Physical Chemistry Laboratory Experiments

Physical Chemistry Laboratory Experiments

John M. White
University of Texas, Austin

Prentice-Hall, Inc., *Englewood Cliffs, New Jersey*

Library of Congress Cataloging in Publication Data

White, John M. (date)
Physical chemistry laboratory experiments.

Includes bibliographical references.
1. Chemistry, Physical and theoretical—Laboratory manuals. I. Title.
QD457.W46 541'.3'028 74-11029
ISBN 0-13-665927-6

10 9 8 7 6 5 4 3 2 1

Printed in the United States of America

Prentice-Hall International, Inc., *London*
Prentice-Hall of Australia, Pty. Ltd., *Sydney*
Prentice-Hall of Canada, Ltd., *Toronto*
Prentice-Hall of India Private Limited, *New Delhi*
Prentice-Hall of Japan, Inc., *Tokyo*

This book is dedicated to

John R. White,
John H. McCormick,
and Thomas C. White,

who,
in love with the soil,
taught me respect
for Nature
and her traditions.

Contents

Chapter 3
A Collection of Experiments Introducing Techniques
150

Chapter 4
Physical Properties of Gases
161

Chapter 5
Thermodynamics
175

Chapter 6
Kinetics
258

Chapter 7 Spectroscopy 338

Chapter 8 Bulk Electric and Magnetic Properties 425

Chapter 9 Electrochemistry 451

Chapter 10 Macromolecular Physical Chemistry 483

Chapter 11 Molecular Structure 511

Appendices 524

A Few Words to the Student

Reports are required for this laboratory course. Their primary purpose is to communicate your experimental results and your analysis of them to your instructor. Their second and quite vital purpose is to provide training in the effective written communication of technical information.

Report writing is required of nearly everyone doing scientific research. It is the means by which findings become permanently available to others. For this reason, the ability to write effectively is very desirable and should be cultivated. Practice over a long period of time is necessary to achieve this objective.

When writing a report keep in mind the following:

1. Who will read the report?
2. Why will they read the report?
3. What is the purpose of this report?

The answers to these three questions will provide the direction the report should take—what material to include and how to present it. In this laboratory, your attention should be focused on the presentation, treatment, and discussion of experimental data.

The writing style should be your own, but it must be clear and concise. Generally speaking, clarity can be achieved with relatively simple sentences and commonly used words. Technical words should be carefully defined if persons reading the report may not be familiar with them. Writing concisely does not mean excluding necessary detail. It does mean expressing a thought in its simplest terms. Information that is not relevant should not be included, but necessary detail must be included.

The experiments outlined in this text include considerably more theoretical background than is common in many laboratory texts. The purpose of including this background is to underscore the relationships existing between the laboratory data being collected and the general principles related to that experiment. It is important that you strive to relate the experimentally measurable quantities to more than just a single formula into which numbers are substituted. The value of the laboratory will increase immensely if you understand not only the apparatus but the theory as well.

In proceeding through this course, it is advisable to familiarize yourself with the contents of this text and with the general layout and requirements for each experiment. Chapters 1 and 2 discuss data handling and experimental techniques. Most of the experiments assume that you have read and understood one or more sections of this material. It is therefore advisable to get very well acquainted with these two chapters early in the course. Several appendices have been included at the end of the text that may be of some use at various times during your work. A brief perusal of the table of contents and a look at the appendices now may save considerable time later. Each experiment has a theory section and one or two sections related to the actual experiment. These are followed by a section describing the procedure for analyzing the data and by a section containing several questions. Finally each experiment has a list of references that may be useful for further study. A comment has been given with each of the references to indicate what you may expect to find in it.

John M. White

Chapter 1

Data Handling and Computation

1.1 The Precision of Experimental Data

INTRODUCTION. Experience in collecting experimental data teaches us that measurements are always subject to some uncertainty. If we make several independent determinations of some experimental quantity, for example, temperature, the values obtained will not be identical to an arbitrary number of significant figures even though we make every effort to maintain identical experimental conditions. Because the values are not identical, we are faced with the problem selecting the "best" value to report and use in any calculations. In this section criteria are developed which with certain assumptions enable one to select the best value and estimate its precision.

DEFINITIONS. Before proceeding it will be helpful to define certain terms and make clear the basic assumptions which underlie the mathematical development of the error analysis concepts given below. We may regard the error in any measurement as arising from three sources: mistakes, systematic error, and random error. Mistakes occur when there is an error in reading an instrument or in recording what is read. Systematic errors affect all measurements in the same way and arise either because of an improperly calibrated measuring device or because of a consistently improper way of using an instrument. In principle, both mistakes and systematic errors can be eliminated by persistent careful experimentation. The third source of error, random error, remains even after the other two have been eliminated. In a single measurement with a certain experimental apparatus, neither the source of this error nor its magnitude can be predicted with certainty. This indeterminate character is a mixed blessing preventing reduction of the error in a single measurement to an arbitrarily small value but permitting the employment of statistical methods and a probabilistic interpretation of the uncertainty in a result. The mathematical development which follows treats only random error, and application of the mathematical conclusions to any experimental data implies, for validity, that the experimental errors are indeed random. While randomness is generally assumed, it is not easy to demonstrate. It arises from a variety of sources including (1) the quantization of light and matter, (2) stray signal pickup, (3) thermal fluctuations, (4) building vibrations, and (5) instrument sensitivity.

Two additional terms need definition: accuracy and precision. These terms have quite different meanings, and the distinction between accurate data and precise data is important. Accurate data have very little systematic error, whereas precise data have very little random error. This distinction may be illustrated with the following example: Suppose that 50 independent measurements are made of the temperature of pure water boiling at 1 atm pressure. If each of the individual measurements is 96.2□ with variations only in the blank column, then the data are quite precise because there is very little variation from measurement to measurement. They are, however, not accurate because the accepted true value for water boiling at 1 atm pressure is 100°C. The difference between the accepted value (100°C) and the measured value (96.2□) must be attributed to systematic error. In this case the sources might include (1) improperly calibrated thermometer, (2) incorrect pressure reading, (3) improper location of the thermometer, and (4) impure water.

A COMMENT ON SIGNIFICANT FIGURES. In any experimental measurement there is, as pointed out above, some uncertainty. In reporting the results of an experiment, the numerical values should reflect this uncertainty by giving a ± figure. Methods for estimating the magnitude of the uncertainty are outlined in Sections 1.2, 1.3, and 1.5. It is absolutely essential to furnish estimates of the uncertainty in a reported value so that others can interpret the utility and reliability of a measurement. In conjunction with the notion of uncertainty, we shall consider the concept of significant figures. It is important to note that there are no set rules regarding significant figures, and thus the following must be regarded as illustrative and hopefully sensible. Suppose that we measure three quantities and combine them in some kind of mathematical relation to calculate a fourth quantity, y. Clearly the uncertainty in the fourth quantity is determined by the uncertainties in the three measured quantities. Section 1.3 shows one way of connecting all these uncertainties. Suppose that we estimate the uncertainty in the calculated quantity, y, to be ±0.001 and that we calculate $y = 1.2137564$ for the quantity of interest. We might be tempted to report that

$$y = 1.2137564 \pm 0.001$$

However, most of the digits in the number reported for y are clearly irrelevant because of the magnitude of the uncertainty. A more reasonable value to report would be

$$y = 1.214 \pm 0.001$$

As a general rule we should not report any more significance than the uncertainty will allow. In some cases we may have good reason for wanting to report morc than one significant figure in the uncertainty. For example, we might determine that the uncertainty for the above is ±0.0012. Then a reasonable y value would be reported as

$$y = 1.2138 \pm 0.0012$$

1.2 The Statistical Treatment of Data

When several measurements of a single quantity are made the random errors may be treated statistically. The rudimentary development of such a treatment is the subject matter of this section.

MEASURES OF PRECISION. Suppose that we make several measurements of a certain experimental variable under equivalent conditions. Each of the determinations will very likely be different. What is the "best" value? Is there any good way to choose a best value of the quantity measured? Once a best value is chosen or calculated, what can we say regarding its precision?

One means, and likely the most intuitive means, of choosing a best value is to calculate the average of a set of measurements. Letting X_i be an individual measurement, the average $\overline{X}$ is defined as

$$\overline{X} = \frac{1}{N} \sum_{i=1}^{N} X_i \qquad (1\text{-}1)$$

where N is the total number of determinations. Later in the discussion it will be shown that for a Gaussian distribution of X_is (events) the largest probability occurs when $X = \overline{X}$. Hence, if we assume that the measurements of the quantity X are distributed according to a Gaussian distribution, the most probable value of X is the average value of that set of measurements and it is taken to be the best estimate of the true value of X.

The precision of a given set of data is generally specified as some function of the deviations, $\{d_i\}$, of the set $\{X_i\}$ from the average value, $\overline{X}$:

$$d_i = X_i - \overline{X} \qquad (1\text{-}2)$$

One measure of precision is the mean deviation, which is defined as

$$\alpha \equiv \frac{1}{N} \sum_{i=1}^{N} |d_i| \qquad (1\text{-}3)$$

where $|d_i|$ is the absolute value of d_i.

Another measure of precision is the standard deviation, σ, which is defined as

$$\sigma \equiv \sqrt{\frac{1}{N-1} \sum_i d_i^2}$$

or (1-4)

$$\sigma = \sqrt{\frac{1}{N-1} \sum_i (X_i - \bar{X})^2}$$

The standard deviation as calculated above is the best estimate of the spread or dispersion of a Gaussian distribution of measurements, as will be discussed later.

Variance is also used as a measure of precision and is defined as

$$\text{var}(X) = \sigma^2 \tag{1-5}$$

It is important to consider the variation of σ with N in Eq. (1-4). Increasing N will in general increase both the value of the summation and $N - 1$. If we assume as a special case that $d_i^2 = K$ a constant for all i measurements, then

$$\sigma = \sqrt{\frac{N}{N-1} K}$$

A plot of σ versus N for this special case shows that σ is a monotonically decreasing function of N, which emphasizes the significance of making several measurements. Even if d_i^2 is not constant for all i, a decrease in the standard deviation is expected with an increasing number of measurements.

Illustrative Example 1 Consider the following set of five temperature measurements presumed to possess only random error: 297°K, 299°K, 296°K, 297°K, 297°K. The estimated variance σ_T^2 of these measurements is calculated as follows:

$$\text{var}(T) = \sigma_T^2 = \frac{1}{N-1} \sum_i (T_i - \bar{T})^2$$

$$\bar{T} = \sum_i \frac{T_i}{N}$$

$$= \frac{1}{5}(1486)$$

$$= 297.2°\text{K}$$

$$\text{var}(T) = \frac{1}{4}(0.04 + 3.24 + 1.44 + 0.04 + 0.04)$$

$$\text{var}(T) = 1.20$$

PROBABILITY. With the above very brief introduction to some of the more intuitive statistical concepts we now proceed to a fuller discussion of probability and the Gaussian or normal error distribution.

Most people have intuitive notions about probability. For example, in flipping a penny we expect a head 50% of the time and a tail 50% of the time and so we say that the probability of obtaining a head is 1/2. However, if we flip a penny 10 times, it is quite likely that we would not get 5 heads and 5 tails. On the other hand, in a very large number of trials, the occurrence of heads would be expected to approach 1/2 of the total number of trials. This leads us to a definition of probability.

> Definition: The probability of the occurrence of a certain event is the limiting value of the ratio of the number of occurrences of that particular event to the total number of trials as the number of trials approaches infinity.

This definition of probability is useful when we are dealing with discrete events such as a head or tail in coin flipping or the probability of drawing a three of hearts from a deck of cards. Most experimentally observable quantities are continuous, not discrete, variables. For example, the pressure of a gas is generally continuously variable in the range of observations. These continuously variable quantities and the probabilities associated with them require a generalization of the above definition.

PROBABILITY DISTRIBUTION FUNCTIONS. Both discrete and continuous observables are often discussed statistically in terms of probability distribution functions. Consider first a discrete distribution such as that associated with the tossing of a die. If we assume as a model that all possible faces are equally probable, then we can construct a discrete probability distribution function f(n) for all the possible events, where n ranges from 1 to 6 depending on the number of spots showing on the face of the die. There are six possible events, each equally probable according to the model. Hence, the probability of observing any face is just 1/6, and the discrete distribution function is shown in Fig. 1-1.

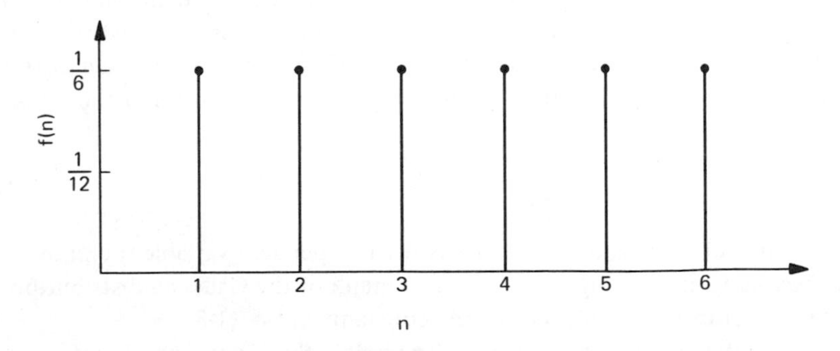

Figure 1-1. Distribution function for six discrete equally probable events.

Note that the sum of all the probabilities is unity because all possible events are included in constructing the distribution function. Therefore, we are absolutely certain of obtaining one of the events in any "experiment." Mathematically then we require the normalization condition:

$$\sum_n f(n) = 1 \tag{1-6}$$

The distribution function shown in Fig. 1-1 is based on a certain model for the system. As such, it assumes a very large number of trials so that limiting values (i.e., probabilities) of all the events may be obtained. This distribution is called a population distribution and is to be distinguished from a sample distribution arising from a limited number of trials. Quantities calculated from sample distributions are only estimates of what these same quantities would be if calculated from population distributions. For example, if we toss a die 12 times, it is unlikely that the observed sample distribution would be a good approximation to the population distribution or vice versa.

For a continuous variable Eq. (1-6) must be rewritten as a continuous, rather than discrete, sum, and the normalization condition becomes

$$\int f(x)\, dx = 1 \tag{1-7}$$

where $f(x)\, dx$ is the probability of observing x in the range x to $x + dx$. $f(x)$ is referred to as the population distribution for the continuous variable.

GAUSSIAN OR NORMAL ERROR DISTRIBUTION. A population distribution which is widely used as a model for observations made on a continuous variable is the Gaussian or normal error distribution. Although a theoretical derivation of the Gaussian distribution is indeed possible, we take the viewpoint here that the real justification for using it to describe the distribution of random errors is that many sets of experimental observations turn out to obey it. We shall treat the Gaussian distribution as an experimental fact, state its formula, and examine its meaning and uses. The Gaussian distribution function is also referred to as the normal error function, and errors distributed according to the Gaussian distribution are said to be normally distributed. The mathematical form of the Gaussian distribution is

$$f(x) = Ae^{-h^2(x - m)^2} \tag{1-8}$$

where A, h, and m are constants and x is the independent variable (i.e., the *experimental* variable). Figure 1-2 shows a graph of the Gaussian distribution function. Notice the significance of the constants in Eq. (1-8).

What significance does the function $f(x)$ hold? Since x is a continuous variable, the probability of x having exactly any particular value is zero. Nonzero

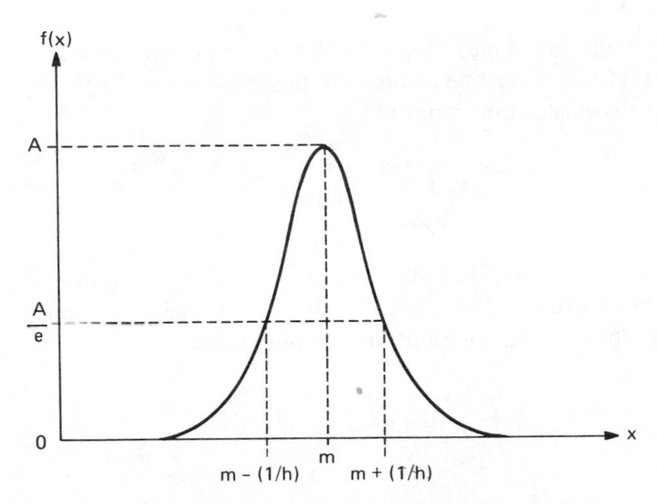

Figure 1-2. Gaussian distributive function.

probabilities exist for the values of the variable x only in an interval, say between x and x + dx. We now arrive at the proper interpretation of the function f(x): For a small interval dx, f(x) dx represents the probability of making a measurement and observing X_i in the interval between x and x + dx. This is graphically interpreted in Fig. 1-3, where the area of the shaded strip on the left represents f(x) dx. For a finite interval,

$$P(a, b) = \int_a^b f(x)\ dx \tag{1-9}$$

is the probability that measurements will fall somewhere in the interval $a \leqslant x \leqslant b$.

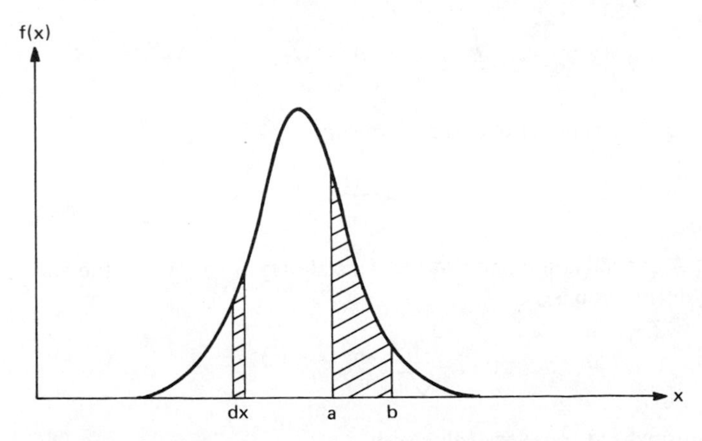

Figure 1-3. Graphical representation of probabilities associated with a Gaussian distribution function.

The probability that a measurement will yield some value of x is unity, since all possible values of x must be included in the definition of f(x). This is expressed as the normalization condition

$$\int_{-\infty}^{+\infty} f(x)\, dx = 1 \qquad (1\text{-}10)$$

and is analogous to Eq. (1-7). Functions for which the condition of Eq. (1-10) is satisfied are said to be normalized; i.e., the total probability is unity. Applying Eq. (1-10) to the Gaussian distribution function yields

$$\int_{-\infty}^{+\infty} Ae^{-h^2(x - m)^2}\, dx = 1 \qquad (1\text{-}11)$$

Equation (1-11) furnishes an algebraic relation between the constants A, h, and m and prevents the arbitrary selection of values for all three. For example, knowing h and m suffices to determine A. By making a change of variable and performing the quadrature of Eq. (1-11), A can be found. Make the variable change

$$h(x - m) = z \qquad (1\text{-}12)$$

Equation (1-11) then becomes

$$A\int_{-\infty}^{+\infty} e^{-z^2} dz = h \qquad (1\text{-}13)$$

The value of the integral in Eq. (1-13) can be shown to be

$$\int_{-\infty}^{+\infty} e^{-z^2}\, dz = \sqrt{\pi} \qquad (1\text{-}14)$$

Inserting Eq. (1-14) into Eq. (1-13) furnishes

$$A = \frac{h}{\sqrt{\pi}} \qquad (1\text{-}15)$$

Note that A depends on h and not m. This allows us to rewrite the Gaussian distribution function as

$$f(x) = \frac{h}{\sqrt{\pi}} e^{-h^2(x - m)^2} \qquad (1\text{-}16)$$

which is normalized for every value of h.

Next let us consider the average or mean value of x for this distribution. f(x) dx represents the probability of occurrence of the measurement in the interval x to x + dx. The mean value of x which we denote as $\bar{x}$ is found by integrating the product of this probability and the value of x corresponding to this interval. That is,

$$\bar{x} = \int_{-\infty}^{+\infty} x f(x) dx \tag{1-17}$$

To calculate $\bar{x}$ for a Gaussian probability distribution we substitute into Eq. (1-17) from Eq. (1-16):

$$\begin{aligned} \bar{x} &= \frac{h}{\sqrt{\pi}} \int_{-\infty}^{+\infty} x e^{-h^2(x-m)^2} dx \\ &= \frac{1}{\sqrt{\pi}} \int_{-\infty}^{+\infty} \left(\frac{z}{h} + m\right) e^{-z^2} dz \end{aligned} \tag{1-18}$$

Integration yields

$$\begin{aligned} \bar{x} &= \frac{m}{\sqrt{\pi}} \int_{-\infty}^{+\infty} e^{-z^2} dz = \frac{m}{\sqrt{\pi}} \sqrt{\pi} \\ \bar{x} &= m \end{aligned} \tag{1-19}$$

We thus conclude that the constant m is the mean value of the independent variable.

The variance is calculated in a similar fashion:

$$\sigma^2 = \overline{(x - m)^2}$$

$$\begin{aligned} \sigma^2 &= \int_{-\infty}^{+\infty} (x - m)^2 f(x)\, dx \\ &= \int_{-\infty}^{+\infty} \frac{h}{\sqrt{\pi}} (x - m)^2 e^{-h^2(x-m)^2} dx \end{aligned}$$

Recalling the change of variable in Eq. (1-12), we obtain

$$\sigma^2 = \frac{1}{h^2\sqrt{\pi}} \int_{-\infty}^{+\infty} z^2 e^{-z^2} dz \tag{1-20}$$

Performing the integration in Eq. (1-20) yields

$$\sigma^2 = \frac{1}{2h^2} \quad \text{or} \quad \sigma = \frac{1}{\sqrt{2}h} \tag{1-21}$$

Note that the standard deviation is inversely proportional to h. Larger values of h result in a more sharply peaked curve and smaller σ values (cf, Fig. 1-4). Since h is large for sharply peaked curves, corresponding to a small spread of errors, h is sometimes called the measure of precision of the distribution (i.e., a large value of h implies precise data).

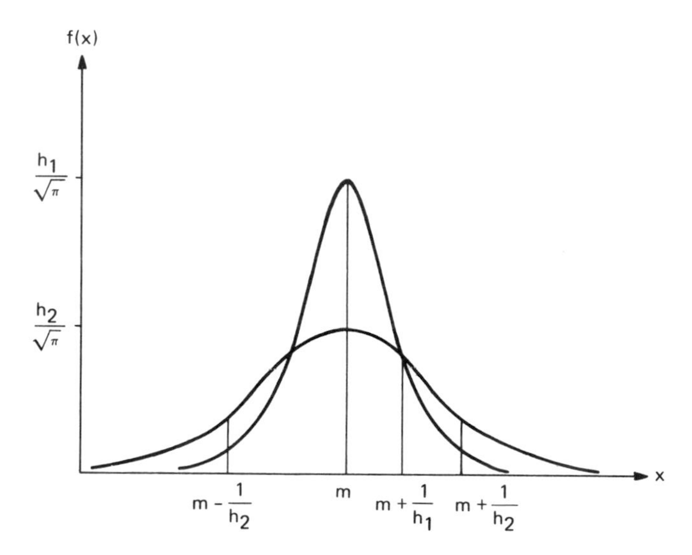

Figure 1-4. Gaussian distribution for two values of h ($h_1 = 2h_2$) and a fixed m.

Writing the Gaussian distribution in terms of σ rather than h furnishes

$$f(x) = \frac{1}{\sqrt{2\pi}\sigma} e^{-(x-m)^2/2\sigma^2} \tag{1-22}$$

Note that f(x) is symmetric about x = m.

The mean deviation of the Gaussian distribution is obtained as follows:

$$\begin{aligned} \alpha &= \int_{-\infty}^{+\infty} |x - m| f(x)\, dx \\ &= \int_{-\infty}^{+\infty} |x - m| \frac{h}{\sqrt{\pi}} e^{-h^2(x-m)^2}\, dx \end{aligned} \tag{1-23}$$

Let y = |x - m|. Making the change of variable $z = h^2y^2$ and integrating yields

$$\sigma = \frac{\pi}{2} \cdot \alpha \simeq 1.25\alpha \tag{1-24}$$

Comparing Eqs. (1-24) and (1-21), we see that

$$\alpha = \frac{1}{\sqrt{\pi h^2}} = \sqrt{\frac{2}{\pi}}\,\sigma \tag{1-25}$$

Another quantity of interest is the probability that a measurement will fall within any specified limits. Of particular interest is the probability of a measurement falling within $\pm\sigma$ of the mean value. This consideration will lend clarity to the significance of the standard deviation. The probability P that a measurement will fall between $m - \sigma$ and $m + \sigma$ is given by

$$P = \int_{m-\sigma}^{m+\sigma} \frac{1}{\sqrt{2\pi}\,\sigma}\, e^{-(x-m)^2/2\sigma^2}\, dx \tag{1-26}$$

Making the change of variable $t = (x - m)/\sigma$, we obtain

$$P = \frac{1}{\sqrt{2\pi}} \int_{-1}^{+1} e^{-t^2/2}\, dt \tag{1-27}$$

which is symmetric about $t = 0$. The quadrature in Eq. (1-27) can be performed by numerical methods only, and tables of values exist for $P(T)/2$ where T is defined by the expression

$$\frac{P(T)}{2} = \frac{1}{\sqrt{2\pi}} \int_{0}^{T} e^{-t^2/2}\, dt = \frac{1}{2\sqrt{2\pi}} \int_{-T}^{+T} e^{-t^2/2}\, dt \tag{1-28}$$

The values of P(T) for a few values of T are

$P(1) = 0.683$	$1 - P(1) = 0.317$
$P(2) = 0.954$	$1 - P(2) = 0.046$
$P(3) = 0.997$	$1 - P(3) = 0.003$

Notice that the probability is about 0.68 that the measured quantity lies within one standard deviation of the mean, while the probability that it will lie outside three standard deviations is only 0.3%. Frequently *95% confidence limits* are specified and are given mathematically as $\pm 2\sigma$. We may be 95% certain that a single measurement of x will lie within $\pm 2\sigma$ of the average value. Figure 1-5 illustrates the probability of being within one standard deviation of the mean.

Keep in mind that Eqs. (1-8)–(1-28) are properties of a Gaussian population distribution. Their use in analyzing a sample distribution of experimental data assumes that the model for the population distribution of the random errors in

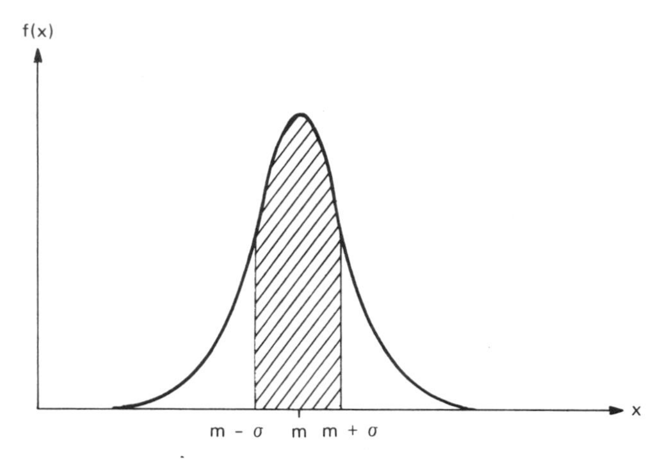

Figure 1-5. Probability of being within one standard deviation of the mean.

the data is indeed Gaussian. With this assumption, properties of the sample distribution, for example, the average, are taken to be the best estimates of those same quantities which characterize the Gaussian population distribution.

As an example of estimation, consider the standard deviation. When a small sample distribution is used as a basis for estimating the standard deviation of the population distribution, the estimate itself is subject to considerable uncertainty, which is specified quantitatively by

$$\sigma_\sigma = \sqrt{\frac{2}{N}}\,\sigma = \sqrt{\frac{2}{N(N-1)}\left(\sum_i d_i^2\right)} \tag{1-29}$$

This expression makes very clear, because of the N dependence, the need for making several independent determinations of x if meaningful estimates of precision are to be obtained.

Illustrative Example 2 An estimate of the uncertainty in the standard deviation σ_T of the mean of T for the data of Illustrative Example 1 is calculated as follows:

$$\sigma_{\sigma_T} = \sqrt{\frac{2}{N}}\,\sigma_T$$

$$= \sqrt{\frac{2}{5} \times 1.2}$$

$$\cong 0.7$$

This makes clear the large uncertainty in the estimate of σ_T.

DATA REJECTION. If we measure a quantity several times, frequently the individual measurements will for the most part lie close to the average value. A few, however, may deviate quite widely from the average value. We then ask whether there is any substantial reason for rejecting them. It is assumed that there is no readily apparent experimental reason for rejecting the particular points in question. For example, known impurities in a particular sample could be a valid experimental reason for rejecting a particular measurement. Barring this kind of reason the only remaining justification for rejecting a measurement must be based on rather arbitrary statistical arguments. The Gaussian probability distribution provides one means of establishing criteria for rejection. Assuming the sample of data to be well described by a Gaussian probability distribution, we can then say, for example, that the probability of an observation deviating from the average by more than $\pm 3\sigma$ is 0.003. Frequently data are rejected if they deviate by more than 3σ from the mean.

It should be kept clearly in mind that the Gaussian distribution applies rigorously only to the population distribution and only approximates the sample distribution.

PROBLEMS

Imagine carrying out an experiment with the apparatus shown in Fig. 1-6 in which the number of drops of water per minute from the capillary is measured as a function of the height of the column of water in the tube above the capillary.

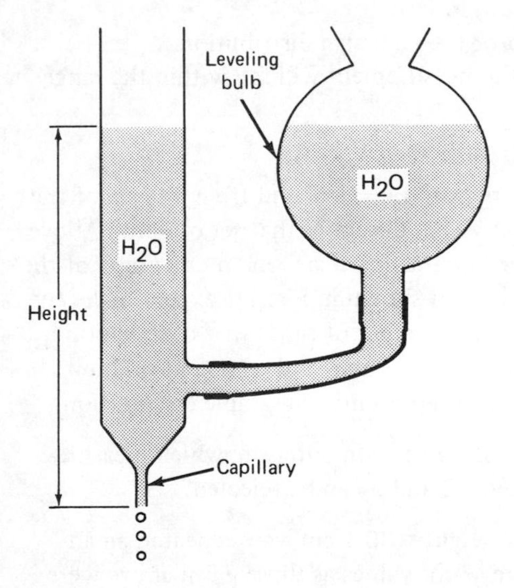

Figure 1-6. Drop rate apparatus.

Assume that the following set of data was obtained. Use these data for statistical analysis:

Table 1. DATA SHEET

Height of water level above the capillary tip (cm)	Number of drops measured over 1 min intervals
80.1*	40, 44, 38, 42, 41, 39, 37, 39, 41, 36, 40
75.2	38, 35, 36, 36, 37
70.0	32, 31, 33, 29, 30
64.9	26, 28, 27, 25, 26
60.1	21, 25, 21, 20, 23
55.0	20, 20, 19, 19, 20
50.0	14, 15, 16, 14, 14
39.0	9, 10, 10, 9, 10
30.1	6, 5, 6, 6, 6,

*Use this set of data for Problems 2 and 3.

1. For each set of drop rate data, calculate the
   (a) Mean deviation.
   (b) Variance.
   (c) Standard deviation.

2. For one set of the drop rate data, assume the errors to be distributed according to the Gaussian distribution and
   (a) Calculate the index of precision, h.
   (b) Calculate the mean deviation for the Gaussian distribution.
   (c) Calculate the probability that a measurement will fall within the range 0.5σ to 1.5σ of the mean.
   (d) Estimate the uncertainty in σ.

3. Using Eq. (1-22) and the parameters m and σ calculated from one set of the data, plot the Gaussian distribution which describes that set of data. Above this graph and along a line representing the abscissa, construct a graph of the actual sample distribution similar to that shown in Fig. 1-1 except make the height of each of the lines equal to the number of times that observation was made. Connect the tips of these vertical lines and comment on how well the Gaussian distribution actually represents the sample distribution.

4. Using 99% confidence limits ($\pm 3\sigma$) as a rejection criterion, which measurements of the sample used in Problems 2 and 3 can be rejected?

5. Suppose that the measurements at height = 80.1 cm were repeated an additional 11 times and that the same set of values as those given above were obtained. Combining the two sets will furnish new estimates for the statistical parameters m and σ. Calculate the new σ and an estimate of its uncertainty, compare them with results of Problem 2, and comment critically.

1.3 Error Propagation and Error Estimates for Small Samples

Frequently because an insufficient number of measurements are made, the use of statistical methods of error analysis cannot be justified. While the minimum number of measurements required is rather arbitrary it is frequently set at 4, and if the number of measurements is less than this minimum and precision estimates are desired, then one is forced to use means other than the statistical methods of the previous section. In this situation experience and intuition can be used to obtain reasonable estimates of the precision of the directly determined experimental quantities. With these estimates a mathematical technique known as the propagation of errors is employed to arrive at an estimate of the uncertainty in the quantity the experiment is designed to measure. This approach is useful in at least two situations:

1. When the number of measurements (sample size) is too small to use statistical methods.
2. When an experiment is being designed an a priori estimate of the uncertainty in the final result can be obtained based on reasonable estimates of the precision of the instruments used in the experiment.

As a simple example suppose that we desire the translational energy of a moving vehicle. The direct measurement of mass, m, and speed, v, suffice to determine the translational energy since $E_T = \frac{1}{2}mv^2$. If only one determination of E_T is made, then statistical estimates of the precision cannot be made, but if by experience we can place reasonable estimates on the precision of our scales and speedometer, then an estimate of the uncertainty in E_T can be obtained by the methods outlined below. Note that this estimate can be made even prior to the measurement of E_T—hence the utility of the error propagation method in designing experiments to give meaningfully precise results.

MATHEMATICAL DEVELOPMENT. Consider the calculation of a quantity F from experimentally observable quantities x, y, and z. We write

$$F = F(x, y, z) \tag{1-30}$$

Notice that the directly measured quantities x, y, and z are the independent variables in this equation.

The total derivative of F is

$$dF = \left(\frac{\partial F}{\partial x}\right)_{y,z} dx + \left(\frac{\partial F}{\partial y}\right)_{x,z} dy + \left(\frac{\partial F}{\partial z}\right)_{x,y} dz \tag{1-31}$$

If the changes in the independent variables are finite but small, then the values of the partial derivatives will not be affected appreciably by the changes in x, y, and z. For example, $(\partial F/\partial x)_{y,z}$ will be nearly constant over the small finite interval Δx. Then we can write the approximate relation

$$\Delta F = \left(\frac{\partial F}{\partial x}\right)_{y,z} \Delta x + \left(\frac{\partial F}{\partial y}\right)_{x,z} \Delta y + \left(\frac{\partial F}{\partial z}\right)_{x,y} \Delta z \tag{1-32}$$

where the partial derivatives are evaluated at the measured values of x, y, and z. Note that the variation of F with variations in x, y, and z is given by this equation and that we may regard the quantities ΔF, Δx, Δy, and Δz as uncertainties, the latter three being reasonable estimates based on experience and intuition, and ΔF the estimate we desire to obtain. For example, suppose that x is temperature = T; then $\Delta x = \Delta T$, and we might estimate, based on our experience with the apparatus at hand, that $\Delta T = \pm 0.1°K$. Similar estimates for Δy and Δz must be made; then Eq. (1-32) can be utilized to estimate ΔF.

There is one serious difficulty with Eq. (1-32) which prevents its direct application. The estimates for the uncertainties in the independent variables will always be plus or minus some quantity. For example, above ΔT was estimated as $\pm 0.1°K$. How is the choice between the plus and minus sign to be made? There is no a priori way to make this choice, and we are thus faced with a dilemma whose resolution requires an additional assumption about Δx, Δy, and Δz; namely, these uncertainties are randomly distributed about the measured values of x, y, and z. With this assumption the problem of choosing the plus or minus sign is avoided, and the following mathematical considerations make this clear.

First, Eq. (1-32) is squared to give

$$\Delta F^2 \cong \left(\frac{\partial F}{\partial x}\right)_{y,z}^2 \Delta x^2 + \left(\frac{\partial F}{\partial y}\right)_{z,x}^2 \Delta y^2 + \left(\frac{\partial F}{\partial z}\right)_{x,y}^2 \Delta z^2 + 2\left(\frac{\partial F}{\partial x}\right)_{y,z} \left(\frac{\partial F}{\partial y}\right)_{z,x} \Delta x\, \Delta y + 2\left(\frac{\partial F}{\partial x}\right)_{y,z}\left(\frac{\partial F}{\partial z}\right)_{x,y} \Delta x\, \Delta z + 2\left(\frac{\partial F}{\partial y}\right)_{z,x} \left(\frac{\partial F}{\partial z}\right)_{y,x} \Delta y\, \Delta z \tag{1-33}$$

Equation (1-33) contains two types of terms in the uncertainties of the independent variables, those like Δx^2 and those like $\Delta x\, \Delta y$ (cross terms). The first type will always be positive irrespective of the choice of the plus or minus sign on Δx. The second type is still troublesome from this point of view, but the assumption of randomly distributed Δx, Δy, and Δz permits, if an average is taken, the circumvention of this difficulty. It can be rigorously shown that the average value of the product of the uncertainties in two randomly distributed independent variables is identically zero. Making use of the results of Section 1.2 and writing the average value of the product of the uncertainties as $\overline{\Delta x\,\Delta y}$ the following relations follow:

$$\overline{\Delta x\,\Delta y} = \overline{(x - \bar{x})\,(y - \bar{y})} = \overline{(x - \bar{x})} \cdot \overline{(y - \bar{y})} = (\bar{x} - \bar{x}) \cdot (\bar{y} - \bar{y}) = 0 \qquad (1\text{-}34)$$

This development follows because x and y are independent variables. The average values of x and y are $\bar{x}$ and $\bar{y}$. In a more intuitive vein the same conclusion is drawn if one considers that a random distribution for Δx implies that in a series of measurements Δx is expected to be positive as often as negative with an average uncertainty in x of zero.

Taking averages in Eq. (1-33) furnishes

$$\overline{(\Delta F)^2} = \left(\frac{\partial F}{\partial x}\right)^2_{y,z} \overline{\Delta x^2} + \left(\frac{\partial F}{\partial y}\right)^2_{z,x} \overline{\Delta y^2} + \left(\frac{\partial F}{\partial z}\right)^2_{y,x} \overline{\Delta z^2} \qquad (1\text{-}35)$$

For the average values appearing on the right-hand side of Eq. (1-35) the squares of the estimated uncertainties in the independent variables are taken. For example, using ΔT given above as $\pm 0.1°C$ furnishes $\overline{\Delta T^2} = 0.01$. The square root of Eq. (1-35) furnishes a root-mean-square estimate of the uncertainty in F which we denote ΔF_{rms}. The fractional uncertainty in F is simply $\Delta F_{rms}/F$.

As a simple example of the application of Eq. (1-35), consider the following gas thermometry experiment using the ideal gas law and pV measurements to determine T. The expression to consider is

$$T = \frac{pV}{nR} \qquad (1\text{-}36)$$

where p, V, and n are directly measured and used with Eq. (1-36) to determine T. If we have four or more independent measurements of p, V, and n, statistical methods may be used to treat the resulting set of calculated temperatures; otherwise the estimation procedure outlined above must be used. For the latter procedure an expression for the root-mean-square uncertainty in the temperature, ΔT_{rms}, can be written in terms of the uncertainties in the independent variables, p, V, and n. By analogy with Eq. (1-35), there follows

$$\Delta T_{rms} = \left[\left(\frac{\partial T}{\partial p}\right)^2_{n,V} \overline{\Delta p^2} + \left(\frac{\partial T}{\partial n}\right)^2_{V,p} \overline{\Delta n^2} + \left(\frac{\partial T}{\partial V}\right)^2_{p,n} \overline{\Delta V^2}\right]^{1/2} \qquad (1\text{-}37)$$

Expressions for the partial derivatives are obtained by straightforward differentiation of Eq. (1-36) with the result

$$\begin{aligned}\Delta T_{rms} &= \left[\left(\frac{V}{nR}\right)^2 \overline{\Delta p^2} + \left(\frac{pVR}{n^2R^2}\right)^2 \overline{\Delta n^2} + \left(\frac{p}{nR}\right)^2 \overline{\Delta V^2}\right]^{1/2} \\ &= \frac{pV}{nR}\left[\frac{\overline{\Delta p^2}}{p^2} + \frac{\overline{\Delta n^2}}{n^2} + \frac{\overline{\Delta V^2}}{V^2}\right]^{1/2}\end{aligned} \qquad (1\text{-}38)$$

To evaluate Eq. (1-38) the measured values of p, V, and n together with the uncertainty estimates are needed. For illustrative purposes, consider the following example:

$$
\begin{aligned}
\Delta p &= \pm 0.1 \text{ mm Hg} = \pm 0.1 \text{ Torr} \\
\Delta V &= \pm 0.1 \text{ cm}^3 \\
\Delta n &= \pm 0.001 \text{ mole} \\
p &= 50 \text{ Torr} \\
V &= 1000 \text{ cm}^3 \\
n &= 0.05 \text{ mole} \\
R &= 62.4 \times 10^3 \text{ cm}^3 \text{ Torr mole}^{-1} \text{ deg K}^{-1}
\end{aligned}
\tag{1-39}
$$

$$\Delta T_{rms} = 80.2[4 \times 10^{-6} + 4 \times 10^{-4} + 10^{-8}]^{1/2}$$

Examination of Eq. (1-39) reveals an important lesson: independent variables contribute different amounts to the overall uncertainty of a result. This is especially important in experimental design and improvement of apparatus. Any marked improvement in ΔT_{rms} in this example obviously entails improving the measurement of n since its uncertainty contributes at least a factor of 10 more to the uncertainty in T than do either the pressure or volume measurements.

Illustrative Example 3 How To Determine the Form of a Root-Mean-Square Uncertainty Relation

Suppose that $y = xz^2 + bt$, where x, z, and t are independent variables and b is a constant. Develop an expression for the root-mean-square uncertainty in y.

We desire to find Δy_{rms} in terms of uncertainties in x, z, and t. Following the outline of the above equations,

$$\Delta y_{rms} = \left[\left(\frac{\partial y}{\partial x}\right)^2 \overline{(\Delta x)^2} + \left(\frac{\partial y}{\partial z}\right)^2 \overline{(\Delta z)^2} + \left(\frac{\partial y}{\partial t}\right)^2 \overline{(\Delta t)^2}\right]^{1/2}$$

$$\frac{\partial y}{\partial x} = z^2, \quad \frac{\partial y}{\partial z} = 2xz, \quad \frac{\partial y}{\partial t} = b$$

$$\Delta y_{rms} = \left[z^4\overline{(\Delta x)^2} + 4x^2z^2\overline{(\Delta z)^2} + b^2\overline{(\Delta t)^2}\right]^{1/2}$$

To use this relation we need to apply it to some set of measured values of (x, z, t) and have estimates for Δx, Δz, and Δt.

PROBLEMS

1. For each of the following expressions, develop a relation for the root-mean-square uncertainty in y.
   a. $y = ax + b$; where x is the independent variable.
   b. $y = xz$; both x and z are independent variables.
   c. $y = x + z$; both x and z are independent variables.

2. The Clausius-Clapeyron equation can be used to obtain heats of vaporization if at least two pressures and temperatures are measured. Mathematically

$$\Delta H_{vap} = R\left[\frac{\ln p_1 - \ln p_2}{(1/T_1) - (1/T_2)}\right]$$

Derive an expression for the rms uncertainty in ΔH_{vap} assuming that p_1, p_2, T_1, and T_2 are each independent variables.

3. From kinetics of the elementary reaction, A + A $\longrightarrow$ products, the rate law is written

$$-\frac{dA}{dt} = k_A [A]^2$$

The integrated form of this rate expression is

$$k_A = \frac{1}{t_2 - t_1} \cdot \frac{A_1 - A_2}{A_1 A_2}$$

a. Assuming that errors in t_1, t_2, A_1, and A_2 are independent and random, derive an expression for

$$\frac{\overline{\Delta k_A^2}}{k_A^2}$$

b. Using $t_2 - t_1 = 1000$ sec, $\Delta t_1 = \Delta t_2 = \pm 1$ sec, and $A_2 = 0.90A_1$, and assuming that the relative error in the analysis of A_2 and A_1 is ±0.1%, calculate a value for

$$\frac{\overline{\Delta k_A^2}}{k_A^2}$$

c. What part of the experiment needs attention if the expected accuracy is to be improved?

4. Consider a situation where the fractional error in x, y, or z of Eq. (1-35) is large. Why is Eq. (1-35) not valid in these circumstances?

1.4 Systematic Error

Even though the previous three sections do not discuss systematic error, its role must never be minimized and neglected because frequently it may be an even more important consideration than random error. Systematic error can arise from a variety of sources and each individual experimental situation must be examined in its own context. Possible sources of systematic error include

1. Improperly calibrated instruments. If a gas thermometer is calibrated using the ice point (0°C) and the steam point (100°C) but the atmospheric pressure is 700 Torr rather than 760 Torr, then the steam point is actually less than 100°C and the calibration is, as a result, incorrect.
2. Personal bias. For example, in reading an ammeter one may view the pointer from a direction other than perpendicular to the scale and thereby obtain readings which are higher or lower than those actually indicated by the meter.
3. Consistently imperfect technique. For example, in titrating one could consistently overshoot the end point.
4. Unsatisfied experimental assumptions. The following example illustrates this point.

Suppose that we set out to measure the boiling point of pure water at 1 atm pressure and in ignorance we use a sample obtained from the Gulf of Mexico. Using accurate temperature and pressure probes, 10 measurements of the boiling point are made with the result $T_{boil} = 102.15 \pm 0.04°C$, where 0.04 is the standard deviation. By most standards this result is quite precise, but to conclude that pure water at 1 atm pressure boils at 102.15°C is, obviously in this case, incorrect because we have systematically used impure starting materials. The point to be made clear is the following: Good precision alone is an insufficient basis for many conclusions.

There is another, and in some respects even more important, point. Systematic errors of the type described above arise because of physical phenomena which are themselves often worthy of study. In the above example we would quite naturally be led to investigate the effect of an ionic solute on the boiling point of water.

From these observations the following conclusions may be drawn: (1) The valid interpretation of data depends on both the proper evaluation of random error and the elimination of systematic error. (2) Those systematic errors which arise because an assumption about the system under study is not satisfied may often lead one to an area of fruitful investigation. (3) In any experiment, considerable insight and thoroughness is required to eliminate systematic error or even reduce it to insignificance.

1.5 Curve Fitting by Least-Squares Procedure

In experimental research relationships among different variables are sought. For example, in kinetics experiments the relationship between a rate coefficient and the reaction temperature is frequently desired. In spectroscopy the extinction coefficient as a function of wavelength is often sought. The data one collects, being always subject to uncertainties of one kind or another, are not expected to fit precisely any theoretical or empirical relationship between variables. Suppose, for example, that a set of data is available relating two variables and that there is supposedly a linear relationship between them. A plot of one variable versus the other will not, in general, furnish a graph in which all the points lie on a single straight line. This does not necessarily mean that the two variables are not linearly related because the collected data possess some uncertainty. This example makes clear the need for some kind of curve-fitting procedure which attempts to answer the question, What is the best relationship of a given type between these variables? For the example just discussed we attempt to find the "best" straight line. Frequently this is accomplished by the so-called *eyeball* technique in which one simply draws in a curve with about the same number of points above as below. More sophisticated approaches are often desirable and we shall describe here one mathematical approach in which *best* is defined in a least-squares sense.

Before setting out the mathematics of this approach we shall discuss it qualitatively using the simplest example, a linear least-squares fit. We assume that data relating two experimental quantities are available and that we desire to find the best linear relation between the variables. A linear relation between two variables contains two parameters, say the slope and intercept, and in the least-squares method, we adjust the values for these parameters until the deviations between the experimentally measured quantities and those predicted by the linear fit obey a certain relationship; namely, the sum of the squares of the deviations is minimized. That slope and intercept which minimizes this sum is by least-squares definition the best linear fit.

Turning now to the general mathematical development we assume a set of experimental data relating two quantities that we denote as $\{Y_i, X_i\}$. Further, we assume some mathematical relationship between these variables, which we write as

$$y = f(x, a, b, c, \ldots) \tag{1-40}$$

where y and x are the counterparts of Y_i and X_i and $\{a, b, c, \ldots\}$is taken to represent a set of parameters which must be chosen to give the best fit. In the linear case two parameters suffice. For any set $\{a, b, c, \ldots\}$the value of y corresponding to a given x can be calculated from (1-40). We assume no error in the experimental values X_i and so we may rewrite (1-40) as

$$y_i = f(X_i, a, b, c, \ldots) \tag{1-41}$$

The deviation $Y_i - y_i$ is written as

$$d_i = Y_i - y_i \tag{1-42}$$

and we seek to minimize the expression

$$g = \sum_{i=1}^{n} d_i^2 \qquad (1\text{-}43)$$

where n is the number of pairs of values. Note that Y_i is measured, whereas y_i is calculated. In expanded form (1-43) may be written

$$g = \sum_i [Y_i - f(X_i, a, b, c, ...)]^2 \qquad (1\text{-}44)$$

Since Y_i and X_i are fixed by the experiment, to minimize g we must minimize with respect to a, b, c, ... Applying the well-known formulas of calculus we thus require that simultaneously $\partial g/\partial a = 0$, $\partial g, \partial b = 0$, etc. Mathematically from (1-44) this becomes a set of simultaneous equations:

$$\frac{\partial g}{\partial a} = 0 = 2 \sum_i [Y_i - f(x_i, a, b, c, ...)] \frac{\partial f}{\partial a}$$

$$\frac{\partial g}{\partial b} = 0 = 2 \sum_i [Y_i - f(x_i, a, b, c, ...)] \frac{\partial f}{\partial b} \qquad (1\text{-}45)$$

$$\frac{\partial g}{\partial c} = 0 = 2 \sum_i [Y_i - f(x_i, a, b, c, ...)] \frac{\partial f}{\partial c}$$

$$\vdots$$

The solution of this set of simultaneous equations furnishes the least-squares values of a, b, c, ...

The use of this procedure may be illustrated for the linear case as follows:

$$y_i = aX_i + b = f(X_i, a, b) \qquad (1\text{-}46)$$

Minimization is with respect to a and b, and using (1-45) the results are

$$\sum_{i=1}^{n} (Y_i - aX_i + b)X_i = 0$$
$$\sum_{i=1}^{n} (Y_i - aX_i + b) = 0 \qquad (1\text{-}47)$$

Expanding Eqs. (1-47) furnishes

$$\sum_i X_iY_i - a \sum_i X_i^2 + b \sum_i X_i = 0$$
$$\sum_i Y_i - a \sum_i X_i + nb = 0 \qquad (1\text{-}48)$$

Solving (1-48) for a and b leads to

$$b = \frac{(\sum_i X_i^2)(\sum_i Y_i) - (\sum_i X_i)(\sum_i X_i Y_i)}{n \sum_i X_i^2 - (\sum_i X_i)^2}$$

$$a = \frac{n(\sum_i X_i Y_i) - (\sum_i X_i)(\sum_i Y_i)}{n \sum_i X_i^2 - (\sum_i X_i)^2} \tag{1-49}$$

The intercept b and slope a both have some uncertainty. An estimate of the standard deviation of the intercept is

$$\sigma_b = \left[\left(\frac{\sum_i d_i^2}{n-2}\right)\left(\frac{\sum_i X_i^2}{n \sum_i X_i^2 - (\sum_i X_i)^2}\right)\right]^{1/2} \tag{1-50}$$

Similarly the standard deviation of the slope is estimated by the expression

$$\sigma_a = \left[\left(\frac{\sum_i d_i^2}{n-2}\right)\left(\frac{n}{n \sum_i X_i^2 - (\sum_i X_i)^2}\right)\right]^{1/2} \tag{1-51}$$

In passing it is worthwhile to note that some expressions may be rendered linear through a change of variable or of mathematical form. A prominent example is the exponential relation $z = e^{bx}$. Taking logarithms of both sides gives a linear form.

We may ask about the validity of least-squares methods in general and in particular about the equations developed here. We have assumed negligible error in X_i, and if this is not valid, then Eq. (1-41) and those following it must be altered. The general validity of least-squares fitting rests on the assumption that errors in experimental measurements are random and therefore distributed normally about the true value.

Illustrative Example 4 Consider a set of experimentally measured pairs of numbers $\{x_m, y_m\}$ which are presumed to be related linearly. Calculate by a least-squares procedure the best slope and intercept (i.e., the best linear fit) of these data. Estimate the uncertainty in the slope and the intercept.

x_m	y_m
0.0514	1.35
0.0954	1.67
0.162	1.76
0.165	1.99
0.233	2.20
0.300	2.65
0.355	2.93

The quantities which must be calculated are related to the above data by Eqs. (1-49), (1-50), and (1-51). The various parts of these expressions are evaluated separately to give

$$n = 7$$

$$\sum_i (x_i)^2 = 0.3355$$

$$\sum_i x_i = 1.3618$$

$$(\sum_i x_i)^2 = 1.8545$$

$$\sum_i y_i = 14.55$$

$$\sum_i x_i y_i = 3.1899$$

Combining these according to Eq. (1-49) furnishes for the intercept

$$b = \frac{(0.3355)(14.55) - (1.3618)(3.1899)}{(7)(0.3355) - 1.8545} = 1.0884$$

and for the slope

$$a = \frac{(7)(3.1899) - (1.3618)(14.55)}{(7)(0.3355) - 1.8545} = 5.0898$$

Using Eq. (1-50) an estimate of the uncertainty in the intercept is given by using the quantities calculated above together with the calculated values of y and y_c, obtained from the linear equation

$$y_c = 5.0898x_m + 1.0884$$

The values of y_c corresponding to the values of x_m are listed in the following table. Also listed are the deviations $y_m - y_c$, which are denoted d. From the following table

$$\sum_i d_i^2 = 0.0817$$

x_m	y_c	d
0.0514	1.1568	0.1932
0.0954	1.5739	0.0961
0.162	1.9129	-0.1529
0.165	1.9281	0.0620
0.233	2.2742	-0.0742
0.300	2.6152	0.0348
0.355	2.8951	0.0349

The estimated uncertainty in the intercept is then calculated as

$$\sigma_a = \sqrt{\frac{(0.0817)(0.3355)}{5[(7)(0.3355) - 1.8545]}} = 0.1053$$

while the estimated uncertainty in the slope is

$$\sigma_b = \sqrt{\frac{(0.0817)(7)}{5[(7)(0.3355) - 1.8545]}} = 0.4811$$

These uncertainties represent estimates of the standard deviation of the intercept and slope, respectively.

A NOTE ON GRAPHICAL ANALYSIS. It should be apparent to anyone having progressed through a science curriculum as far as physical chemistry that graphs are constantly employed to display and analyze data. A few notes are in order regarding the construction and use of graphs. First, regarding the construction, the following recipe may be helpful:

1. Draw a set of axes.
2. Determine the maximum and minimum values of both the ordinate (y axis) and the abscissa (x axis) of the data to be plotted.
3. Using the values from item 2, mark off the axes with hash marks and number the marks.
4. Label the axes and include units.
5. Plot the data and encircle each point.
6. Connect the data points with a smooth curve. Do not pass the curve through the surrounding circles.
7. Number the figure and add any necessary captions.

Refer to Fig. 1-7 as an illustration of these points. One point not illustrated in Fig. 1-7 is the problem of data points "falling off the curve"—this is intimately related to the least-squares problem discussed above. Suppose, for example, that we have a set of quantities (x_i, y_i) which are linearly related according to theory. If these pairs are determined experimentally and plotted as y_i versus x_i, they will not lie precisely on a straight line. This problem is addressed by the least-

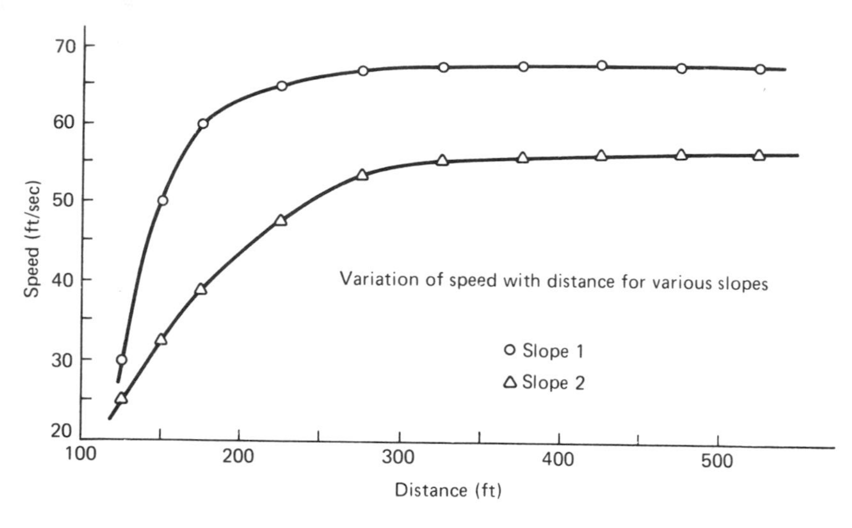

Figure 1-7. Example of graphical presentation of data.

squares curve-fitting method. However, we often skip the least-squares analysis and *eyeball* in what appears to be a good linear fit of the data. In many cases this is adequate, but it is worth keeping in mind that this procedure is not particularly systematic and can lead in some cases to misleading conclusions, especially when comparing two *eyeballed* straight lines.

It is worth noting that integrals can be evaluated by determining graphically (count the squares) the area under the curve of the integrand versus the variable of integration. Suppose, for example, that we wish to evaluate the integral

$$\int_a^b f(x)\, dx$$

To accomplish this graphically we would need a graph of f(x) versus x from at least x = a to x = b. The area under this curve, expressed in the units of the ordinate and abscissa of the graph, is the numerical value of the definite integral.

Finally, we should note that the numerical value of the first derivative of the function f(x) in the above problem can be evaluated at any $x = X_0$ in the range x = a to x = b by drawing a tangent to f(x) at X_0 and determining its slope. This procedure is not particularly reliable, and numerical differentiation methods such as those outlined in Section 1.6 are to be preferred for quantitative results.

PROBLEMS

1. Problem 2 of Section 1.7 uses Eqs. (1-46) to develop a program for linear curve fitting.
2. Suppose that the data set $\{Y_i, X_i\}$ is assumed to be related according to

$$y = ax^2 + bx + c$$

Derive the necessary relations for finding the best a, b, and c in the least-squares sense. The experiment entitled "Fluorescence of Molecular Iodine" (see Section 7.6) makes use of the solution to this problem. If there is sufficient time, write a computer program for fitting to a second-order polynomial.

1.6 Numerical Analysis

The use of numerical analysis in chemical research has become especially widespread since the development of sophisticated digital computers. This situation arises because the present state of theoretical physics does not permit analytical solutions to be found for a wide variety of important and very interesting problems. In the absence of analytical solutions (equations) an approximate but hopefully very accurate solution may be found by numerical analysis. The solution appears as a table of numbers. For example, the dynamics of chemical reactions may be studied with numerical techniques for integration of differential equations; this approach is taken in Section 6.5. Another example is the theoretical development of wave functions for molecular electronic structure in which many complicated definite integrals and matrices are evaluated using numerical methods. This list could be extended almost indefinitely, and because of the widespread applicability of numerical techniques to problems in science and engineering, one is sure to be well repaid for studying numerical methods of analysis. In the material presented here, we have attempted to furnish a brief, but reasonable, survey of several commonly encountered problems and methods. Some of these are used later in the text, especially in Section 1.7. For further study, refer to the texts listed in the References.

Numerical analysis is often described as the art and science of solving mathematical problems numerically. The science of numerical analysis is concerned with the development of methods for the numerical solution of problems. The art lies in choosing the most effective method for the solution of a given problem Given a function f(X), the need for numerical analysis generally arises in two ways: First, f(X) may be known analytically but we may wish to perform operations on f(X) which have no analytic solutions. Consider the following example using the analytic function:

$$f(X) = X + \ell n\,(X)$$

Given a value of X we can find f(X), but the inverse problem—given f(X), find X—has no solution in terms of a finite number of elementary functions. Hence, we must solve this problem numerically. Second, f(X) may be known only at a finite number of points $\{X_i\}$. If we wish to evaluate f(X) between two of the X_i, integrate f(X), or differentiate f(X), we are required to use numerical methods.

BASIC METHOD. The basic approach in numerical analysis is to replace the true solution f(X) by an approximation g(X), usually a linear combination of elementary functions, in terms of which the given problem can be solved. By far the most common functional form for g(X) is a polynomial (Fourier approximations being a major exception). The reason for the overwhelming use of polynomial approximations can be understood as follows.

First it must be remembered, although it is easy to forget, that digital computers can perform only the most elementary arithmetic operations; namely, addition, subtraction, multiplication, and division. Thus, any function which cannot be evaluated by performing a finite number of these four operations must be evaluated in approximate form. This is true for a large number of very elementary functions, for example, sin (X). Since for arbitrary X, sin (X) cannot be evaluated by a finite number of arithmetic operations, a computer must use some approximation such as a truncated Taylor expansion. Thus, to form a Fourier approximation [which is in terms of sin (X) and cos (X)] we would first have to approximate sin (X) and cos (X). Polynomials, on the other hand, are among the most general functions and can be evaluated directly on a computer. They are also easily integrated and differentiated. These features account for their widespread use in approximations. In the remainder of this chapter we shall assume that the approximation function is a polynomial of degree n, $P_n(X)$.

There are three common criteria for the determination of $P_n(X)$:

1. Exact approximation: We require that f(X) and $P_n(X)$ agree exactly at a sequence of points $\{X_i\}$. That is, we require that $f(X_i) = P_n(X_i)$ for all X_i in the selected sequence.
2. Least-squares approximation: We require that

$$\sum_i [f(X_i) - P_n(X_i)]^2$$

be a minimum. When n = 1, that is, when $P_n(X)$ is linear, this is the familiar linear least-squares fitting procedure.
3. Minimax approximation: We require that max $|f(X) - P_n(X)|$ be as small as possible over some interval. That is, we require $P_n(X)$ to be such that the maximum absolute deviation from f(X) in the interval is minimized. This approximation is difficult to obtain and will not be considered further.

Restricting our approach to polynomials does not fix precisely the method of solving a given numerical analysis problem. Generally speaking, several methods of approach to a given problem will still be available, and the selection of one of them depends on consideration of two factors: error and speed. The final selection of a method will frequently represent a compromise between the two.

Errors in numerical methods arise in a variety of ways. Gross errors such as arithmetical errors can occur, but the use of a computer significantly reduces their probability and we shall assume that they do not occur. Random error in empirical data will also propagate through numerical calculations, but such error is always present and cannot be eliminated by numerical analysis. The errors of interest here are those caused by the numerical methods themselves, and they fall into three categories:

1. *Truncation error* is caused by replacing an infinite process by a finite one. For example, most functions have polynomial representations of infinite degree; hence, the use of finite polynomial approximation will inherently introduce error. Other operations such as integration and differentiation are defined as limit processes which cannot be performed on a computer so that an approximation must be used.
2. *Roundoff error* arises because a digital computer can carry only a finite number of decimal places. Since most numbers have infinite decimal representations, we must use approximations of the actual numbers. Roundoff error is most important in long computations where it can build up and introduce substantial inaccuracy in the final result.
3. *Loss of significance* is closely related to roundoff error and generally occurs when the difference between two approximately equal numbers is calculated. Consider the difference

$$\begin{array}{r} 6.2316778 \\ -6.2316754 \\ \hline 0.0000024 \end{array}$$

Although the original numbers have eight significant figures each, the result has only two. If the result is used in further calculations, the accuracy of those calculations will be greatly diminished. Loss of significance can generally be avoided by the use of double-precision arithmetic.

The inherent error in numerical methods is usually the criterion for the choice of one method over another. We first decide what error we can tolerate in the results and then look for a method which will meet or surpass the resulting criterion.

If two methods have equivalent error properties, we would naturally choose the faster of the two for use on a computer. The speed depends on the number of arithmetical operations which must be performed to obtain a desired degree of accuracy. An approximate determination of the speed of a method can be accomplished by considering the number of multiplications, divisions, and library routine evaluations the method requires. Additions and subtractions can generally be neglected because they are very much faster than other operations.

Note that speed can be improved by careful programming. In particular watch for:

1. Repeated calculations: Do not repeat the same calculation over and over; do it once and store it. For example, in finding the roots of a quadratic equation do not evaluate $\sqrt{b^2 - 4ac}$ twice since it is the same for each root. This may seem trivial, but if we are solving 1000 quadratics, it could cut computation time in half.
2. Elementary relations between functions: If your program requires $\cos^2 X$ and $\sin^2 X$, do not calculate both with the library routines. Calculate one and use $\sin^2 X + \cos^2 X = 1$ to find the other.
3. Derivatives: The derivatives of many functions repeat all or part of the functions themselves. Suppose that you need f(X) and f′(X), where

$$f(X) = X^2 \exp\{X\} - X$$
$$f'(X) = X^2 \exp\{X\} + 2X \exp\{X\} - 1$$

   Do not evaluate $X^2 \exp\{X\}$ twice and $\exp\{X\}$ three times. Do each only once.
4. Periodicity: A function is periodic of period p if $f(X + p) = f(X)$. Notable examples are the trigometric functions. Do not evaluate both $\sin(\pi/4)$ and $\sin(9\pi/4)$ since they are the same number.

There are many other opportunities to improve speed, and efficient programming arises from thought and experience.

INTERPOLATION. We shall now consider interpolation as our first example of numerical analysis. Given m + 1 values of a function f(X) (unknown generally) at the points $\{a_i\}$, i = 0, 1, ..., m, we wish to find a suitable polynomial approximation to f(X). We require that $P_n(a_i) = f(a_i)$; hence, we seek an exact approximation. To determine $P_n(X)$ we have to find its degree n and n + 1 coefficients. It can be shown that m + 1 points are sufficient to determine a unique polynomial of degree m. This is just a generalization of the rule that two points determine a first-order polynomial (straight line). Further, if f(X) were actually a polynomial of degree m, then $P_n(X)$ and f(X) can be made identical (no error) if we choose n = m. We can do no better than this and so we set n = m. To determine the coefficients we write $P_m(X)$ as

$$\sum_{j=0}^{m} \ell_j(X) f(a_j)$$

where the $\ell_j(X)$ are also polynomials. From our exactness criterion we require that

$$\ell_j(a_i) = \delta_{ij} = \begin{cases} 0, & \text{if } j \neq i \\ 1, & \text{if } j = i \end{cases} \tag{1-52}$$

We can satisfy the first requirement $\ell_j(a_i) = 0$ by including in $\ell_j(X)$ a factor

$$[X - a_0]\,[X - a_1]\,\ldots\,[X - a_{j-1}]\,[X - a_{j+1}]\,\ldots\,[X - a_m] \tag{1-53}$$

Clearly, this factor is zero for $X = a_i$, $i \neq j$. The requirement $\ell_j(a_j) = 1$ can be accomplished by dividing (1-53) by

$$[a_j - a_0]\,[a_j - a_1]\,\ldots\,[a_j - a_{j-1}]\,[a_j - a_{j+1}]\,\ldots\,[a_j - a_m] \tag{1-54}$$

or

$$\ell_j(X) = \frac{[X - a_0]\ \ldots\ [X - a_{j-1}]\,[X - a_{j+1}]\ \ldots\ [X - a_m]}{[a_j - a_0]\ \ldots\ [a_j - a_{j-1}]\,[a_j - a_{j+1}]\ \ldots\ [a_j - a_m]} \tag{1-55}$$

Then

$$P_m(X) = \sum_{j=0}^{m} f(a_j) \prod_{\substack{i=0 \\ i \neq j}}^{m} \frac{[X - a_i]}{[a_j - a_i]} \tag{1-56}$$

where Π means product. On a computer $P_m(X)$ can be evaluated easily using DO loops (see Section 1.7). Equation (1-56) is known as the Lagrangian interpolation formula. The error is given by

$$E = \frac{f^{m+1}(\alpha)}{[m + 1]!} \prod_{j=0}^{m} (\alpha - a_j) \tag{1-57}$$

where $f^{m+1}(\alpha)$ is the $(m + 1^{th})$ derivative of f(X) evaluated at an unknown point in the interval $[a_o, a_m]$. This error term illustrates the uncertainty inherent in numerical methods when we have no knowledge of f(X). However, we would hope that if we make m large enough, the $[m + 1]!$ term will make the error small. With no knowledge of f(X), a useful procedure is to calculate several $P_m(X)$, increasing m (the number of points −1) each time until the required number of decimal places remains unchanged by the inclusion of additional points. This procedure is often carried over into other numerical methods. Note that if $X = a_j$ for $j = 0, \ldots m$, the error becomes zero, which agrees with our original criterion for determining $P_m(X)$. Also, the error is zero if f(X) is a polynomial of degree $\leqslant m$ since $f^{m+1}(\alpha)$ is zero.

Illustrative Example 5 Consider the following example of interpolation. Calculate by interpolation $y = \ell n\,(0.85)$ given

$\ell n\,(0.80) = -0.22314, \quad \ell n\,(0.90) = 0.10536, \quad \ell n\,(1.00) = 0.00000$

Since we have three values of f(X), we set m = 2. Then from Eq. (1-55)

$$\ell_0 = \frac{(0.85 - 0.90)(0.85 - 1.00)}{(0.80 - 0.90)(0.80 - 1.00)} = 0.375$$

$$\ell_1 = \frac{(0.85 - 0.80)(0.85 - 1.00)}{(0.90 - 0.80)(0.90 - 1.00)} = 0.750$$

$$\ell_2 = \frac{(0.85 - 0.80)(0.85 - 0.90)}{(1.00 - 0.80)(1.00 - 0.90)} = -0.125$$

and from Eq. (1-56) $\ell n\,(0.85) \simeq P_2(0.85) = (0.375)(-0.22314) + (0.750)(-0.10536) + (-0.125)(0.00000) = -0.16269$. The correct answer is -0.16251, and the relative error is 0.13%. You can verify that the actual error is in the range predicted by the error formula.

QUADRATURE. The numerical evaluation of the definite integral

$$I = \int_a^b f(X)\, dX$$

is called numerical quadrature, the term numerical integration being reserved for the integration of differential equations. Again we seek to approximate f(X) with $P_n(X)$ since we can integrate a polynomial easily.

Suppose that f(X) is known at a and b. Then we can approximate f(X) by a straight line joining a and b. The result is shown in Fig. 1-8. The approximate value of I is just the area of the hatched trapezoid: $I \simeq (b - a)\,\{\frac{1}{2}f(a) + \frac{1}{2}f(b)\}$ and will very likely be quite inaccurate. If we can evaluate f(X) at points intermediate on [a, b], we can use a higher-order polynomial approximation. However, there are strong reasons for not doing this. A better procedure is to divide [a, b] into a number, m, of subintervals of length h = (b - a)/m. We then approximate f(X) by a straight line over each subinterval. The result for m = 6 is shown in Fig. 1-9. The approximate value of I is just the sum of the areas of the small trapezoids. If we denote f(a + nh) by f_n, we obtain, for the general case of m subintervals,

$$I \simeq h\left[\frac{1}{2}f_0 + f_1 + f_2 + \ldots + f_{m-2} + f_{m-1} + \frac{1}{2}f_m\right]; \; h = \frac{[b - a]}{m} \qquad (1\text{-}58)$$

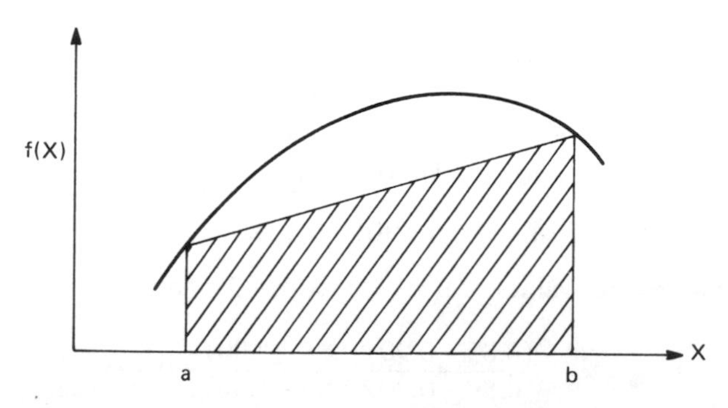

Figure 1-8. First approximation for quadrature on [a, b].

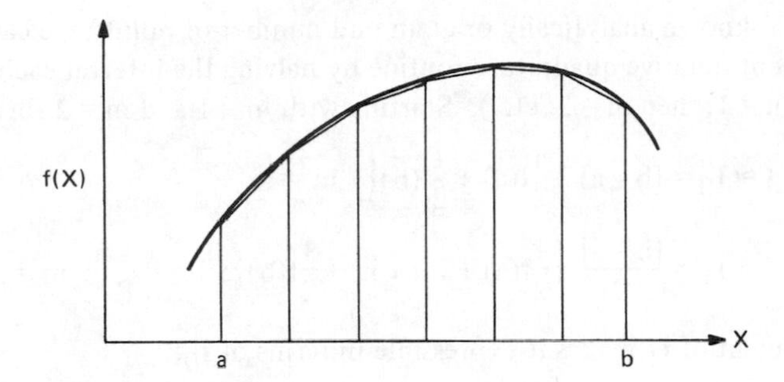

Figure 1-9. Extended approximation for quadrature on [a, b].

This formula is called the trapezoidal rule. The error is given by

$$E = -\frac{mh^3 f''(X)}{12} = -\frac{(b - a)^3}{12m^2} f''(X) \tag{1-59}$$

As $m \longrightarrow \infty$, E falls to zero so we can get arbitrarily close to the true I by making m large enough, provided f(X) is continuous on [a, b]. Again, a good procedure is to calculate several approximations with increasing m until the required number of decimal places stabilizes. This stabilization is known as convergence.

Convergence in this case can be accelerated considerably. Consider I_1 and I_2, two trapezoidal approximations calculated with m_1 and m_2 subintervals, respectively. Then we can write the true value I as

$$I = I_1 - \frac{(b - a)^3}{12m_1^2} f''(X_1) \tag{1-60}$$

$$I = I_2 - \frac{(b - a)^3}{12m_2^2} f''(X_2)$$

Although X_1 and X_2 may be different points, let us assume that $f''(X_1)$ and $f''(X_2)$ are approximately equal. Then we can solve approximately for I and obtain

$$I \simeq \frac{(m_2^2/m_1^2)(I_2 - I_1)}{(m_2^2/m_1^2) - 1} \tag{1-61}$$

If the above assumption is valid, this result should be better than either I_1 or I_2. This process is called Richardson extrapolation.

If f(X) is known analytically or at an odd number of points, we can obtain a very efficient iterative quadrature routine by halving the interval each time (i.e., choosing m = 1, then m = 2, etc.). Starting with m = 1 and m = 2 furnishes

$$I \simeq I_0 = [b - a]\left\{\frac{1}{2}f(a) + \frac{1}{2}f(b)\right\}; \quad m = 1$$

$$I \simeq I_1 = \frac{[b - a]}{2}\left\{\frac{1}{2}f(a) + f(a + h) + \frac{1}{2}f(b)\right\}; \quad h = \frac{b - a}{2}; \quad m = 2 \tag{1-62}$$

Rearrangement of I_1 makes it expressible in terms of I_0:

$$I_1 = \frac{[b - a]}{2}\left\{\frac{1}{2}f(a) + \frac{1}{2}f(b)\right\} + hf(a + h)$$

$$I_1 = I_0/2 + hf(a + h) \tag{1-63}$$

Hence, we do not have to repeat any computations in order to get I_1 if we already have I_0. If we let I_k be the trapezoid rule for 2^k subintervals, we obtain, for the general case,

$$I_k = \frac{1}{2}I_{k-1} + h \sum_{\substack{n=1 \\ n \text{ odd}}}^{2^k} f(a + hn); \quad h = \frac{[b - a]}{2^k} \tag{1-64}$$

Also note that h for 2^k subintervals is just half of h for 2^{k-1} subintervals. Starting a DO loop at n = 1 and incrementing by 2 is a convenient way to pick up the f(a + nh) terms for odd n.

Since $m_k^2 = \frac{1}{4}m_{k+1}^2$ (i.e., $m_k = \frac{1}{2}m_{k+1}$), the Richardson extrapolation becomes

$$I \simeq \frac{1}{3}[4I_k - I_{k-1}] \tag{1-65}$$

This procedure is quite efficient since calculation of successive I_k and the extrapolation require relatively few operations. A generalization of this procedure, called Romberg integration, is one of the fastest and most versatile quadrature routines in existence.

Illustrative Example 6 As an example of quadrature, consider the following definite integral. Evaluate $\int_1^3 dx/x$ to an accuracy of three decimal places with and without Richardson extrapolation.

Repeated application of Eq. (1-64) yields successively

$$I_0 = (3 - 1)\left[\frac{1}{2}(1) + \frac{1}{2} \cdot \frac{1}{3}\right] = 1.3333$$

$$I_1 = \frac{I_0}{2} + 1\left[\frac{1}{2}\right] = 1.1667$$

$$I_2 = 1.1166$$

$$I_3 = 1.1032$$

$$I_4 = 1.0998; \; 2^4 = 16 \text{ subintervals}$$

$$I_5 = 1.0989$$

$$I_6 = 1.0987; \; 2^6 = 64 \text{ subintervals}$$

Using Richardson extrapolation [Eq. (1-65)] we get successively

$$I \simeq \frac{1}{3}(4I_1 - I_0) = \frac{1}{3}(4.6668 - 1.3333) = 1.1112$$

$$I \simeq \frac{1}{3}(4I_2 - I_1) = 1.1000$$

$$I \simeq \frac{1}{3}(4I_3 - I_2) = 1.0987$$

$$I \simeq \frac{1}{3}(4I_4 - I_3) = 1.0986$$

Hence, with Richardson extrapolation we get the same accuracy with only 16 subintervals.

DIFFERENTIATION. To differentiate a function f(X), we could approximate it with a polynomial and then differentiate the polynomial. If we differentiate the Langrangian interpolation polynomial $P_m(X)$, the error becomes

$$E(X) = \frac{f^{(m+2)}(\alpha_1)}{(m+2)!} \sum_{j=0}^{m} (X - a_j) + \frac{f^{(m+1)}(\alpha_2)}{(m+1)!} \frac{d}{dx} \sum_{j=0}^{m} (X - a_j) \tag{1-66}$$

Evaluating derivatives by this exact approximation method is very risky, especially if f(X) is known only from experiment. The reason can be shown as follows.

Consider a function f(X) which is actually a straight line. We make three measurements of f(X) and, because of random experimental error, they do not fall precisely on a straight line. If we then approximate f(X) by $P_2(X)$, the result is shown in Fig. 1-10. Although $P_2(X)$ is not a bad approximation of f(X) itself, the derivative of $P_2(X)$ is a very poor approximation of the derivative of f(X). The fluctuations caused by random experimental error cause $P_2(X)$ and the derivatives to fluctuate in a manner which does not at all resemble the derivatives of f(X). As a result numerical differentiation of exact approximation to experimental data can often be disastrous.

A much better procedure, especially if the form of f(X) is known, is to use a least-squares approximation. Least-squares approximations tend to smooth out the random fluctuations in experimental data; hence, their derivatives are likely to approximate the derivatives of experimental functions better than the derivatives of exact approximations.

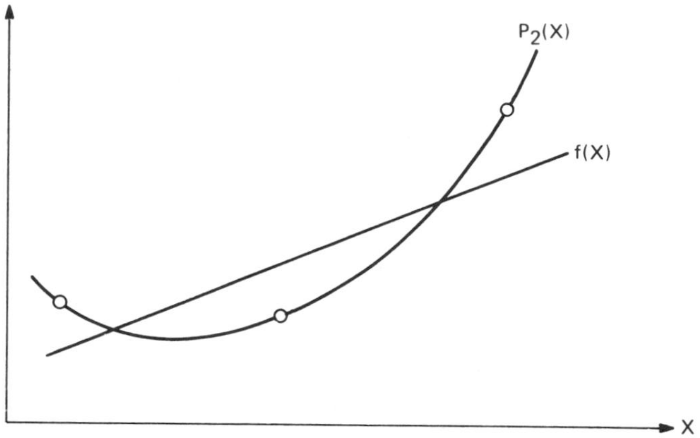

Figure 1-10. Exact approximation of three points.

NUMERICAL INTEGRATION. The solution of many differential equations cannot be found analytically, and in such a situation the ready availability of digital computers makes a numerical solution attractive. In what follows we shall limit ourselves to a single first-order equation; the results, however, may be readily generalized to any number of simultaneous equations or to equations of an order that is higher than first order. There is also a wide variety of methods. We shall discuss only two: the second-order predictor-corrector and the second-order Runge-Kutta methods. With these considerations in mind we shall consider the following problem: Given the differential equation

$$\frac{dy}{dx} = f(x, y) \tag{1-67}$$

and the boundary condition

$$y(x_0) = y_0 \tag{1-68}$$

and assuming the existence of a solution of (1-67) on the interval $[x_0, b]$, find a solution $y(x_i)$ at the points $\{x_i\}$, $i = 1, 2, \ldots$ with x_i equally spaced a distance h apart.

The general method of attack is to approximate by numerical means a solution of (1-67) at x_1—call it Y_1—which makes use of y_0 and $f(x_0, y_0)$ and then to obtain an approximate solution Y_2 at x_2 using x_0, x_1, $f(x_0, y_0)$, and $f(x_1, Y_1)$, and so forth. The result, illustrated in Fig. 1-11, is a stepwise solution of the differential equation in which the solution at x_i is based on approximate solutions at previous points. Because the solution at a given point is based partially on values calculated for previous points, errors can arise not only because of roundoff and truncation at the particular point being evaluated but also because of error in pre-

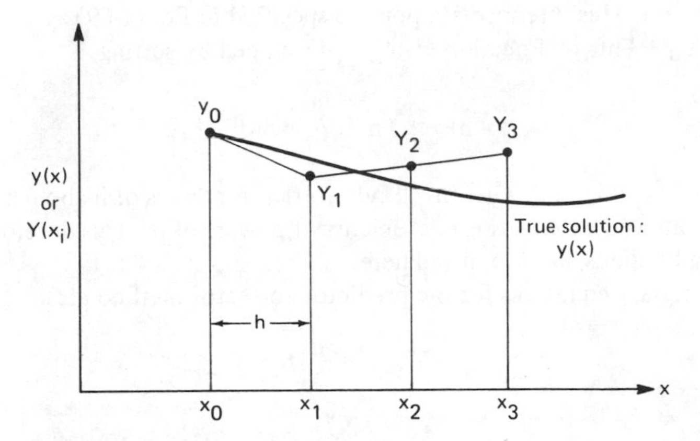

Figure 1-11. Schematic numerical integration.

vious points. We say that error has propagated when the latter occurs. It is important in numerical integration to be able to estimate how the error propagates as well as to be able to estimate the speed, roundoff error, and truncation error of the method. Figure 1-11 illustrates the step size h, and it would appear intuitively that a more accurate solution could be obtained if h were decreased. Of course, decreasing h will increase the time required to find the solution on the interval $[x_0, b]$ and so we are forced to compromise, at some point, speed and accuracy.

Two other considerations are the order of the method and how to start the solution. The order of the method is n if the formula used to approximate the solution is an exact approximation when the true solution is a polynomial of degree n. To start a solution we must choose a method which is self-starting [i.e., given (1-67) and (1-68), Y_1 can actually be generated]. The Runge-Kutta method is self-starting. Other methods such as the predictor-corrector method have features which may be desirable but are not self-starting. For example, the predictor-corrector method is generally faster and has better error properties than the Runge-Kutta method. In the second-order predictor-corrector formulas the quantities y_0, and Y_1 are needed to calculate Y_2, then Y_1 and Y_2 are needed for Y_3, etc., and so one commonly used approach is to obtain Y_1 by the Runge-Kutta method and then switch to the predictor-corrector method for the rest of the solution. This is accomplished quite readily on a computer.

The necessary equation for the second-order Runge-Kutta method is

$$Y_{n+1} = Y_n + hf\left(x_n + \frac{1}{2}h, Y_n + \frac{1}{2}hf_n\right) \tag{1-69}$$

This equation depends only on the approximate solution at the previous point, Y_n, the step size, h, and the magnitude, f, of the differential equation at an intermediate point. This intermediate point is specified in Eq. (1-69) as $x = [x_n + \frac{1}{2}h]$ and $y = [Y_n + \frac{1}{2}hf_n]$. Equation (1-69) is developed by setting

$$Y_{n+1} - Y_n = \sum_{i=1}^{m} a_i g_i(h, f) \qquad (1\text{-}70)$$

and then expanding both sides in a Taylor series in powers of h about the point (x_n, Y_n) followed by equating coefficients of powers of h. This development is lengthy and tedious and is omitted here.

The necessary equations for the predictor-corrector method are

$$Y_{n+1} = Y_{n-2} + \frac{3h}{2}[f_n + f_{n-1}] \qquad (1\text{-}71)$$

and

$$Y_{n+1} = Y_n + \frac{h}{2}[f_{n+1} + f_n] \qquad (1\text{-}72)$$

Equation (1-71) is called the predictor while (1-72) is known as the corrector. Both of these equations estimate Y_{n+1} and they are used in an iterative fashion as follows: First, (1-71) is used to estimate Y_{n+1}; then (1-72) is evaluated, the differential equation magnitude f_{n+1} being estimated using x_{n+1} and the estimate of Y_{n+1} just calculated. This furnishes a new estimate for Y_{n+1}, which can then be inserted in a second cycle through Eq. (1-72). This iterative process is repeated until the change in Y_{n+1} from one cycle to the next is less than some predetermined value. At this point the solution is said to have converged. We are then ready to obtain the solution at Y_{n+2}.

Illustrative Example 7 Consider the following example of numerical integration: Integrate numerically the equation $dy/dx = xy$ using $h = 0.01$ and $y(0) = 1.0$. First we make use of the Runge-Kutta second-order equation to find Y(0.01) as

$$Y(0.01) = 1.0 + 0.01\,[(0.005)(1.0) + 0.0]$$

$$= 1.000050$$

The true solution is $y(0.01) = 1.000050001$. With both y(0) and Y(0.01) at hand we turn to the second-order predictor-corrector method in continuing the solution. The predictor becomes

$$Y(0.02) = 1.0 + \frac{3}{2}(0.01)(1.00005)(0.01) + 0$$

$$= 1.00015$$

The first cycle (iteration) through the corrector furnishes

$$Y(0.02) = 1.00005 + (0.005)(0.02)(1.00015) + (1.00005)(0.01)$$

$$= 1.00020$$

The second iteration yields

$$Y(0.02) = 1.00005 + (0.005)(0.02)(1.00020) + (1.00005)(0.01)$$
$$= 1.00020$$

implying convergence to five-decimal accuracy. The true solution is $y(0.02) = 1.000200019$.

In most calculations higher-order methods will normally be used because they have better convergence and error properties. For example, Section 6.5 makes use of fourth-order Runge-Kutta and predictor-corrector methods. A further problem in some cases is lack of convergence. However, each problem has its own convergence considerations and so we omit further discussion and at this point refer the reader to the References.

ROOTS OF NONLINEAR EQUATIONS. Here we are looking for real solutions to the equation $f(X) = 0$, where $f(X)$ may be almost any function including a polynomial. Note that we also cover equations of the type $h(X) = g(X)$, since if we define $f(X)$ as $h(X) - g(X)$, then $f(X) = 0$ is equivalent to $h(X) = g(X)$. We denote the true root as β, that is, $f(\beta) = 0$.

All the methods we shall cover take an initial approximation and seek repeatedly to improve it. Such methods are called iterative methods, and the calculation of a new approximation from the previous one (ones) is called one iteration. If $\text{limit}_{i\to\infty}|\beta - X_i| = 0$, where X_i is the ith approximation to β, the method is convergent and would give the true answer (zero error) if we could perform an infinite number of iterations. Again our method will be to continue iterating until the required number of decimal places stabilizes. If two methods are convergent, we are interested in their rates of convergence, that is, the number of iterations required to obtain a given accuracy.

Iterative methods fall naturally into two classes. Always-convergent methods converge no matter how bad our initial guess at β is. Conditionally convergent methods may or may not converge depending on how bad our initial guess is. Always-convergent methods generally converge slowly, while conditionally convergent methods are much faster, if they converge at all.

We shall consider an always-convergent method. Suppose we know that $f(X)$ has opposite sign at two points X_1 and X_2. Then continuity tells us that $f(X)$ has at least one zero on $[X_1, X_2]$. We approximate $f(X)$ by $P_1(X)$ and, letting $f(X_1) = f_1$ and $f(X_2) = f_2$, we have

$$f(X) \simeq P_1(X) = \frac{[X - X_2]}{[X_1 - X_2]} f_1 + \frac{[X - X_1]}{[X_2 - X_1]} f_2 \tag{1-73}$$

Since we seek X such that $f(X) = 0$, we set $P_1(X) = 0$ and solve for X, calling it X_3:

$$X_3 = X_2 \frac{f_1}{[f_1 - f_2]} + X_1 \frac{f_2}{[f_2 - f_1]} \tag{1-74}$$

where X_3 is an approximation of β. Now we compute $f(X_3)$ and compare its sign with $f(X_1)$ and $f(X_2)$. If $f(X_3)$ and $f(X_2)$ have the same sign, we replace f_2 with f_3 and X_2 with X_3. If f_1 and f_3 have the same sign, f_1 becomes f_3 and X_1 becomes X_3. Then we repeat the above calculation and obtain a new X_3. The successive X_3s represent better and better approximations of β. This is called the method of false position, or false position for short.

False position always converges provided that f(X) changes sign on either side of β. This will be true unless β has even multiplicity. The multiplicity of a root is the number of times the same root appears. For example, the roots of $X^2 + 2X + 1 = 0$ are both the same and so the multiplicity (i.e., 2) is even and false position fails. There are ways out of this difficulty, but we shall not consider them here.

Although false position is absolutely convergent (for roots of odd multiplicity), it converges slowly. We shall now consider a faster, although conditionally convergent, method. Define h such that $\beta = X_i + h$, where X_i is the ith approximation to β. We expand $f(X_i + h)$ in a Taylor series and keep only two terms:

$$f(X_i + h) = f(X_i) + hf'(X_i) \tag{1-75}$$

where $f'(X_i)$ is the first derivative of f evaluated at the point $x = X_i$. Since $f(X_i + h) = f(\beta) = 0$, we have

$$h \simeq -\frac{f(X_i)}{f'(X_i)} \tag{1-76}$$

Our iteration scheme is

$$X_{i+1} = X_i - \frac{f(X_i)}{f'(X_i)} \tag{1-77}$$

This is the Newton-Raphson method, and it is much faster than false position, although it may not converge. If we have very poor knowledge of the location of β, a good procedure is to start with false position, get a good approximation to β, and then switch to the Newton-Raphson method as shown in the following example.

Illustrative Example 8 Solve $\ln(X) + X = 0$ by false position with $X_1 = 0.40000$ and $X_2 = 1.00000$ to an accuracy of five decimal places. We have $f(X) = \ln(X) + X$:

$$f_1 = f(0.400000) = -0.51629$$

$$f_2 = f(1.00000) = 1.00000$$

$$\beta \simeq X_3 = \frac{(1.00000)(0.40000)}{(1.00000 + 0.51629)} + \frac{(-0.51629)(1.00000)}{(-0.51629 - 1.00000)}$$

$$X_3 = 0.60430, \qquad f(X_3) = 0.40000$$

Since $f(X_3)$ and $f(X_2)$ have the same sign, we replace X_2 with X_3 and f_2 with f_3 and continue:

$$X_3 = \frac{(0.10061)(0.40000)}{(0.10061 + 0.51629)} + \frac{(-0.51629)(0.60430)}{(-0.51629 - 0.10061)}$$

$$X_3 = 0.57098$$

and continuing we get for successive X_3

$$X_3 = 0.56755$$

$$X_3 = 0.56719$$

$$X_3 = 0.56715$$

$$X_3 = 0.56714$$

$$X_3 = 0.56714 \quad \text{(seven iterations)}$$

If instead we take our second estimate of β (0.57098) and switch to the Newton-Raphson method, we get

$$X_0 = 0.57098$$

$$f'(X) = \frac{1}{X} + 1$$

$$X_1 = 0.57098 - \frac{0.01058}{(1/0.57098 + 1)} = 0.56714$$

$$X_2 = 0.56714 \quad \text{(two iterations)}$$

The faster convergence of the Newton-Raphson method is obvious.

SYSTEMS OF LINEAR EQUATIONS. It is not uncommon in physical problems to encounter a system of linear equations of the form

$$\begin{array}{l} a_{11}X_1 + a_{12}X_2 + \cdots + a_{1n}X_n = a_{1,n+1} \\ a_{21}X_1 + a_{22}X_2 + \cdots + a_{2n}X_n = a_{2,n+1} \\ \vdots \\ a_{n1}X_1 + a_{n2}X_2 + \cdots + a_{nn}X_n = a_{n,n+1} \end{array} \tag{1-78}$$

where our goal is to find the set of X_i values which will satisfy this simultaneous set of equations. An analytical solution to this problem exists and is furnished by by Cramer's rule for determinants. Since the evaluation by the normal method of an nth order determinant requires more than n! multiplications, even a small system requires a fantastic amount of computation if Cramer's rule is used. We seek a faster method.

We start by multiplying the first row by a_{i1}/a_{11} and subtracting the result from the ith row for i = 2, n. The result is

$$\begin{array}{l} a_{11}X_1 + a_{12}X_2 + \cdots + a_{1n}X_n = a_{1n+1} \\ 0 \quad + a_{22}^{(1)}X_2 + \cdots + a_{2n}^{(1)}X_n = a_{2n+1}^{(1)} \\ \cdot \qquad \cdot \qquad \cdot \qquad \cdot \\ \cdot \qquad \cdot \qquad \cdot \qquad \cdot \\ \cdot \qquad \cdot \qquad \cdot \qquad \cdot \\ 0 \quad + a_{nn}^{(1)}X_2 + \cdots + a_{nn}^{(1)}X_n = a_{nn+1}^{(1)} \end{array} \tag{1-79}$$

We have eliminated all elements in the first column except $a_{11}X_1$. The result is called the first derived system, denoted by superscript 1. We now multiply the second row by $a_{i2}^{(1)}/a_{22}^{(1)}$ and subtract from the ith row, i = 3 ... n. This gives the second derived system:

$$\begin{array}{r} a_{11}X_1 + a_{12}X_2 + a_{13}X_3 + \cdots + a_{1n}X_n = a_{1n+1} \\ a_{22}^{(1)}X_2 + a_{23}^{(1)}X_3 + \cdots + a_{2n}^{(1)}X_n = a_{2n+1}^{(1)} \\ a_{33}^{(2)}X_3 + \cdots + a_{3n}^{(2)}X_n = a_{3n+1}^{(2)} \\ \cdot \qquad \cdot \qquad \cdot \\ \cdot \qquad \cdot \qquad \cdot \\ \cdot \qquad \cdot \qquad \cdot \\ a_{n3}^{(2)}X_3 + \cdots + a_{nn}^{(2)}X_n = a_{nn+1}^{(2)} \end{array} \tag{1-80}$$

If we continue, we get

$$\begin{array}{r} a_{11}X_1 + a_{12}X_2 + a_{13}X_3 + \cdots + a_{1n}X_n = a_{1n+1} \\ a_{22}^{(1)}X_2 + a_{23}^{(1)}X_3 + \cdots + a_{2n}^{(1)}X_n = a_{2n+1}^{(1)} \\ a_{33}^{(2)}X_3 + \cdots + a_{3n}^{(2)}X_n = a_{3n+1}^{(2)} \\ \cdot \qquad \cdot \qquad \cdot \\ \cdot \qquad \cdot \qquad \cdot \\ \cdot \qquad \cdot \qquad \cdot \\ a_{nn}^{(n-1)}X_n = a_{nn+1}^{(n-1)} \end{array} \tag{1-81}$$

where

$$a_{ij}^{(k)} = a_{ij}^{(k-1)} - \frac{a_{ik}^{(k-1)}}{a_{kk}^{(k-1)}} a_{kj}^{(k-1)}; \qquad \begin{array}{l} k = 1, n-1 \\ j = k+1, \ldots, n+1 \\ i = k+1, \ldots, n \end{array} \tag{1-82}$$

Then from Eq. (1-81) we have

$$X_n = \frac{a_{nn+1}^{(n-1)}}{a_{nn}^{(n-1)}} \tag{1-83}$$

and

$$X_i = \frac{1}{a_{ii}^{(i-1)}} \left[a_{in+1}^{(i-1)} - \sum_{j=i+1}^{n} a_{ij}^{(i-1)} X_j \right]$$

This procedure is called Gaussian elimination. The calculation of the X_is is called back substitution. The X_is are the true X_is except for roundoff. If at any stage

$$a_{kk}^{(k-1)} = 0$$

two rows or columns can be interchanged to get a nonzero element in kk position. This procedure requires on the order of n^3 multiplications and so it is much faster than Cramer's rule.

Many of the problems in the solution of linear equations occur when the system is ill-conditioned. Ill-conditioning is best illustrated by giving an example. Following Ralston (1965) we consider the system

$$2X_1 + 6X_2 = 8$$

$$2X_1 + 6.00001X_2 = 8.00001$$

where $X_1 = 1$ and $X_2 = 1$ are the true solutions. Consider a second system

$$2X_1 + 6X_2 = 8$$

$$2X_1 + 5.99992X_2 = 8.00002$$

where $X_1 = 10$ and $X_2 = -2$ are the true solutions. Here a gross change in the solution is caused by a tiny change in the coefficients. This system is ill-conditioned, and if, for example, the coefficients are experimental data with limited accuracy, it will be nearly impossible to obtain an accurate solution for this system. Because of ill-conditioning and the large number of calculations required to obtain solutions, roundoff error is very important in the solution of linear systems. If the coefficients are known accurately enough to warrant it, it is always a good idea to use double-precision arithmetic when solving linear systems.

There are endless variations of Gaussian elimination. The simple form given here is satisfactory for small systems ($n < 15$). For larger systems, shifting the derived systems in and out of the computer memory can occupy as much time as actually doing the calculations. Consult the References for more efficient methods.

Illustrative Example 9 Consider the following example of Gaussian elimination: Solve the system

$$
\begin{aligned}
X_1 + X_2 + X_3 + X_4 &= 4 \\
X_1 + X_2 - X_3 + 2X_4 &= 3 \\
2X_1 + X_2 - 2X_3 - X_4 &= 0 \\
X_1 - X_2 - X_3 - X_4 &= -2
\end{aligned}
$$

The first derived system is

$$
\begin{aligned}
X_1 + X_2 + X_3 + X_4 &= 4 \\
-2X_3 + X_4 &= -1 \\
-X_2 - 4X_3 - 3X_4 &= -8 \\
-2X_2 - 2X_3 - 2X_4 &= -6
\end{aligned}
$$

Since $a_{22}^{(1)}$ is zero, we interchange rows 2 and 3:

$$
\begin{aligned}
X_1 + X_2 + X_3 + X_4 &= 4 \\
-X_2 - 4X_3 - 3X_4 &= -8 \\
-2X_3 + X_4 &= -1 \\
-2X_2 - 2X_3 - 2X_4 &= -6
\end{aligned}
$$

Continuing, we get for the final system

$$
\begin{aligned}
X_1 + X_2 + X_3 + X_4 &= 4 \\
-X_2 - 4X_3 - 3X_4 &= -8 \\
-2X_3 + X_4 &= -1 \\
7X_4 &= 7
\end{aligned}
$$

Back substitution gives $X_1 = X_2 = X_3 = X_4 = 1$.

1.7 Computer Programming

The increased use of digital computers in our society is self-evident. On every hand it seems we are presented with information processed by a digital computer; our bank statements, school records, financial obligations, and hour-exam scores are evident examples. Scientific research is no exception; both experimental and theoretical research, more often than not, involve computer calculations in some way. With a little thought and experience a scientist can readily appreciate the importance of understanding the elements of writing computer programs. The discussion which follows is intended as an introduction to writing simple programs. For further information the reader should consult the References and the systems manuals describing the particular computer being used.

PROGRAM OUTLINE. A digital computer is useless until we are able to communicate with it, and computer programs are the medium by which this is accomplished. For the purposes of this discussion we shall use only one of the many languages of computer communication, FORTRAN. We shall further assume that the communication is achieved by means of a deck of cards, as contrasted to teletype, magnetic tape, etc. With these assumptions a typical program outline is that shown in Fig. 1-12. Each division in Fig. 1-12 represents a certain sequence of cards, each of which has information encoded on it in language the computer system "understands." A typical card is shown in Fig. 1-13.

The first group of cards in the program, control cards, furnishes the system with information about who is programming, the size of the program, the time required to complete the computation, the programming language, etc. Each system has its own requirements; hence, the reader is referred to the appropriate systems manuals which are written for each computation system.

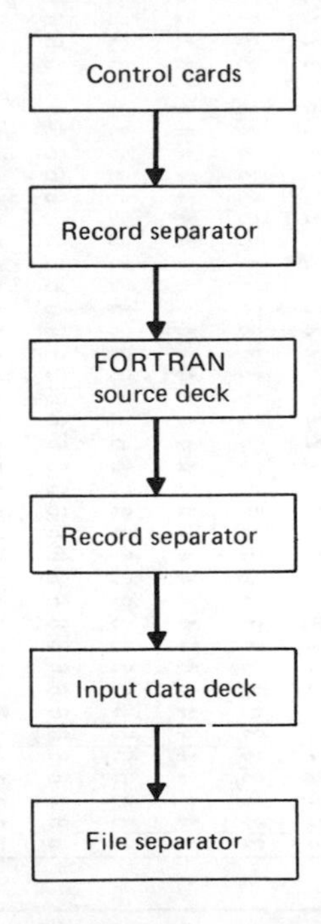

Figure 1-12. Computer program outline.

129 X=A+B/C

C ← FOR COMMENT

STATEMENT NUMBER

CONTINUATION

FORTRAN STATEMENT

IDENTIFICATION

0 0 0 0 0 0 ■ 0 0 0 0 ■ 0 0 0 0 0 0 0 0 0 0 0 0 0 0 0 0 0 0 0 0 0 0 0 0 0 0 0 0 0 0 0 0 0 0 0 0 0 0 0 0 0 0 0 0 0 0 0 0 0 0 0 0 0 0 0 0 0 0 0 0 0 0 0 0

1 2 3 4 5 6 7 8 9 10 11 12 13 14 15 16 17 18 19 20 21 22 23 24 25 26 27 28 29 30 31 32 33 34 35 36 37 38 39 40 41 42 43 44 45 46 47 48 49 50 51 52 53 54 55 56 57 58 59 60 61 62 63 64 65 66 67 68 69 70 71 72 73 74 75 76 77 78 79 80

1 1 ■ 1 1 1 1 1 ■ 1 1 ■ 1 1 1 1 1 1 1 1 1 1 1 1 1 1 1 1 1 1 1 1 1 1 1 1 1 1 1 1 1 1 1 1 1 1 1 1 1 1 1 1 1 1 1 1 1 1 1 1 1 1 1 1 1 1 1 1 1 1 1 1 1 1 1 1 1 1 1 1

2 2 2 ■ 2 2 2 2 2 2 ■ 2 2 2 2 2 2 2 2 2 2 2 2 2 2 2 2 2 2 2 2 2 2 2 2 2 2 2 2 2 2 2 2 2 2 2 2 2 2 2 2 2 2 2 2 2 2 2 2 2 2 2 2 2 2 2 2 2 2 2 2 2 2 2 2 2 2 2 2 2

3 3 3 3 3 3 3 ■ 3 3 3 3 ■ 3 3 3 3 3 3 3 3 3 3 3 3 3 3 3 3 3 3 3 3 3 3 3 3 3 3 3 3 3 3 3 3 3 3 3 3 3 3 3 3 3 3 3 3 3 3 3 3 3 3 3 3 3 3 3 3 3 3 3 3 3 3 3 3 3

4 4 4 4 4 4 4 4 4 4 4 4 4 4 4 4 4 4 4 4 4 4 4 4 4 4 4 4 4 4 4 4 4 4 4 4 4 4 4 4 4 4 4 4 4 4 4 4 4 4 4 4 4 4 4 4 4 4 4 4 4 4 4 4 4 4 4 4 4 4 4 4 4 4 4 4 4 4 4 4

5 5 5 5 5 5 5 5 5 5 5 5 5 5 5 5 5 5 5 5 5 5 5 5 5 5 5 5 5 5 5 5 5 5 5 5 5 5 5 5 5 5 5 5 5 5 5 5 5 5 5 5 5 5 5 5 5 5 5 5 5 5 5 5 5 5 5 5 5 5 5 5 5 5 5 5 5 5 5 5

6 6 6 6 6 6 6 6 6 6 6 6 6 6 6 6 6 6 6 6 6 6 6 6 6 6 6 6 6 6 6 6 6 6 6 6 6 6 6 6 6 6 6 6 6 6 6 6 6 6 6 6 6 6 6 6 6 6 6 6 6 6 6 6 6 6 6 6 6 6 6 6 6 6 6 6 6 6 6 6

7 7 7 7 7 7 ■ 7 7 7 7 7 7 7 7 7 7 7 7 7 7 7 7 7 7 7 7 7 7 7 7 7 7 7 7 7 7 7 7 7 7 7 7 7 7 7 7 7 7 7 7 7 7 7 7 7 7 7 7 7 7 7 7 7 7 7 7 7 7 7 7 7 7 7 7 7 7 7 7 7

8 8 8 8 8 8 8 ■ 8 8 8 8 8 8 8 8 8 8 8 8 8 8 8 8 8 8 8 8 8 8 8 8 8 8 8 8 8 8 8 8 8 8 8 8 8 8 8 8 8 8 8 8 8 8 8 8 8 8 8 8 8 8 8 8 8 8 8 8 8 8 8 8 8 8 8 8 8 8 8 8

9 9 9 9 ■ 9 9 9 9 9 9 9 9 9 9 9 9 9 9 9 9 9 9 9 9 9 9 9 9 9 9 9 9 9 9 9 9 9 9 9 9 9 9 9 9 9 9 9 9 9 9 9 9 9 9 9 9 9 9 9 9 9 9 9 9 9 9 9 9 9 9 9 9 9 9 9 9 9 9 9

1 2 3 4 5 6 7 8 9 10 11 12 13 14 15 16 17 18 19 20 21 22 23 24 25 26 27 28 29 30 31 32 33 34 35 36 37 38 39 40 41 42 43 44 45 46 47 48 49 50 51 52 53 54 55 56 57 58 59 60 61 62 63 64 65 66 67 68 69 70 71 72 73 74 75 76 77 78 79 80

888157 BSC

Figure 1–13. FORTRAN statement card.

Once the computer reads and approves what it finds in the control cards division it is ready to process the program itself (contained in the FORTRAN source deck in our example). A record separator card is placed between the control cards and the source deck. This is usually a card with 7, 8, and 9 encoded in column 1. The program contains instructions, in the FORTRAN language, which describe in a sequential and logical way what the computer is being asked to do. For example, statements which put numerical data into the system and which retrieve numerical data for print-out are encoded on cards in this division. Also included are statements describing algebraic and logical operations. It is important in passing to mention that the computer translates FORTRAN into its own machine language using a compiler before the actual computations are carried out.

Following the FORTRAN source deck division, i.e., after we have encoded all the operations we desire the computer to perform, is another record separator, which essentially informs the computer that the FORTRAN is completed. In the simple program outlined in Fig. 1-12 there follows the data deck, which consists of a group of cards like those in Fig. 1-13 which have numerical data encoded on them. No FORTRAN occurs in this part of the program. After the data deck comes a file separator that informs the computer that the last of the cards has been reached. A file separator is generally a card with 6, 7, 8 and 9 encoded in column 1.

FORTRAN STATEMENTS. Assuming that we have the correct sequence of control cards, the next sequence consists of the FORTRAN source deck. In this section various types and examples of FORTRAN statements will be discussed. The material presented is not exhaustive but will give sufficient information to permit the writing of relatively straightforward programs. The reader should consult the References for more complicated programming. The books by McCracken are very popular.

As a starting point an explanation of the FORTRAN statement card is in order. This card, outlined in Fig. 1-13, consists of 80 columns numbered consecutively across the card. Each of these columns can in principle contain information in the form of 0 to 12 rectangular punches; i.e., each of the columns has the sequence 0 through 9 vertically and any number of these may be punched out (encoded) using a card punch. In addition each column may have two additional punches above the 0 row. In the FORTRAN part of the deck of cards, different groups of columns play special roles, whereas in the control card and data deck divisions each group of columns has no special significance. This point is often confusing when writing programs. The point to be remembered is that FORTRAN statements occur in a particular part of the program, the FORTRAN source deck, and in this part there are different groups of columns which have special significance.

The first group of significance consists of columns 7 through 72. The FORTRAN statement itself *must* appear in these columns. Information outside this group will not be treated as FORTRAN by the computer system. Generally the FORTRAN statement will begin in column 7, although it is not necessary to do so.

Knowing where to put the FORTRAN statement, we now proceed to discuss the statements themselves. The three other groups of columns are 1 through 5, 6, and 73 through 80. They will be discussed as the need arises.

The first card in the FORTRAN source deck is the program card. The card consists of the word PROGRAM followed by an alphanumeric identifier, called the program name. An alphanumeric identifier is any group of 7 or fewer letters and numbers which begins with a letter—with one exception: The letter O followed by six digits is not an alphanumeric identifier. For example,

1. A13B: alphanumeric
2. 2BCD: not alphanumeric
3. 0123456: not alphanumeric

Returning to a description of the program card, after the program name there is a set of parentheses inside of which are placed certain quantities called files. Examples are INPUT, OUTPUT, PUNCH, and READ INPUT TAPE I. These refer to various modes of getting information into and out of the computer system. For beginners, only INPUT and OUTPUT are needed. An example of an acceptable program card is, beginning in column 7, PROGRAM APOLLO (INPUT, OUTPUT)

Following the program card is a wide variety of cards, their sequence being dictated, with the exception of FORMAT statements, by the order in which operations are to be carried out. We shall discuss a few of these types below and then consider a particular problem.

ARITHMETIC OPERATIONS. FORTRAN permits arithmetic operations using both variables and constants. There are two types of each: fixed point and floating point. Variables and constants which have no decimal point are called fixed point, while those needing a decimal point are called floating point. Keeping in mind the definition of an alphanumeric identifier, we define a fixed-point variable as an alphanumeric identifier which starts with one of the letters I, J, K, L, M, or N. As examples,

1. I21 fixed-point variable
2. 21 floating-point constant
3. 21 fixed-point constant
4. A21 floating-point variable

Arithmetic operations called for by various FORTRAN statements are outlined below:

X = A + B	addition
X = A - B	subtraction
X = A * B	multiplication (A × B)
X = A/B	division
X = A ** B	exponentiation A^B

The variables A and B above are floating-point variables. Fixed-point variables may also be used in the same way. In more advanced types of FORTRAN, fixed- and floating-point variables may be mixed on the right-hand side of the above expressions; the type of result is determined by the variables on the left, which is floating point in our case. Throughout the discussion here we assume that it is unwise to mix fixed- and floating-point variables in FORTRAN statements.

Many expressions occurring in scientific work are complex; the corresponding FORTRAN statements are also complex. For example, the expression

$$X = \frac{A^2 + 3B^3}{2D + E^4}$$

calls for a mixture of four arithmetic operations. The corresponding FORTRAN statement is

```
X = (A**2 + 3.* B**3) / (2.*D + E**4)
```

Note that exponents may be expressed as fixed-point constants. With this statement we have introduced the use of parentheses, which are a vital part of FORTRAN statements. The computer follows the following rules: (1) Evaluate the expressions inside the innermost parentheses, then the next innermost, etc. (2) Evaluate expressions inside each group of parentheses before evaluating any operations not contained within parentheses. (3) Combine the groups of parentheses according to the arithmetic operations called for. In the above example we have two groups of parentheses in the form

```
X = (    ) / (    )
```

The computer will evaluate the expressions inside each set prior to carrying out the final division.

At this point it is important to consider the use of parentheses in more detail, as illustrated in the following examples:

FORTRAN statement	Algebraic equivalent
X = A + B / C – E	$X = A + \frac{B}{C} - E$
X = (A + B) / C – E	$X = \frac{A + B}{C} - E$
X = (A + B) / (C – E)	$X = \frac{A + B}{C - E}$

These FORTRAN expressions differ only in their use of parentheses, and each has, as shown above, a different algebraic equivalent. The above examples illustrate the computer's hierarchy of arithmetic operations: First, exponentiations are performed; second, multiplications and divisions; and finally, additions and subtractions. These are, of course, performed in a manner compatible with the parentheses. In the first statement above, the first operation is B/C followed by the addition of A and the subtraction of E. In the second statement the interior of the parentheses is evaluated first, then the division, and finally the subtraction of E. These comments and examples serve to underscore the importance of carefully translating the arithmetic expression at hand into its FORTRAN equivalent.

In passing, it is important to note that the computer must have numerical values for all the variables on the right of any expression before it can make the evaluation. Hence, the numerical value of any variable must be calculated or read as input before that variable can appear on the right-hand side of an expression.

LIBRARY FUNCTIONS. Digital computer systems normally have several mathematical functions incorporated in them which you may use without programming a numerical method for their evaluation. The following partial list gives examples:

FORTRAN	Numerical
X = COS (Y)	Cosine of Y; Y in radians
X = SIN (Y)	Sine of Y; Y in radians
X = SQRT (Y)	Square root of Y
X = EXP (Y)	e to Yth power
X = ALOG (Y)	Natural log of Y
X = ALOG10 (Y)	Base 10 log of Y

These library functions may be used in arithmetic expressions just as variables can. For example,

```
X = A * COS (Z) * EXP (Y)
```

Furthermore, arithmetic expressions may be written inside the parentheses as in

```
X = EXP (Y + Z**2.)
```

The utility of these and other library functions is obvious.

CONTROL MANIPULATIONS. It is often believed that the advantage of using a computer is in the speed with which it can carry out an arithmetic operation. This is only partially true. The real advantage accrues because a digital computer system can carry out a repetitious calculation in a logical way with great speed. For example, if you want to multiply 2 X 2, the computer is a waste of time because by the time you have written the program to get the result you could have written down the result. If for some reason, however, you wanted a table for multiplication by 2 of integers from 0 to 10,000, you could probably save time by using a computer because it can be logically programmed to carry out the whole calculation and good use can be made of the computer's speed. In this section the manipulations which control the way in which a calculation is carried out are discussed.

Of the many control operations we shall discuss only a few. Many control operations make use of statement numbers. Statement numbers are sequences of 1 to 5 digits which are encoded in columns 1 to 5 of the FORTRAN statement card and serve as an identifier for that particular statement. For example,

Column	4 5 6 7 8 9 10 11
Statement	1 5 X = Y + Z

In this case the identifier of the statement is 15, whereas the FORTRAN operation is the statement X = Y + Z. Note: Column 6 is blank; its use will be discussed later.

One of the most useful control operations is the DO statement, which has the form

DO 101 I = 1, 50

The number 101 is an identifier reference and I is a fixed-point variable. The statement says the following: "Do all operations between here and statement 101 using I = 1. When statement 101 is reached, set I = 2 and repeat all operations between here and 101 again, increase I to 3, etc. After the I = 50 cycle is complete, cease the repetition and continue beyond statement 101."

Another often-used control statement is GO TO. For example, GO TO 101 instructs the computer to proceed directly to the statement whose identifier (address) is 101.

There are several IF statements:

1. If (X) I, J, K. Here X is a variable which is tested for its sign by this statement. If it is negative, the computer goes to address I; if zero, to J; and if positive, to K. To actually be executed I, J, and K must be specified explicitly as fixed-point constants.
2. IF (A. EQ. B) GO TO 327
   IF (A. LT. B) GO TO 327
   IF (A. GT. B) GO TO 327
   IF (A. GE. B) GO TO 327
   IF (A. LE. B) GO TO 327

Each of these statements is a true-false test. The first is executed as follows: "If A equals B, go to address 327; otherwise continue with the next FORTRAN statement," The other examples are similar; LT is less than, GT is greater than, GE is greater than or equal to, and LE is less than or equal to. The expression GO TO 327 may be replaced by another logical or arithmetic operation. For example,

IF (A. LT. B) Z = 29.362

Here, if A is less than B, then Z is set equal to 29.362; otherwise Z retains its value and the next statement is executed.

3. IF (A. LT. B) 77, 78. This third type of IF statement is also a true-false test. Executed, it is translated as "If A is less than B, proceed to statement number 77; if not, proceed to 78." Similar statements with GT, EQ, etc., are valid.

Refer to the References for additional true-false tests.

Another control statement is the CONTINUE statement. It must contain a reference address and is most often used as the last card in a DO loop, serving only to locate the desired end of the loop. Otherwise it is a do-nothing statement. Finally we have the END statement. This must appear as the last card in the FORTRAN source deck. In summary we have discussed briefly the control statements DO, GO TO, IF, CONTINUE, and END.

INPUT-OUTPUT. With the information at hand, one can, with practice, program a wide variety of expressions for evaluation by a computer system. We shall now turn our attention to statements by means of which we can get information into and out of the computer. For our purposes, we assume that input data are always encoded on cards and that output is always printed. With these assumptions input is achieved by READ statements and output by PRINT statements. Each of these statements in general requires a FORMAT statement in order to be executed. The forms are

```
 READ  41, A, B, C
 READ  15, I
PRINT  15, J
PRINT  41, X
```

Each of these statements is interpreted in essentially the same way. The first instructs the computer to read, according to a method specified by statement number 41, the quantities A, B, and C. The actual numerical values of A, B, and C will be located on a data card in the input data deck division (Fig. 1-12). The last statement above instructs the computer to print, according to statement number 41, the quantity X. This assumes, of course, that the actual numerical value of X is at hand.

For the proper execution of the above statements the statements numbered 41 and 15 must be FORMAT statements, which inform the computer how the input data are arranged on the data cards and how the output is to be printed. As an example we write

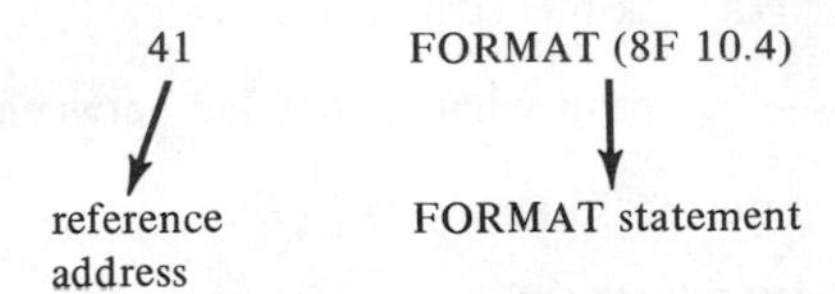

This statement could be used in conjunction with the first and last statements above. With the first statement it would achieve the reading of A, B, and C from the data card. The instructions telling the computer where to look on the data card for A, B, and C are located within the parentheses. The number 8 specifies 8 fields (numbers) on the card, F specifies a floating-point form, 10.4 specifies 10 columns per field, and if the decimal point is not explicitly encoded, it will be placed so that the last four columns of the field are to the right of it. Generally the decimal is included, and if so, the 4 has no meaning. With this format the numerical value of A should be located in the first 10 columns (first field) of the card, the value of B in columns 11 through 20, and the value of C in columns 21 through 30. We have specified 8 fields with 8F 10.4 but have read only 3 values. This is permitted; you can overspecify but not underspecify. Note that we have set 8 fields with 10 columns each or 80 columns total. This, of course, is equal to the total number of columns on a single card. We could have called for fewer total columns but not more.

There are several other types of format; two often used are the I and E formats. For example,

41 FORMAT (8E 10.4)

This format has the same interpretation as 8F 10.4 except that the numbers are in the E (exponential) format. This means that they would be written in the form 1.3211E+04, which with no spaces takes up 10 columns (1 field) and is equivalent to the floating-point number (13211.). The utility of the exponential format becomes clear when one considers numbers whose significant figures are far to the left or right of the decimal point; the E format compacts them.

The I format is used when reading or printing fixed-point quantities. For example,

15 FORMAT (1 I 5)

would set aside one fixed-point field 5 columns wide (the first 5 columns on the card). With fixed-point quantities, the number must be encoded as far to the right-hand side of the field as possible. Blank columns are treated as zeros by the computer.

Other kinds of information may also be included in format statements. As examples we include

N/: skip N lines: N is an integer

NX: skip N spaces

* ABC ------ *: print what is encoded between the stars

The combination of statements below

```
15  FORMAT (5/, 25X, * ZEBRA = *, F 10.2)
    PRINT     15,   Z
```

would move down the output paper 5 lines and then over 25 spaces from the left-hand margin. The word ZEBRA = would then be typed, followed by the print-out of Z in floating-point format. The output page of the computer is 120 columns wide, and column 1 should not be used for printing out data.

THE MEANING OF = IN FORTRAN. The symbol = has a meaning in FORTRAN which is somewhat different from its usual meaning. In FORTRAN, the word *replace* best describes the symbol =. For example, a statement such as X = X + A is perfectly valid in FORTRAN and does not imply that A = 0.0. Rather the statement instructs the computer to replace the quantity on the left by the quantity on the right. In this case X is on the left and so we are calculating a new X. On the right we have X + A and so the old X + A furnishes the new X.

SUBSCRIPTED VARIABLES AND ARRAYS. Very often in a computational problem, a situation arises in which we desire to calculate and store away for later use several values of a particular quantity. For example, if $Y = AX^2$, we might wish to have at hand the values of Y corresponding to X = 0, 1, ---(N - 1). Since at any given instant each location in the computer memory can contain only one quantity, we must set aside N locations in the memory for the sequence of Y values. This sequence of locations is called an array and is set aside with a DIMENSION statement such as

```
DIMENSION  Y(N)
```

This statement must appear prior to any calculation of Y values, and the value of N must be given explicitly as a fixed-point quantity in the above statement. For example, DIMENSION Y(100) would set aside an array of 100 locations for Y values.

Whenever a variable is dimensioned as above, it must be written as a subscripted variable everywhere it appears in a FORTRAN statement except in a PRINT statement where one wants to print the complete array. A subscripted variable is written in the same form as it appears in the DIMENSION statement.

For the present case we would rewrite our Y equation as $Y(I) = AX^2$. In general a subscripted variable may have as many as three subscripts Y(I, J, K), requiring in turn a DIMENSION statement with the same number of subscripts to set aside memory locations for all the quantities.

A PROBLEM. We now shall consider a simple illustrative problem.

Given the polynomial form $y = ax^3 + bx^2 + cx + d$, calculate y for x = 0, 0.1, 0.2, ..., 10.0 and print out values of x and y in a table. Let a = 5.1, b = 2.8, c = 3.1, and d = 1.1. We write only the FORTRAN source deck and data deck, omitting all the control cards shown in Fig. 1-12. First we develop a simple flow chart (Fig. 1-14) and then write a FORTRAN program.

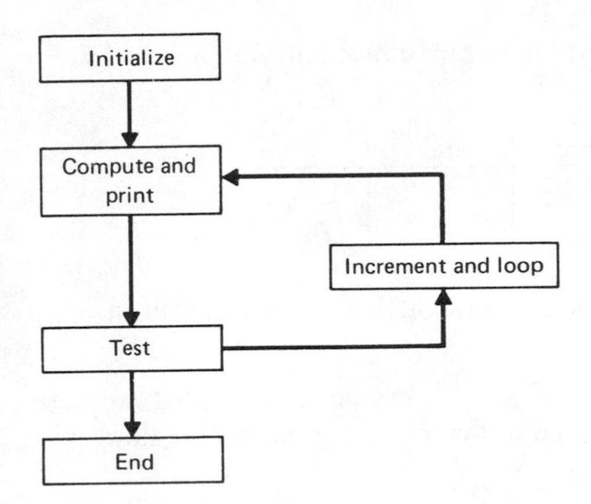

Figure 1-14.

The FORTRAN program corresponding to this flow chart might be as follows (the numbers at the top are FORTRAN statement card column numbers):

```
1        5   6   7                                    72
                 PROGRAM POLY (INPUT, OUTPUT)
         1       FORMAT (8F 10.2)
         2       FORMAT (5/, 25X, *X*, 25X, *Y*)
         3       FORMAT (20X, E 10.3, 20X, E 10.3)
                 READ 1, A, B, C, D
                 X = 0.0
                 PRINT 2
                 DO 4 I = 1, 101
                 Y = A*X**3. + B*X**2. + C*X + D
                 PRINT 3, X, Y
                 X = X + 0.1
         4       CONTINUE
                 END
```

Initialization is a very important part of programming and consists of setting the initial values of all the variables which are needed to begin the calculation. In our case we need to know values for X, A, B, C, and D. In the program above, initialization is achieved by setting X = 0.0 and reading in A, B, C, and D. After initialization, PRINT 2 provides a heading for the table of output. Then the repetitious part of the calculation is performed using a DO loop terminated by the CONTINUE statement at reference address 4. Inside the DO loop the polynomial is evaluated and the results are printed in exponential form according to format 3. X is then increased to the next value and the whole process is repeated again until 101 cycles have been performed. This completes the program. Note that the speed could be improved by factoring an X out of the first three terms of Y. Why is this helpful?

The data card for A, B, C, and D will have the form specified in format 1:

First 10 columns	Second 10 columns	etc.	etc.
5.1	2.8	3.1	1.1

The numbers may be anywhere within their proper 10-column field if the decimal point is encoded.

The problem may be coded in other ways. For example, subscripted variables could be used and a different type of end test employed. Consider the flow chart in Fig. 1-15.

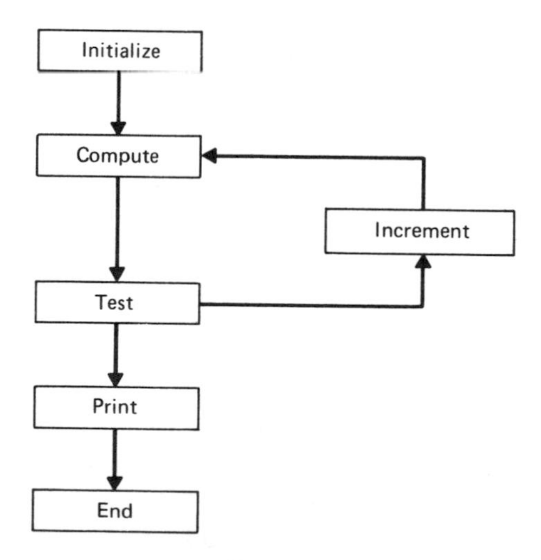

Figure 1-15. Flow chart.

The corresponding FORTRAN code could be written as

```
1   5 6 7                                                   72
      PROGRAM POLY (INPUT, OUTPUT)
    1 FORMAT (8F 10.2)
    2 FORMAT (5/, 25X, *X*, 25X, *Y*)
    3 FORMAT (20X, E 10.3, 20X, E 10.3)
    4 FORMAT (1H1)
      DIMENSION Y(200), X(200)
      READ 1, A, B, C, D, G, XMAX
      X(1) = 0.0
      K = 1
    5 Z = X(K)**2
      W = Z*X(K)
      Y(K) = A*W + B*Z + C*X(K) + D
      IF (X(K).GE.XMAX) GO TO 6
      K = K + 1
      X(K) = X(K - 1) + G
      GO TO 5
    6 CONTINUE
      KMAX = K
      PRINT 4
      PRINT 2
      DO 7 K = 1, KMAX
      PRINT 3, X(K), Y(K)
    7 CONTINUE
      END
```

This code is longer than the previous one but is also considerably more general in that nearly all the parameters are read in or calculated. Thus, with data card changes we could carry out any calculation having the same form. You should study the FORTRAN statements in this program carefully.

MISCELLANEOUS NOTES.

1. If you desire to print something on a new page, use the following combination of statements:

```
I  FORMAT (1 H 1)
   PRINT I
```

   where I is an explicitly given address.

2. If you have a FORTRAN statement which is too long to go between columns 7 and 72 on a single card, put a digit in column 6 of the next card and continue the FORTRAN statement.
3. Comment cards are often useful. Encoding a C in column 1 denotes a comment card, and anything may be typed in the remaining columns. The computer ignores such cards insofar as FORTRAN is concerned. The message on the card will, however, appear in the print-out of the FORTRAN source deck. Such cards are of use in reminding one what a certain program or statement is designed to do.

An illustrative completely encoded program including control cards, FORTRAN statements, and data deck is shown in Fig. 1-16. Also given, in Fig. 1-17, is a copy of the actual computer-printed output. The program is equivalent to program POLY discussed above except the input data have been changed. Note the use of comment cards.

PROBLEMS

Many of the following problems assume familiarity with the material in the preceeding sections, especially Section 1.6.

1. a. Using the trapezoidal rule, numerically perform the following quadrature:

$$I = \int_{x=1}^{x=3} \frac{dx}{x}$$

Iterate the estimate for I until the value of I changes by less than 0.1% of its previous value. Print out a table of values with the headings

Number of iterations	Quadrature	Fractional change
1	Estimate for I	Change from previous estimate for I
2		
3		

b. How near the true result is the final estimate?
c. What troubles are encountered if the lower limit of integration extends toward x = 0?

Listing of control cards, Fortran cards, and data cards.

```
POLY.CMDJ0028,WHITE.
RUN(S)
MAP.
LGO.

      PROGRAM POLY(INPUT,OUTPUT)
C   FLOATING POINT FORMAT USED TO READ IN INPUT DATA.
    1 FORMAT(8F10.2)
C   FORMAT FOR OUTPUT TABLE HEADING.
    2 FORMAT(5/,25X,*X*,25X,*Y*)
C   FORMAT FOR OUTPUT OF DATA.
    3 FORMAT(20X,E10.3,20X,E10.3)
    4 FORMAT(1H1)
C   DIMENSION STATEMENT RESERVING COMPUTER STORAGE FOR X AND Y ARRAYS.
      DIMENSION Y(200),X(200)
C   READ IN ACCORDING TO STATEMENT NUMBER ONE, SIX QUANTITIES. THESE
C   SHOULD APPEAR ON THE FIRST DATA CARD.
      READ 1,A,B,C,D,G,XMAX
C   INITIALIZING X(K)
      X(1)=0.0
C   K IS FIXED POINT COUNTER FOR SUBSCRIPT ON X AND Y.
      K=1
C   Z AND W ARE USED TO SPEED UP COMPUTATION.
    5 Z=X(K)**2
      W=Z*X(K)
      Y(K)=A*W+B*Z+C*X(K)+D
C   TEST FOR STOPPING CALCULATION.
C   XMAX IS THE LARGEST VALUE OF X DESIRED.
      IF(X(K).GE.XMAX) GO TO 6
C   INCREMENT K BY ONE.
      K=K+1
C   CALCULATE X(K) FROM PREVIOUS X(K) AND THE INCREMENT G.
      X(K) =X(K-1) + G
      GO TO 5
    6 CONTINUE
C   SET UP TO PRINT OUT X AND Y ARRAYS.
      KMAX=K
C   SKIP TO A NEW PAGE IN THE OUTPUT.
      PRINT 4
C   PRINT TABLE HEADING.
      PRINT 2
      DO 7 K=1,KMAX
C   PRINT X(K) AND Y(K).
      PRINT 3,X(K),Y(K)
    7 CONTINUE
      END
(data card)
6.26      1.28      3.47      5.22      0.10      4.0
```

Figure 1-16. Listing of control cards, FORTRAN cards, and data cards.

X	Y
0.	5.220E+00
1.000E-01	5.586E+00
2.000E-01	6.015E+00
3.000E-01	6.545E+00
4.000E-01	7.213E+00
5.000E-01	8.057E+00
6.000E-01	9.115E+00
7.000E-01	1.042E+01
8.000E-01	1.202E+01
9.000E-01	1.394E+01
1.000E+00	1.623E+01
1.100E+00	1.892E+01
1.200E+00	2.204E+01
1.300E+00	2.565E+01
1.400E+00	2.976E+01
1.500E+00	3.443E+01
1.600E+00	3.969E+01
1.700E+00	4.557E+01
1.800E+00	5.212E+01
1.900E+00	5.937E+01
2.000E+00	6.736E+01
2.100E+00	7.613E+01
2.200E+00	8.571E+01
2.300E+00	9.614E+01
2.400E+00	1.075E+02
2.500E+00	1.197E+02
2.600E+00	1.329E+02
2.700E+00	1.471E+02
2.800E+00	1.624E+02
2.900E+00	1.787E+02
3.000E+00	1.962E+02
3.100E+00	2.148E+02
3.200E+00	2.346E+02
3.300E+00	2.556E+02
3.400E+00	2.779E+02
3.500E+00	3.014E+02
3.600E+00	3.264E+02
3.700E+00	3.527E+02
3.800E+00	3.804E+02
3.900E+00	4.096E+02
4.000E+00	4.402E+02
4.100E+00	4.724E+02

Figure 1-17. Output of Program POLY.

2. Write a computer program to determine a least-squares linear fit of a set of data and determine the slope and intercept. Print out in tabular form the input abscissas and ordinates, the deviation of the input ordinates from the computed ordinates, the slope, and the intercept.

3. Write a program which combines the false position method and the Newton-Raphson method to find the roots of the third-order polynomial

$$f(x) = Ax^3 + Bx^2 + Cx + D$$

which lie in the interval $-5 \leqslant x \leqslant +5$. Use your own ingenuity to develop a general method for finding regions where f(x) has different signs. Assume that A, B, C, and D are to be read in as data.

4. a. Write a program to evaluate the third-order polynomial of Problem 3 for $x = -5.0$, $x = -4.9$, $x = -4.8$, etc., up to $X = +5.0$. Assume that A, B, C, and D are to be read in as data. Print out in tabular form the constants A, B, C, D, x, and f(x).
   b. Generalize part a so that the program could make the same calculation for several polynomials, each with a different set of constants.

REFERENCES

Section 1.2

Y. Beers, *Introduction to the Theory of Error* (Addison-Wesley, Reading, Mass., 1957).

P. R. Bevington, *Data Reduction and Error Analysis for the Physical Sciences* (McGraw-Hill, New York, 1969).

W. E. Deming, *Statistical Adjustment of Data* (Chapman & Hall, London, 1943).

D. A. S. Fraser, *Statistics, An Introduction* (Wiley, New York, 1958).

L. G. Parratt, *Probability and Experimental Errors in Science* (Wiley, New York, 1961).

Section 1.3

Y. Beers, *Introduction to the Theory of Error* (Addison-Wesley, Reading, Mass., 1957).

P. R. Bevington, *Data Reduction and Error Analysis for the Physical Sciences* (McGraw-Hill, New York, 1969).

Section 1.5

Y. Beers, *Introduction to the Theory of Error* (Addison-Wesley, Reading, Mass., 1957).

P. R. Bevington, *Data Reduction and Error Analysis for the Physical Sciences* (McGraw-Hill, New York, 1969).

H. Margenau and G. M. Murphy, *The Mathematics of Physics and Chemistry* (Van Nostrand-Reinhold, New York, 1956).

Section 1.6

R. A. Buckingham, *Numerical Methods* (Pitman, New York, 1962).

R. W. Hamming, *Numerical Methods for Scientists and Engineers* (McGraw-Hill, New York, 1962).

F. B. Hildebrand, *Introduction to Numerical Analysis* (McGraw-Hill, New York, 1956).

Z. Kopal, *Numerical Analysis* (Chapman & Hall, London, 1955).

A. Ralston, *A First Course in Numerical Analysis* (McGraw-Hill, New York, 1965).

Section 1.7

D. M. Anderson, *Computer Programming, FORTRAN IV* (Appleton, New York, 1966).

T. R. Dickson, *The Computer and Chemistry* (W. H. Freeman, San Francisco, 1968).

D. D. McCracken, *A Guide to* FORTRAN *Programming* (Wiley, New York, 1961).

D. D. McCracken, *A Guide to* FORTRAN *IV Programming* (Wiley, New York, 1965).

D. D. McCracken, FORTRAN *with Engineering Applications* (Wiley, New York, 1967).

K. B. Wiberg, *Computer Programming for Chemists* (Benjamin, Reading, Mass., 1965).

FORTRAN *Reference Manual, Control Data® 6400/6500/6600 Computer Systems* (Control Data Corp., Minneapolis, 1969).

Chapter 2

Experimental Apparatus and Methods

2.1 Vacuum Systems and Pressure Measurement

The purpose of the following discussion is familiarization with some of the basic components of a high-vacuum system. Several of the laboratory experiments involve evacuated apparatus, and it is very important to become familiar with the equipment used in these systems and also with the terminology one uses in discussing high-vacuum equipment.

Figure 2-1 shows a block diagram of a simple high-vacuum system and includes the essential components. The first component, labeled A, is a mechanical pump. The usual mechanical pump is of the rotary type in which the gas is removed from the system by an eccentric rotating piston. Figure 2-2 shows a sketch of a rotary pump and illustrates the mechanical means by which it operates. The eccentric piston (1) rotates, and the chamber in which it rotates is divided into two subchambers (2) and (3) by a vane (4) which is in continuous contact with the eccentric piston. As the piston rotates, volume (2) is decreased, thereby increasing the pressure in this volume. When volume (2) becomes very small the valve (5) is opened and the compressed gas is exhausted to the atmos-

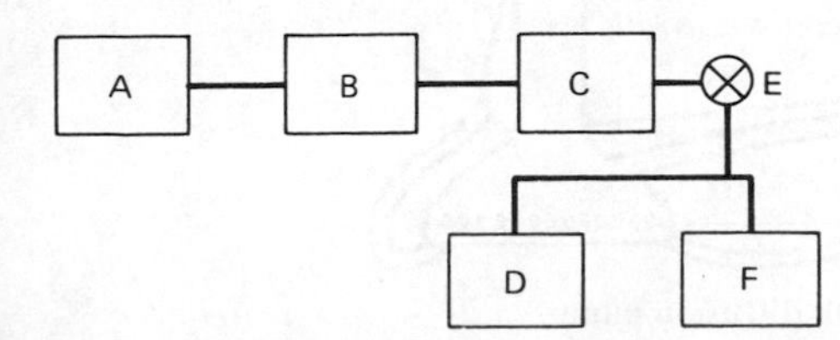

Figure 2-1. Block diagram of a simple vacuum system.

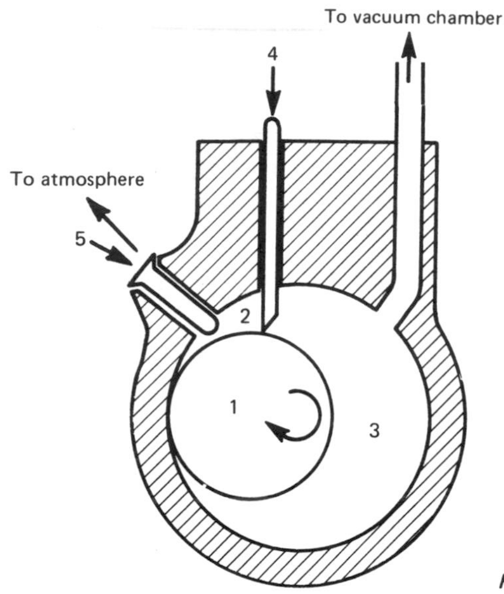

Figure 2-2. Rotary mechanical pumps.

phere. These mechanical pumps can reduce the pressure in a system to approximately 10^{-3} Torr or 1 μ (1 Torr = 1 mm Hg, 1 $\mu = 10^{-3}$ Torr).

A pressure of 10^{-3} Torr corresponds to 3×10^{13} molecules/cm^3 at room temperature (see Table 2-1). It is frequently desirable to achieve lower pressures than this in order to avoid contamination of a sample by residual gases in the system. To achieve this goal, use is made of a diffusion pump, designated in Fig. 2-1 as B. With these pumps pressures of 10^{-6} - 10^{-8} Torr can be achieved with ease. The only moving part in a diffusion pump is a high-speed jet of vapor (oil or mercury) which is directed away from the vacuum chamber (component D in Fig. 2-1). Figure 2-3 shows a diagram of an oil diffusion pump constructed

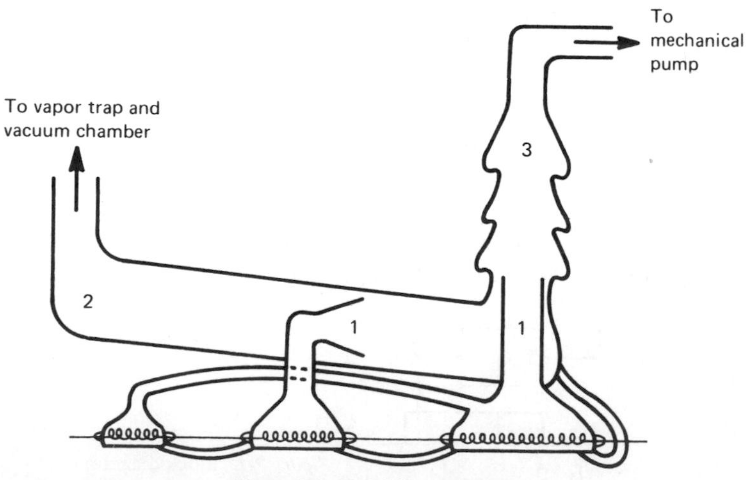

Figure 2-3. Oil diffusion pump.

of glass (other types are made of metal). The oil is heated by an electrical filament and evaporated. It then is directed away from the vacuum chamber by the jets (1) and in doing so transfers momentum to gas molecules diffusing out of the apparatus (region 2) being evacuated. This process carries them to region 3 where they are pumped away by the mechanical pump. The mechanical pump is necessary for the operation of the diffusion pump because the pressure must be reduced to only a few microns before the diffusion pump will operate. A mechanical pump being used in this capacity is referred to as a forepump or backing pump.

Table 2-1. CHARACTERISTICS OF VACUUMS

Pressure region	Atmosphere	Low vacuum	High vacuum
Pressure (Torr)	760	10^{-3}	10^{-6}
Number of molecules striking wall (per cm^2)	3×10^{23}	4×10^{17}	4×10^{14}
Mean free path (cm)	6.5×10^{-6}	5	5×10^{3}
Density of molecules (cm^{-3})	2×10^{19}	3×10^{13}	3×10^{10}
Pump used		Mechanical	Diffusion

The lowest pressure achievable by a diffusion pump is limited by the diffusion of pumping fluid vapor from the diffusion pump back into the vacuum chamber. To prevent the diffusion of vapor back into the vacuum chamber a vapor trap (component C in Fig. 2-1) is installed. Figure 2-4 shows a commonly used vapor trap (sometimes called a cold trap). This trap is surrounded by a dewar filled with liquid nitrogen (T = 77°K), which serves to condense the vapor diffusing from the pump toward the vacuum chamber. This vapor trap also acts as a pump for the vacuum chamber because any gas whose pressure in the vacuum chamber is greater than its equilibrium vapor pressure at liquid nitrogen temperature will move from the chamber to the vapor trap and be condensed there. This is an example of cryogenic pumping.

To isolate the vacuum chamber from the pumping system, use is made of a valve called a stopcock (part E in Fig. 2-1). Figure 2-5 shows one type of stopcock which is commonly used. The joint is lubricated with a grease that has a very low vapor pressure. For example, good silicone greases have vapor pressures on the order of 10^{-6} Torr at room temperature. Good hydrocarbon greases are available having vapor pressures as low as 10^{-9} Torr. The plug (C) in the stopcock shown in Fig. 2-5 is hollow, and the stopcock body has cup B. By turning the stopcock so that the exhaust hole (A) in the plug is aligned with the exhaust tube, one can evacuate the plug and the cup. This feature makes such stopcocks very useful for high-vacuum work since a very good seal between the two surfaces is easily achieved with the plug volume and the cup evacuated.

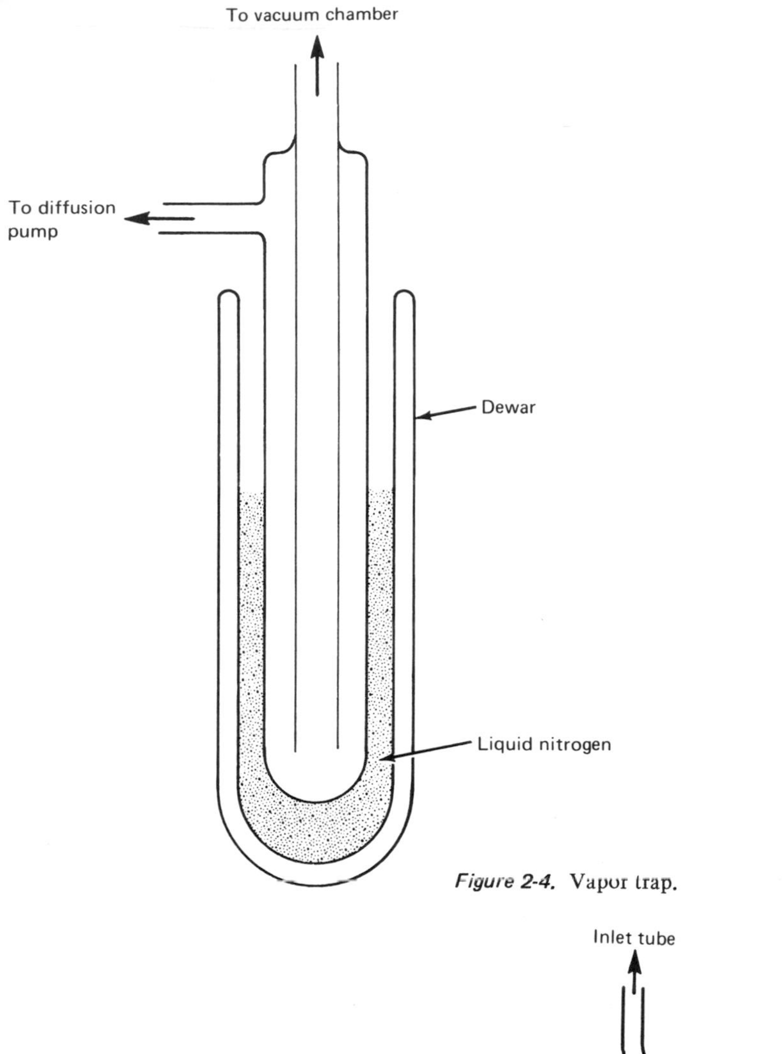

Figure 2-4. Vapor trap.

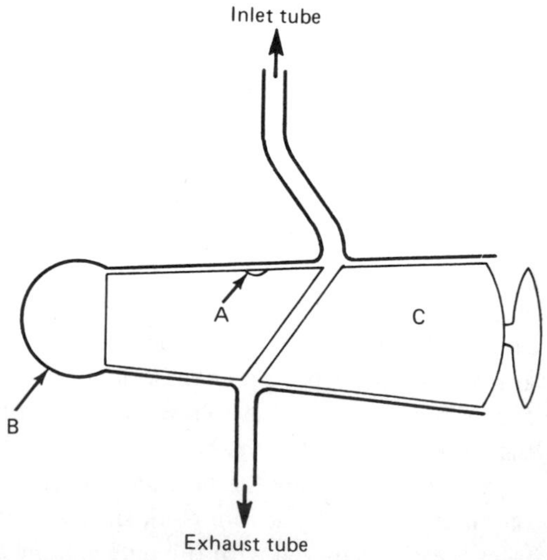

Figure 2-5. High-vacuum stopcock. A, exhaust hole; B, cup. C, plug.

Part F in Fig. 2-1 is a pressure-measuring device and may consist of one or more of the following instruments: mercury manometer, thermocouple gauge, ionization gauge, McLeod gauge, spoon gauge, and pressure transducer. There are many varieties of these as well as other instruments; refer to the references for further information. In the following discussion only the essential features of these devices and a few points about operating them are given. Table 2-2 lists the nominal pressure range of each of these instruments.

Probably the most familiar pressure gauge is the mercury manometer consisting of a glass U-tube partially filled with mercury. One tube is connected to the sample system whose pressure is to be measured, while the other is connected to a reference system whose pressure is known. Measuring the difference in heights of the two mercury columns furnishes only the pressure difference between the sample and the reference. To obtain the sample pressure the following relationship should be used:

$$p_{sample} = \Delta h + p_{ref} \tag{2-1}$$

where Δh is the height of the reference mercury column less the height of the sample column. Generally speaking, the reference pressure will be either atmospheric pressure or very nearly zero (evacuated reference). From a practical point of view there is a strong tendency for mercury to become slightly contaminated, resulting in a nonuniform interaction between it and the walls of the glass tubing. This tendency toward sticking may lead to erroneous pressures but may be minimized by gently tapping the manometer before reading column heights and by obtaining a correction factor through measurement of the difference in column heights when the pressures in the sample and reference systems are known to be equal. Another way of minimizing the tendency toward sticking is to frost the inside of the glass manometer tubing. Frosting of glass can be accomplished by rubbing it with wet carborundum.

Table 2-2. PRESSURE-MEASURING DEVICES

Gauge	Useful range
Mercury manometer	1 Torr to atmosphere
Bourdon tube	1 Torr to atmosphere
Thermocouple gauge	10 μ to 1 Torr
McLeod gauge	10^{-5} - 10 Torr
Ion gauge	10^{-11} - 10^{-3} Torr
Spoon gauge	0.05 Torr upwards
Pressure transducer	0.01 Torr upwards

Another gauge commonly used for pressures above atmospheric which also can be used for pressures in the range 1 - 760 Torr is the Bourdon tube gauge. The operation of this gauge depends on the mechanical motion of a thin-walled, closed-end, curved tube across which a pressure difference exists. This tube, whose spatial orientation depends on the magnitude of the pressure difference, is mechanically linked to an indicator whose position is read from a dial. Many of these gauges are relatively inexpensive but are useful only as qualitative indicators of pressure. More accurate, and consequently more expensive, versions are available.

A diagram of a McLeod gauge is shown in Fig. 2-6. It is one of the most useful of all vacuum gauges because it furnishes an absolute measurement of low pressures and may be used to calibrate the other gauges mentioned above, which

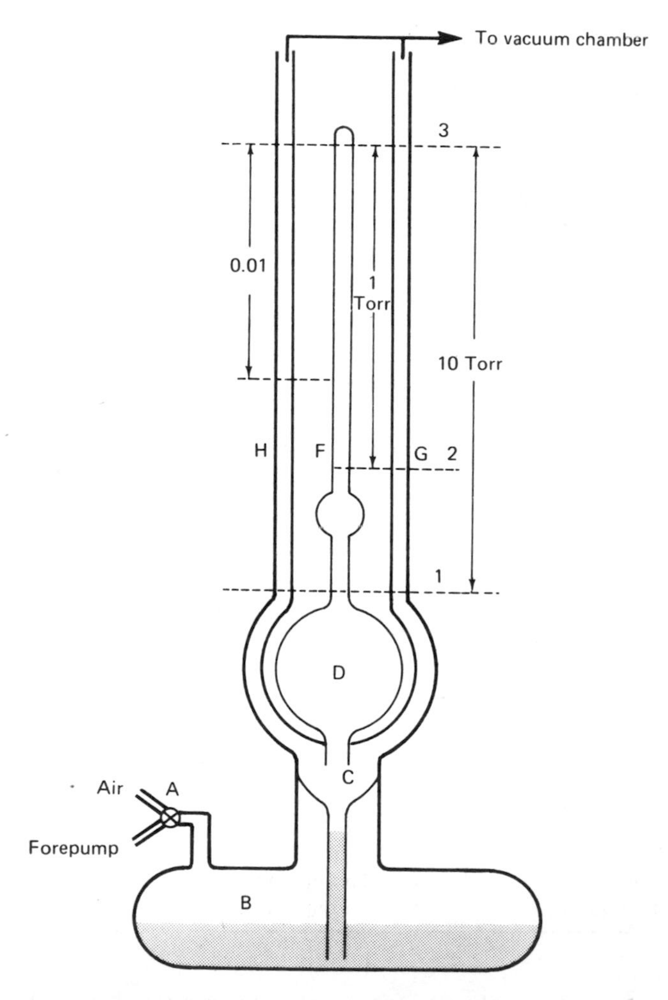

Figure 2-6. McLeod gauge.

are not absolute gauges. Figure 2-6 illustrates a gauge with three pressure ranges covering the overall range 0.01 μ to 10 Torr. To operate it when the vacuum chamber pressure is in one of the gauge ranges, one raises the mercury level by turning stopcock A to the "air" position. This admits air to the mercury reservoir (B), which in turn causes the mercury to rise. When the level reaches point C, the volume (D) and the capillary (F) are cut off. Therefore, a known volume (known by previous calibration) of gas is trapped whose pressure one desires to determine. This volume is then compressed to one of three points (1, 2, or 3) depending on the pressure range. This compression of the volume increases the pressure of the trapped gas. If the final volume is known and the final pressure is measured, one can calculate (assuming the ideal gas law) that

$$pV = (Ah)b \tag{2-2}$$

where p is the unknown initial pressure, V the known initial volume, A the cross-sectional area of the capillary, h the height of the final gas column, and b the difference in heights of the final mercury columns in the capillary (F) and the evacuated reference. The product Ah is thus the final volume, while b is the final pressure. The gauge shown in Fig. 2-6 has been constructed and calibrated so that when the pressure is in the range 1.0 - 10.0 Torr the mercury level is raised until the center column is at point 1. The pressure is then read by comparing the height of the mercury column in tube G with a scale attached to the gauge. For pressures in the 100-μ to 1-Torr range, the level in tube G is raised to point 2 and the pressure is read by comparing the mercury height in tube H with an attached scale. In the lowest pressure range (0.01–100 μ) the level in tube G is raised to point 3 and the pressure is obtained from the mercury height in the capillary. Once the pressure has been determined the mercury level is slowly lowered by turning stopcock A to the forepump position.

An ionization gauge of the common Bayard-Alpert type is shown in Fig. 2-7. It operates like a vacuum tube with the resistance-heated filament emitting electrons which are accelerated toward the grid by applying to it a positive voltage (~ 150 V) with respect to the cathode (filament). The grid is constructed of a widely spaced wire coil and so most of the electrons pass through it and make collisions with gas molecules in the inner region. These collisions lead to positive ions which are collected by the negatively biased center electrode (plate) where they pick up an electron and thus furnish a measurable electron current in the external circuit. Since the number of collisions and thus the number of positive ions formed depend linearly on the pressure of gas, we expect the measured ion current to increase with pressure. This kind of gauge is inoperable at pressures above 10^{-3} Torr because the positive ions formed undergo additional collisions on their path to the plate. These secondary collisions lead to a glow discharge in the gas. Below 10^{-11} Torr the positive ion current collected at the plate becomes quite small and comparable in magnitude with the small X-ray photoelectric current. The latter arises when electrons strike the grid, an X-ray photon is

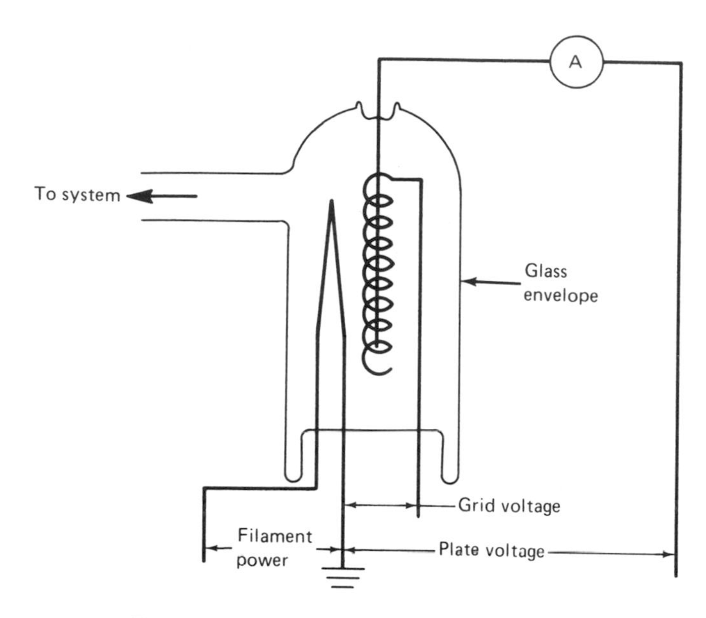

Figure 2-7. Bayard-Alpert type of ionization gauge.

emitted, and the X-ray, in striking the plate, ejects an electron from it. These two phenomena, gas discharge and X-ray photoelectron current, limit the range of any type of ionization gauge.

Since at a fixed electron energy the ionization cross sections or probabilities of an electron-molecule collision leading to ionization vary from molecule to molecule, the ionization gauge must be calibrated (usually with a McLeod gauge) for each gas or gas mixture whose pressure is to be determined. Most commercial gauges have an approximate calibration based on pure nitrogen.

Another type of pressure gauge which must be calibrated before it can be used to measure absolute pressures is the thermocouple gauge depicted in Fig. 2-8. The operation of this gauge depends on the thermal conductivity of the gas in the gauge tube. An ac voltage is applied across a resistance divider containing a thermocouple (TC), and the resulting current heats the thermocouple junction

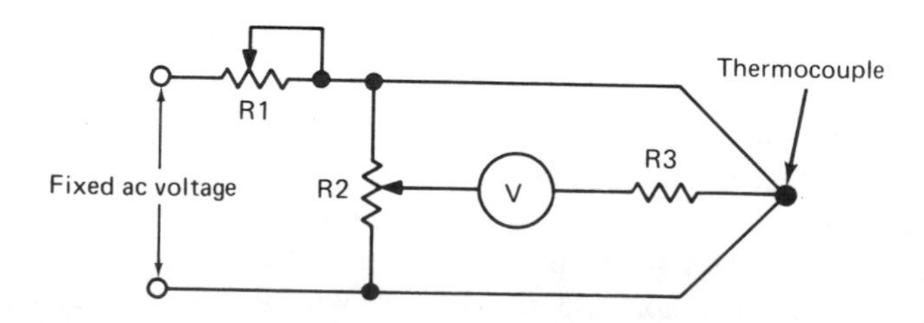

Figure 2-8. Thermocouple gauge circuit.

and produces a dc voltage. This voltage is measured by the voltmeter, and its magnitude depends on the thermocouple temperature, which in turn depends, at constant current, on the rate at which energy is conducted away from the thermocouple. The rate of energy transfer is dependent on the number of gas molecule-thermocouple collisions per unit of time and on the ability of different molecules to gain energy in a collision and to transport it to the walls of the thermocouple gauge tube. The latter property is known as the thermal conductivity and generally decreases with the mass and increases with the complexity of molecules. For a fixed gas composition, as the pressure increases, the rate of transfer of energy from the thermocouple to the gas will increase, the temperature of the thermocouple will decrease, and the thermocouple voltage will drop. The mathematical relationship between pressure and voltage cannot be derived for the tube geometries normally used. Suffice it to say that voltage does not vary linearly with pressure. The range of most thermocouple gauges is 10–1000 μ.

A few comments may be made about the circuit of Fig. 2-8 used in conjunction with the tube. R1 adjusts the ac current through the tube, which normally ranges from 100 to 200 mA and is specified by the manufacturer. R2 is a balancing resistor which adjusts the ac voltage across the meter to zero. Since this is a dc voltmeter, it will oscillate rapidly if any ac voltage is applied across it.

An interesting gauge, but one not widely used because of its fragile nature, is the glass spoon gauge shown in Fig. 2-9. Its operation depends on the properties of a very thin glass bulb made by collapsing a very thin sphere. As indicated in the figure one side is connected to a reference system and the other to the sample system. The gauge is constructed so that when there is no pressure difference between the two systems the two pointers are aligned. If the sample system is at a higher pressure than the reference system, the difference in force across the thin-walled spoon tends to make it straighten and deflect the attached pointer to the right and vice versa. Clearly this gauge must be either calibrated or used as a null indicator. The null method is described in detail below.

Another pressure-measuring instrument which can be used over the same range as the glass spoon gauge and which is mechanically much more reliable is the magnetic reluctance pressure transducer whose cross section is shown in Fig. 2-10. The body consists of two stainless steel blocks separated by a very thin stainless steel diaphragm which is magnetically permeable. Notice that the two stainless steel halves are provided with ports, one side leading to the reference system and the other to the sample system. In the interior of the block are two magnetic cores which are separated from the diaphragm by a small gap. In addition, inductance coils labeled L are located within the magnetic core. The mutual inductance of these coils depends on the length of the gap between the diaphragm and the magnetic core (this distance determines the magnetic reluctance, which is analogous to resistance in electrical current flow). When the pressures are balanced, the gap size is the same on both sides of the transducer, the induc-

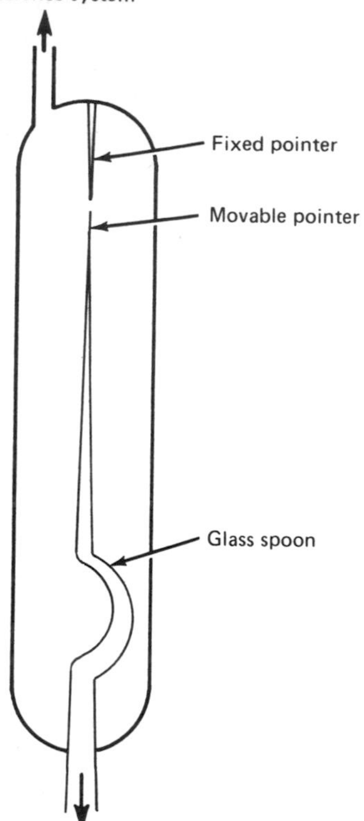

Figure 2-9. Glass spoon gauge.

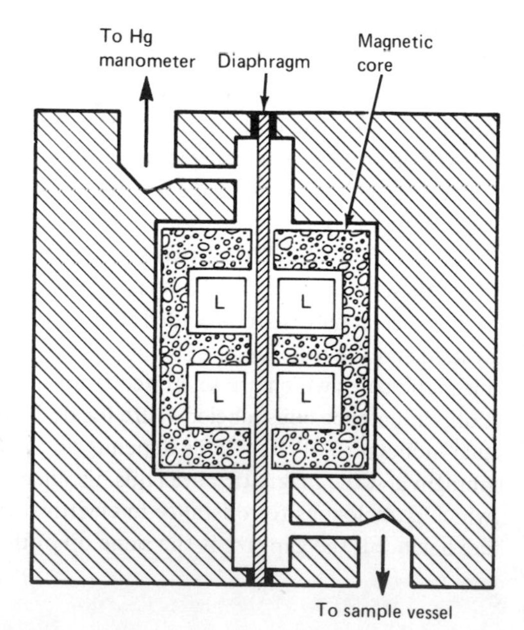

Figure 2-10. Cross section of pressure transducer.

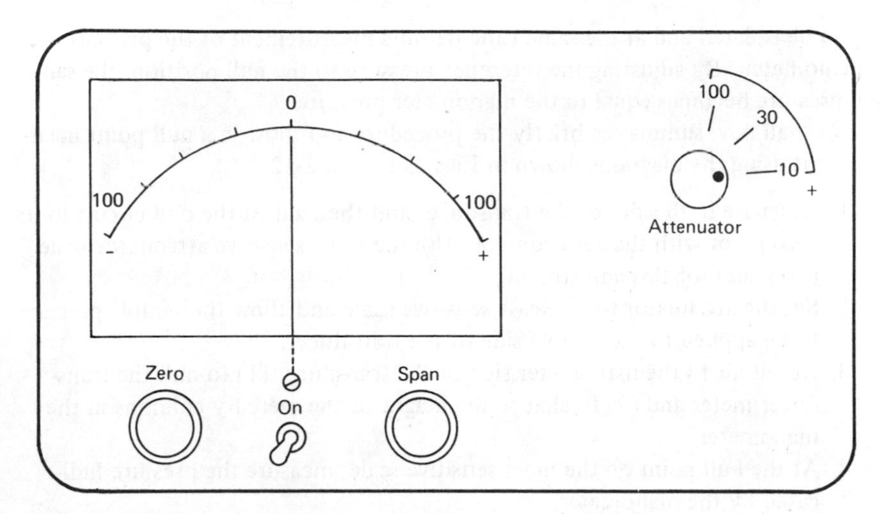

Figure 2-11. Transducer control circuit.

tance values are all the same, and the detector circuit shown in Fig. 2-11 indicates zero or null position. The sensitivity of these devices varies over a very wide range and depends on the thickness of the diaphragm. Generally speaking the devices are more reliable as null rather than as absolute pressure indicators.

Figure 2-12 illustrates the use of a pressure transducer as a null-measuring instrument in a system where it is desirable to exclude mercury vapor from the sample system but where the sample pressures may otherwise be adequately measured with a mercury manometer. The pressure transducer permits the sam-

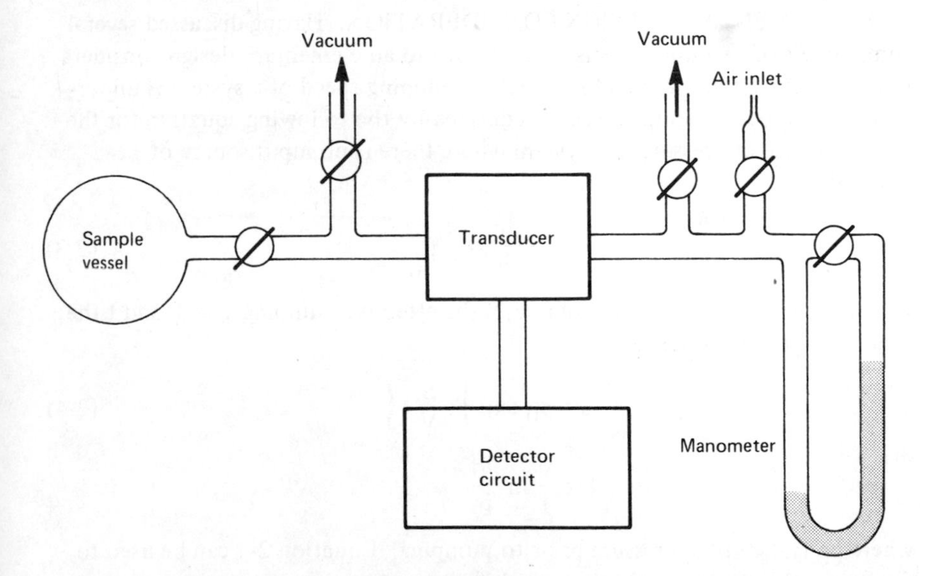

Figure 2-12. Schematic of null point pressure measurement apparatus.

ple to be isolated and at the same time permits measurement of the pressure with a manometer. By adjusting the reference pressure to the null position, the sample pressure becomes equal to the manometer pressure.

We shall now summarize briefly the procedure to follow in a null point measurement using the diagrams shown in Figs. 2-11 and 2-12.

1. Evacuate both sides of the transducer and then adjust the null circuit to its zero point with the zero control. Use the most sensitive attenuator scale (10 scale) for this adjustment.
2. Set the attenuator to the least sensitive scale and allow the sample pressure to be applied to the sample side of the transducer.
3. Admit air to the manometer side of the transducer (1) to null the transducer meter and (2) to change the height of the mercury columns in the manometer.
4. At the null point on the most sensitive scale, measure the pressure indicated by the manometer.
5. The overall sensitivity may be increased by increasing the span control potentiometer reading.
6. The meter is of the zero-center type so that both positive and negative deviations from the null point may be observed.

In summary the previous discussion has pointed out the basic components of a glass high-vacuum system, namely, mechanical pump, diffusion pump, vapor trap, high-vacuum stopcocks, and a pressure-measuring device. With these components one constructs a chamber for an experiment requiring low pressures of air or other material.

AN ELEMENTARY DESIGN CONSIDERATION. Having discussed several components of vacuum systems we now turn to an elementary design consideration. In some circumstances the effective pumping speed of a system is important. The effective pumping speed is defined by the following equation for the rate of change of pressure in a system where there is no input source of gas molecules:

$$-\frac{dp}{dt} = \frac{S}{V}p \qquad (2\text{-}3)$$

where p is the pressure, V the volume, S the effective pumping speed, and t the time. Integration furnishes

$$p = p_0 \exp\left\{-\frac{S}{V}t\right\} \qquad (2\text{-}4)$$

or

$$S = \frac{V}{t}\ln\frac{p_0}{p}$$

where p_0 is the initial pressure prior to pumping. Equation 2-4 can be used to determine the effective pumping speed if the pressure versus time is measured.

An obvious question concerns the magnitude of S and how it can be varied by components in the system. First we note that pumps having greater capacity than others will also have greater pumping speeds. The pump may not, however, be the determining factor in the effective pumping speed. Consider, for example, Fig. 2-13, which shows a pump connected by a piece of cylindrical tubing to a system which is to be evacuated. Intuitively we expect the rate at which the pressure drops in the system to depend on both the pump and the length and diameter of the connecting tube. These effects are summarized in the effective pumping speed S for the apparatus. Focusing on the tube, the effective pumping speed will drop off as the diameter decreases and/or the length increases. This observation makes clear the desirability of using large-diameter connecting tubing and eliminating unnecessary length.

Figure 2-13. Effect of connecting tubing on effective pumping speed.

PRECAUTIONS.

1. Never turn off a mechanical pump unless you immediately allow the pressure on both sides of it (inlet and exhaust) to reach atmospheric pressure. When a pump is turned off and not exhausted, the vacuum on the inlet side causes the oil in the pump to be drawn up into the evacuated region.
2. Always turn stopcocks slowly, applying a constant torque. A sudden torque will frequently break the glass tubing.
3. Admit air or any other gas to the vacuum system slowly in order to avoid sudden pressure bursts which will cause the mercury to "bump" and break the glass tubing and/or scatter mercury throughout the system.
4. Keep your mind on what you are doing. If you don't know, ask.

LEAKS. Leaks are anathemas to anyone working with vacuum systems, and they always appear at inopportune times. Finding and eliminating them is sometimes a real headache, but with experience in vacuum practice the problem becomes less severe. For the vacuum systems described in this text, checking for leaks will generally mean evacuating the system and then isolating the pump and watching a pressure-measuring device on the system. If after isolating the pumps the pressure remains constant, no leaks are present. If the pressure rises, either the system is leaking or there is some part of the system that is outgassing (material originally stuck to the walls is coming off at a rate faster than it is resticking to the walls). Outgassing can often be distinguished from a leak by noting how the pressure varies with time. If the pressure increases at first but then levels off at some constant value well below atmospheric pressure, the system is outgassing, whereas if it continues to rise to atmospheric pressure, a leak is present.

In a glass system leaks in glass-blown joints can often be found with a Tesla high-voltage coil. With the system evacuated and the coil operating, its probe tip is passed over the glass surface. If a pinhole leak is present, the high voltage at the coil will excite the molecules passing through the pinhole and it will "light up" as these molecules radiate. Small leaks can be repaired with epoxy or hard wax if reblowing the glass is inconvenient.

In metal vacuum systems the conventional method of leak detection (for small leaks) is the helium mass spectrometer method. The system is connected to a mass spectrometer tuned for helium ions and is evacuated. Helium gas is then sprayed sparingly over the outside of the system. When sprayed near a leak the helium enters the system and is detected by the mass spectrometer.

2.2 Temperature Measurement and Control

One of the variables which nearly always has some effect on experimental measurements is temperature; hence, its accurate measurement and control is a must in most situations. Intuitively we all connect temperature and heat in at least a qualitative fashion, and for some nonscientific situations this is quite satisfactory. However, a quantitative basis is necessary in scientific work, and we shall proceed to give a brief resumé of the absolute temperature scale.

At the outset it should be noted that temperature may be specified by a number for a body in thermal equilibrium and that this number is independent of the composition of the body. Having accepted this, the only problem remaining is agreement on a scale of numbers which may be used to represent the temperature. William Thomson (Lord Kelvin) used the second law of thermodynamics to define the so-called thermodynamic temperature scale. His definition is very satisfying since it is completely devoid of any reference to the material used as the thermometer. Kelvin's development utilizes the results of Carnot concerning relations between power and heat. His results show that the maximum work W derivable from an amount of heat, Q, in a process where the initial state is characterized by a temperature T_1 and the final state by T_2 is given by

$$W = Q\left(\frac{T_1 - T_2}{T_1}\right) \tag{2-5}$$

The thermodynamic temperature scale has a lower bound, namely absolute zero. Thus, the size of the degree on the absolute temperature scale can be fixed by arbitrarily choosing the temperature of some system. The presently accepted standard is the triple point of water (i.e., where liquid, gas, and solid coexist at equilibrium), which is defined as 273.16°K.

By international agreement the constant-volume hydrogen gas thermometer is the primary standard. In many circumstances, however, it is desirable to have a less cumbersome means of calibration, and out of this need the fixed points of the International Temperature Scale have been defined and are given in Table

2-3. The scale thus established is, in an absolute sense, completely arbitrary; however, the points are chosen to approximate very closely the thermodynamic temperature scale. Actual measurements of temperature from −183° to 630.5°C are made using a platinum resistance thermometer and two interpolation formulas together with the data from Table 2-3. From −183° to 0°C the interpolation formula is

$$R_t = R_0[1 - At + Bt^2 + C(t - 100)t^3] \tag{2-6}$$

whereas from 0° to 630.5°C the formula is

$$R_t = R_0[1 + At + Bt^2] \tag{2-7}$$

R_0, A, and B are determined by calibration at the ice, steam, and sulfur points, respectively, and C is determined by the resistance of the thermometer at the oxygen point. From 630.5°C to the gold point a thermocouple of platinum and platinum-rhodium is used to make measurements. As it stands, the International Temperature Scale covers a restricted range, and above and below its limits other scales must be used; several have been proposed. Using relations (2-6) and (2-7) the platinum resistance thermometer may be used to calibrate other devices.

Turning our attention now to thermometers we first enumerate criteria for useful ones. In principle, any physical property which is a sensitive function of temperature in a certain range may be utilized as a thermometer in that range provided it is calibrated by comparison with the standards for the thermodynamic or international scale. Other criteria, including cost, ruggedness, and simplicity, are also vital.

Table 2-3. FIXED POINTS ON THE INTERNATIONAL TEMPERATURE SCALE*

Substance	Condition	Temperature (°C)
Oxygen	Boiling point	−182.97
Water	Freezing point	0.00
Water	Boiling point	100.00
Sulfur	Boiling point	444.60
Antimony	Freezing point	630.50
Silver	Freezing point	960.5
Gold	Freezing point	1063

*Atmospheric pressure = 760 Torr in every case.

The most common thermometer is the mercury-in-glass thermometer, which utilizes the temperature dependence of the density of mercury and glass or equivalently the differential-volume thermal expansion of these materials. The latter is simply the fractional change in volume of a given weight of material when the temperature is changed by 1 deg. The calibration is based on the change in length of a mercury column when the temperature is increased from a known reference point near the lower end of the Hg thermometer scale to a second reference point near the upper end. A standard Pt resistance thermometer could be used as the reference, Between the two reference points a linear function for the length of the Hg column in the capillary is assumed. The error introduced with this assumption varies depending on the range of temperatures covered. There are two broad categories of mercury-in-glass thermometers: (1) the total immersion type in which the whole thermometer is assumed to be at the sample temperature and (2) the partial immersion type where only the lower part (indicated by a line on the stem) is at the sample temperature while the remainder is at room temperature. The range of temperatures accessible with the mercury thermometer is limited, roughly speaking, by the freezing and boiling points of mercury. Below the freezing point of mercury, fluid-in-glass thermometers using alcohol may be used. At higher temperatures some other type of thermometry is generally employed.

A special mercury-in-glass thermometer known as a Beckmann thermometer is illustrated in Fig. 2-14. It is of the differential type in that its total range is only 5° or 6°C but the lower end of this range is variable. With this restriction the Beckmann thermometer is useful for measuring accurately the temperature difference (hence, differential thermometer) between two samples where $\Delta T \lessapprox 6°C$. The lower end of the temperature range is adjusted by changing the amount of mercury in the reservoir of Fig. 2-14. To accomplish this, the thermometer bulb is warmed to drive mercury by expansion into the reservoir. When the temperature of the bulb is slightly greater than the temperature desired, the column is broken by tapping the thermometer rather sharply to break the column of mercury into two portions at the reservoir entrance. Once calibrated the thermometer must not be turned upside down or the calibration may be lost.

Another type of thermometer is the gas thermometer, which depends on the variation with temperature of the pressure of a fixed volume of gas. The cumbersome nature of the gas thermometer apparatus renders this device inconvenient for most purposes. Further discussion of gas thermometry is given in Section 4.1.

One other kind of thermometer involving gases is sometimes used, especially at low temperatures. These thermometers depend for their operation on the variation with temperature of the vapor pressure of a solid or liquid. For example, oxygen thermometers are sometimes used to measure temperatures near 80°K. To construct an oxygen thermometer one would generally connect a U-tube manometer to a closed tube containing pure oxygen, the tip end of which is placed in good thermal contact with the body whose temperature is required. The manometer then furnishes the vapor pressure of pure oxygen, the latter being controlled by the temperature of the body being studied.

We have already indicated the utility of resistance thermometers with the mention above that the International Temperature Scale relies upon the platinum resistance thermometer. Basically this kind of thermometer depends on

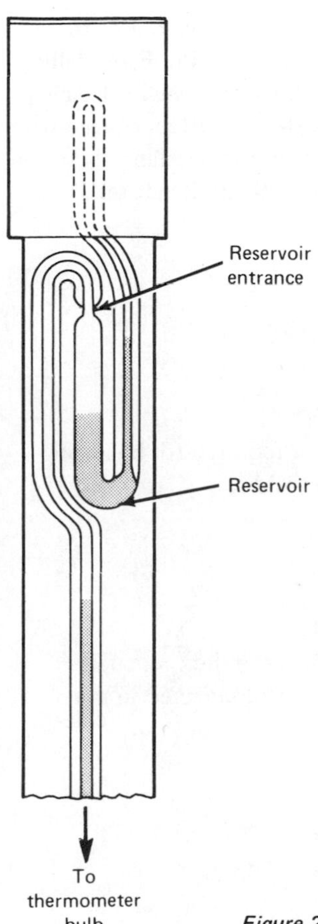

Figure 2-14. Beckman thermometer.

the temperature variation of electrical resistance. At the microscopic level, resistance arises when, under the influence of a voltage, electrons move through a substrate and encounter (collide with) atoms of the substrate lattice. They are thereby impeded, and the extent of this phenomena is measured by the resistance. As the temperature increases so does vibrational motion of the substrate lattice, and in general the tendency is toward increased resistance. Platinum is especially useful because it has, for metals, a high resistance and is chemically inert.

Physically, platinum thermometers are normally constructed by winding the wire on a dielectric material in a way which minimizes strain and its variation with temperature. The length of wire is generally selected to make the resistance be either 25.5 or 2.55Ω at 0°C with attending sensitivities of 0.1 and 0.01Ω deg^{-1}, respectively. Since the conductors connected to the platinum wires also have resistance which varies with temperature, caution must be exercised in designing the circuitry. A typical arrangement is shown in Fig. 2-15. Essentially the circuit employs four leads from the thermometer, a Wheatstone bridge circuit, and a switching arrangement, which makes negligible the effect of lead wire resistance on the measurement. At the balance condition the resistance between A and D and that between B and D must be equal. With the switches set as shown in Fig. 2-15

$$R' + R_1 = R_t + R_2 \tag{2-8}$$

Switching all the switches to their alternate position furnishes

$$R'' + R_2 = R_t + R_1 \tag{2-9}$$

where R' and R'' are the settings of the variable resistor required for balancing. Equations (2-8) and (2-9) are readily solved for R_t to yield

$$R_t = \frac{R' + R''}{2} \tag{2-10}$$

Clearly this arrangement minimizes lead resistance effects.

Another widely used temperature-measuring device is the thermocouple, which consists of a pair of dissimilar electrical conductors connected in the

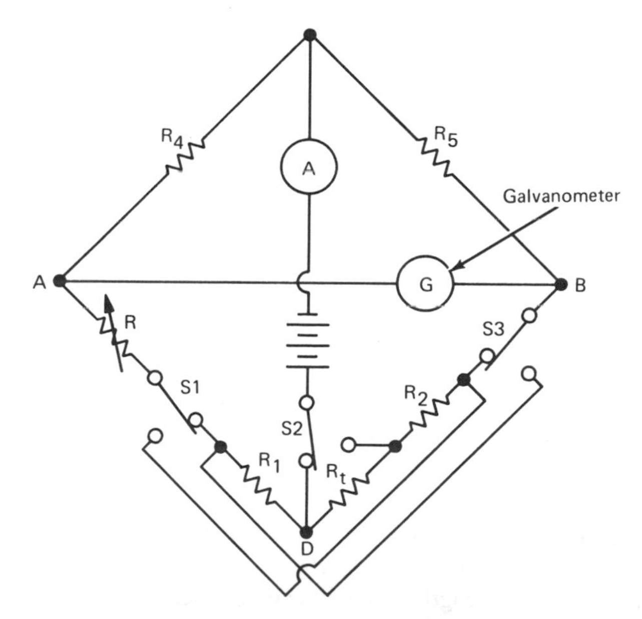

Figure 2-15. Platinum resistance thermometer. R_1, R_2, lead wire resistance; R_t, thermometer resistance; R, balancing resistor.

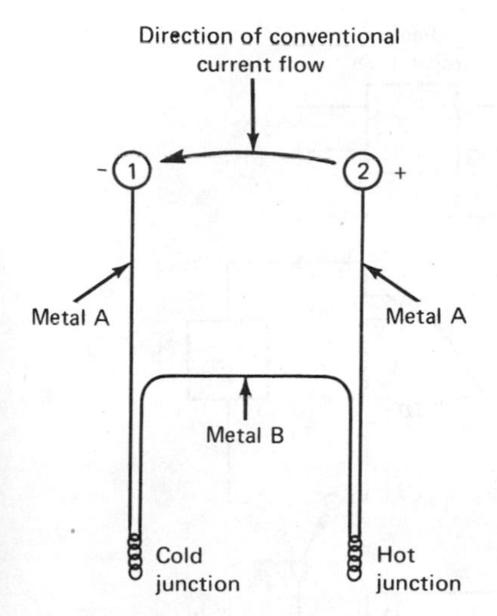

Figure 2-16. Thermocouple.

fashion shown in Fig. 2-16. When the junctions are at different temperatures, an emf is developed. The phenomenon was discovered by Seebeck in 1821. Use of a thermocouple as a temperature-measuring device depends on the fact that the emf varies monotonically with temperature. For some pairs of metals a monotonic relation between emf and temperature does not exist, and as a result some caution must be exercised in selecting pairs of metals. If in Fig. 2-16 we connect points 1 and 2, a current will flow because of the emf. The direction of conventional (+) current flow depends on the two metals, A and B. In Fig. 2-16, A is assumed positive with respect to B, and in this case the conventional current will flow from the hot to the cold junction. The theoretical development of these experimentally observed phenomena is a problem in solid-state physics and is beyond the scope of this treatment.

A variety of common thermocouples exists and a few are listed in Table 2-4 together with their useful temperature range. The standard arrangement is to have one junction, the reference, at °C with the second junction at the temperature of the sample. For precise work, thermocouples must be carefully made by spot welding. In addition, contact potentials between the thermocouple and the external circuitry must be avoided or accounted for. For the work requiring thermocouples in the experiments in this book the apparatus shown in Fig. 2-17 is satisfactory. The potentiometer circuit employed is described in greater detail in Section 2.4. Here we shall describe the procedure for use of the whole circuit.

Prior to measurement of the thermocouple emf, the apparatus must be standardized. The following procedure is suggested:

1. Check the wiring of the circuit.
2. Connect the power supply and filament transformer to 115-V outlets. The power supply should be about 3 V dc.

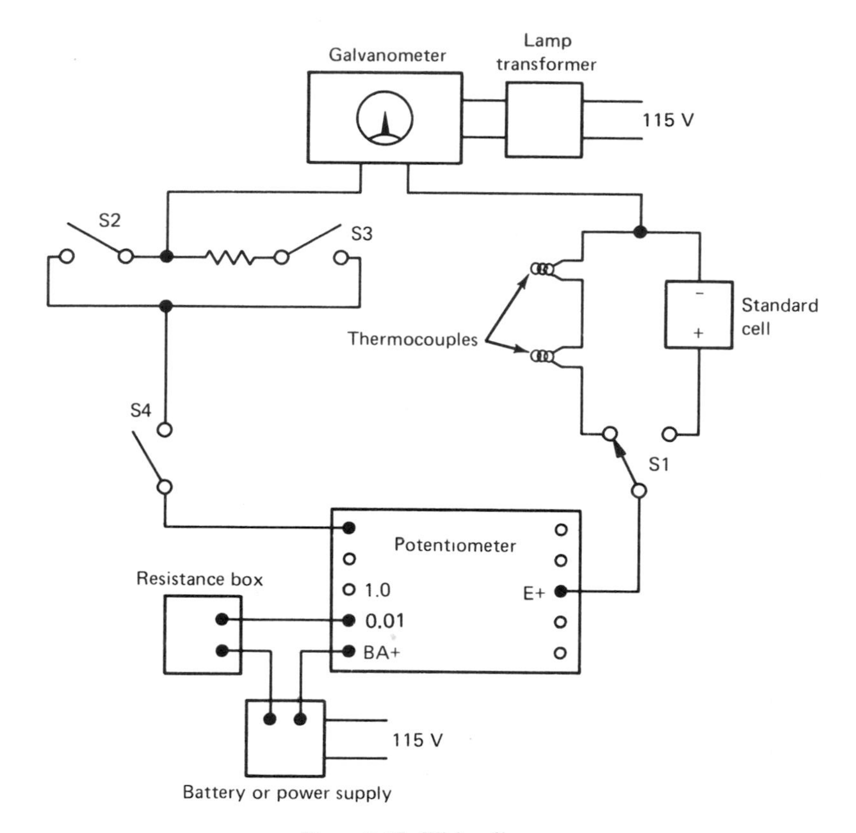

Figure 2-17. Wiring diagram.

3. Disconnect the lead to the 0.01 post of the potentiometer and connect it to the 1.0 post. This changes the range of emf values which can be measured. When the power supply is connected to the 1.0 post the range is 0–1.6 V (the standard cell range); when connected to the 0.01 post it is 0–16 mV (the thermocouple range).
4. Move switch S1 to the standard cell position.
5. Set the potentiometer to read the standard cell emf, which is written on the outside of its container.
6. Set the resistance box at 100 Ω, close switch S3, momentarily depress the key switch (S4), and note the direction of the galvanometer deflection.
7. Reset the resistance box at 200 Ω and repeat step 6. If the direction of deflection has changed, the balance point (the point at which no current flows through the galvanometer) is between 100 and 200 Ω. Keep adjusting the resistance box until no deflection is observed on the galvanometer when the fine switch (S2) is closed.

At this point the emf applied to the potentiometer has been made both directly proportional and equal to the resistance of the potentiometer. This completes the standardization. The resistance box setting should remain the same throughout the remainder of the experiment unless a check of the standardization reveals a change in the system.

To measure an emf the following procedure is followed:

1. Rewire the potentiometer as in Fig. 2-17 and turn switch S1 to the thermocouple position (i.e., reconnect the lead to the 0.01 post).
2. Measure the emf of the thermocouple by adjusting the potentiometer until no deflection of the galvanometer is observed using the fine switch (S2). The reading on the potentiometer is the emf of the thermocouple.

To convert from emf to temperature a calibration curve is needed. For chromel-alumel the values are given in the appendices. In Section 3.2 the copper-constantan thermocouple is calibrated.

Table 2-4. COMMON THERMOCOUPLES

Thermocouple	Temperature range (°C)
Copper-constantan	−250–300
Iron-constantan	−200–1000
Chromel-alumel	−190–1300
Platinum–platinum/rhodium	0–1700

There is a variety of other temperature-measuring devices including thermistors, optical pyrometers, bolometers, and thermopiles. They are described in some of the References.

We shall now turn our attention to means by which temperature may be controlled. In many experiments it is not only desirable to measure the temperature but also to vary and control it precisely. Broadly speaking, the method by which this is accomplished is to immerse the system in a heat bath whose heat capacity is large compared to the system of interest. Such a device is known as a thermostat.

For temperatures in the range 10° to 90°C a common heat bath is the circulating water bath in which heat is dissipated at a controllable rate through resistance heaters into the water which is stirred to remove thermal gradients. The temperature is controlled by the rate at which energy is put into the system and by the rate at which it is dissipated into the surroundings. Through the use of thermal switches (for example, like those used in homes to control furnaces and air conditioners) the heating rate can be increased when the temperature drops below the set point and vice versa. Above 90°C up to about 250°C hydrocarbon oils can be used in the same way as water.

Air thermostats are also widely used because apparatus tend to be complicated and not readily submergible in a liquid. Ovens of various types are available commercially for the range from room temperature to 300°C, and special types are available for reaching 1000°C. For the experiments described herein the oven can be readily constructed simply as a box with high-quality insulation in the walls and a fan to circulate the air. These ovens are subject to greater thermal gradients than the liquid devices.

Figure 2-18 shows schematically the basic ingredients of a small oven whose temperature is regulated solely by the operator. It consists of an insulating wall structure made of Transite or some other mechanically strong and insulating material ($\frac{1}{4}$ -in.-thick Transite is quite satisfactory for temperatures up to 100°C). Heating is achieved using a cartridge heater or a length of resistance wire connected to an ac voltage control (Variac or Powerstat, for example). A small fan is provided to circulate the air and thus minimize temperature gradients. Temperature is measured using a thermometer or some other device. The heater characteristics are essentially determined by the volume of the thermostat and the highest temperature required. For a box with a volume near 1 or 2 ft^3 and temperatures below 100°C, a 500-W cartridge heater should be adequate unless rapid heating schedules are called for. Higher temperatures can be reached in two ways: heaters with more power capability and a box with additional insulation around the walls. Operation of the thermostat unattended and at a constant temperature requires the addition of a thermostat which is connected into the heater power circuit and controls the voltage applied to the heater.

Below room temperature the cooling of the system can be achieved by immersion in some liquid contained in a dewar flask. Some readily available systems are listed in Table 2-5. The slushes listed are equilibrium mixtures of the solid and liquid phases of the material listed. They can be made by adding dry ice or liquid nitrogen slowly to the liquid material with stirring.

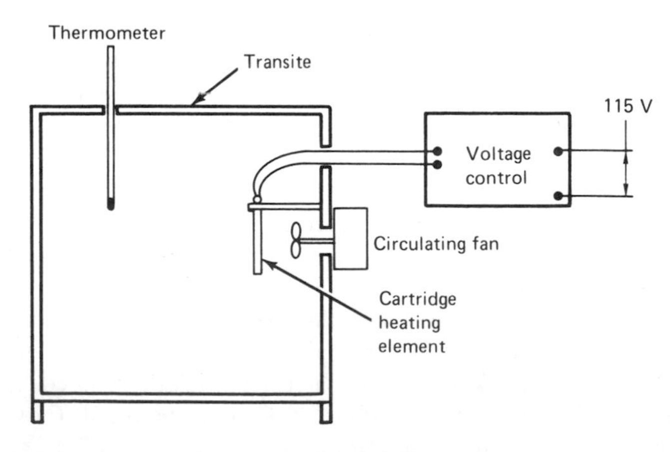

Figure 2-18. Small air thermostat.

Table 2-5.

System	Temperature (°C)
Ice water	0
CCl_4 slush	−22.9
Bromobenzene slush	−30.6
$CHCl_3$ slush	−63.5
Dry ice-acetone (flammable)	−78
Dry ice-trichloroethylene (nonflammable)	−78
C_2H_5Br slush	−119
N-pentane slush	−131.5
Isopentane slush	−160.5
Liquid nitrogen	−196
Liquid hydrogen	−253
Liquid helium	−269

PROBLEMS

1. What error is incurred if a total immersion thermometer is used as a partial immersion device?

2. Using handbook tables, calculate the emf developed across two chromel-alumel thermocouples connected in series as hot-cold-hot-cold. Would such a device be more sensitive or less sensitive than a single thermocouple? What is its sensitivity near room temperature?

2.3 Optical Instruments

Chemical research relies heavily on electromagnetic radiation of various kinds. For example, the field of photochemistry by definition makes use of the absorption of light by matter. Usually the wavelengths range from 6000 Å downward. At longer wavelengths microwave spectroscopy, electron spin resonance, and nuclear magnetic resonance involve transitions induced by electromagnetic radiation. Spectroscopy of various sorts often plays a crucial role in analytical chemistry. Even when absorption does not occur, light can be used as an analytical tool through such phenomena as refraction, reflection, and optical rotatory dispersion. Because of the fundamental role played in experimental research by electromagnetic radiation having wavelengths in the vacuum ultraviolet through the infrared region of the spectrum, it is of importance to become acquainted with various pieces of apparatus which are of use in this kind of research. To that end we shall discuss here devices for use in the ultraviolet, visible, and infrared regions of the spectrum. It is neither possible nor desirable to discuss the available apparatus thoroughly. For additional information, consult the References.

BASIC COMPONENTS. Any optical system contains a set of components which may be arranged in various ways to suit the needs of the experiment being performed. These components may be categorized as source, filter, sample, and detector, and one method of arranging them is shown in Fig. 2-19. This arrangement, which shows a source, two filters, a sample, and a detector, is typical of apparatus used in photochemical experiments. Other experiments may require different numbers of each of the components, and as mentioned above they are often arranged differently.

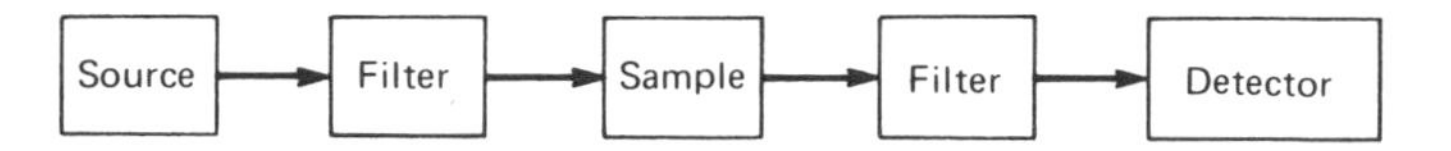

Figure 2-19. Basic components of an optical system.

As the name suggests, a source is any device from which useful electromagnetic radiation emanates. Sources vary depending on the wavelength region of interest, the power required, and the configuration of the system.

Filters are devices used to alter the wavelength distribution of the light incident upon them. In this sense a grating or prism monochromator is a filter which accepts many wavelengths but transmits only a narrow band, the nominal wavelength of which is predetermined by the geometric position of the grating or prism. Other filters commonly used are various glasses, plastics, and other chemicals which absorb at some wavelengths but transmit at others.

The sample is that part of the apparatus which is of particular interest. It must be contained in a vessel of some sort, and the walls of the vessel must transmit at least the radiation of interest.

The detector may be any one of a wide variety of devices including the phototube, photocell, photomultiplier, photographic film, human eye, thermopile, bolometer, and chemical reaction. An obvious requirement is its sensitivity to the wavelengths of interest in a given experiment. Different detectors respond with differing sensitivities in different spectral regions. Hence, experience is required in order to make a wise selection of devices.

With this general introduction to the basic features of optical systems we shall proceed to examine in turn the two diverse spectral regions: visible-ultraviolet and infrared. The visible-ultraviolet region includes wavelengths from about 8000 Å downward to 2000 Å, while the infrared ranges from 8000 Å upward to 1 mm.

VISIBLE-ULTRAVIOLET. At the outset it is important to consider materials out of which suitable lenses, windows, and cell walls may be constructed. Any material chosen for this purpose must readily transmit light in the wavelength region of interest. Two of the most commonly used materials are Pyrex and fused silica (quartz). Pyrex is satisfactory for wavelengths above 3000 Å but absorbs strongly below 3000 Å. Quartz is satisfactory for the entire visible-ultraviolet region. Devices made from quartz are more expensive than those of

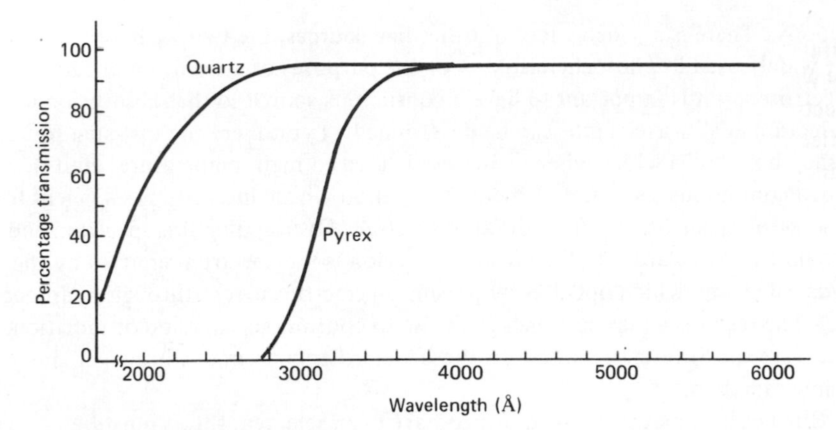

Figure 2-20. Percent transmission of Pyrex and quartz as a function of wavelength.

Pyrex because of both material cost and difficult glass blowing. Figure 2-20 shows the percentage transmission as a function of wavelength for both. Other materials are sometimes used, especially in sample containers. The book by Calvert and Pitts listed in the References contains a fairly complete list.

Sources for use in the ultraviolet-visible spectral region are of two types: line sources and continuous sources. Line sources arise from electronic transitions in atoms which are excited by electrical or microwave discharge. The atoms must be in the gas phase and at fairly low pressures (a few Torr). Typical examples include the low-pressure mercury arc which is shown in Fig. 2-21, with lines at 1849, 2537, 2653, 2967, 3130, 3340, 3660, 4046, 4358, and 5769 Å as well as other weaker lines and the cadmium source with lines at 2288, 2980, and

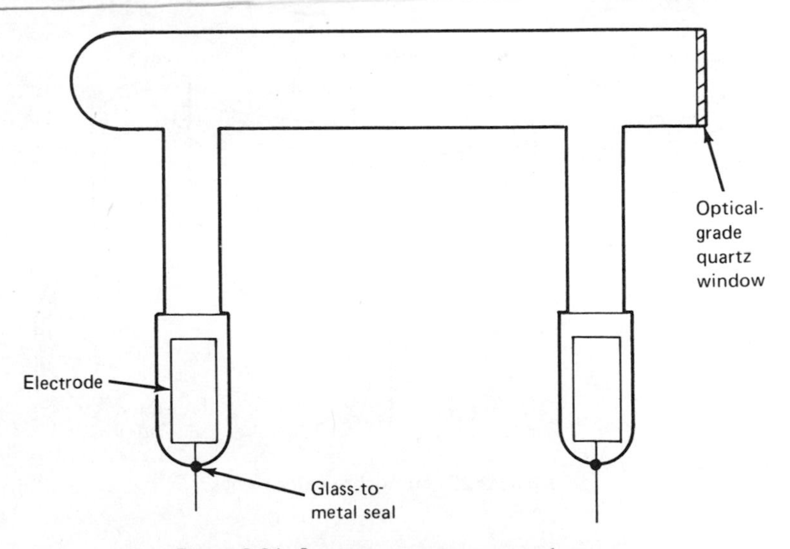

Figure 2-21. Low-pressure mercury arc lamp.

3466 Å. There is a wide variety of other line sources; the two we have mentioned are widely used in photochemistry. For the purposes of absorption or reflection spectroscopy it is important to have a continuous source so that absorption or reflection at all wavelengths can be determined. Typical are the tungsten source (fancy light bulb) which when resistively heated to high temperatures emits a broad continuous spectrum of thermal radiation whose intensity is sufficient for absorption spectroscopy from 3000 to 8000 Å. Overlapping this spectrum and extending downward into the ultraviolet region is the spectrum emitted by the hydrogen lamp, which operates by passing an electric current through hydrogen gas. The result is a plasma which gives rise to continuous emission of radiation. West (see the References) discusses these and other sources of ultraviolet and visible radiation.

Once optical materials and a source have been selected, filters must be selected. In absorption or emission spectroscopy the filter will normally be a monochromator which uses either a prism or a grating together with appropriate focusing optics to disperse the incoming radiation into geometrically different regions of space. With a suitable detector, the wavelength distribution of the transmitted light can be analyzed and used in elucidating molecular structure or concentration. Figure 2-22 illustrates a simple grating monochromator. Light from a source is gathered by a condensing lens system; as a result light diverging from the source is rendered converging toward the collecting lens, forming an image of the source on the entrance slit. The collecting lens creates a bundle of light rays which uniformly illuminate the grating. The grating disperses the incident radiation, and a concave mirror brings a certain wavelength into focus at the exit slit after correction for aberration has been made by the corrector lens. Rotating the grating about an axis perpendicular to its face brings different wavelengths into focus at the exit slit.

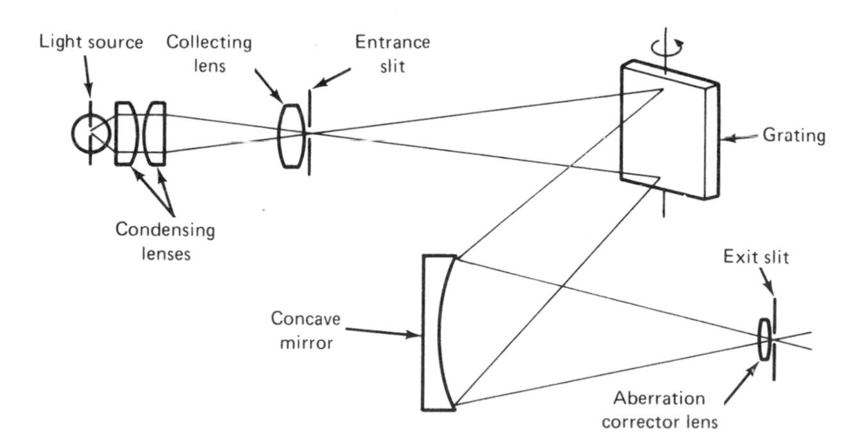

Figure 2-22. Monochromator optics.

The grating consists of a set of parallel rulings evenly spaced across the surface of a mirror. A schematic of the surface and its important dimensions is shown in Fig. 2-23. The wavelength of the diffracted rays emanating in a particular direction is given by

$$n\lambda = a(\sin \alpha \pm \sin \beta) \tag{2-11}$$

where n is an integer known as the order, a is the distance between grooves, and the other parameters are shown in Fig. 2-23. The minus sign is used in Eq. 2-11 if β is on the same side of the surface normal as α. The blaze angle θ is determined by the wavelength range of interest and is constructed so that at a wavelength centrally located within the range, specular reflection occurs in the same direction as diffraction for that wavelength. The further a wavelength is removed from this blaze condition, the poorer is the transmission by the grating. According to Eq. (2-11), if a grating and a direction are specified by $\{a, \alpha, \beta\}$, then λ', $\lambda'/2$, $\lambda'/3$, . . . all appear in the same direction. For example, if the value of n is fixed at 1 and the grating is adjusted to transmit 6000-Å radiation, then 3000 Å, 1500 Å, etc., will also be transmitted. To avoid the lower wavelengths (higher orders) an additional filter transmitting at 6000 Å but strongly absorbing at 3000 Å and below must be used.

The grating may be replaced by a prism in which light is refracted rather than diffracted. The dispersion then depends on the variation with wavelength of the index of refraction. The geometric arrangement of the associated collection and transmission optics will also differ from that shown in Fig. 2-23.

Filters other than monochromators are also in common use. These filters include chemical filters in which solutions of various salts are used. By adjusting their composition, various wavelength bands can be isolated. A typical example

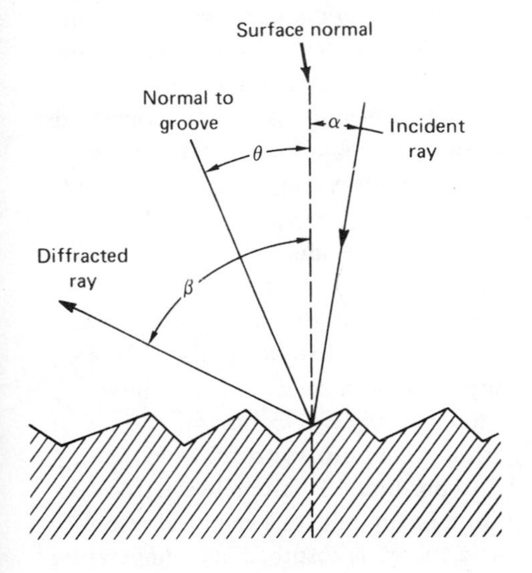

Figure 2-23. Grating surface.

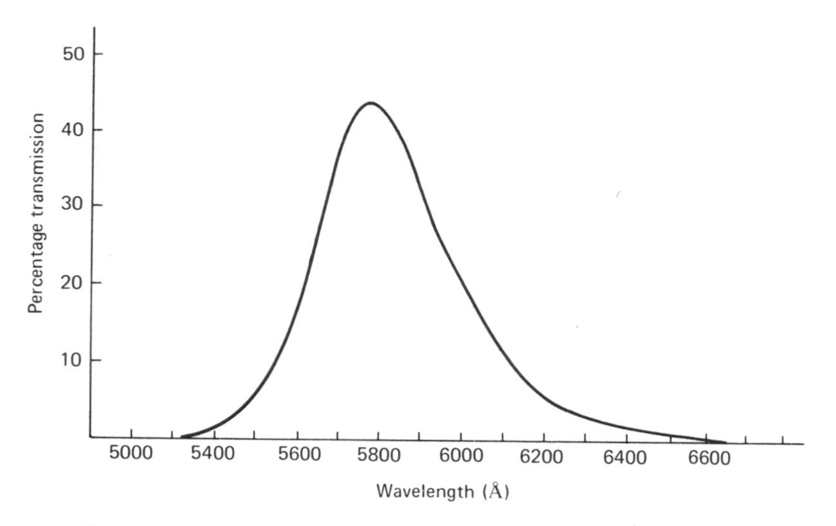

Figure 2-24. Transmission spectrum of a combination of two liquid filters: (1) a slightly acidic aqueous solution containing 10 g of $CuCl_2 \cdot 2H_2O$ and 30 g of $CaCl_2$ per 100 cc of solution and (2) a solution containing 3 g of $K_2Cr_2O_7$ per 100 cc of solution. A 1-cm-long cell is used for (1) and a 5-cm cell for (2).

is shown in Fig. 2-24. A lengthy list is given in Calvert and Pitts (see the References). Interference filters have come into common use with the advent of accurate thin-film deposition techniques. These filters, as their name suggests, depend on interference patterns arising as a result of the reflectance and transmittance properties of thin metal films. A wide variety of these filters is available commercially. Various glasses are also useful; a rather complete set is available from Corning. Nearly any material whose optical properties vary from opaque to clear as the wavelength is changed is a suitable candidate for a filter. The limiting factors seem to be the perseverance and ingenuity of the experimentalist.

Once the proper wavelengths, derived from a suitable source, have been isolated and passed through the reaction vessel, two tasks remain: dispersion and detection of the light emanating from the cell. In some experiments the light may be of the same wavelength distribution as that incident on the cell. In other cases, fluorescence, for example, it may be of a different character. Finally, in many experiments the source and first filter shown in Fig. 2-19 may be absent, the sample vessel itself then serving as a source. Regardless of these considerations, the filtering can be accomplished with devices such as those mentioned above and so the only problem remaining is detection.

Detectors fall into three broad categories: (1) photoelectric in which the photon flux interacts with some material to produce a measurable electron current, (2) photochemical in which the photon beam induces a chemical reaction, and (3) thermoelectric in which the photon energy is absorbed in a temperature-sensitive medium.

The first category includes photomultipliers, phototubes, and photocells. A circuit diagram of a typical photomultiplier is shown in Fig. 2-25. As previously suggested, the photomultiplier is an electronic device for converting a photon current into an electron current. The apparatus consists of an evacuated tube containing several amplifying stages, a resistance network which divides the high voltage applied to the tube, a high-voltage power supply, and an electrometer. The photon current is incident upon a photosensitive surface (electrode 1) which emits electrons in proportion to the magnitude of the photon current. These electrons are accelerated to electrode 2 (a dynode) because of the potential applied between electrodes 1 and 2. At the second dynode each incident electron ejects several electrons (a multiplication) which are accelerated to dynode 3. At each dynode the same multiplication process occurs. Finally, after the last acceleration, the electrons are collected and the current is measured with an electrometer. Note the polarity; the last electrode is at ground potential, and the first is at a very negative potential. A typical photomultiplier has 13 multiplying stages, and with approximately 1400 V applied across the tube the multiplication factor is about 10^5. As with all detectors the response is wavelength-dependent and the manufacturer's specifications should be checked for details.

Photochemical detectors include photographic techniques and a wide variety of liquid- and gas-phase reactions called actinometers (photon counters). Photographic detection is of widespread use, and with this technique the light dispersed from a grating similar to that shown in Fig. 2-22 is allowed to strike a strip of photosensitive emulsion. The exit slit shown in Fig. 2-22 is removed.

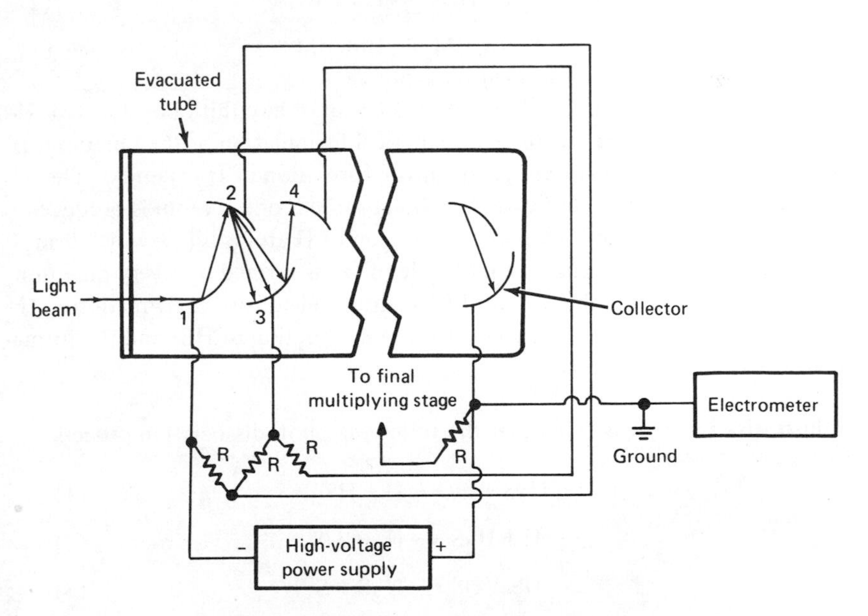

Figure 2-25. Diagram of photomultiplier.

The resulting exposure in which different wavelengths strike the emulsion at different positions provides, on development, a spectrum of the light incident on it. By suitable calibration in terms of wavelength and film darkening, the wavelength distribution of the light can be determined. This technique finds use in both absorption and emission spectroscopy. Emulsions of various types are readily obtainable and must be selected chiefly on the basis of their speed (how much light is required to give a detectable result), grain (the size of the photosensitive silver halide crystals used in the emulsion), and spectral sensitivity (the wavelength region where the emulsion is photosensitive). For accurate work great care must be exercised in the development of the emulsion and in the evaluation of the intensity of bands using a densitometer, which measures the extent of darkening.

Chemical actinometers involve gases or liquids which on photon absorption lead to a well-defined chemical transformation of known quantum yield (number of molecules of product per photon absorbed). A typical liquid actinometer is potassium ferrioxalate, which is useful over the range 2500–5770 Å. On photon absorption ferric ion is reduced to ferrous ion whereas the oxalate ion is oxidized. The quantum yields vary with wavelength but have been accurately determined. A typical gas-phase actinometer is HBr, which shows a quantum yield of unity for formation of H_2 and is useful from 2000 to 2500 Å. The photochemistry of the HBr reaction is mechanistically well understood and may be written as

$$HBr + h\nu \longrightarrow H + Br \tag{1}$$

$$H + HBr \longrightarrow H_2 + Br \tag{2}$$

$$Br + Br + M \longrightarrow Br_2 + M \tag{3}$$

where $h\nu$ denotes a photon and M denotes the wall or any third body such as H_2 or HBr. Without working out the mathematical formulation of the kinetics it is possible to see that the quantum yield for the formation of H_2 is unity. The reasoning goes as follows: For each photon absorbed, one H atom is produced in reaction (1) and the only subsequent reaction of H atoms is (2), which leads to one molecule of H_2. Thus, one molecule of H_2 is formed for every quantum of light absorbed, or, in other words, the quantum yield for the formation of H_2 is unity. What are the quantum yields for the destruction of HBr and the formation of Br_2?

Illustrative Example 1 Consider the following photodissociation process

$$H_2S + h\nu \longrightarrow H + HS \tag{1}$$

$$H + H_2S \longrightarrow H_2 + HS \tag{2}$$

$$HS + HS \longrightarrow H_2S + S \tag{3}$$

The quantum yield for the formation of H_2, which we denote as ϕ_{H_2}, is calculated as follows. The rate of absorption (photons per second) is denoted as I_a and with this and the above mechanism we can write

$$\frac{d[H]}{dt} = I_a - k_2[H][H_2S]$$

Under steady-state conditions the rate of production of H atoms equals the rate of their destruction (loss); thus $d[H]/dt = 0$ and

$$k_2[H][H_2S] = I_a$$

But

$$\frac{d[H_2]}{dt} = k_2[H][H_2S] = I_a$$

If I_a is constant throughout the experiment and if $[H_2] = 0$ at the beginning of the experiment, then at any time t

$$[H_2] = I_a t = Q$$

where Q is the total number of quanta absorbed in time t. By definition

$$\phi_{H_2} = \frac{[H_2]}{Q}$$

and for this case, then,

$$\phi_{H_2} = \frac{Q}{Q} = 1$$

Notice that we could obtain the same result by defining the quantum yield in terms of the ratio of two rates: the rate of production of H_2 divided by the rate of photon absorption.

The third type of detector may be illustrated by the thermopile shown in Fig. 2-26. This device is constructed of a series of thermocouples alternately exposed

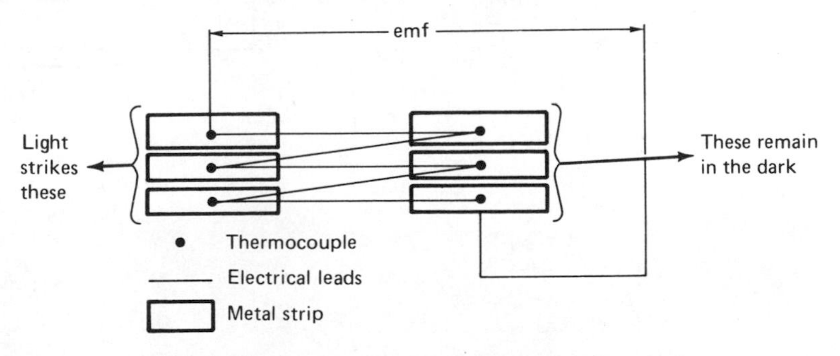

Figure 2-26. Thermopile schematic (three-junction).

and not exposed to the incident radiation. By suitable preparation, usually lampblack coating, those thermocouples which are exposed to light are made very absorbing. The energy of the light is thus deposited in the material, raising its temperature and giving rise to an emf difference between the hot and cold junctions. A commonly used thermopile has 13 junctions, with the hot junctions made of thin ribbons of metal arranged to intercept a beam of light 1.5 mm wide and 15 mm high. Common thermocouple materials are bismuth-silver or copper-constantan.

As typical instrumentation, we shall now present two complete devices, one suitable for photochemistry and the other for absorption spectroscopy. Figure 2-27 illustrates a typical photochemical arrangement (see Comtet in the References). S is a 1000-W Hg-Xe high-pressure lamp operated at approximately 32 V dc and 32 A, which provides a continuous spectrum in the 2000–6000-Å region. The lamp is located inside an air-cooled housing equipped with a spherical mirror in order to collect more light. The light beam is rendered parallel by lens L_1 and focused by lens L_2 on the entrance slit of a grating monochromator (M) with linear reciprocal dispersion of 16 Å/mm. Linear reciprocal dispersion is a measure of how effective the prism or grating is in dispersing the light. The figure 16 Å/mm implies that if the exit slits are adjusted to a width of 1 mm, the band of wavelengths transmitted will be 16 Å wide. The almost monochromatic light is passed through a mercury vapor filter (F) to eliminate by absorption any possible mercury resonance radiation from the lamp and is then collimated by a quartz lens L_3. To monitor variations in the intensity of the lamp, a plate of quartz (Q) is placed at 45° to the incident beam, and the intensity of reflected light ($\simeq$ 10% of the total beam) is monitored by a phototube P_1 connected to a

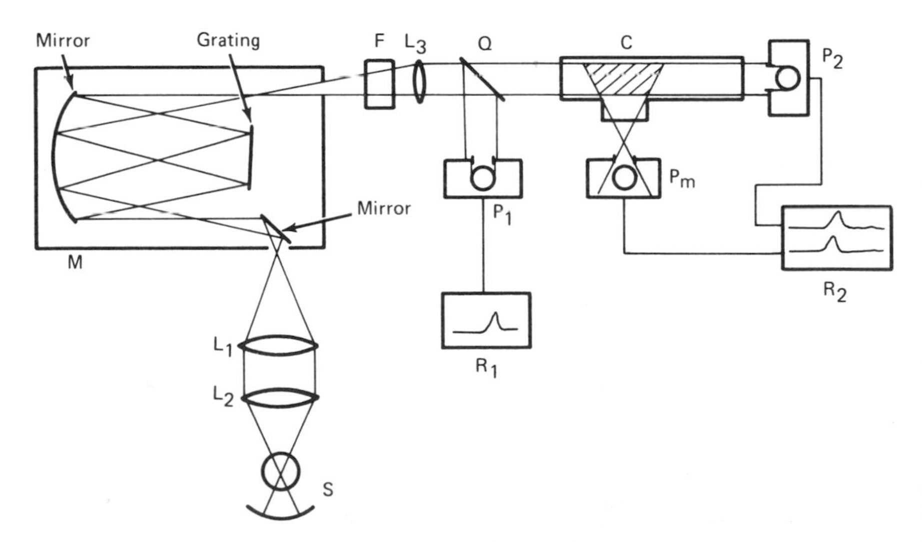

Figure 2-27. Schematic of the optical arrangement suitable for photochemistry.

recorder R_1. The beam transmitted through Q is passed through a T-shaped quartz cell (C). The light transmitted by the cell is monitored with a phototube P_2, while the light emitted as fluorescence or phosphorescence is measured at 90° with a photomultiplier (P_m) operated with a high-voltage power supply. Recorder R_2 measures simultaneously the signals from P_2 and P_m. All optical materials through which light must pass are constructed of high-quality quartz.

As a typical example of an absorption spectrophotometer, consider Fig. 2-28. This apparatus has two sources, a tungsten lamp and a hydrogen lamp, and so the complete ultraviolet-visible region is accessible. The source is selected by rotating the mirror (a) in or out of the beams. There are two dispersing elements in the monochromator, a prism (F) and a grating (J), and so we may regard the system as having two monochromators in series. The internally reflecting prism (F) has high transmission properties, while the grating (J) furnishes high resolving power (low linear reciprocal dispersion). E, G, I, and K are spherical mirrors, while D, H, and L are the entrance and exit slits of the monochromators. At the exit of the second monochromator the transmitted light is made parallel by lens M and is chopped at 30 Hz by a motor-driven semicircular mirror. The light is alternately reflected through the sample cell and the reference cell by mirrors O and P, respectively. The light transmitted by either cell is mirrored to the photocell (X) by mirrors (V, V′, W, and W′). As a result the phototube alternately senses the light intensity from the sample and reference cells, and an electron current proportional to the light intensity is generated. The two signals are compared with appropriate electronic circuitry and recorded as the net optical density,

$$\log_{10} \frac{I_{ref}}{I_{sample}},$$

where I_{ref} is the reference cell signal and I_{sample} is the sample cell signal. Chopping accomplishes a twofold purpose: First, signal-to-noise (S/N) ratios are improved in this way. Second, by using a matched cell in the reference beam, absorption of light by the windows of the sample cell is automatically accounted for, and the recorded optical density is attributable solely to the contents of the sample cell.

INFRARED. With the foregoing cursory examination of apparatus suitable for visible-ultraviolet use, we shall proceed to a similar consideration of the infrared region. While the logical considerations remain, for the most part, the same, the infrared and visible-ultraviolet apparatus differ primarily in the types of materials used and the types of useful sources and detectors.

The most widely used source in infrared spectrometers is the Nernst filament, made of rare-earth oxides in the form of small-diameter tubes. Passage of electric current through this device heats it, and a typical operating temperature is about 1800°C. The thermally emitted radiation is continuous and very rich in

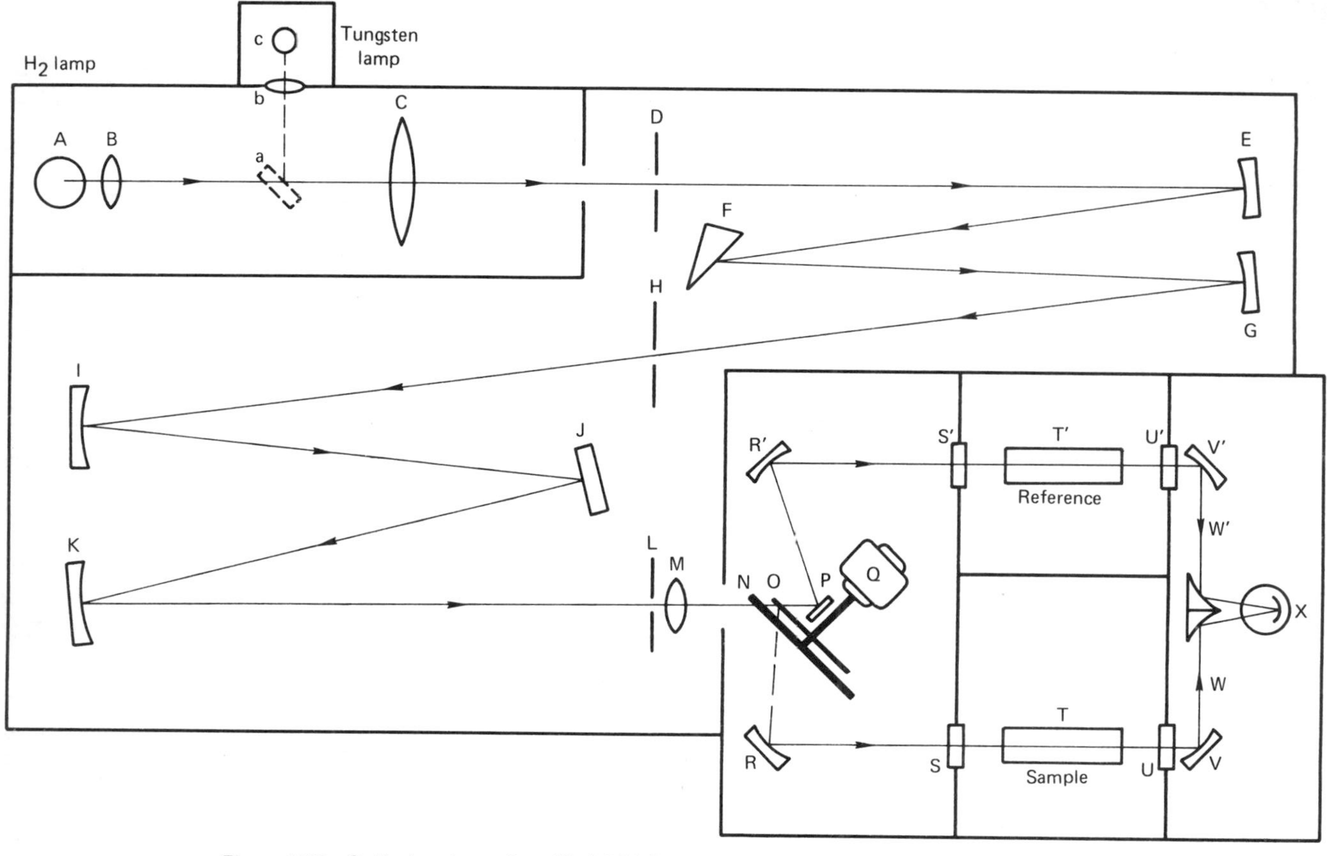

Figure 2-28. Optical system. Cary Model 14 Spectrophotometer (omitting infrared section). (With permission of Varian Instruments.)

infrared wavelengths. The Nernst tube has a large negative coefficient of resistivity (resistance decreases rapidly as the temperature increases). As a result the tube must be preheated by an external source before it can actually be started. Furthermore, resistance, other than the tube itself, must be placed in series with the filament to limit the current drawn when the filament gets hot.

Filters for the infrared are of the same types as those used for the visible-ultraviolet region except different materials are used. Alkali and alkaline earth halide crystals are widely used. Sodium chloride is useful to wavelengths of 16 μ ($1\ \mu = 10^{-6}$ m) and CsI is useful to 50 μ. Other materials are also used, for example, quartz (3.6 μ), sapphire (5.9 μ), and a host of polycrystalline materials. Again, the References should be consulted for more complete listings and further sources of information.

Detectors for the infrared include photocells, thermopiles, bolometers, and pneumatic devices. As usual, the appropriate device must be selected on the basis of its sensitivity in the wavelength region of interest.

Photocells depend for their operation on the production of conduction electrons by absorption of infrared radiation into a semiconducting material. An excellent photoconductive detector is made of germanium doped with gallium and cooled to 4°K. Such devices are sensitive through the infrared to wavelengths as long as 130 μ. Lead sulfide is also a popular photoconducting material, useful to 3.6 μ.

Thermopiles have been discussed earlier and so we turn to bolometers, which depend on the absorption of radiation into a metal (a thin platinum strip, for example). Essentially a constant current is passed through the thin metal strip, and the voltage drop across it is measured. Absorbed radiation causes the temperature to increase, thereby increasing the resistance and increasing the measured voltage drop.

Typical of the pneumatic devices is the Golay detector, sketched in Fig. 2-29. Its operation depends on detecting pressure changes in the gas cell. This cell is filled with gas which absorbs infrared radiation of particular wavelengths. On absorption, the temperature, and thus the pressure, increases. The pressure change is detected as motion of the thin diaphragm.

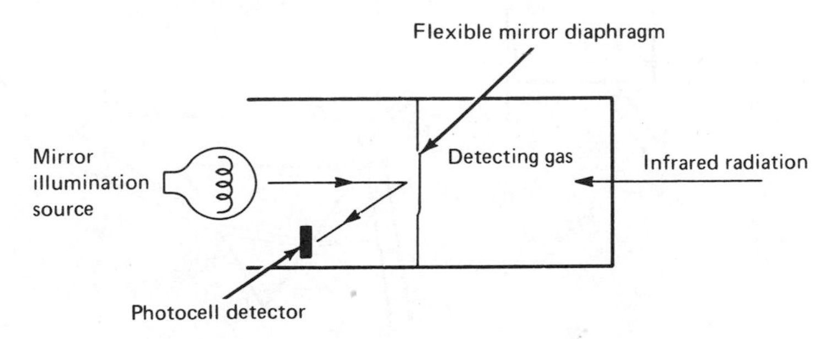

Figure 2-29. Schematic of Golay pneumatic detector.

We shall now turn our attention to the description of typical infrared spectrometers: one of the prism type and the other of the grating type. Figure 2-30 illustrates a single-beam prism type of infrared spectrometer. Radiation from the source (S) is chopped at frequency ω by a rotating disc (D). The resulting divergent radiation is focused by mirrors M1 and M2 through the sample (X) onto the entrance slit of the monochromator. Diverging through the monochromator to mirror M3 the beam is then reflected through the dispersing prism (P) to mirror M4 and back through the prism. The resulting radiation comes to a focal point intermediately located between M5 and M6 and finally is focused on the thermopile detector by M7. The resulting signal has a periodic waveform whose frequency is equal to the chopping frequency. Rotating the prism brings different wavelengths to focus on the detector. As is typical with infrared spectrometers the radiation passes through the sample before it is dispersed. The single-beam operation makes quantitative analysis difficult because an accurate measure of the intensity of the signal in the absence of sample is required. Although this can in principle be obtained in a separate measurement, its accuracy is limited as a result of instrument drift.

A more sophisticated double-beam grating spectrometer is schematically shown in Fig. 2-31. Light from the source (S) is passed through both a sample and reference cell and is focused on two separate entrance slits of the monochromator. The beams are sampled alternately as time passes by the beam switch (rotating mirror). After dispersion by the grating the signals (first sample, then reference, etc.) arrive at the detector. The associated electronics connected to the thermocouple detector are arranged so that no net output is recorded un-

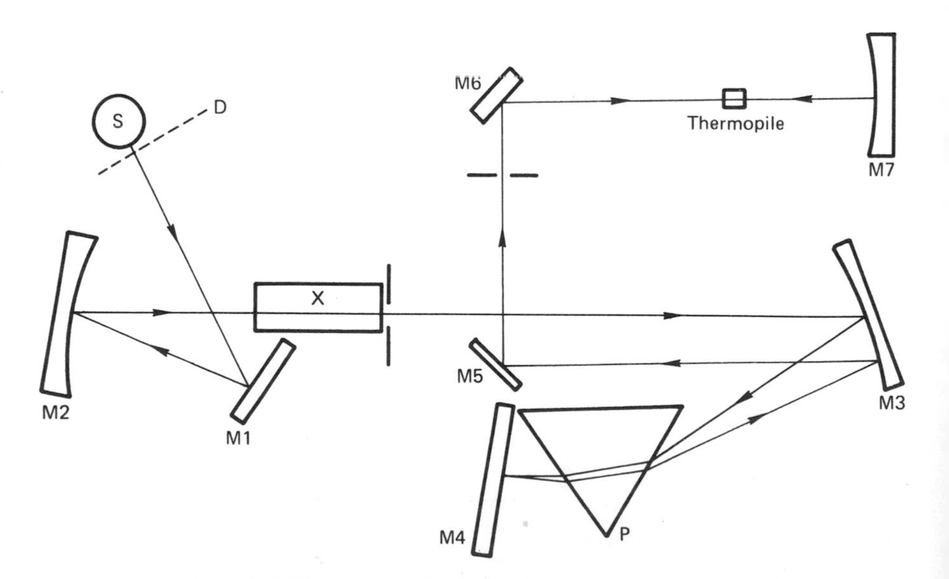

Figure 2-30. Single-beam prism infrared spectrometer.

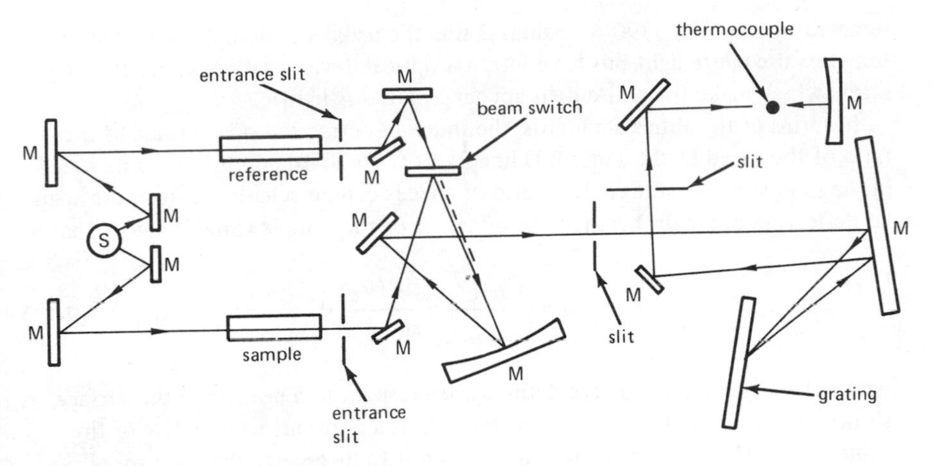

Figure 2-31. Schematic of a double-beam grating infrared spectrometer.

less the sample and reference signals have different magnitudes. When a difference exists the output (usually a deflection on a strip chart recorder) is proportional to it. The double-beam operation permits a continuous monitoring of the signal levels by means of the reference beam (no sample absorption), thus eliminating most of the effects of instrument drift.

REFRACTOMETRY AND LIGHT SCATTERING. Thus far we have given attention to optical systems which are useful in investigations involving absorption or emission of radiation. In the material which follows we shall describe other optical systems which are used in the experiments described in this text. To be specific, refractometry and light-scattering apparatus are discussed. Many other kinds of optical devices are in common use: those used in polarimetry, optical rotatory dispersion, streaming birefringence, Faraday effect, Kerr effect, and ellipsometry, to name a few. A discussion of them is beyond the needs and scope of this text.

Refractometry. It is common knowledge that visible rays of light are bent on passing from air into water and that ordinary sunlight is separated spatially into its different-colored components on passage from air into water and back into air. These phenomena are characterized and discussed in terms of indices of refraction. Refractometry is an experimental method which distinguishes between different species on the basis of their refractive index. The refractive index of a substance is defined in terms of the angular deflection which a light ray undergoes as it passes from the substance into another of known refractive index. The observed angular deflection is a consequence of the fact that light travels with different speeds in different media. The refractive index (or the speed) is a function of both the temperature of the sample and the wavelength of light being used. For analytical work all indices of refraction must employ a common wavelength and a common temperature, 25°C being the usual temperature and the 5890-Å sodium D line the usual wavelength. Many refractometers use *white* light but have internal optical devices called Amici compensators which make the white light appear as 5890-Å light.

In terms of the above standards, the index of refraction n is defined as the ratio of the speed of the sodium D line in air at standard conditions to its speed in the sample under study. This ratio of speeds can be calculated from the angular deflection of a light beam at the interface of the sample (Snell's law). That is,

$$n = \frac{v_{air}}{v_{sample}} = \frac{\sin(\theta_a)}{\sin(\theta_s)} \tag{2-12}$$

where the angles θ_a and θ_s are defined with respect to a normal to the surface, as shown in Fig. 2-32. For a given substance, n is a constant irrespective of the value of θ_s. The speed of light in air turns out to be greater than the speed of light in most other environments and so, generally, $n > 1$. Figure 2-32 shows schematically a single ray passing from the sample into air. Assuming that n is indeed greater than unity, Eq. (2-12) requires $\sin(\theta_s) < \sin(\theta_a)$ or $\theta_a > \theta_s$. In passing note that neither θ_a nor θ_s are well defined for angles greater than $\pi/2$. Examining Fig. 2-32 we consider the effect of increasing θ_s. As required by Eq. (2-12), θ_a will remain larger than θ_s until θ_a becomes equal to $\pi/2$. This is the so-called critical ray, and we denote the incident angle as θ_s°. For all values of θ_s between θ_s° and $\pi/2$, light rays will not pass from the sample into the air but rather are internally reflected at the interface, as illustrated in Fig. 2-33. The actual refractometer measurement is discussed below, but Fig. 2-34 points up the essential principle. In this figure the rays are passing from the air into the sample. All the rays between $\theta_a = 0$ and $\theta_a = \pi/2$ become angularly compressed on passage into the sample and are limited to the angular region $\theta_s = 0$ to $\theta_s = \theta_s^\circ$.

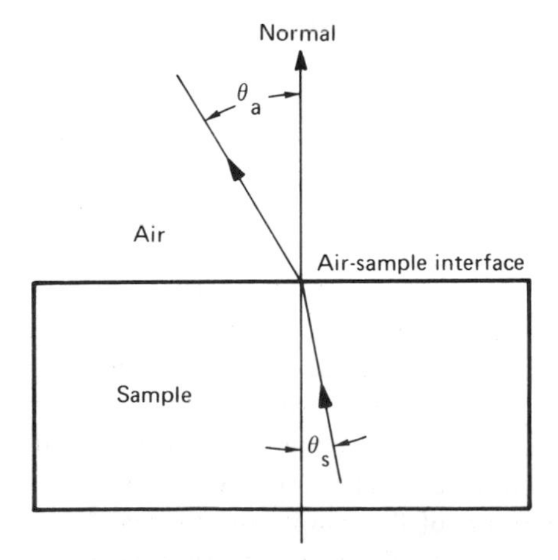

Figure 2-32. Refraction at an air-sample interface.

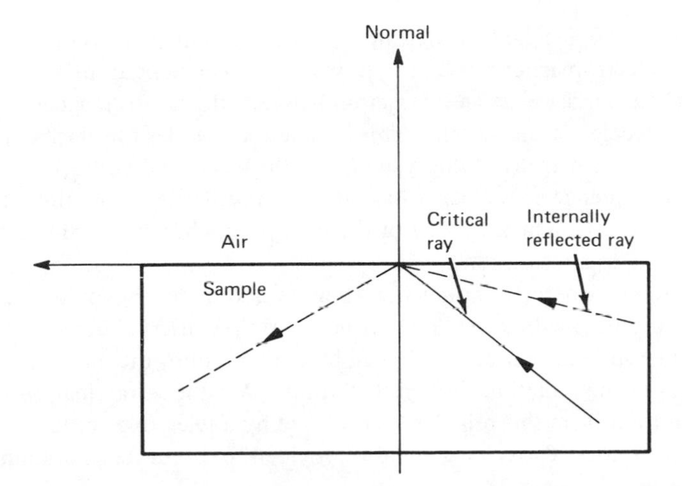

Figure 2-33. Critical and internally reflected rays.

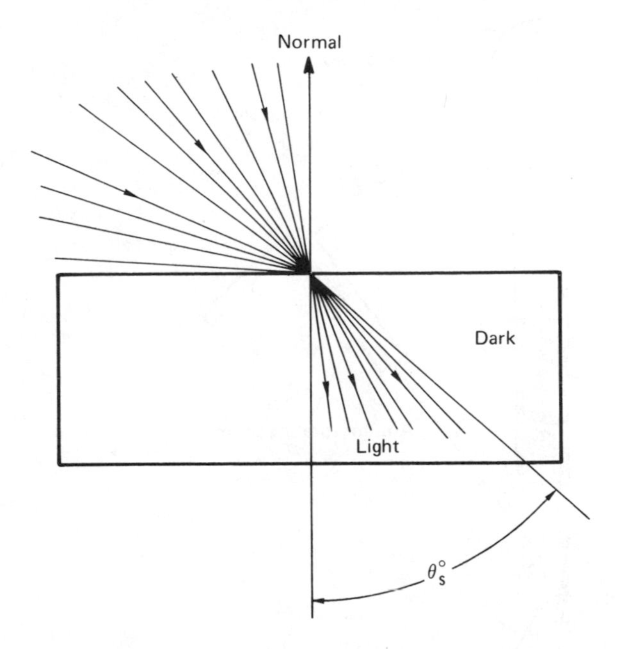

Figure 2-34. Light and dark regions within sample.

Therefore, an observer inside the sample would find a sharp transition from lightness to darkness on passing through the critical angle θ_s°. The magnitude of θ_s° depends on the sample because refractive index varies from sample to sample. It is this variation which renders refractometry useful as an analytical tool.

Qualitatively, the molecular level description of refraction arises as follows. Experimentally, electromagnetic radiation is observed to be bent at interfaces. By implication, there must exist an interaction between the electromagnetic radiation and the molecules of the system. This interaction couples the electromagnetic radiation to the electrons of the system, and the electron distribution moves under the influence of the perturbation. As a result atoms and molecules are polarized. Retardation of the speed of the light occurs because of this kind of interaction—and, as a result, refraction.

As a typical refractometer, consider the Abbe refractometer shown in Fig. 2-35. It uses two prisms with a thin layer of liquid sample inserted between the two. Light from a suitable source is reflected by a mirror onto the illuminating prism, which passes the radiation through the sample into the refracting prism. The light transmitted from this prism is then viewed by a telescope containing the Amici compensators. Cross hairs are also provided for accurate positioning of the field of view.

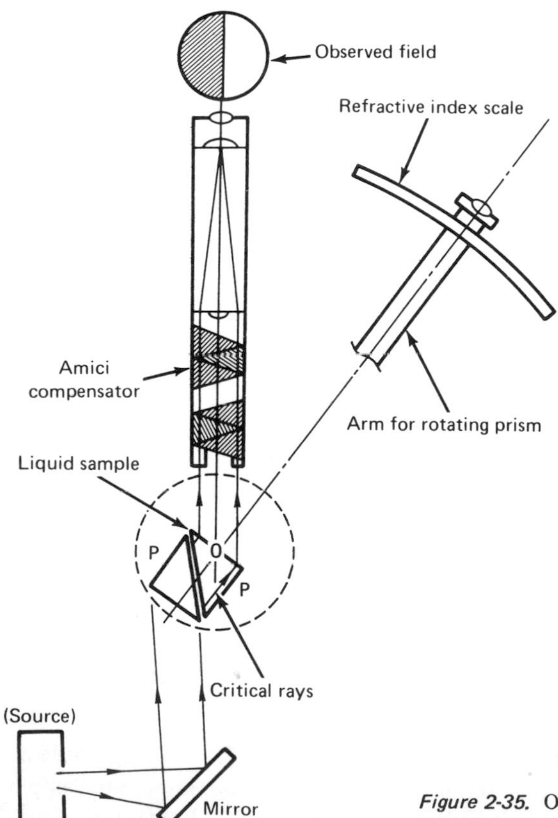

Figure 2-35. Optical path in an Abbe refractometer.

The refractometer makes use of the critical ray phenomena discussed above, but it is the critical ray inside the prism surface, not the sample, that is important, because the refractive index of the prism is larger than that of samples which can be measured. Mathematically, at the critical angle

$$\frac{1}{\sin \theta'_{\text{prism}}} = \frac{v_{\text{sample}}}{v_{\text{prism}}} = \left(\frac{v_{\text{sample}}}{v_{\text{air}}}\right)\left(\frac{v_{\text{air}}}{v_{\text{prism}}}\right) = \frac{n_{\text{prism}}}{n_{\text{sample}}} \tag{2-13}$$

Figure 2-36 illustrates the phenomena. As the light passes from the sample into the refracting prism, its rays are bent. For some particular θ' in the prism, the light will have entered at 90° with respect to the surface. Light will appear at angles less than θ' but not greater than θ'. An observer, as shown in Fig. 2-36, will note a dark and light area. As shown by Eq. (2-13), θ' depends on both the prism and the sample. Thus, θ' depends on the sample and can be used to measure the refractive index of the sample if the position of the prism is adjusted by rotation about point 0 (Fig. 2-35) so that the boundary between the light and dark parts of the field of view is centered on the cross hairs of the telescope. This rotation also changes a scale pointer, which moves along an index of refraction scale. The scale pointer directly furnishes the index of refraction when these adjustments are completed provided the instrument has been calibrated.

In operation the refractometer prisms should never be touched and they should be clean and dry prior to use. A few drops of sample between the prisms

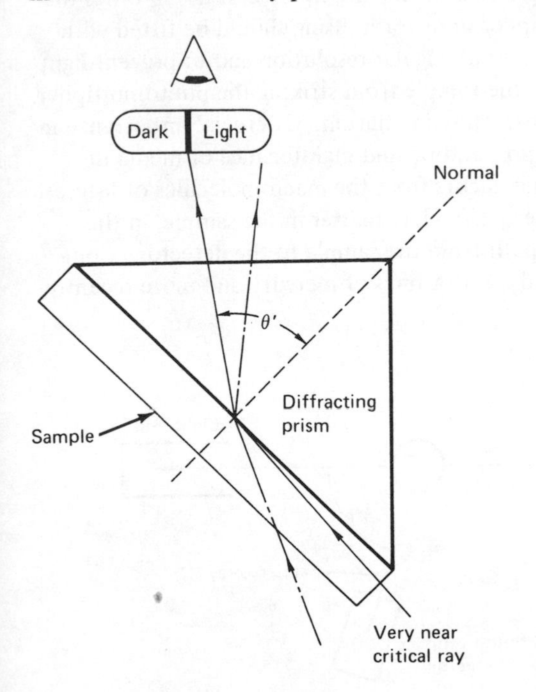

Figure 2-36. Critical rays in refractometer.

are sufficient. The rotatable mirror and light source should be adjusted so that the lower prism is adequately illuminated. When the light-dark interface is located in the telescope the Amici compensators should be adjusted until the interface is sharp and colorless. With the cross hairs centered on the light-dark dividing line, the index of refraction can be read from the calibrated scale. Generally it is wise to recalibrate the instrument using samples with known refractive indices.

Light Scattering. Apparatus designed for the purpose of measuring scattering by aerosols, macromolecules, dust, etc., are conceptually identical with spectrophotometers except in one respect: The detector is normally movable in a plane perpendicular to an axis through the sample. This added degree of freedom permits measurement of the angular dependence of the scattered light. From a theoretical point of view the angular dependence is of fundamental significance. Figure 2-37 illustrates schematically the essential features of a light-scattering measurement system. The incident radiation must be highly monochromatic because scattering intensity depends on the fourth power of the wavelength. The incident radiation must also be well collimated so that the scattering angle θ can be defined with little uncertainty. The photomultiplier detector should be located a large distance from the sample so that the sample may be considered approximately as a point-scattering source. Large in this context means that the distance from sample to detector should be at least 10 times the sample thickness. The photomultiplier detector housing should be fitted with narrow entrance slits to provide for good angular resolution and to prevent light scattered from particles other than the sample from striking the photomultiplier. A light trap is provided for the nonscattered radiation. Of crucial importance in light-scattering experiments is the preparation and maintenance of media in which the only significant scattering occurs from the macromolecules of interest. This implies avoiding dust and other particulate matter in the sample, in the sample container, and in the light path from the sample to the detector. Common light sources are the 4360- and 5460-Å lines of mercury and more recently lines from visible-region lasers.

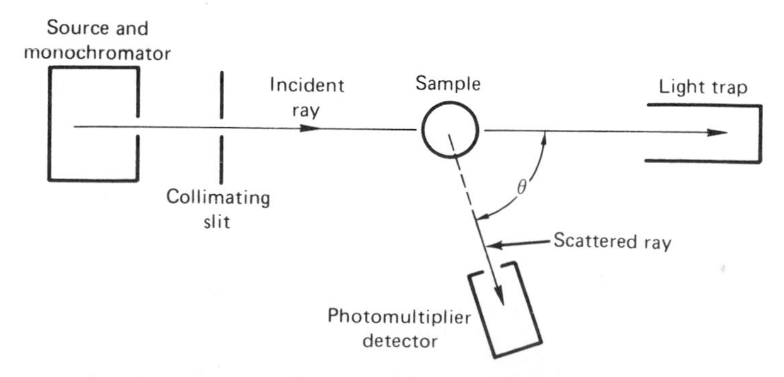

Figure 2-37. Schematic of light-scattering apparatus.

2.4 Electronics

The use of complex electronic apparatus in chemical research seems to increase constantly, and today most physical chemistry laboratories utilize a considerable variety of electrical systems. While it is obviously impossible to discuss electronics thoroughly in one section, it is nevertheless worthwhile to describe some of the fundamental concepts and a few of the more common pieces of equipment.

In electrical circuit analysis and description, the quantities of basic interest are currents and voltages. Various devices are used to supply and alter both these quantities to achieve some desired end, for example, amplification, rectification, and heating. Current refers to the flow of charged particles; most often in electronics the charged particles are electrons, with a notable exception being electrochemical cells. Current arises because voltage differences (electrical potential-energy differences) exist within the circuit. In many cases we are interested in how a certain part of a complete circuit responds to various voltages and/or currents. That is, we apply some known voltage or current to the input of a device (input signal) and ask for the characteristics of the output voltage and current (response or output signal). In the following paragraphs attention is focused on methods for describing mathematically the input and output signals and characterizing the functions of several electrical devices.

TIME DEPENDENCE OF VOLTAGES AND CURRENTS. At the outset it is of considerable importance to set down in a mathematical way the methods by which the time-varying properties of electrical circuits are described. Basically, there are two circuit types: (1) dc (direct-current) circuits in which the algebraic sign of neither current nor voltage varies with time and (2) ac (alternating-current) circuits in which the sign of both the current and voltage varies with some characteristic frequency. The latter type of circuit is generally described by one or more sinusoidal functions. Even though the actual time dependence may not be sinusoidal, it may, nevertheless, be constructed mathematically by linearly combining sinusoidal functions whose frequencies are multiples of a fundamental frequency (Fourier synthesis). For this reason we shall discuss only sinusoidal ac waveforms and the response of circuit elements to them.

An arbitrary time-dependent sinusoidal function can be written as

$$V(t) = V_0 \sin(\omega t + \phi) \tag{2-14}$$

where $V(t)$ is the instantaneous value of the function, V_0 the maximum value or amplitude of the function, t the time, ω the angular frequency in radians per second, and ϕ the phase angle in radians. Sinusoidal waves of the same frequency and amplitude may show different time dependences because of variations in ϕ.

For example, consider the two functions

$$V'(t) = V_0 \sin(\omega t)$$

$$V''(t) = V_0 \sin(\omega t + \frac{\pi}{4})$$

In this case V'' is said to lead V' by $\pi/4$ radians. The function $V'''(t) = V_0 \sin[\omega t - (\pi/2)]$ lags V' by $\pi/2$ radians.

A decomposition of $V(t) = V_0 \sin(\omega t + \phi)$ can be achieved by using the trigonometric relation for the sum of two angles; namely

$$V(t) = V_0(\sin \omega t \cos \phi + \cos \omega t \sin \phi) \tag{2-15}$$

Illustrative Example 2 Consider the common 60-cycle ac voltage available commercially. While it is not representable with high accuracy as a pure 60-cycle sine wave, for our purposes this approximation will suffice. Therefore in Eq. (2-14) we use $\omega = 60$. If the voltage is 110 V, this generally refers to the root-mean-square voltage defined in Eq. (2-48) and shown there to be $0.707V_0$ or $V_0 = 155$ V in Eq. (2-14). Sometimes the term peak-to-peak voltage is used and is defined as $2V_0$, or 310 V in the present case. Pictorially, then, Eq. (2-14) says that if one sits at a fixed point in an electrical circuit, the voltage at that point is not the same at every instant. Rather it varies in an oscillatory (sinusoidal) fashion from 0 to $+V_0$ to 0 to $-V_0$ to 0 to $+V_0$, etc. Construct a plot of V(t) from Eq. (2-14) assuming that $\phi = 0$.

The features of the trigonometric formula for the sum of two angles are incorporated in an even more useful representation of the time-dependent waveforms: the complex number representation. In this method the function $V(t) = V_0 \sin(\omega t)$ is represented by the complex equation

$$\mathcal{V}(t) = V_0 \exp\left\{j(\omega t - \frac{\pi}{2})\right\} = V_0 \left\{\cos(\omega t - \frac{\pi}{2}) + j \sin(\omega t - \frac{\pi}{2})\right\} \tag{2-16}$$

where $j = \sqrt{-1}$. With this representation the methods of complex algebra are used when making calculations, and the real part of the resulting complex function is taken as being physically meaningful. In the present example the real part of $\mathcal{V}(t)$ is $V_0 \cos[\omega t - (\pi/2)] = V_0 \sin \omega t$, which is the function that $\mathcal{V}(t)$ *represents* in the physical world. Similarly $V = V_0 \cos \omega t$ would be represented by $\mathcal{V}(t) = V_0 \exp\{j\omega t\}$. The utility of this representation will become clear when the analysis of ac circuits containing resistance, capacitance, and inductance is discussed. The complex representation may be used for voltage, for currents, and for impedance. The impedance is calculated, as described later on, from the resistance, capacitance, and inductance of a circuit. Appendix VIII discusses some of the elementary relations employed in complex variables. There the imaginary quantity is denoted i rather than j.

PASSIVE CIRCUIT ELEMENTS. Circuit constituents may be divided into two categories: active elements and passive elements. Passive elements are capable of dissipating and storing, but not generating or supplying, electrical energy. Resistors, capacitors, and inductors are included. Batteries, power supplies, vacuum tubes, and transistors are active elements having the capability of supplying electrical energy to a circuit. Briefly, we shall now turn to a discussion of passive circuit elements.

Resistors. One of the most common elements encountered in circuit diagrams is the resistor represented with the symbol ⩘⩘ . A current flowing through a resistor gives rise to a voltage drop in accord with the well-known Ohm's law:

$$V = IR \tag{2-17}$$

with I denoting the current in amperes, R denoting the resistance in ohms, and V being the resulting potential drop in volts across the resistor. The resistance itself is analogous to mechanical friction in that resistance changes electrical energy into heat. At the atomic level resistance is described in terms of collisions between moving current carriers and the lattice of atoms making up the substrate in which the change carriers move. The effect of these collisions is to transfer energy to the vibrational motion of the lattice from the translational energy of the current carrier with an attending amount of power dissipated as heat according to Joule's law:

$$P = I^2R \tag{2-18}$$

where P is the power in watts. Remember, 1 W is equal to 1 J/sec. The above equation makes clear the necessity of considering power requirements when choosing resistors. Common resistors in low-power circuits have $\frac{1}{2}$-, 1-, and 2-W power ratings. A wide variety of power resistors is available for applications requiring higher-wattage components.

For many purposes of circuit analysis it is both satisfactory and helpful to replace a network of resistors connected in series and/or parallel by a single equivalent series resistors. Figure 2-38(a) and (b) illustrates the general principle, while Fig. 2-38(c) and (d) gives the results for series and parallel connection of resistors. This approach will be helpful in any circumstance where the analysis involves the network as a whole and does not depend particularly on one or more elements of the network. The use of an equivalent circuit simplifies both calculations on and understanding of the performance of the whole circuit of which the network forms a part.

Capacitors. When an electrical energy source is connected to two conductors separated by an insulating medium, the conductors develop charges of opposite sign and energy is stored in the resulting electric field. Such an arrangement of conductors and a dielectric is called a capacitor. Capacitance is defined as the ratio of the charge on either conductor to the voltage applied. That is,

$$C = \frac{Q}{V} \tag{2-19}$$

where Q is the charge in coulombs, V the potential in volts, and C the capacitance in farads. Figure 2-39 shows how capacitors in series and parallel are combined to furnish equivalent series circuits (note the symbol). It should be noted

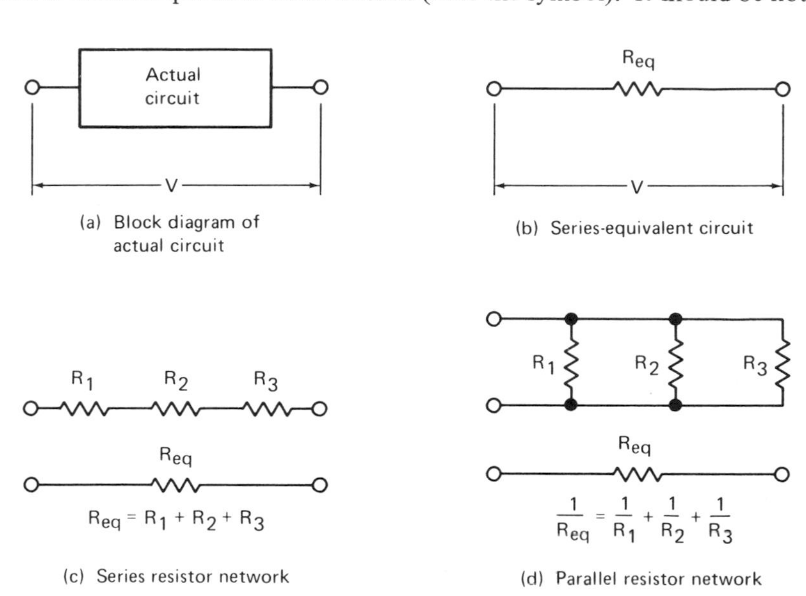

Figure 2-38. Series and parallel networks and their equivalents.

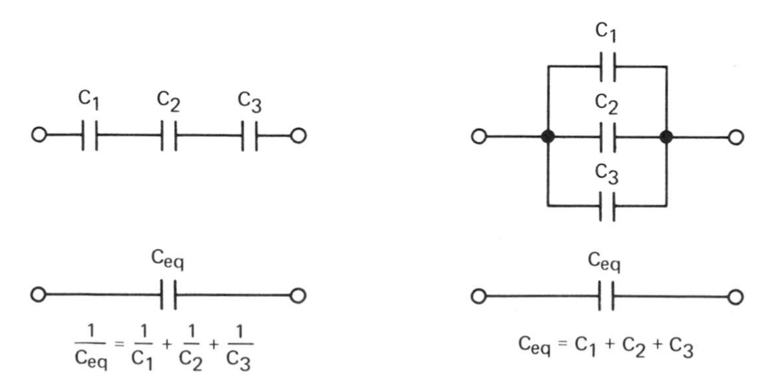

Figure 2-39. Series and capacitor networks.

that capacitance occurs between any two conductors (electrodes) separated by an insulating medium (dielectric). For example, capacitance between hookup wires or the electrodes in a vacuum tube may become crucial in certain circumstances (very fast timing circuits, for example). However, in many circuits interelectrode capacitance can be neglected in comparison with the capacitance of the capacitors wired into the circuit. Generally the latter consist of two thin metal foils separated by an insulating material of high dielectric constant. This sandwich is often rolled into a compact device. A widespread exception to the above is the tuning capacitor in a radio receiver in which the plates are flat and air is the dielectric.

Like resistors, capacitors must be selected using criteria in addition to their capacitance, C. In this case the voltage requirements are important because there is an upper limit to the voltage which can safely be applied across a capacitor without causing breakdown. Breakdown in this context means that the high electric field in the dielectric becomes sufficient to make the normally insulating dielectric become a conductor. The result is a short circuit. Of course in any application the value of C must be chosen appropriately; generally speaking, capacitance values range from a few micro microfarads ($\mu\mu F = 10^{-12}$ F) up to a few hundred microfarads ($\mu F = 10^{-6}$ F) but may go higher in special applications (for example, the power supply of a flash photolysis apparatus). It is a matter of gaining experience to have a feeling for the type and size of capacitor required in various types of appiications.

Although a purely capacitive circuit element has not been developed, it is important to recognize that, by definition, ideal capacitors are not dissipative elements. This implies that they do not release energy as heat (like resistors) but rather store it in the accompanying electric field and release it by means of a discharge current whenever the voltage drops. Furthermore, the effect of capacitance on the current and voltage characteristics of a given circuit is frequency-dependent. This will become clear later in this discussion when ac circuit analysis is described.

Since currents and voltages are of interest in circuit analysis, the differential form of the equation defining capacitance is very useful because it relates current and the time derivative of the voltage rather than charge and voltage. It may be written

$$C = \frac{dQ/dt}{dV/dt} = \frac{i}{dV/dt} \qquad (2\text{-}20)$$

where the time rate of change of charge Q is, by definition, the current i.

Illustrative Example 3 A typical capacitor has $C = 50\ \mu F$ and can withstand up to 250 V. The charge on the capacitor when $V = 200$ V is calculated as follows:

$$Q = CV = (50 \times 10^{-6})(200) = 10^{-3} \text{ coulomb}$$

If each charged particle has one electronic unit of charge associated with it, the total number of negatively charged particles (electrons) involved is

$$N_e = \frac{10^{-3}\ \text{coulomb}}{1.6 \times 10^{-19}\ \text{coulomb/electron}}$$

$$= 0.63 \times 10^{+16}\ \text{electrons}$$

Inductors. The third basic type of passive circuit element is the inductor, denoted in circuits by ⌒⌒⌒ or ⌒⌒⌒ (with straight lines above), where the straight lines denote an iron core. Although inductors generally consist of coils of wire wound either on a metal core such as a transformer core or a dielectric core, inductance effects arise in any circuit in which the current is changing with time. Currents and magnetic fields are inextricably connected and so wherever there is a current there is also a magnetic field which can store electrical energy. Whenever the current changes, the field must follow, and energy thus flows into and out of the field. Furthermore, as Faraday first discovered, whenever there is a changing magnetic field, and thus a changing current, an induced voltage arises in the conductor which opposes the change. This induced voltage is intimately connected with the power required to do the work–the work in this case being done by the source of current (and power) in increasing or decreasing the magnetic field. It is important to recognize that the induced voltage drop has nothing to do with resistance and further that inductors (ideal) are not dissipative circuit elements. The form of Faraday's law which is useful in electronics is

$$V = -L\frac{di}{dt} \tag{2-21}$$

This equation furnishes the voltage drop across any conductor in which the current is changing according to di/dt and in which the inductance is L H (henries). Note the minus sign, suggesting that the induced voltage drop opposes the direction of current change.

Every conductor has an inductance regardless of its configuration, but a coil has a higher inductance than a straight piece of wire. A coil wound on an iron core has an even higher inductance. In actual circuit applications, high values of inductance on the order several henries are often used in power supply circuits and these so-called chokes are always iron-core-type inductors. Low values of inductance often find application in radio-frequency (rf) circuits and range in value from a few microhenries (μH) to millihenries (mH). These inductors, called coils, either are wound on a dielectric material or are simply wound in the configuration of a stretched spring. Figure 2-40 illustrates the combination of inductances in series and in parallel.

Thus far our discussion of inductance has included only those effects induced in the same conductor and has not included those effects called mutual inductance involving two or more electrically insulated but spatially close conductors. For example, suppose that two coils have spatially the same axis and are close

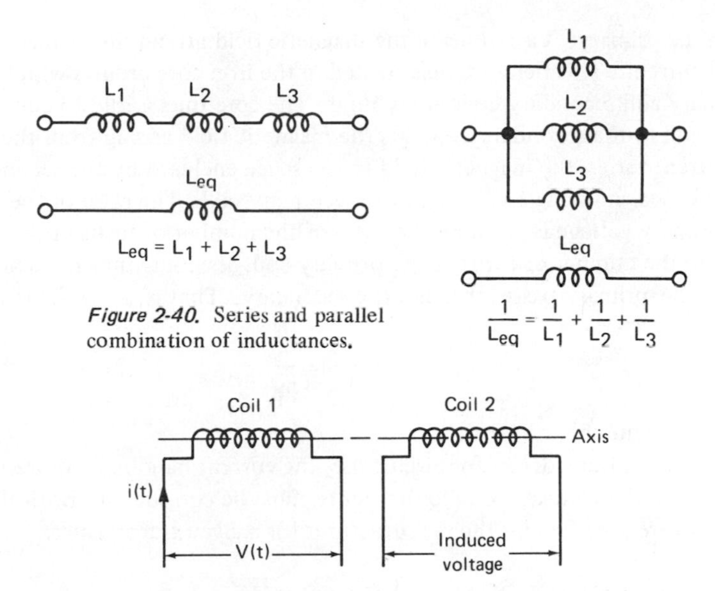

Figure 2-40. Series and parallel combination of inductances.

Figure 2-41. Mutual inductance between two coils with the same spatial axis.

together, as shown in Fig. 2-41. An alternating voltage applied to coil 1 will cause an alternating current to flow with an attending variation of both magnetic field (essentially confined within the coil) and opposing voltage. The space enclosed by coil 2 overlaps somewhat the space where the magnetic field of coil 1 is strong. As this magnetic field changes it induces a voltage drop across and current flow through coil 2; this describes mutual inductance, and its effect will occur whenever the magnetic field associated with one conductor changes and is spatially in the region of a second conductor. The importance, in a given circumstance, of mutual inductance depends on how the induced voltage compares in magnitude with other voltages in the circuit.

Mutual inductance is the essential ingredient of transformer operation where two electrically insulated coils, generally wound on an iron core, are used to change the amplitude of an ac voltage. A simple diagram and its circuit symbol are shown in Fig. 2-42. An alternating current from a power source flows

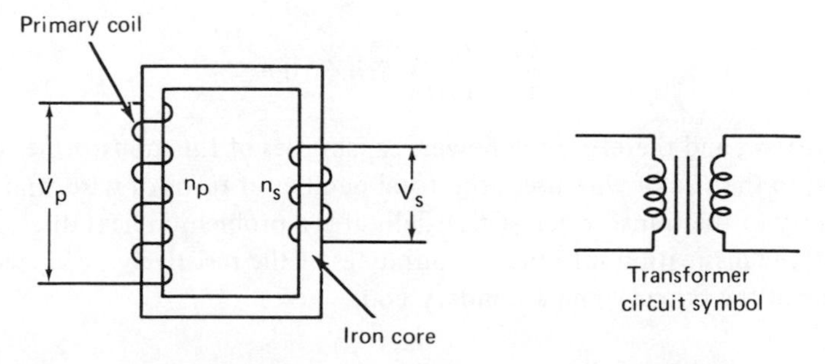

Figure 2-42. Iron-core transformer and circuit symbol.

through the primary. Variations in the magnetic field attend the changes in primary current. This field is concentrated in the iron core around which both the primary and secondary coils are wound. The core thus is said to couple the primary coil to the secondary coil. As the magnetic field arising from the primary current varies, the magnetic field in the space enclosed by the secondary coil also varies, inducing a voltage in the secondary coil. The ratio of the secondary to primary voltage is given by the ratio of the number of turns in the secondary coil to the number of turns in the primary coil, assuming that the whole field set up by the primary passes through the secondary. That is,

$$\frac{V_s}{V_p} = \frac{n_s}{n_p} \tag{2-22}$$

where n is the number of turns.

In selecting a transformer for circuit use, the current handling requirements, in addition to the voltage level requirements, must be considered. Both these capabilities are specified by the manufacturer for a given transformer.

Illustrative Example 4 Suppose that a sinusoidal current $i = I_o \sin \omega t = 10 \sin(60t)$ is applied to a 5.0-H choke (coil). The induced voltage is given by

$$\begin{aligned} V(t) &= (-L)(\omega \cos \omega t) \\ &= (-50)(60) \cos(60t) \\ &= -300 \cos(60t) \end{aligned}$$

which at t = 0 furnishes

$$V(0) = -300 \text{ V}$$

This voltage (induced) opposes that voltage which gives rise to the sinusoidal current flow.

Illustrative Example 5 A common transformer which often appears in circuits containing electronic tubes has a 110-V primary and a 6.3-V secondary. The turns ratio is then

$$\frac{n_s}{n_p} = \frac{6.3}{110} = 5.7 \times 10^{-2}$$

The current, and therefore the power, capabilities of this transformer are related to the size of wire used, the total number of turns of wire, and the geometry of the transformer. Essentially it is a problem in heat dissipation, that is, the dissipation into the surroundings of the resistively produced heating of the primary and secondary coils.

BASIC RULES OF CIRCUIT ANALYSIS. As implied in the previous section it is possible to analyze many circuits by simplifying them to equivalent resistors, capacitors, and inductors. For other circuits this is not possible and two very important rules known as Kirchhoff's rules are utilized. These may be stated as follows:

> Kirchhoff's first rule: The algebraic sum of the currents is equal to zero at any point in a circuit where three or more conductors come together. This law is essentially a statement of the conservation of electric charge and may be thought of as requiring that the current entering point is equal to the current leaving that point. Mathematically, we can write for any *point*
>
> $$\sum_i I_i = 0 \tag{2-23}$$
>
> Kirchhoff's second rule: The algebraic sum of the potential differences around any closed loop in a circuit is equal to zero. This is simply a statement of the law of conservation of energy in electrical circuits. Mathematically,
>
> $$\sum_i V_i = 0 \tag{2-24}$$
>
> This rule requires that the potential change between any two points in a circuit be the same regardless of the path taken between the two points.

As an example of the use of Kirchhoff's rules, consider the Wheatstone bridge circuit shown in Fig. 2-43, which can be used to measure an unknown resistance.

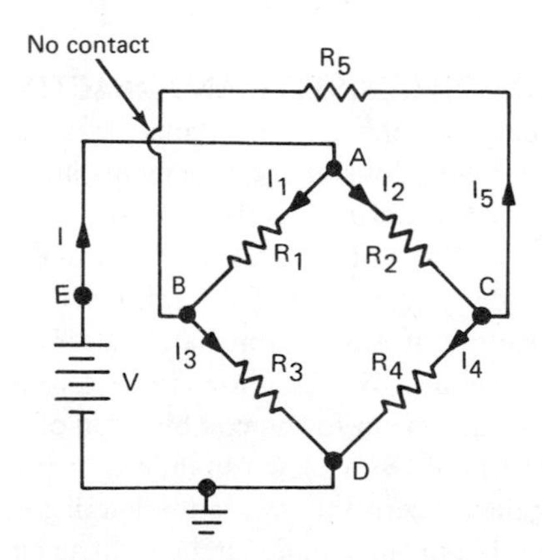

Figure 2-43. Wheatstone bridge circuit.

From Kirchhoff's first rule, three independent equations arise:

$$\begin{aligned} I - I_1 - I_2 &= 0 \quad \text{(point A)} \\ I_1 + I_5 - I_3 &= 0 \quad \text{(point B)} \\ I_2 - I_5 - I_4 &= 0 \quad \text{(point C)} \end{aligned} \tag{2-25}$$

There is an additional equation corresponding to branch point D but it is not independent of the other three equations.

From Kirchhoff's second rule three additional equations arise, one from each of the loops in the circuit:

$$\begin{aligned} V - I_1R_1 - I_3R_3 &= 0 \quad \text{(loop EABDE)} \\ I_1R_1 - I_2R_2 - I_5R_5 &= 0 \quad \text{(loop BACB)} \\ I_4R_4 - I_5R_5 - I_3R_3 &= 0 \quad \text{(loop DCBD)} \end{aligned} \tag{2-26}$$

Equations (2-25) and (2-26) contain six unknown currents, and in principle this set of six simultaneous equations can be solved for any current that is desired. Generally, in a Wheatstone circuit, R_5 is a current meter, R_1 and R_2 are in a known fixed ratio, R_3 is continuously variable, and R_4 is unknown. R_3 is adjusted until no current flows through R_5; at this point the bridge is said to be balanced.

Solving the above set of equations for I_5 furnishes

$$I_5 = \frac{V(R_1R_4 - R_2R_3)}{R_5R_3R_2 + R_1R_2R_5 + R_1R_2R_3 + R_1R_3R_4}$$

At the balance point $I_5 = 0$, which implies that $R_1R_4 - R_2R_3 = 0$. Hence,

$$R_4 = \frac{R_2}{R_1} R_3 \tag{2-27}$$

THE COMBINATION OF RESISTIVE AND REACTIVE CIRCUIT ELEMENTS. In many circuits, resistance, capacitance, and inductance are all important. A treatment of combinations of these elements involves the concept of impedance. The impedance Z is defined so that the form of Ohm's law is preserved when reactive elements are present. Both inductance and capacitance influence current flow, but the underlying source of this influence is not the same as in the case of resistance. Therefore, it is not immediately clear how to relate current and voltage in a reactive circuit. We shall now proceed to develop the concept of impedance, giving special attention to the combination of resistive and reactive elements and how the form of Ohm's law can indeed be preserved.

We begin by noting that except for transient (charging up) effects inductors and capacitors are of no importance in dc circuits. In ac circuits, however, the effects of both capacitance and inductance become important because the associated fields are now time-dependent, resulting in induced currents and voltages in the circuit.

First consider capacitance. By definition, $C = Q/V$ for the case of a fully charged capacitor under the potential V. During charging and discharging the same relation holds for differential increments and the time dependence may be introduced using Eq. (2-20). Thus, in a capacitor to which an alternating (time-dependent) source of voltage is attached, the voltage and current are connected by the differential equation

$$\frac{dV}{dt} = \frac{i}{C} \tag{2-28}$$

Suppose that the voltage on the capacitor varies sinusoidally according to the expression $V = V_0 \sin \omega t$. Solving Eq. (2-28) for the current and substituting for the time derivative of the voltage, we obtain

$$i(t) = \omega C V_0 \cos (\omega t)$$

$$= \omega C V_0 \sin \left(\omega t + \frac{\pi}{2}\right)$$

We conclude that the application of a sinusoidal voltage to an ideal capacitance gives, in the capacitor, a sinusoidal current of the same form but leading the voltage by a phase shift of $\pi/2$ radians. Note that the amplitude, I_0, of the current equation is a function of the frequency, the voltage amplitude, and the capacitance. That is,

$$I_0 = \omega C V_0$$

which may be rearranged to the form of Ohm's law as

$$V_0 = I_0 X_c$$

where

$$X_c = \frac{1}{\omega C} \tag{2-29}$$

plays the role of a resistance and is called the capacitive reactance. Note that for a given applied V_0, I_0 increases as ω and/or C increase.

If an alternating voltage of the same form is applied to an inductor, the following relations arise:

$$V(t) + V_{induced} = 0$$

$$V(t) = L \frac{di}{dt}$$

We write as a trial function for the current

$$i(t) = I_0 \sin (\omega t + \phi)$$

and obtain

$$V(t) = \omega L I_0 \cos (\omega t + \phi)$$

which is of the proper form provided $\phi = -\pi/2$ and $\omega L I_0 = V_0$.

Hence, the following equations arise:

$$i(t) = I_0 \sin\left(\omega t - \frac{\pi}{2}\right)$$

$$V_0 = I_0 X_L = I_0 \omega L \tag{2-30}$$

The first of these two equations points up the phase lag of $\pi/2$ radians by which the current follows the applied voltage. Equation (2-30) has the form of Ohm's law and the quantity $X_L = \omega L$ is called the inductive reactance. Note its dependence on frequency and inductance.

To develop the proper method for combining resistance and reactance (capacitive or inductive) to furnish the impedance, consider the circuit in Fig. 2-44 in which the voltage $V = V_0 \sin \omega t$ is applied to a series combination of R and C.

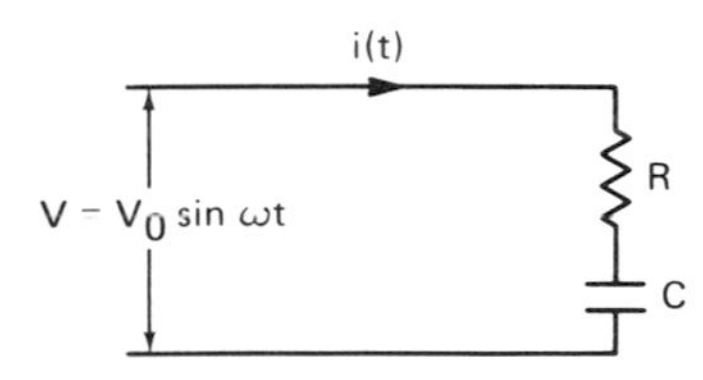

Figure 2-44. Series circuit with resistance and capacitance.

At any instant of time

$$V = iR + \frac{Q}{C}$$

and differentiating gives

$$\frac{dV}{dt} = \frac{di}{dt} R + \frac{i}{C} \tag{2-31}$$

As a trial solution we take the current to be $i = I_0 \sin(\omega t + \phi)$, which furnishes, when substituted into Eq. (2-31), the trigonometric equation

$$\omega V_0 \cos \omega t = \omega I_0 R \cos(\omega t + \phi) + \frac{I_0}{C} \sin(\omega t + \phi)$$

which must be solved for ϕ.

Expanding $\cos(\omega t + \phi)$ and $\sin(\omega t + \phi)$ as

$$\cos(\omega t + \phi) = \cos \omega t \cos \phi - \sin \omega t \sin \phi$$

$$\sin(\omega t + \phi) = \sin \omega t \cos \phi + \cos \omega t \sin \phi$$

furnishes

$$\omega V_0 \cos \omega t = \omega I_0 R [\cos \omega t \cos \phi - \sin \omega t \sin \phi]$$
$$+ \frac{I_0}{C} [\sin \omega t \cos \phi + \cos \omega t \sin \phi]$$

We now obtain two equations by substituting $\omega t = 0$ and $\omega t = \pi/2$. When we use $\omega t = 0$ the result is

$$\omega V_0 = \omega I_0 R \cos \phi + \frac{I_0}{C} \sin \phi$$

and when $t = \pi/2$ (2-32)

$$O = -\omega I_0 R \sin\phi + \frac{I_0}{C}\cos\phi$$

Solving the latter for ϕ furnishes the desired solution, which can be substituted into the assumed current equation, namely

$$\phi = \tan^{-1}\left(\frac{1}{\omega CR}\right) = \tan^{-1}\left(\frac{X_C}{R}\right)$$

With this expression we may obtain sin ϕ and cos ϕ from the right triangle shown in Fig. 2-45 in which the two sides are R and $1/\omega C$, respectively:

$$\sin\phi = \frac{X_C}{(R^2 + X_C{}^2)^{1/2}}$$

$$\cos\phi = \frac{R}{(R^2 + X_C{}^2)^{1/2}} \qquad (2\text{-}33)$$

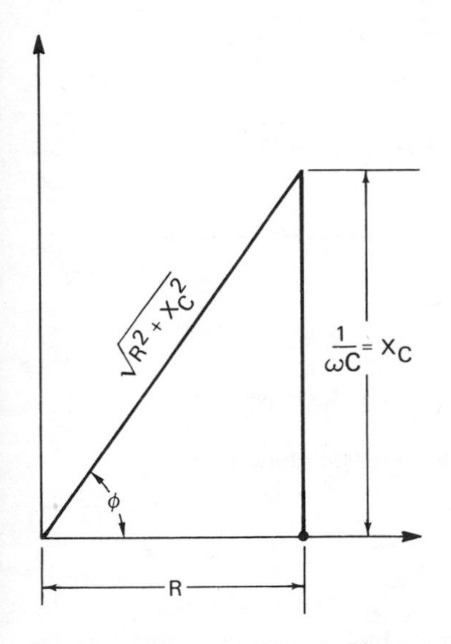

Figure 2-45. Right triangle relation for resistance and reactance used to determine sine and cosine of phase angle ϕ.

Using (2-33) in the first part of (2-32) and rearranging, we obtain a relation between the amplitudes of the current and voltage:

$$V_0 = \sqrt{R^2 + X_C{}^2}\, I_0 = Z I_0$$

A similar analysis of a series connection including resistance, capacitance, and inductance (compare Problem 7 at the end of this section) gives rise to a relation between the amplitudes of the current and voltage of the form

$$V_0 = I_0 Z \qquad (2\text{-}34)$$

where the impedance Z represents the combined series effect of R, C, and L or any combination thereof. Its magnitude is given by the following combination of resistance R and reactance X:

$$Z = \sqrt{R^2 + X^2} \tag{2-35}$$

where

$$X \equiv X_L - X_C \tag{2-36}$$

A few comments are now in order. We have shown that in series ac circuits containing resistive and reactive elements, the form of Ohm's law may be preserved insofar as the current and voltage amplitudes are concerned but only if R and X are combined in the way described by Eq. (2-35). This expression may be viewed as the magnitude of a complex number in which resistance appears along the real axis and reactance along the imaginary axis, as shown in Fig. 2-46.

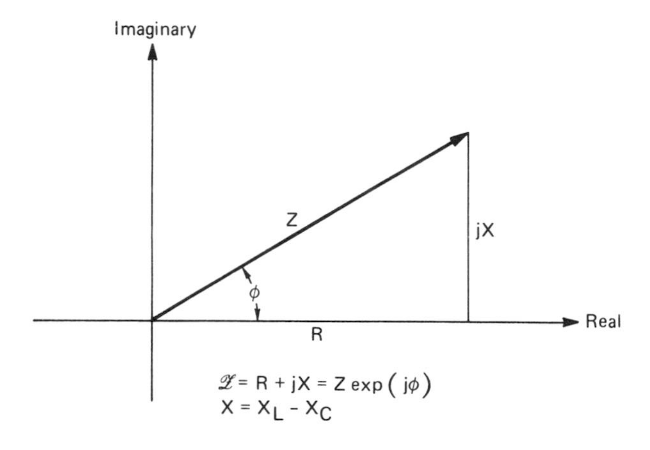

Figure 2-46. Real and imaginary parts of complex impedance.

This leads naturally to the notion of complex impedance, defined as

$$\mathscr{Z} = R + jX \tag{2-37}$$

or the equivalent

$$\mathscr{Z} = Z \exp\{j\phi\}$$

This is the complex representation of impedance which is used in calculations involving complex current and complex voltage representations.

Keep in mind that the above expressions are valid only in series circuits, and to be useful in analyzing complex networks, there must first be a reduction to an equivalent series inductor, capacitor, and resistor. In series, reactances combine according to

$$X_{eq} = \sum_i X_i \tag{2-38}$$

while in parallel

$$X_{eq} = \frac{1}{\sum_i (1/X_i)} \tag{2-39}$$

Once the series-equivalent elements have been determined they are combined according to Eq. (2-35) to furnish the circuit impedance, Z.

Recalling that X_L and X_C are both frequency-dependent it becomes clear that Z and I_0 are also frequency-dependent. Problem 9 (at the end of this section) illustrates how this dependence can be utilized advantageously to develop resonant circuits.

Having established the notion of a complex impedance we shall proceed to illustrate the utility of the complex number representation for the solution of ac circuit problems. As an example, consider the solution of the circuit in Fig. 2-44. The following equations arise:

$$\mathscr{V} = V_0 \exp\{j(\omega t - \frac{\pi}{2})\}$$

$$\mathscr{Z} = R - jX_c = Z \exp\{j\phi\}$$

where $Z = \sqrt{R^2 + X_C^2}$ and $\phi = \tan^{-1}(X_C/R)$. The complex current may be obtained directly by solving the ac complex form of Ohm's law:

$$\mathscr{I} = \frac{\mathscr{V}}{\mathscr{Z}} = \frac{V_0}{Z} \exp\{j(\omega t - \frac{\pi}{2} - \phi)\} \tag{2-40}$$

The real part has physical significance and it is

$$i(t) = \frac{V_0}{(R^2 + X_C^2)^{1/2}} \cos(\omega t - \frac{\pi}{2} - \phi)$$

$$i(t) = \frac{V_0}{(R^2 + X_C^2)^{1/2}} \sin(\omega t + \eta)$$

This equation is equivalent to that obtained previously except it is obtained here without recourse to solving differential equations. It should be noted that the validity of the present approach rests upon the *form* of the circuit differential equations and, in fact, that the complex number method is simply a time-saving computational device. It may be used in any circuit for which the equivalent complex impedance is worked out.

Illustrative Example 6 Consider a series circuit of a resistor ($R = 5 \times 10^3\ \Omega$), an inductor ($L = 3 \times 10^{-3}$ H), and a capacitor ($C = 5 \times 10^{-6}$ F). The complex impedance of this arrangement depends on the frequency of the voltage signal applied. Suppose that we choose 10^2 Hz as the frequency. The complex impedance is then calculated in the following way by first finding the reactance

$$X_{eq} = \frac{1}{10^2 \times 5 \times 10^{-6}} + 10^2 \times 3 \times 10^{-3}$$
$$= 2 \times 10^3 + 3 \times 10^{-1}$$
$$\simeq 2 \times 10^3$$
$$\mathscr{Z} = 5 \times 10^3 + j(2 \times 10^3)$$

At this frequency the capacitive reactance clearly dominates the total reactance.

If we choose a frequency of 10^7 Hz, the equivalent reactance is

$$X_{eq} = 2 \times 10^{-2} + 3 \times 10^4$$

and the complex impedance is

$$\mathscr{Z} = 5 \times 10^3 + j(3 \times 10^4)$$

At these relatively high frequencies the reactance is clearly dominated by the inductive element.

ACTIVE CIRCUIT ELEMENTS. Earlier, circuit elements were divided into two categories: those which act as sources of electrical energy and those which do not. The former category includes batteries, vacuum tubes, solid-state diodes, and transistors, while the latter includes resistors, capacitors, and inductors. Having discussed the latter we shall now turn our attention to a description of vacuum tubes and solid-state devices.

Vacuum Tubes. The vacuum tube is still widely used in electronic circuits despite encroachment by solid-state devices. While there are a wide variety of vacuum tubes and a wide variety of applications for them, we shall discuss only two tubes and two applications here: the diode rectifier and the triode amplifier.

Diode. The vacuum diode, or rectifier tube, is shown schematically in Fig. 2-47. It contains two working electrodes, the cathode and the plate, together with a cathode heater. (The cathode and heater are the same element in many

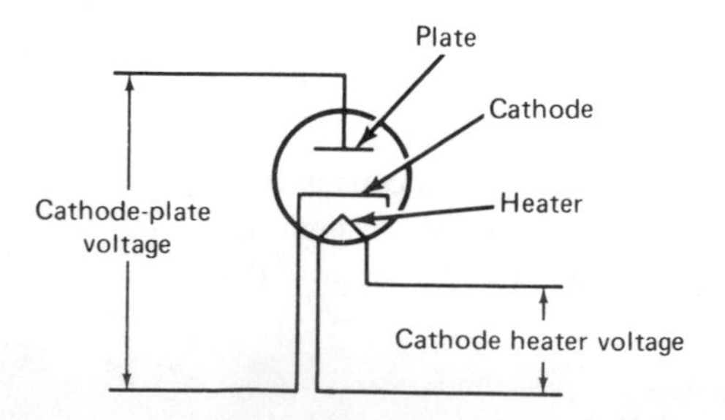

Figure 2-47. Vacuum diode.

diodes.) The power supplied to the heater raises the temperature of the cathode until electrons are emitted. The thermally emitted electrons will travel to the plate provided the voltage on the plate is more positive than the voltage on the cathode. When the plate voltage is lower than the cathode voltage the thermally emitted electrons will be repelled from the plate, and in an ideal situation no current will flow in the cathode-plate circuit. However, because the thermally emitted electrons from the cathode have some initial translational energy, the cathode-plate current does not drop sharply to zero when the plate voltage drops below the cathode voltage; rather a certain negative voltage is required to completely stop conduction (see Fig. 2-48). Thus, the vacuum diode acts like an on-off switch for electric current, the switching being controlled by the voltage applied between the cathode and the plate. Note that the magnitude of the current which flows in the cathode-plate circuit is limited by the number of electrons emitted thermally from the cathode. Since this circuit will also contain the load in any application, power available to the load is intimately related to the current capabilities of the diode. This is discussed further in connection with dc power supplies.

Triode. In many electronic circuits, amplification is an important feature, and the vacuum triode is often used in such applications. Its elements are diagrammed in Fig. 2-49. The functions of the cathode, plate, and heater are similar to the diode. The additional electrode, the grid, is the element which controls the cathode-to-plate current and thus the amplification.

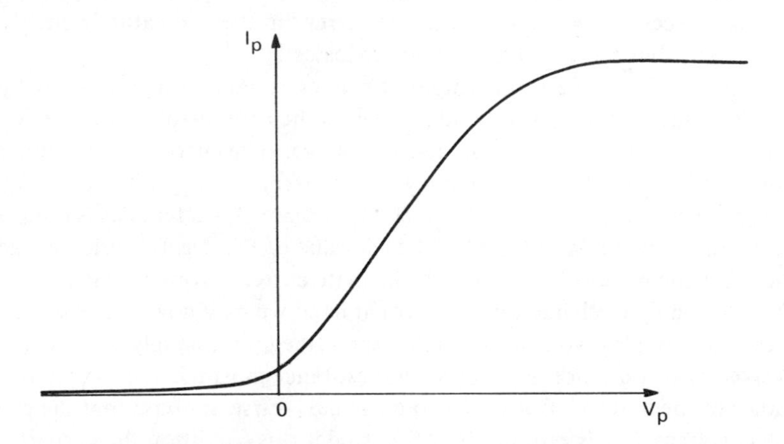

Figure 2-48. Plate current versus plate voltage for a vacuum diode.

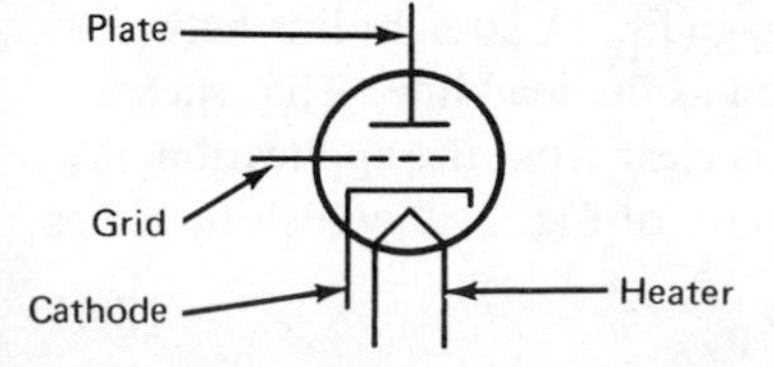

Figure 2-49. Vacuum triode.

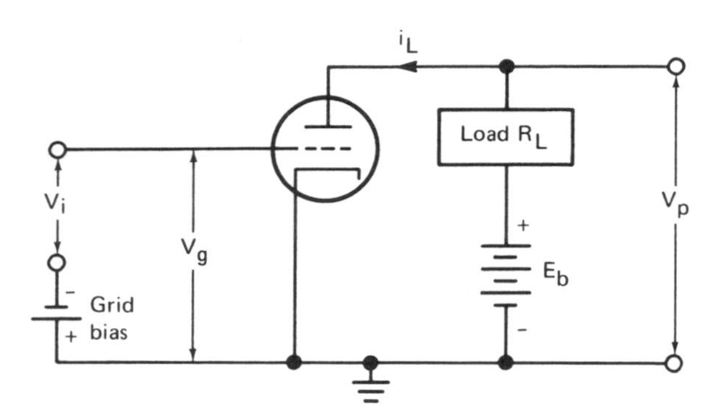

Figure 2-50. Block diagram of circuit capable of voltage amplification.

Figure 2-50 shows a typical triode amplifier in the common cathode mode [i.e., the cathode is electrically in both the input and output circuits (via ground)]. The thermally emitted electrons from the cathode give rise to a current in the plate circuit provided they are conducted to the plate. If the voltage on the grid is zero, the electrons will travel to the plate because the power supply (battery) provides the necessary positive plate potential E_b. The grid voltage with respect to the cathode can be changed in two ways: first, through variations in the applied input voltage V_i and, second, by means of the grid bias power supply. Normally, the grid bias is adjusted so that the grid is always negative with respect to the cathode even when the input voltage is positive. This is necessary in order to keep the current in the grid-cathode circuit very small and thus avoid undesirable power losses.

The characteristics of a triode amplifier may be summarized graphically by plotting the current through the load, i_L, versus the cathode-plate voltage, V_p, for various fixed values of the cathode-grid voltage. The curves may be obtained experimentally by fixing E_b and varying R_L to effect changes in the plate voltage and plate current. A family of curves called plate characteristics is obtained similar to those shown in Fig. 2-51. At each value of the negative grid voltage V_g there is some plate voltage at which the plate current drops to zero.

With the family of characteristic curves at hand we may now ask how the plate current and plate voltage are related for a fixed plate supply voltage, E_b, and a fixed load resistance, R_L. This load resistance must include the internal resistance of the battery. Consider two extremes: First, suppose that the cathode-plate voltage, V_p, is zero in Fig. 2-50. Under this condition the current through the load, i_L, is given by E_b/R_L. At the other extreme suppose that V_g is very negative so that no current can flow in the plate circuit (i.e., $i_L = 0$). Under this condition V_p is just the battery voltage E_b. A straight line between these two points as shown in Fig. 2-51 is defined as the load line. That such a linear relation between i_L and V_p should exist is clear from the application of Kirchhoff's second law to the cathode-plate circuit of Fig. 2-50, which furnishes

$$E_b = i_L R_L + V_p \tag{2-41}$$

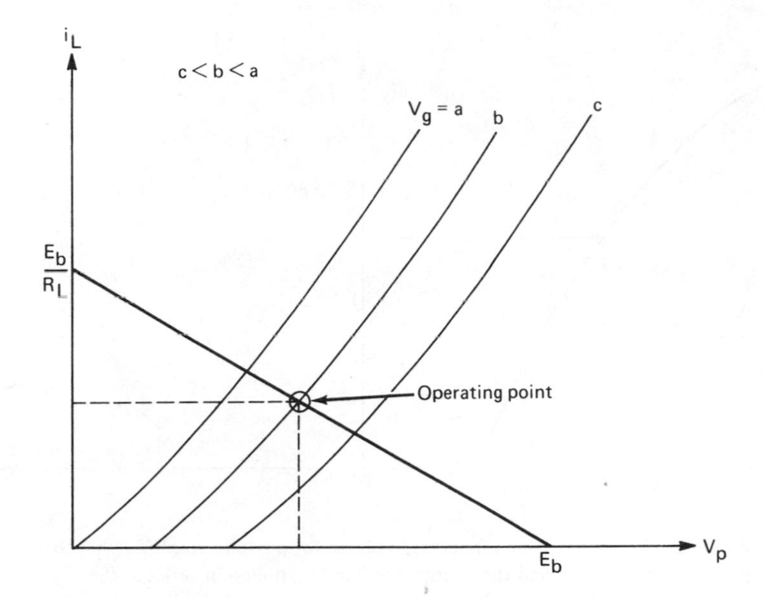

Figure 2-51. Plate characteristics of common cathode amplifier.

We conclude therefore that any changes in V_g will affect i_L in such a way that the above relation and the load line of Fig. 2-51 are satisfied.

If the grid bias (when the input voltage of Fig. 2-50 is equal to zero) is chosen as $V_g = b$ (a negative number), the plate current and voltages are fixed. This point is called the operating point or quiescent point. If the ac signal to be amplified is now placed in series with the grid bias power supply, the instantaneous grid bias voltage oscillates about the dc value b. As a result the plate current, i_L, and plate voltage, V_p, oscillate about their quiescent point values.

A clearer picture of how a vacuum triode amplifies may be gained by studying Fig. 2-52, which shows the plate voltage V_p versus the grid voltage V_g. If the grid bias voltage is b as in Fig. 2-51 and a sinusoidal input voltage of the form $V'_g = V_{g0} \sin(\omega t)$ is applied, the instantaneous grid voltage is

$$V_g = b + V_{g0} \sin(\omega t)$$

Thus, the grid voltage oscillates sinusoidally about the quiescent point, b. The upper and lower limits of this oscillation are shown as dotted lines perpendicular to the abscissa of Fig. 2-52. The plate voltage is controlled according to Eq. (2-41) by the grid bias voltage $E_b = b$, the current through the load i_L, and the equivalent resistance of the load R_L. For a given E_b and R_L, the load line of the triode is fixed, and for a given grid bias (say b) Fig. 2-51 shows how to connect the plate voltage and load current. For example, in the case we are considering the grid voltage oscillates between $b + V_{g0}$ and $b - V_{g0}$ and corresponds in Fig. 2-51 to shifts from the $V_g = b$ curve to neighboring curves along the load line. The implication, of course, is that the grid voltage controls the load current and

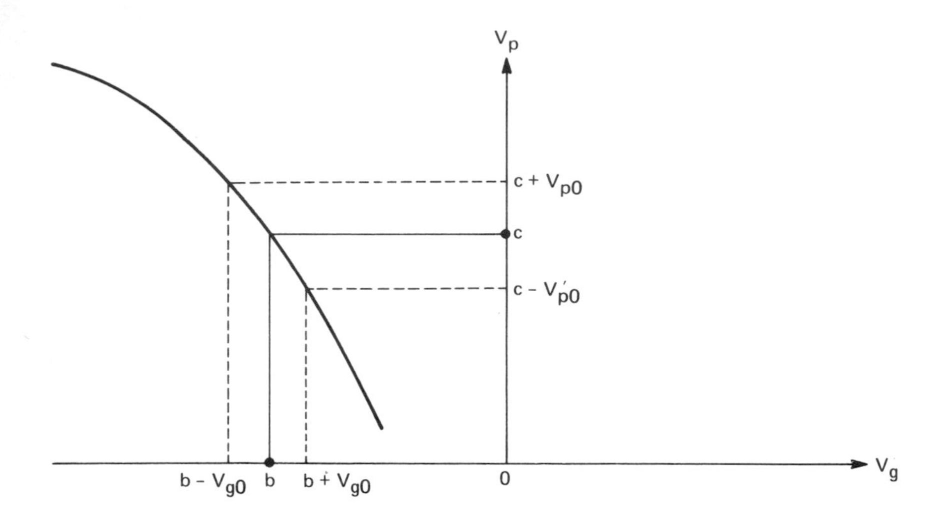

Figure 2-52. Plate voltage versus grid voltage for a triode amplifier. The dc grid bias voltage is b, and the ac input voltage is applied in series in the cathode-grid circuit as $V_g = V_{g0} \sin \omega t$. The output voltage is given by the reflection of the grid voltage onto the characteristic curve. The output voltage will be of the form $V_p = V_{p0} \sin (\omega t - \pi)$ as long as the characteristic curve is linear throughout the operating region.

the plate voltage. This control of the plate voltage by the grid voltage for a given bias and load resistance is shown in Fig. 2-52. The grid voltage varies as described above and as shown on the abscissa in Fig. 2-52. The reflection of these limits in the characteristic curve and then onto the ordinate provides the limits of the plate voltage oscillation. If the characteristic curve is very nearly linear in the region of this reflection, then $V_{p0} \approx V'_{p0}$ and the plate voltage is also sinusoidal. If the characteristic curve is nonlinear in this region, then $V_{p0} \neq V'_{p0}$ and we say that the output is distorted (e.g., the output voltage does not accurately reproduce the *form* of the input voltage). Assuming that $V_{p0} = V'_{p0}$, the voltage gain or amplification G_v is defined as the ratio of two amplitudes: the amplitude of the sinusoidal (ac) part of the output voltage to the amplitude of sinusoidal part of the input voltage. From Fig. 2-52 we have

$$G_v = \frac{V_{p0}}{V_{g0}} \tag{2-42}$$

Note that in Fig. 2-52 an increase in grid voltage is accompanied by a decrease in plate voltage and vice versa. This implies that although the forms of the input and output voltages are the same, there is a phase shift of π radians between the two signals. With $V'_g = V_{g0} \sin \omega t$, the ac part of the output voltage may be written as

$$V'_p = G_v V_{g0} \sin (\omega t - \pi)$$

Frequently an equivalent form is used, namely

$$V'_p = -G_v V_{g0} \sin \omega t$$
$$= -G_v V'_g$$

In the latter representation the phase lag of π radians is accounted for by the negative sign.

Solid-State Devices. In recent years solid-state elements have become increasingly popular because of their small size, long lifetimes, and relatively low cost. The scope of this text prevents a discussion of the solid-state physics of these devices. Suffice it to say that two devices of widespread use are the diode and transistor, which are the solid-state counterparts of the vacuum diode and triode, respectively. Two general types of solid-state material are employed: (1) p type in which the current is conducted by so-called holes and (2) n type in which current is conducted by electrons. A diode is formed at the junction connecting a p-type piece of material with an n-type piece of material. A transistor generally has two p-n junctions arranged either p-n-p or n-p-n. The central region is called the base and the two end regions are called the emitter and the collector.

Transistors, like triode vacuum tubes, are useful as amplifiers. A block diagram of a common emitter transistor amplifier is shown in Fig. 2-53, and comparison with the triode amplifier of Fig. 2-50 reveals their similarities with base and grid, emitter and cathode, and collector and plate playing similar roles. Other transistor amplifiers have either the collector or the base in both the input and output circuits. There are other transistor amplifier circuits as well as many other solid-state devices. The reader is referred to texts on electronics for a description of these.

In circuit diagrams the p-n-p transistor is denoted by ↘/ , while the n-p-n is denoted by ↖/ . The arrow always denotes the emitter and points toward n-type and away from p-type material.

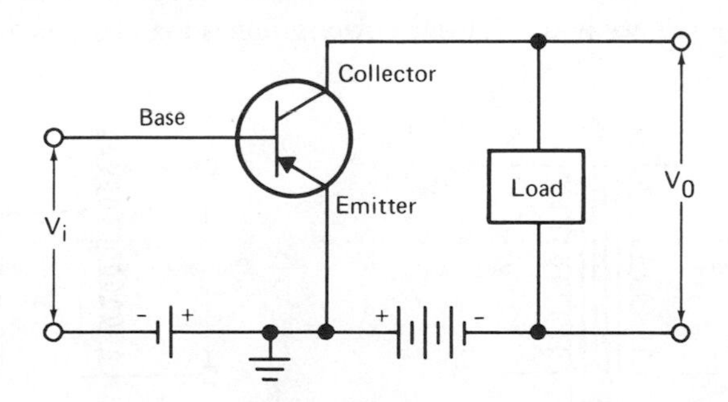

Figure 2-53. Common emitter transistor amplifier.

BASIC ELECTRONIC CIRCUITS. Having discussed several basic circuit components we shall now turn our attention to actual circuits employed in various types of electronic equipment. This is a useful approach because in analyzing any complex circuit, simplification is obtained by decomposing the circuit into sections such as power supply and amplifier. In the discussion which follows we shall present actual circuits for a few of these parts.

Power Supplies. Electronic apparatus contain a power supply (source of electrical energy) as an integral part. With some devices, the 115-V ac commercial voltage source is satisfactory. More often, however, the device may require, for proper operation, ac voltages other than 115 V and/or dc voltages. In this section we shall discuss devices which can serve as sources of electrical energy.

AC power supplies. The standard laboratory source of ac voltage and power is the 115-V commercial supply. Often the voltage required by an instrument is other than 115 V and the supply voltage must be adjusted upward or downward to meet the requirements. This may be accomplished with one of two kinds of transformer: fixed turns ratio or variable turns ratio. Symbolically these transformers are denoted as shown in Fig. 2-54. We have discussed previously the fixed turns ratio transformer; in the variable turns ratio transformer the primary and secondary coils are one and the same. Commercial devices of this type are quite common and are named Variac, Powerstat, Varitran, etc. Their secondary voltage is continuously variable from zero to about 1.2 times the primary voltage. Repetitiously we note that in any transformer application power requirements must be considered. The power-handling capabilities of any transformer depend on how it is constructed, i.e., how many turns of wire, the size of the wire, and the kind of core. The capabilities are given in the manufacturer's specifications.

DC power supplies. One of the most familiar dc sources of electrical energy is the electrochemical cell, which is of two general types: primary or nonrechargeable and secondary or rechargeable.

Rectifier power supplies. Except for special situations, laboratory electrical systems rely more heavily on rectifier power supplies than battery power sup-

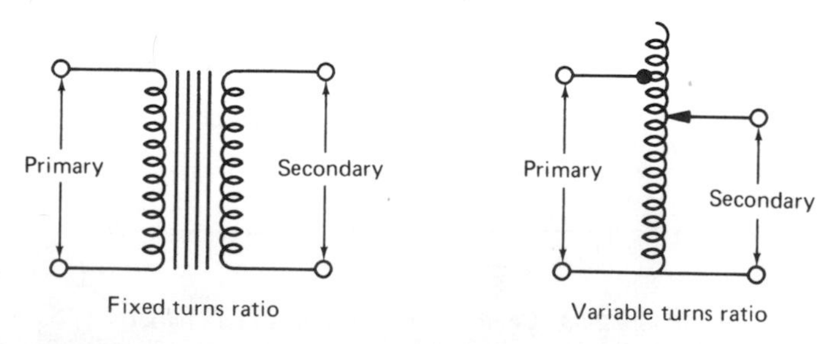

Figure 2-54. Fixed and variable turns ratio transformers.

plies because, with the advent of solid-state devices, the rectifier supplies have become quite reliable, stable, and inexpensive. There are several types of rectifier supplies depending on how accurately the dc output voltage must be maintained, how much ac ripple is permitted, the amount of power required, and the current-voltage levels. In high-voltage, high-power supplies, vacuum tubes are used in rectifier circuits because they have generally higher-current-handling capabilities than solid-state devices. Advances in solid-state technology are, however, rapidly changing this situation. Two types of rectifier circuits are common: half-wave and full-wave circuits. These circuits are discussed below. The symbol —▶|— is used in circuit diagrams to denote rectifiers, and conventionally the device is conductive (on) when the polarity is +▶|− .

A simple tube-type half-wave rectifier circuit is shown in Fig. 2-55. Remember, the rectifier tube acts like an on-off switch, being conductive when the plate voltage is positive with respect to the cathode. Insofar as the load is concerned it receives power from the supply only when the rectifier is conductive. Thus, power is delivered to the load on alternate half-cycles of the input voltage.

Just as with transformers, rectifiers must be chosen with both the voltage and power requirements of the load in mind. These rectifier parameters are specified by the manufacturer and must meet or exceed the requirements of the device drawing power from the supply. Figure 2-48 illustrates how the current conducted by the rectifier varies with plate voltage. In the limit of high plate voltage, saturation is reached where essentially all the electrons emitted by the cath-

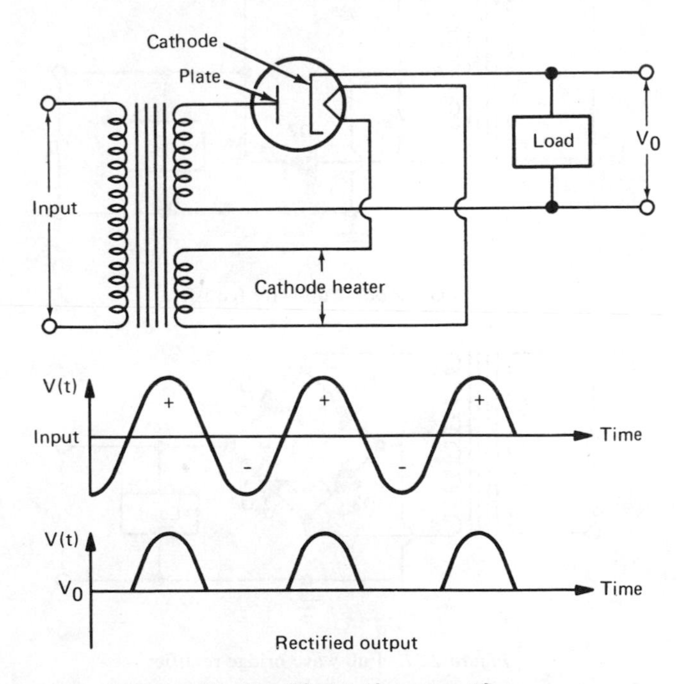

Figure 2-55. Simplified half-wave rectifier.

ode are conducted to the plate. Hence, there is a limit to the current available, implying that the rectifier must be chosen so that sufficient load current is available at the operating voltage.

Some features of the half-wave rectifier should be pointed out. First, note that the ac input is derived from a transformer; second, the cathode heater of the diode is heated by an ac power supply different from that supplying the plate voltage; and, finally, the output is not of constant magnitude but is dc in the sense that the sign of the output voltage is constant.

Two schematic full-wave rectifiers are shown in Figs. 2-56 and 2-57. The circuit of Fig. 2-56 is essentially equivalent to the combination of two half-wave rectifiers, one conducting on one half-cycle of the input and the second on the other half-cycle. The key feature of this circuit is the grounded center-tapped transformer which establishes the sign of the voltage across the diodes and essentially introduces a phase shift of π radians between V_1 and V_2. The net result is that D1 conducts on the positive half-cycle of V_i and D2 on the negative half-cycle, resulting in a full-wave-rectified output voltage.

An alternative type of full-wave rectifier which does not require a center-tapped transformer is the bridge rectifier shown in Fig. 2-57. In this circuit the path taken by the current through the diode arrangement depends on the sign of

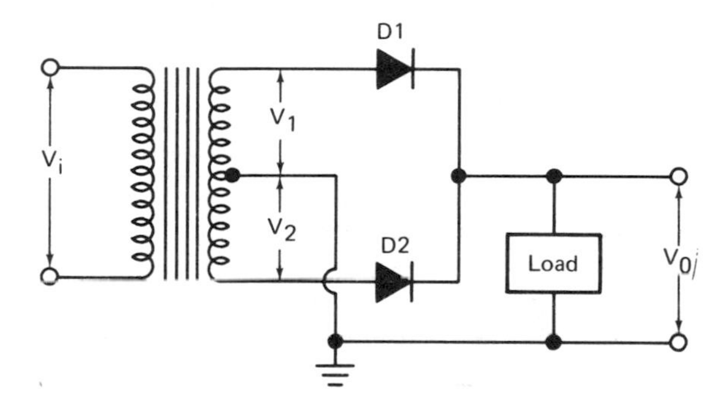

Figure 2-56. Full-wave rectifier.

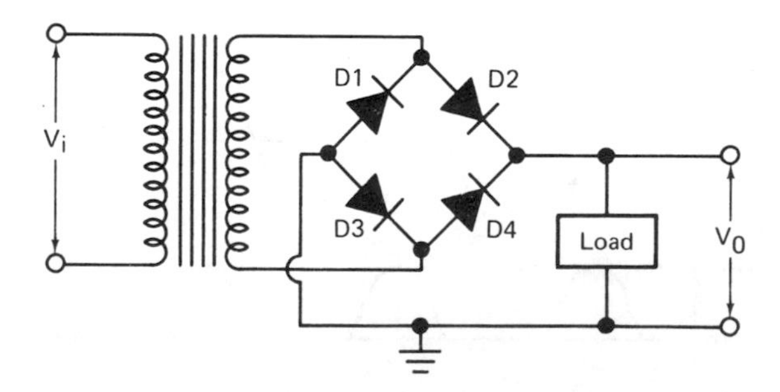

Figure 2-57. Full-wave bridge rectifier.

the applied voltage, V_i. When V_i is positive, D2 and D3 conduct, and when negative, D1 and D4 conduct. In either case the direction of current flow through the load is the same, and thus the load voltage maintains a fixed sign while varying in magnitude.

Filters. In both the half- and full-wave rectifier circuits the output varies in magnitude with time and this ripple is very undesirable in many applications. To reduce it and thus stabilize the magnitude of the output voltage, use is made of filters of either the capacitance input or choke input variety. Choke input means that the first element encountered after the diodes is a choke, while capacitor input means that a capacitor is the first element. Capacitor input filters furnish an output which is larger in magnitude than choke input filters but do not provide as stable an output voltage under varying load conditions as do choke input filters. The stability of the power supply output voltage under variable load conditions is referred to as voltage regulation.

The capacitance-type filter is employed as shown in Fig. 2-58. The magnitude of the capacitance is generally large enough to make the reactance $X_c = 1/\omega C$ small compared to the load resistance R_L, where ω is the frequency of the voltage applied to the rectifier. Under these circumstances the capacitor stores charge as the voltage increases and releases it when the voltage decreases. As shown in Problem 11 (at the end of this section) the rate of both the increase and the decrease in charge depends on the time constant of the capacitor. With filtering the voltage drop across the load is rendered more nearly constant, as shown schematically in Fig. 2-58.

Another filter type uses a combination of L and C in either the L or π configurations, which are shown in Fig. 2-59. Both these filters employ a rather large inductance on the order of a few henries to develop time-dependent opposing voltages when any changing voltages are applied across them. As before, the capacitor has low reactance at the frequency of the line voltage. The net result

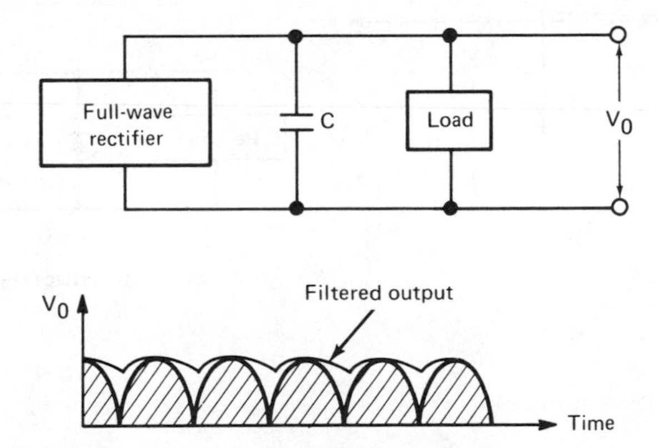

Figure 2-58. Capacitance filtered full-wave rectifier. The hatched area of graph represents the unfiltered output of the full-wave rectifier.

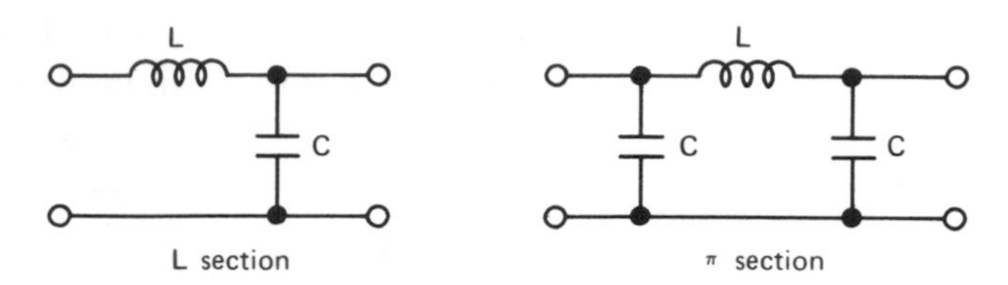

Figure 2-59. L-section and π-section filters.

is a lower-magnitude, but much more constant, voltage supply than is available from the rectifier itself or the capacitor filtered supply.

In all the above filters, the magnitude of the supply output voltage varies somewhat with the current drawn from the supply. As a result, the ripple may be reduced effectively with filtering, but the magnitude of the output voltage may not be constant. If a constant voltage output is required at all current levels, use is often made of zener diodes or voltage regulator tubes. These devices are characterized by very large impedance up to a certain voltage at which point the impedance drops rapidly and they become quite conductive. The zener diode characteristic curve is shown in Fig. 2-60 along with a schematic circuit description. Clearly the output of the rectifier must be greater than or

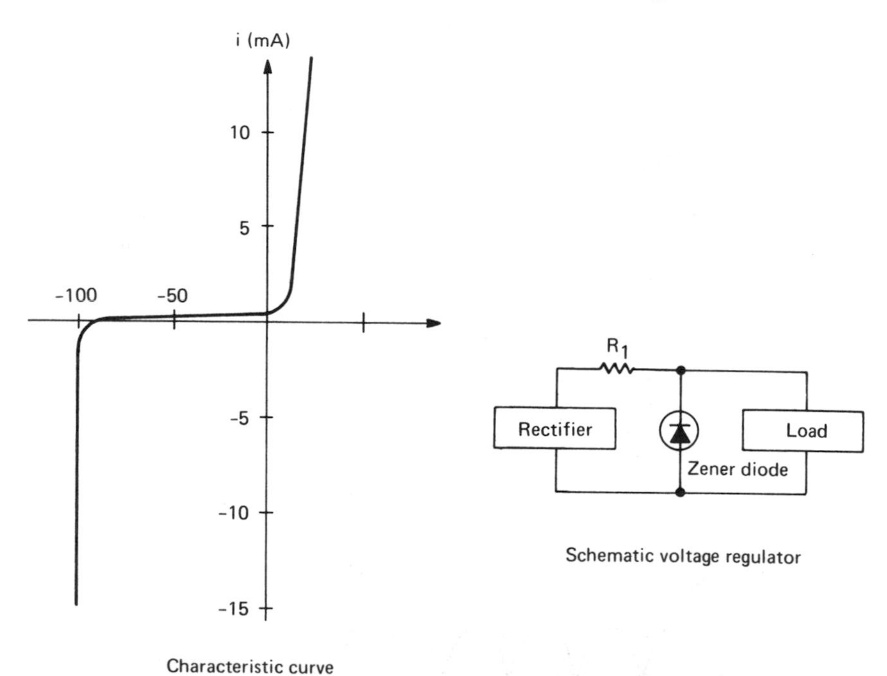

Figure 2-60. Zener diode characteristic curve and schematic voltage regulator circuit. The control voltage is 100 V.

equal to the breakdown voltage of the diode in order for the latter to exert control over the voltage. The current through R_1 is divided between the diode and the load so that the voltage drop across the load is identical to that across the diode, the latter voltage drop remaining constant. As the load draws more current, less current flows through the diode and vice versa. Voltage regulator tubes perform the same function except that the breakdown voltage is about 10% larger than the voltage at which they regulate. They consist of gas-filled diodes, and the change from high to low impedance occurs when the voltages are sufficient to ionize the gas and maintain a gaseous discharge. In circuits, voltage regulator tubes are denoted by the symbol —⊙—.

Another important feature of power supplies is the bleeder resistor which should be connected across the output terminals of any power supply. It provides a path by which the capacitors in the filter section may be discharged when the load is removed for some reason. Properly selected it may also serve to improve voltage regulation by providing a constant minimum load resistance.

In summary, Fig. 2-61 illustrates a practical full-wave rectifier power supply which is both filtered and regulated. The resistor R_2 can serve as the bleeder resistor in this circuit.

Amplifiers. We have already discussed those properties of the vacuum triode and transistor which make them useful as amplifiers. Other devices such as tetrodes, pentodes, and field-effect transistors (FET) are also used in amplifier circuits, but we shall not discuss them here. In actual circuitry, consideration must be given to parameters other than the amplification if a suitable instrument is to be designed. For example, consideration must be given to the power requirements, the frequency range of the signals to be amplified, the coupling to the next stage, the impedance of the amplifier, and the impedance of the circuitry which precedes and follows it. In the material which follows we shall discuss

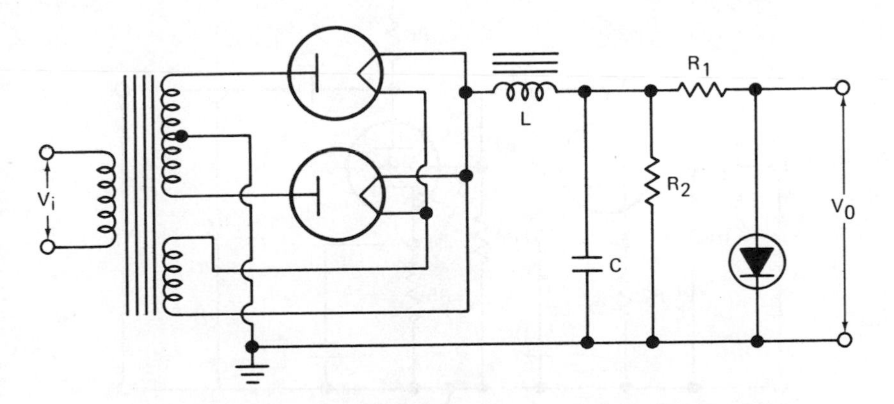

Figure 2-61. Filtered and regulated full-wave rectifier power supply.

these parameters briefly and for the most part in a qualitative fashion. In block diagrams of circuits amplifiers are sometimes represented by the symbol —▷—.

Since the current in the plate circuit of a triode or the collector circuit of a transistor is limited by the current emitted at the cathode or injected at the emitter-base junction, the power which either of these devices can deliver to a load is limited. Any load will require a certain level of power for its proper operation, and the amplifier elements must be chosen accordingly. Characteristic curves given by the manufacturer are the source of this information. We have already indicated that both triodes and transistors require certain dc voltages on their elements, and superimposed on them are the input and output ac signals. For ac signal amplification the voltage of interest at the output is the ac portion alone, and the output circuit must be arranged in a way which prevents application of the dc voltage to the load. Output circuits are generally of three types: resistance coupling, impedance coupling, and transformer coupling. A typical resistance-coupled circuit is shown in Fig. 2-62, where the coupling resistors are R_3 and R_6. In this circuit two triode amplifiers are cascaded to achieve greater amplification than is possible with a single tube. Many of the properties of this circuit have already been discussed, and we shall point out here those features which differ from those in Fig. 2-50. Beginning at the left with the input we choose capacitor C1 to have low impedance at the frequency of the ac input while blocking the constant magnitude dc voltage. This blocking occurs because the capacitor prevents the dc input voltage from causing any current flow in the grid circuit resistor R1, and as a result there is no dc voltage drop across it resulting from dc sources in the circuitry preceding the amplifier. The ac portion of the input voltage does give rise to current and voltage variations in R1 and hence variations at the grid of V1. The quiescent voltages of the cathode, grid, and plate of both V1 and V2 are determined by the magnitude of the B+ voltage

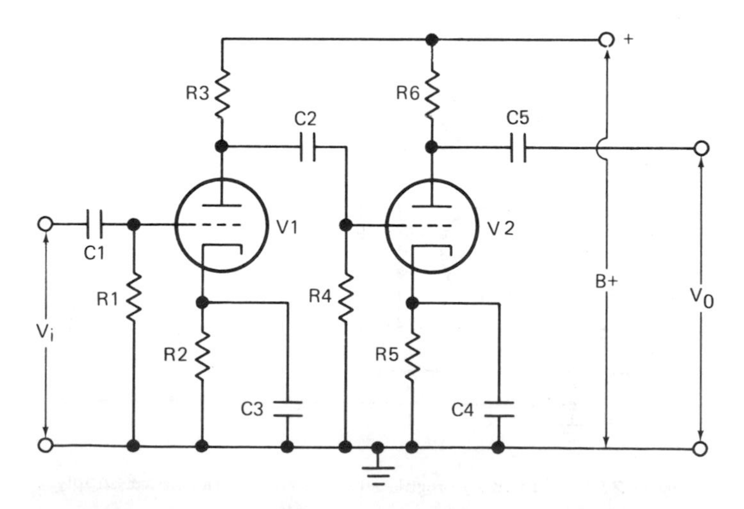

Figure 2-62. Two-stage resistance-coupled triode amplifier.

supply, the load resistors R3 and R6, and the self-biasing resistors R2 and R5. Resistors R2 and R5 give rise to a voltage drop when plate current is drawn, and as a result the cathodes are positive with respect to the grids in both V1 and V2. This self-biasing arrangement makes the grid bias power supply of Fig. 2-50 unnecessary. The output of V1 is applied to the load resistor R3, and the ac portion is coupled to the grid of V2 by applying it across R4. The dc voltage at the plate of V1 causes no voltage drop across R4 because capacitor C2 blocks dc current. The cathode circuits of each tube contain both a biasing resistor and a bypass capacitor (C3 and C4). The function of the latter is to prevent ac plate current from flowing through the cathode resistor and thus varying the grid-cathode bias voltage. With the capacitor there is essentially no voltage drop resulting from ac across the parallel pair, C3 and R2. The output of V2 also employs a capacitor C5 to block dc while readily passing ac to the following stage of the circuit.

An impedance-coupled amplifier looks exactly the same as Fig. 2-62 except R3 and R6 are replaced by large inductors. These have high impedance for ac and low impedance for dc, and as a result the dc voltage drop across them is maintained at a much lower value than in the case of resistance coupling. The net result is a much smaller power requirement for the dc B+ supply.

Because of space limitations we shall not discuss the transformer-coupled amplifier in detail here. It is used whenever relatively large amounts of power must be coupled from one part of a circuit to another. A good example is in the output stage of an audio amplifier which is connected to some speakers.

Another important consideration in any circuit design, including amplifiers, is impedance matching. Every section of a circuit, for example, an amplifier section, has two important impedances: an input impedance and an output impedance. The input impedance is that impedance "seen" by a signal applied at the input terminals of a section, while the output impedance is that impedance seen by a signal applied at the output terminals. The term impedance matching refers to the process in which the input impedance of one circuit section is adjusted to some particular value by means of additional circuit elements. This adjustment is necessary because many devices require a specific value of load impedance for effective operation. For example, suppose that a transistor amplifier requires a load impedance of 1000 Ω and that the actual load has an impedance of 10 Ω. Impedance matching is clearly necessary and one method is to use transformer coupling in which the transistor output is connected through a transformer to the actual load. We desire that the transformer input have an impedance of 1000 Ω while its output impedance should be 10 Ω (see Problem 13 at the end of this section). The input impedance Z_p and the output impedance Z_s of a transformer are related by the square of the turns ratio in the formula

$$Z_p = Z_s \left(\frac{n_p}{n_s}\right)^2 \tag{2-43}$$

For our problem we desire $Z_p/Z_s = 100$ or $n_p/n_s = 10$.

The frequency range of signals to be amplified may dictate the kind of coupling to use between stages of an amplifier. For example, if we desire to amplify only a very narrow range of frequencies while excluding others, what is called a narrow-band-tuned amplifier circuit may be utilized. This device, as illustrated by Fig. 2-63, makes use of an LC parallel circuit as the load for the output of the first stage. As illustrated in Problem 10 (at the end of this section), the impedance of the LC parallel connection depends on the frequency of the applied signal. At the resonant frequency $\omega = (1/LC)^{1/2}$, the impedance passes through a maximum value, and the voltage drop between point A and ground passes through a maximum. For off-resonant frequencies the impedance of the LC combination decreases, the voltage drop decreases, and the ac signals are effectively shorted to ground by capacitor C4. The overall result is a very small voltage drop between point A and ground or effectively no gain at these frequencies. There is therefore a frequency-dependent gain which is strongly peaked at the resonant frequency. The sharpness of the peak about the resonant frequency is determined by the resistance of the parallel combination. There is always some resistance because of the wires in the coil even if a resistor is not directly wired into the circuit. The ratio of the inductive reactance to the resistance is called the quality factor Q and may be expressed as

$$Q = \frac{X_L}{R} = \frac{L\omega}{R} \tag{2-44}$$

High-Q circuits amplify only a very narrow range of frequencies, while low-Q circuits amplify a much broader range.

A variety of other kinds of amplifiers has been designed for various purposes. Among them are the cathode follower and the difference amplifier. Space requirements prevent their discussion here.

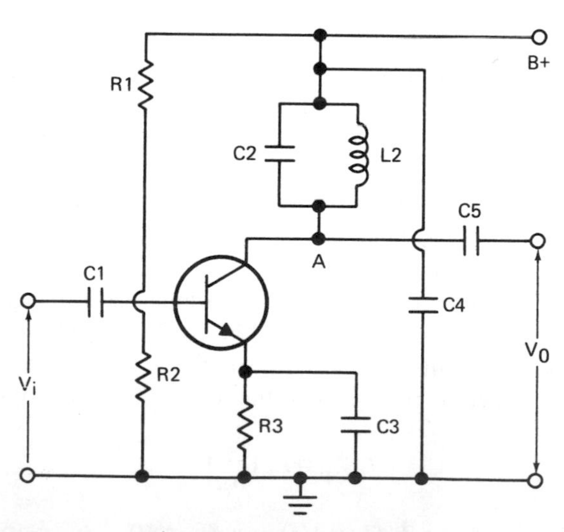

Figure 2-63. Tuned amplifier.

FEEDBACK EFFECTS. The term feedback refers to the process by which a portion of the output signal of an amplifier is applied (fed back) and combined with the input signal (refer to Fig. 2-64). Basically there are two types of feedback: positive and negative. The former is widely used in oscillator circuits, while the latter is used in nonoscillating amplifiers to improve frequency response and to reduce both distortion and effects of tube or transistor aging. In some cases both positive and negative feedback are used. The terms positive and negative refer to the phase relation between the input signal and the feedback signal, with positive implying that the two are in phase and negative implying that they are 180° out of phase. Thus, in positive feedback the two signals add constructively to increase the total input signal, while in negative feedback the signals interfere destructively and the total signal is reduced. Figure 2-64 illustrates schematically a feedback amplifier whose gain is α and whose fractional feedback is β. The input signal v to the amplifier is given by

$$v = v_i + \beta v_0 \tag{2-45}$$

where v_i is the applied voltage and v_0 the output voltage. The signal v_0 is given by αv so that

$$v_0 = \frac{\alpha v_i}{1 - \alpha\beta} \tag{2-46}$$

If the amplifier gain α is quite large and the fraction fed back is selected to make $|\alpha\beta| \gg 1$, Eq. 2-46 becomes

$$v_0 \simeq -\frac{v_i}{\beta} \tag{2-47}$$

This relation illustrates how the output of an amplifier may be made very nearly independent of variations in the amplifier gain. This conclusion takes for granted the constancy of β over long periods of time and will be valid provided the feedback network is constructed of high-quality passive circuit elements. Because Eq. 2-47 shows no dependence on amplifier characteristics, it is obvious that using feedback will reduce distortion between the input and output waveforms.

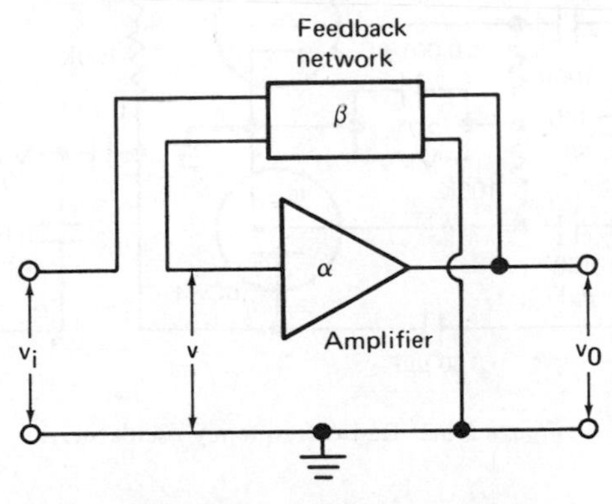

Figure 2-64. Feedback amplifier.

To illustrate the application of Eq. (2-46) we shall consider only one example of what is called an oscillator. This kind of circuit is constructed so that β is positive and the product $\alpha\beta$ approaches unity. Under these conditions the net gain approaches infinity, implying that the output voltage can be maintained without the presence of an input voltage. The positive feedback condition $\alpha\beta = 1$ is known as the Barkhousen condition, and when it is met an amplifier becomes an oscillator. Since oscillators require no input signal other than an initial pulse in order to function, they may be regarded as converting dc power derived from the bias supply into ac power having certain characteristics which reflect the design of the oscillator circuit. For the Barkhousen condition to be met, the phase shift of the feedback signal must be 360°.

One interesting oscillator is the marginal oscillator used, for example, in some nuclear magnetic resonance studies. A block diagram is shown in Fig. 2-65 and a practical circuit is shown in Fig. 2-66. Essentially the sample coil L and the capacitor C form a frequency-selective network which is tuned to the radio-

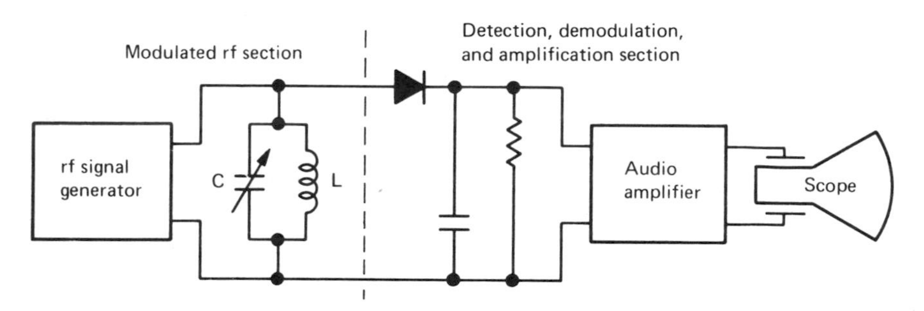

Figure 2-65. Marginal oscillator block diagram.

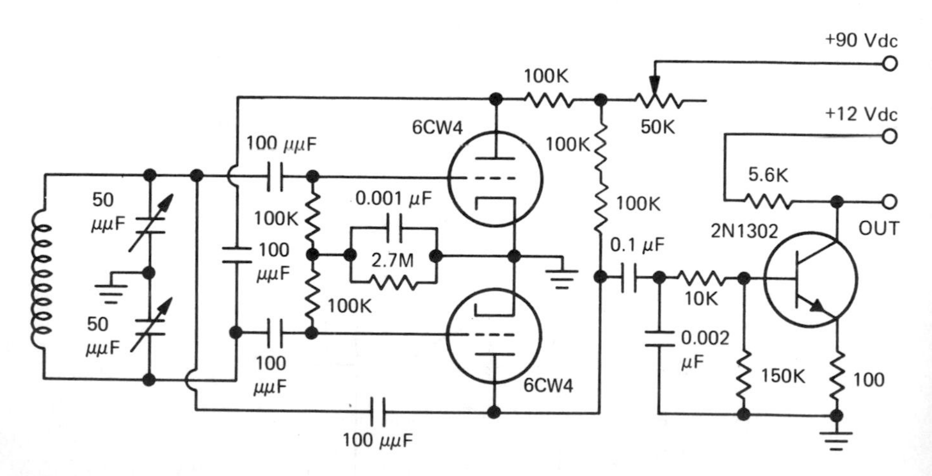

Figure 2-66. Radio-frequency oscillator.

frequency (rf) region of the electromagnetic spectrum. This network is coupled to an rf amplifier through feedback, and when $\alpha\beta = 1$ the system comprises an rf oscillator. Either the feedback β or the amplifier gain α is variable so that the oscillator can be adjusted until it is just able to sustain oscillations with the sample to be studied within the coil L but in a nonresonating condition (i.e., the magnetic field applied to the sample is not of the proper magnitude to give absorption). With this adjustment completed, the magnetic field can be changed to bring the nuclear spins of the sample into resonance. When this occurs, the sample absorbs rf energy from the coil, changing its load and thereby the impedance of the LC circuit. The net effect is a change in voltage drop across the LC combination which can be detected. If the magnetic field applied to the sample is varied repetitiously over a small range with a sawtooth wave in the audio-frequency (af) region, the nuclear spins of the sample will pass through resonance twice each sawtooth cycle, once as the field is increased and again as it decreases. As a result the output of the rf oscillator is modulated at the frequency of the sawtooth generator, the latter also serving to trigger the horizontal sweep on an oscilloscope. The modulated rf voltage has an amplitude which varies periodically with the sawtooth frequency. By demodulating, the envelope of the rf voltage may be detected, amplified, and applied to the vertical input of the oscilloscope. Figure 2-67 illustrates both the variable magnitude rf voltage and the envelope. This kind of circuit may be used in the experiment entitled "Nuclear Magnetic Resonance". (See Section 7.7.)

BASIC MEASURING INSTRUMENTS. Having discussed various kinds of circuits we shall now turn our attention to measuring instruments which find widespread use in experimental chemistry. Nearly all these devices involve meters of one kind or another and we shall begin this section with a description of them.

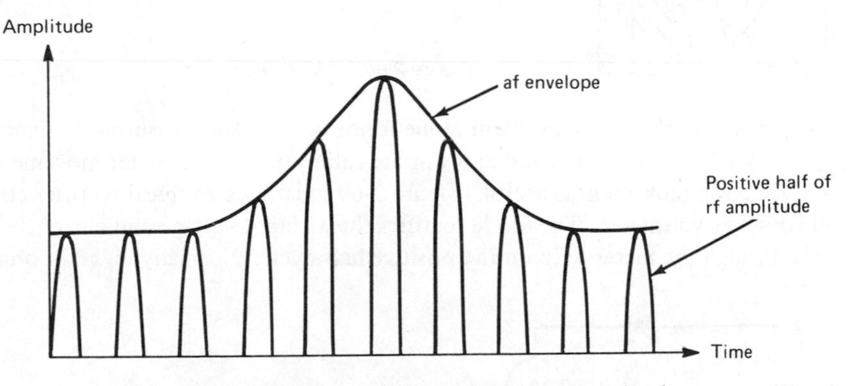

Figure 2-67. Modulated rf nuclear absorption signal.

Meters. A galvanometer of the d'Arsonval type is the basic element of most meters. It is a current-measuring instrument in which a rotatable coil of fine wire suspended by small spiral springs is placed between the poles of a horseshoe magnet. When current is passed through the coil, the associated magnetic field interacts with the field of the permanent magnet, giving rise to a torque tending to wind or unwind the spiral suspension springs. A pointer attached to the coil is thus rotated until equilibrium is achieved. By calibration the pointer position indicates the current through the coil. Figure 2-68 illustrates this meter. The current range of an ammeter can be varied by placing a resistor in parallel with the d'Arsonval movement, thus shunting a certain fraction of the current around the meter. The d'Arsonval meter may also be calibrated as a voltmeter if its internal resistance is measured and Ohm's law is used. The range of the d'Arsonval voltmeter can be extended by placing a resistor in series with the meter. Problems 15 and 16 (at the end of this section) illustrate the procedure for achieving the proper range and why it is desirable to have low resistance in an ammeter and high resistance in a voltmeter.

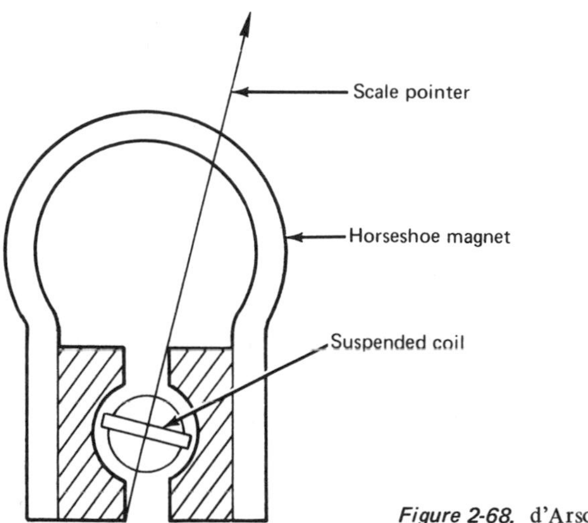

Figure 2-68. d'Arsonval meter.

The d'Arsonval meter movement alone is not suitable for measuring ac signals. By rectification of ac signals and appropriate calibration of the meter movement, the d'Arsonval movement is useful. Figure 2-69 illustrates a typical rectifier circuit for an ac voltmeter. The diode rectifies the ac input voltage and current flows through the meter only on the positive half-cycle. Normally an additional

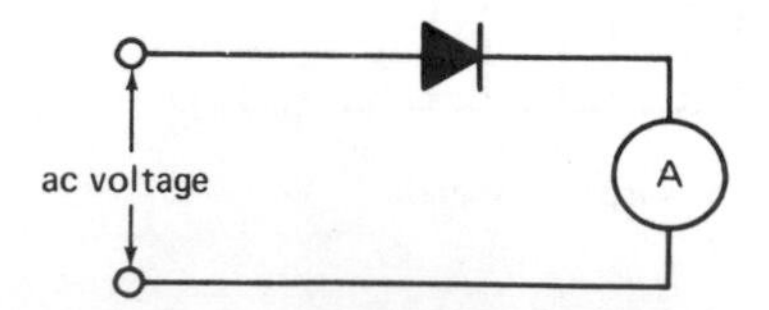

Figure 2-69. Alternating-current voltmeter.

diode is connected in parallel with the diode-meter circuit to conduct on the negative half-cycle, thereby equalizing the load on each half-cycle. When the second diode is used, the device can also, with proper calibration, be used as an ammeter.

Turning specifically to the ac voltmeter, the meter deflection is proportional to the average current flowing in the meter. However, the meter scales are generally calibrated in terms of root-mean-square voltages, which implicitly assumes sinusoidal ac input. The root-mean-square voltage is defined as the following integral over one period T of the sine wave

$$\begin{aligned} V_{rms} &= \left[\frac{1}{T}\int_0^T (V_0 \sin \omega t)^2 \, dt\right]^{1/2} \\ &= 0.707 V_0 \end{aligned} \tag{2-48}$$

This quantity also finds use when comparing the power dissipated in a load by ac and dc sources. $T = \pi/\omega$ for one cycle.

The appropriate use and calibration of a dc voltmeter will make it useful as an ohmmeter. Figure 2-70 illustrates the principles. First with the test leads shorted, the variable resistor, R_V, is adjusted so that the meter reads full scale or $R_x = 0$. At this point the voltage drop across R_1 is V. Connecting in the unknown resistor, R_x, lowers the voltage drop across R_1, and assuming negligible current flow in the meter circuit we have the following equations:

$$\begin{aligned} V &= i_1 R_1 && \text{with leads shorted} \\ V_x &= \frac{R_1}{R_1 + R_x} V && \text{with unknown in circuit} \end{aligned} \tag{2-49}$$

By adjustment when $R_x = 0$ we have made V correspond to a full-scale reading on the meter. When $R_x \neq 0$ the corresponding meter deflection will be less, and the fraction of full-scale deflection f is

$$f = \frac{V_x}{V} = \frac{R_1}{R_1 + R_x} \tag{2-50}$$

Using Eq. (2-50) the scale can be calibrated, albeit nonlinearly, in terms of R_x.

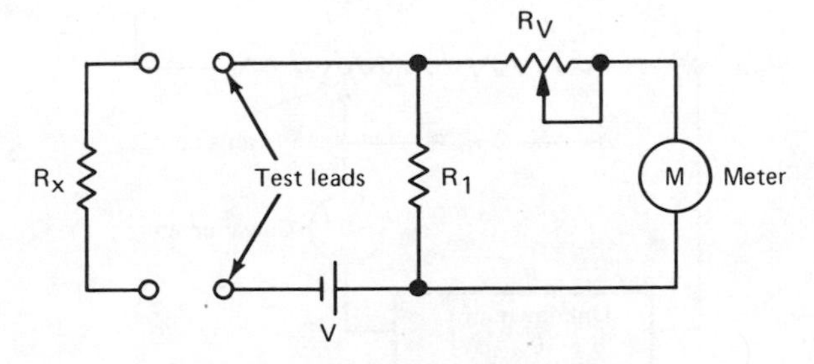

Figure 2-70. Ohmmeter circuit.

The combination of multirange ac and dc ammeter and voltmeter circuitry together with an ohmmeter circuit forms what is called a multimeter or VOM (volt-ohm-milliammeter) if the circuit involves no vacuum tubes. Multimeters introduce errors in voltage readings because the source of the voltage drop must supply power to the voltmeter circuit; hence, the voltage drop when the multimeter is connected differs from the voltage drop when it is not connected. A more sophisticated voltmeter is called a VTVM or vacuum tube voltmeter, and essentially it makes use of a triode amplifier circuit in which the voltage to be measured is supplied to the grid and the power to deflect the meter is supplied by the B+ supply in the grid-cathode circuit. To the extent that no current flows in the grid-cathode circuit the loading of the unknown voltage source is avoided —a very desirable feature. To avoid the complications introduced by the aging of vacuum tubes, which introduces changes in tube characteristics, most VTVM circuits employ a single-ended difference amplifier circuit.

Potentiometers. Laboratory experiments frequently involve the use of potentiometers. These devices measure unknown electrical potentials (voltages) by comparing them with a known potential. The known potential is usually provided by a standard cell. The Weston cadmium cell is the most widely used standard voltage source.

Figure 2-71 shows a simplified potentiometer which illustrates the principles of their operation. Note the polarities of the known and unknown emfs. They are arranged to oppose one another as they develop a voltage across the resistor. That is, the voltage arising from the unknown emf is arranged to oppose the voltage arising from the known emf. The resistor shown in Fig. 2-71 is a length of uniform conductor, L, which has a fixed resistance per unit length. Moving the contact along the wire varies the value of E′, the voltage drop between the end of the resistor and the contact point. The galvanometer measures the current which flows in the circuit containing the unknown emf. By adjusting E′, it is possible to make the galvanometer give a zero reading, implying that the voltage drop across R′ is identical to the voltage of the unknown. This is called the null

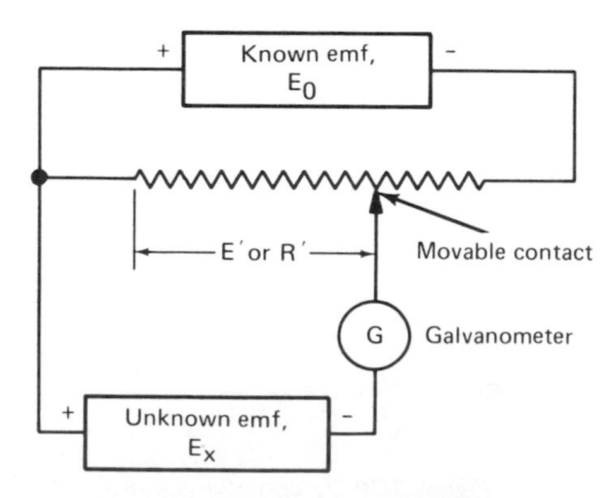

Figure 2-71. Simplified potentiometer circuit.

point. By reaching this condition, it is possible to calculate the unknown emf as shown below.

At the balance condition,

$$E' = E_x$$

From the known length of resistor, X, required to develop E′ we have the following:

$$\frac{E'}{E_0} = \frac{IX\rho}{IL\rho} = \frac{X}{L}$$

where I is the current through the resistor and ρ is the resistance per unit length.

Since $E' = E_x$ at the null point, we can write

$$\frac{E_x}{E_0} = \frac{X}{L}$$

Solving for E_x furnishes the desired relation:

$$E_x = E_0 \frac{X}{L} \tag{2-51}$$

Conclusion: Measurement of L and X together with a known emf provides a means of determining unknown voltages.

Laboratory potentiometers generally are more complicated than those shown in Fig. 2-71. However, the principles of operation remain the same. The added complexity is needed to make the instrument read the unknown voltage directly. Figure 2-72 shows a schematic of one type of laboratory potentiometer. The dc

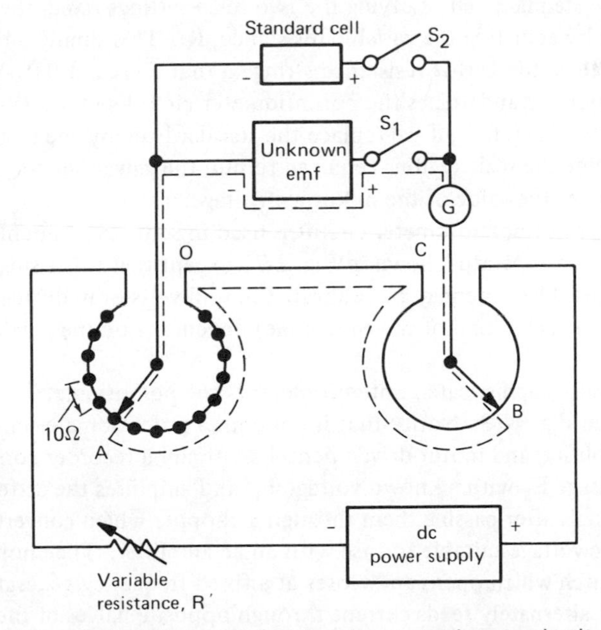

Figure 2-72. Schematic of a laboratory potentiometer circuit.

power supply furnishes a constant voltage across a series of precision resistors. These resistors are of two types. In region A, there is a series of 15–10-Ω resistors which can be tapped (shown schematically by the arrow) so that the resistance between 0 and A is variable, but only in steps of 10 Ω. In region B there is a single resistor of 10 Ω which can be tapped at any value and so the resistance between C and B can have any value between 0 and 10 Ω. Keep in mind that the dc voltage developed by the power supply appears across the whole resistance string (160 Ω). In actual practice both regions A and B are scaled. Region A varies from 0 to 1.5 in steps of 0.1, and region B varies continuously but the smallest scale division is 0.0005.

The circuit path of importance is shown in Fig. 2-72 with dotted lines. As you follow this path there is one emf and two resistances, R_{0A} and R_{BC}. There is also a galvanometer which reads the current in the unknown emf circuit. If R_{0A} and R_{BC} are adjusted so the galvanometer gives a null reading, then no current flows, and the potential drop across the resistors in the path A0CB must be equal in magnitude to the unknown emf but of opposite sign.

Knowing the resistance is insufficient. In addition we must know the voltage drop across the entire 160-Ω string. This can be determined by using the standard cell shown in Fig. 2-72 in place of the unknown emf. By adjusting things properly it is possible to make the potentiometer resistance scale read directly in volts. This is accomplished as follows. Opening S_1 and closing S_2 replaces the unknown emf with the standard cell. We now adjust the potentiometer resistance scale to the value of the voltage of the standard cell. This implies that the actual resistance along A0CB has a magnitude in ohms that is 100 times the voltage of the standard cell. Leaving the two scale settings fixed, the galvanometer is nulled by adjusting the variable resistance, R'. This simply adjusts the voltage applied to the 160-Ω resistance string so that there is a 0.01-V/Ω drop. This measurement standardizes the potentiometer circuit so that its scale reading will be in volts. Therefore, if we replace the standard cell by the unknown emf and redetermine the scale reading required to null the galvanometer, its scale reading furnishes the value of the unknown voltage.

Another type of potentiometer circuit is used in some experiments. It is equivalent to the apparatus shown in Fig. 2-72 except that it is a single self-contained unit. The scale may be calibrated in millivolts or in degrees Centigrade (for a particular kind of thermocouple) depending on the particular instrument.

An even more sophisticated potentiometer is the potentiometric recorder diagrammed in Fig. 2-73. Notice that it contains a potentiometer circuit with a chopper, amplifier, and motor-driven pen. Essentially a recorder compares the unknown voltage E_x with a known voltage E_0 and amplifies the difference in these two signals after passing them through a chopper which converts the dc input to an ac voltage suitable for use with an ac amplifier. The chopper is an electronic switch which opens and closes at a fixed frequency generally 60 Hz. In doing so it alternately feeds current through opposite halves of the primary of the transformer. This change in direction through the primary induces an ac

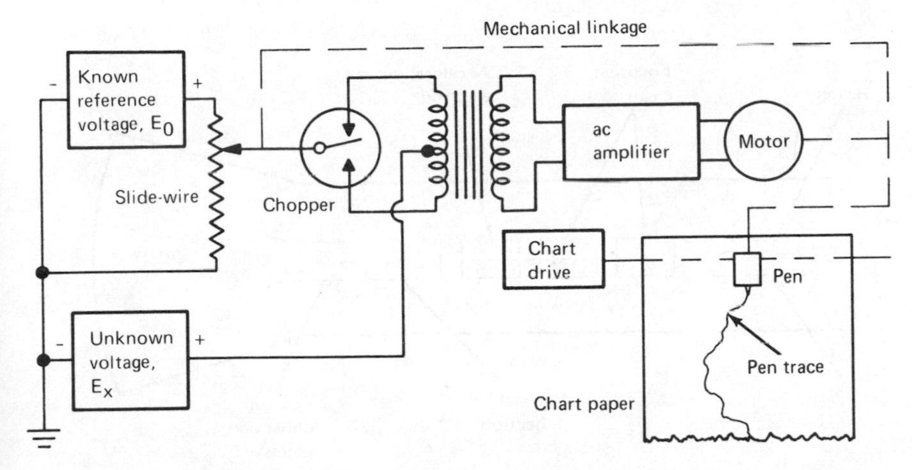

Figure 2-73. Block diagram of a potentiometric recorder.

voltage in the secondary. Notice that the center tap of the primary "floats" at the dc voltage of the unknown. The output of the secondary is then proportional to the difference, $E_0 - E_x$. Depending on the sign of the difference the amplified signal turns a motor one direction or another. A mechanical linkage to both a slide-wire and the recorder pen adjusts $E_0 - E_x$ toward zero and changes the position of the pen. When $E_0 - E_x = 0$, the null point is reached, the amplifier output goes to zero, and the position of the pen becomes constant. By suitable calibration the position of the recorder pen may be linearly related to E_x, and a plot of E_x versus time is obtained through the use of a constant speed chart drive motor.

Oscilloscope. Oscilloscopes are prevalent electronic measuring instruments whose heart is the cathode-ray tube, CRT. This sophisticated vacuum tube contains a thermal source of electrons, an acceleration and focusing section, and two sets of deflection plates—horizontal and vertical, as shown in Fig. 2-74. The versatility of an oscilloscope depends on the sophistication of electronic circuits associated with the CRT, which are diagrammed in Fig. 2-75 for a simple scope. The power supply drives the thermal electron source and the accelerating and focusing circuits. External signals may be applied to both the vertical and horizontal inputs, and after amplification the voltages are applied to the vertical and horizontal deflection plates of the CRT where they serve to deflect the electron beam from its "straight-through" trajectory to the luminescent screen. The vertical deflection is proportional to the voltage applied to the vertical deflection plates; the same is true for the horizontal deflection. In less expensive scopes the amplifiers can handle only ac input signals. Frequently signals are periodic and we desire to observe only one or at most a few periods. This is accomplished by triggering the horizontal amplifier either internally or externally while applying the signal of interest to the vertical input. Triggering implies the repetitious sweeping of the electron beam across the face of the CRT such that the electron

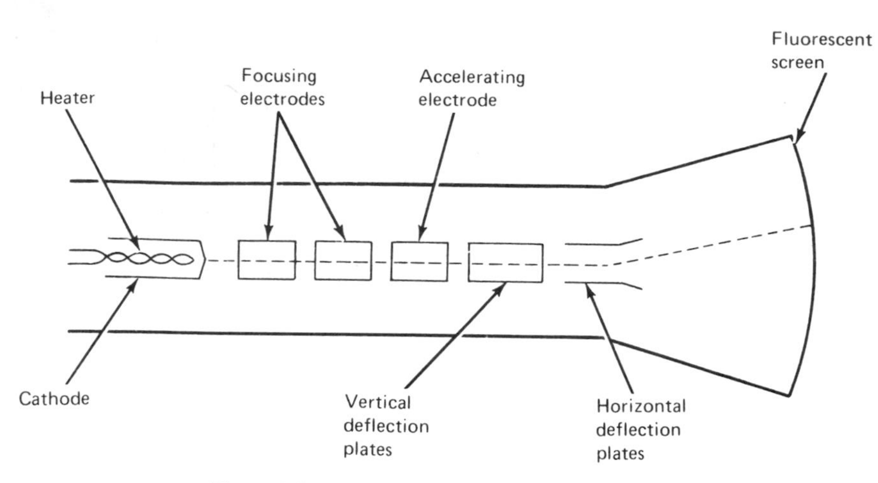

Figure 2-74. Block diagram of a cathode-ray tube.

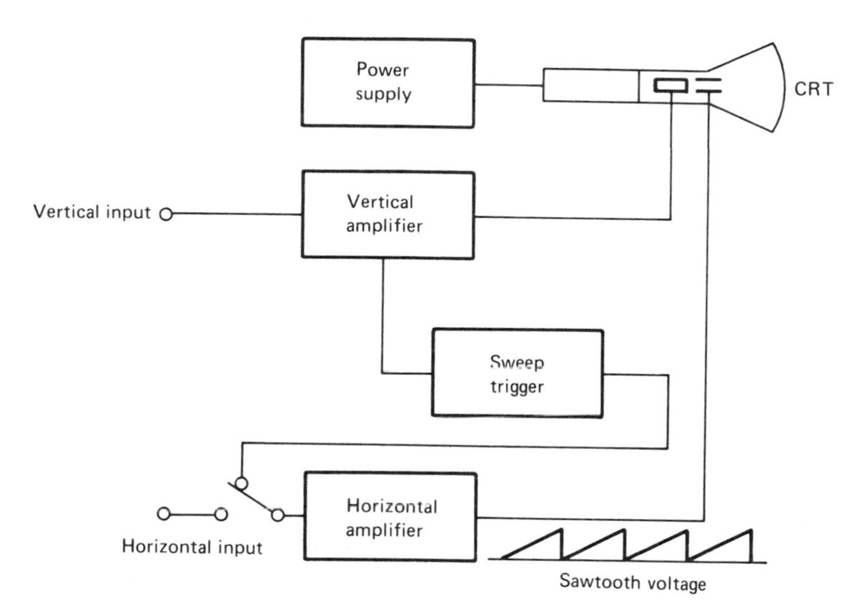

Figure 2-75. Block diagram of an oscilloscope.

beam moves from right to left across the face of the tube much more rapidly than it moves from left to right. This is accomplished with a ramp or sawtooth generator, as shown schematically in Fig. 2-75. By adjusting the time for one traversal to be equal to an integral multiple of the period of the signal of interest, the luminescent trace on the CRT screen is a single trace and is an accurate reproduction of vertical input voltage versus time. This process is called synchronization. In Fig. 2-75 a portion of the vertical input signal is used to trigger the horizontal sweep.

PROBLEMS

1. Prove that

$$\frac{1}{R_{eq}} = \frac{1}{R_1} + \frac{1}{R_2} + \frac{1}{R_3}$$

in the parallel connection shown in Fig. 2-38.

2. Prove the relations illustrated in Fig. 2-39 for both series and parallel connections of capacitors. State your arguments in terms of Kirchhoffs' laws.

3. Prove the relations illustrated in Fig. 2-40 using Kirchhoffs' laws.

4. Suppose that standard 115-V ac power is available and that an application calls for the use of 400-V ac power at the level of 50 W. Design an iron-core transformer (turns ratio, number of turns, wire size) assuming the same size wire for primary and secondary and copper wire. Tables in the *Handbook of Chemistry and Physics* (Chemical Rubber Publishing Co., Cleveland, any edition) furnish wire size and resistance.

5. The Wheatstone bridge circuit of Fig. 2-43 may be solved in a more direct fashion than outlined in the text if one notes that, at balance, the voltages at B and C with respect to ground are equal, thus implying that the voltage drop across R_1 is identical to the voltage drop across R_2, etc. Use this notion to develop Eq. (2-27). Note that in this procedure I_5 is never calculated and that no information away from balance is obtained.

6. Find the voltage drop, in the circuit diagram in Fig. P6, across R_4, in terms of R_1, R_2, R_3, and the applied voltage, making use of a series-equivalent resistor R_{eq} for that part of the network for which it can be usefully applied.

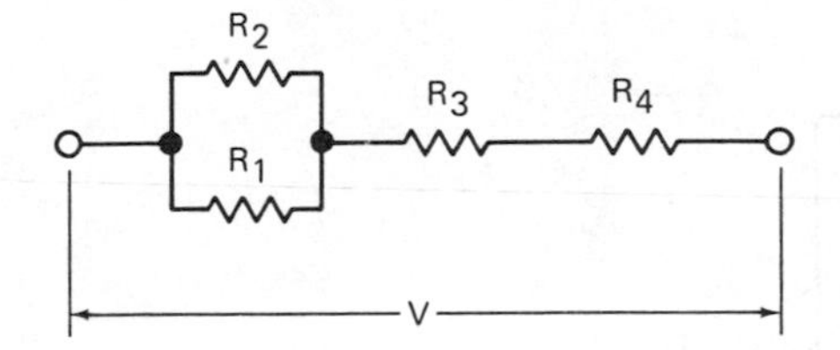

Figure P6.

7. Derive Eq. (2-34) using the circuit diagram in Fig. P7.

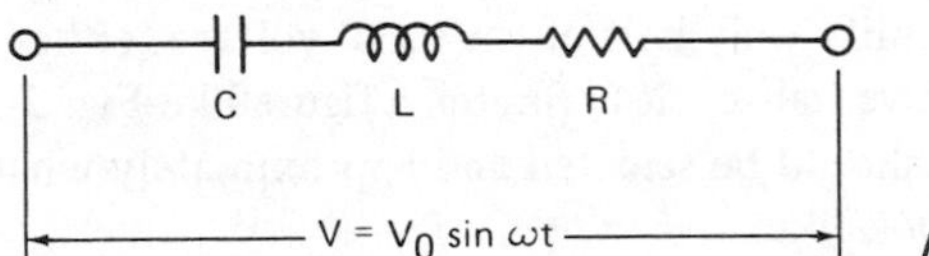

Figure P7.

8. Prove that the current and voltage remain in phase when an ac voltage is applied across a resistor.

9. Consider the circuit shown in Fig. P9 and develop an equation for the impedance as a function of R, L, C, and ω. Make a plot of Z versus ω showing that the function has a minimum value. Comment on how the current in the circuit will vary with ω. When Z is minimized the circuit is said to be resonant. What can be said about the current under these circumstances?

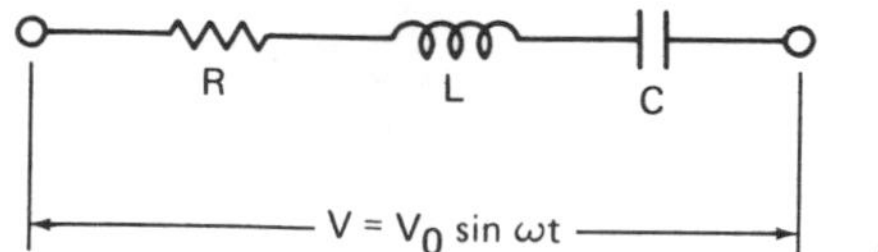

Figure P9.

10. Work out the details of the current, impedance, and resonance conditions for the parallel circuit shown in Fig. P10. Note carefully the frequency dependence of the impedance. Resonance is said to occur when $X_L = X_c$. *Hint:* Calculate complex impedance for each of the three parallel components, and then calculate the inverse of the sum of their reciprocals to find the series-equivalent complex impedance.

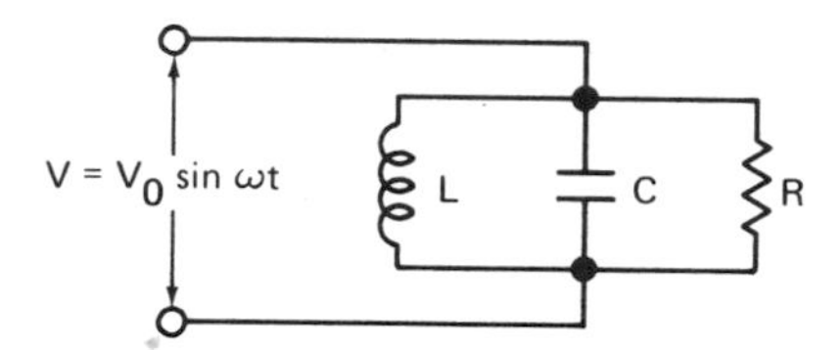

Figure P10.

11. Consider the dc circuit shown in Fig. P11 and find an equation for the current as a function of time, after the switch is closed, which has the form $i(t) = A \exp\{-(t/\tau)\}$. τ is called the time constant and is related to R and C in a way established by the solution of the problem.

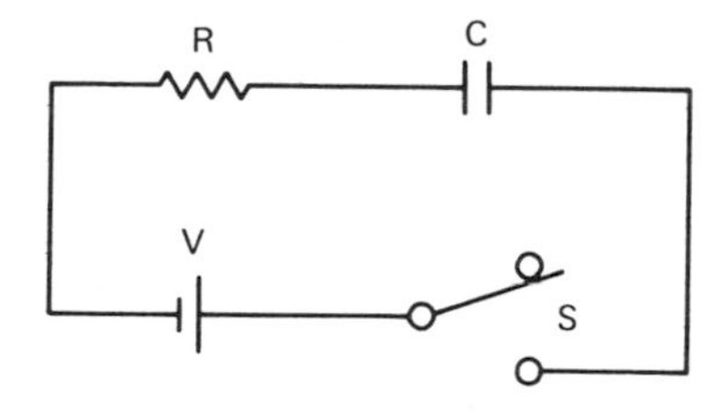

Figure P11.

12. Suppose that you wanted to distort a sinusoidal input to a triode so that the output voltage appeared to amplify only half of the input voltage (either the positive half-cycle or the negative half-cycle). Sketch a figure like Fig. 2-52 showing what operating point should be selected and approximately what the output waveform would look like.

13. Consider, as shown schematically in Fig. P13, a source of current which has an impedance of R_S Ω when viewed from a load R_L connected in series with it. Show that the power dissipated in the load is greatest when $R_s = R_L$.

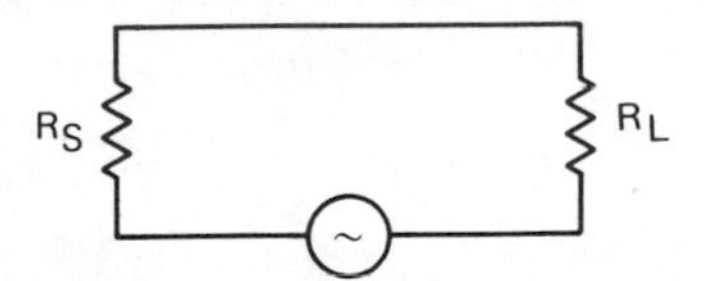

Figure P13.

14. Using the complex number representation as a method of circuit analysis, develop an expression for the output voltage as a function of the input voltage, frequency, and the magnitudes of the resistances and capacitances in the bridged-T network shown in Fig. P14. The resulting expression has an imaginary part which falls to zero at some frequency called the characteristic frequency. Find the characteristic frequency and determine whether, at this frequency, the impedance is a maximum or minimum.

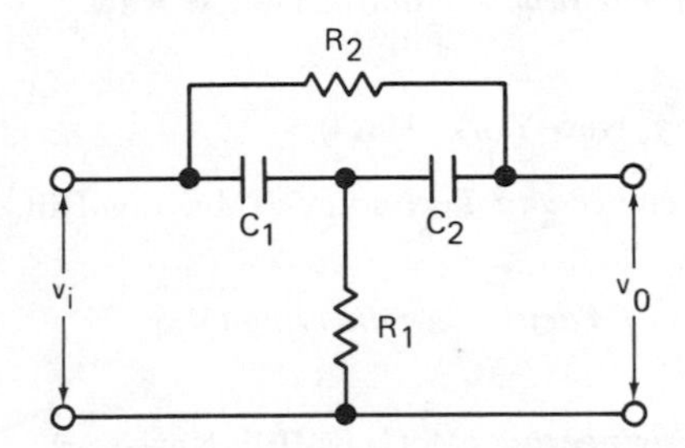

Figure P14.

15. Suppose that a d'Arsonval meter movement has a full-scale deflection of 1 mA without any shunt resistance. If the internal meter resistance is 1000 Ω, what size shunt resistor should be used to make the full-scale deflection correspond to 10 mA? If this meter were connected in series fashion in the circuit shown in Fig. P15, what effect would the meter itself have on the current in the circuit?

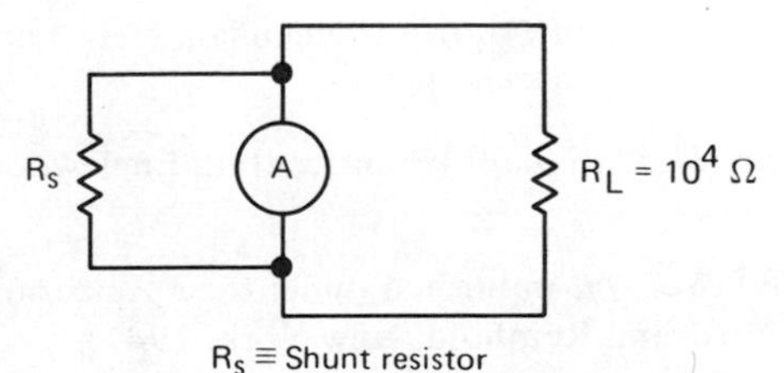

Figure P15.

16. If a meter movement has an internal resistance of 1000 Ω and a full-scale current deflection of 1 mA, what size series resistor should be used to extend the range to 10 V full scale? If this 10-V meter is used to measure the voltage drop across R in the circuit shown in Fig. P16, what effect does the meter itself have on the voltage drop?

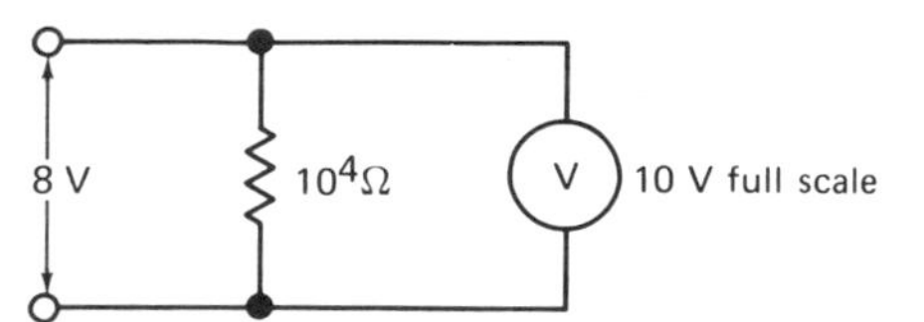

Figure P16.

REFERENCES

Section 2.1

S. Dushman, *Scientific Foundations of Vacuum Technique* (Wiley, New York, 1962).

A. Guthrie, *Vacuum Technology* (Wiley, New York, 1963).

G. Lewin, *Fundamentals of Vacuum Science and Technology* (McGraw-Hill, New York, 1965).

M. S. Pirani and J. Yarwood, *Principles of Vacuum Engineering* (Van Nostrand Reinhold, New York, 1961).

C. M. VanAtta, *Vacuum Science and Engineering* (McGraw-Hill, New York, 1965).

J. Yarwood, *High Vacuum Technique* (Chapman & Hall, London, 1967).

Section 2.2

E. U. Condon and H. Odishaw (eds.), *Handbook of Physics,* 2nd ed. (McGraw-Hill, New York, 1967), pp. 5–34.

W. F. Coxon, *Temperature Measurement and Control* (Macmillan, New York, 1960).

R. L. Weber, *Heat and Temperature Measurement* (Prentice-Hall, Englewood Cliffs, N.J., 1950).

Temperature, Its Measurement and Control, published under the auspices of American Institute of Physics (Van Nostrand Reinhold, New York, 1941).

Section 2.3

D. H. Anderson, N. B. Woodall, and W. West in A. Weissberger (ed.), *Physical Methods of Organic Chemistry,* Vol. I (Wiley-Interscience, New York, 1960), Part III, Ch. XXIX.

J. G. Calvert and J. N. Pitts, Jr., *Photochemistry* (Wiley, New York, 1966).

M. Comtet, Ph.D. dissertation, *I. Photochemistry of trans-1-phenyl-2-butene; II. The Photoisomerization of Azulene,* University of Texas, Austin, 1970.

D. N. Kendall, *Applied Infrared Spectroscopy* (Van Nostrand Reinhold, New York, 1966).

W. West in A. Weissberger (ed.), *Physical Methods of Organic Chemistry,* Vol. I (Wiley-Interscience. New York, 1960), Part III, Ch. XXVIII.

Section 2.4

H. V. Malmstadt, C. G. Enke, and E. C. Toren, Jr., *Electronics for Scientists* (Benjamin, Reading, Mass. 1962).

J. J. Brophy, *Basic Electronics for Scientists* (McGraw-Hill, New York, 1966).

T. B. Brown, *Electronics* (Wiley, New York, 1954).

C. W. Cox and W. L. Reuter, *Circuits, Signals and Networks* (Macmillan, New York, 1969).

P. Cutler, *Semiconductor Circuit Analysis* (McGraw-Hill, New York, 1964).

J. W. Friedman, H. G. Rice, and G. McGinty (eds.), *Basic Electronics* (Prentice-Hall, Englewood Cliffs, N.J., 1965).

W. R. Hill, *Electronics in Engineering* (McGraw-Hill, New York, 1961).

Radio Amateur's Handbook, published under the auspices of the American Radio League, any edition.

Chapter 3

A Collection of Experiments Introducing Techniques

3.1 Empiricism in Experimentation. An Empirical Gas Dynamics Experiment

The goal of this experiment is to empirically relate certain variables in a gas dynamics experiment.

THEORY. In most of the experiments in this book the data gathered are analyzed in terms of theoretically derived expressions, and we shall seek to illuminate this theory during the course of the experiment. While this is a perfectly legitimate and useful goal of laboratory exercises, it may leave the impression that all experimental science has this orientation. Such is absolutely not the case, and, especially in cases where there exists no theory or at least no satisfactory theory, an empirical or semiempirical approach must be used. For many experimenters this is the most exciting kind of research because out of it one hopes to develop phenomenological relations among variables which can later be compared with various theoretical models proposed to account for the phenomena observed.

Success in the empirical approach is largely a matter of persistence, patience, experience, and intuition. Some helpful ground rules can, however, be laid down. First, whatever experiment is being done, make a list of all the variables that could possibly enter into it. Second, design experiments in which many of the variables are held constant while the influence of one on the measurement is studied. This one variable must, of course, be controlled in a known fashion. In most instances it will be impossible to fix all the variables except one. This is a fact of life and must be lived with by attempting to minimize the influence of all the extra variables. Third, carry out a systematic investigation of the variables in the original list. Fourth, attempt to correlate and cross-correlate these results to arrive at a mathematical or graphical relation among the observed variations. Such a relation will summarize the data in a phenomenological expression. Fifth, check the utility of your phenomenological expression by using it in a predictive sense. That is, use it to predict what should be observed in some situation not examined in the course of developing the empirical expression. Then make measurements under these conditions and compare the observation with the prediction.

EXPERIMENT. The apparatus for this experiment is shown in Fig. 3-1 and consists of a timer, a pump, and a vacuum line with several outlets. Eight capillary tubes will be provided which have known lengths and diameters. Using these eight capillaries and connecting parts to hook them to the vacuum line, design a gas dynamics experiment (i.e., measure pressure and time) using the apparatus in any fashion other than materially altering it. Perform the same experiment with all eight capillaries. Record all the data called for in your design. It may be wise to record more than is called for by the design in case hindsight suggests the usefulness of additional data.

Get an unknown capillary from the instructor and measure its length with a ruler. Then run the same experiment run for the knowns.

ANALYSIS. Plot the data obtained with the various known capillaries such that a dependent variable appears on the ordinate and an independent variable, length or diameter, on the abscissa. Families of curves will be helpful, each mem-

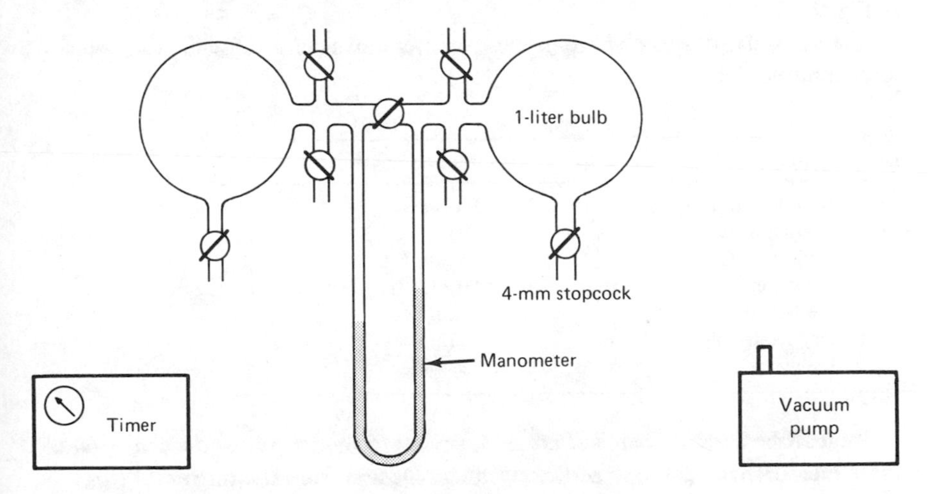

Figure 3-1. Apparatus for capillary calibration.

ber of which holds some variable constant. Try to linearize these curves by plotting a dependent variable (the measured variable) against the independent variable raised to some power or as the argument in an exponential or logarithmic function.

None of these correlations may seem to be entirely satisfactory, and quite likely from a perusal of your results it will become clear that additional experiments are needed. This is the way of empiricism. Report your results in their best empirical form, describe your experimental design, suggest additional experiments, and use the data for the unknown with the correlation to predict its diameter. Remember the intent of this experiment and do not run to gas dynamics texts for some theoretical model of this problem.

3.2 Thermocouple Calibration Study

The purpose of this experiment is to calibrate a copper-constantan thermocouple over the range 77°–373°K. The calibration procedure will involve both numerical and statistical analysis.

Calibration studies are a vital part of scientific research because without them measurements performed in different laboratories and with different types of apparatus have no common basis and cannot therefore be compared. There is a tendency to minimize the importance of careful and adequate calibrations because they are not at all intellectually stimulating. Nevertheless, they are absolutely crucial and must be properly performed and analyzed. In this experiment methods of analysis are emphasized.

EXPERIMENT. The apparatus is described in Section 2.2 and is shown there in Fig. 2-17.

After standardization, obtain potentiometer emf readings for the following bath combinations:

Hot junction	Cold junction
1. Boiling water	Ice water
2. Boiling water	Dry ice-trichloroethylene
3. Boiling water	Liquid N_2
4. Ice water	Dry ice-trichloroethylene
5. Ice water	Liquid N_2
6. Dry ice-trichloroethylene	Liquid N_2
7. Ice water	Ice water

Repeat the measurements of combinations 1, 5, and 4 twice to give a total of three measurements at each of these points. Obtain from the instructor three additional values for combinations 1, 5, and 4.

Experimental Notes.

1. Exercise great care in determining the balance point of the galvanometer. Use the same balancing procedure for each measurement. Check the standardization of the potentiometer during and at the close of the measurements.
2. Make certain that the thermocouple is at the bath temperature when making the emf measurements.
3. To prepare a dry ice-trichloroethylene bath, first pulverize the dry ice. Then add small amounts of the pulverized dry ice slowly to the dewar of trichloroethylene. The dry ice will evaporate immediately and cause considerable foaming. Equilibrium is reached when dry ice begins to accumulate at the bottom of the dewar and there is a slow but steady stream of CO_2 bubbles rising from the bottom.
4. Use distilled water for the ice water and boiling water baths.

ANALYSIS. Prior to doing the analysis it will be wise to review Sections 1.2 and 1.5.

1. Assign the following bath temperatures:
   a. Liquid nitrogen = 77°K.
   b. Dry ice-trichloroethylene = 195°K.
   c. Ice water = 273°K.
   d. Boiling water = 373°K.
2. Plot one set of data (baths 1–7) in the form emf versus ΔT, where ΔT is the temperature difference in the two baths. Find, by a least-squares procedure, the best linear fit to these points. A computer program would be helpful. Compute the deviation of the measured emfs from the values predicted by the linear fit.
3. Using all the data for pairs 1, 5, and 4 (six measurements at each point), compute the average value of the emf and the standard deviation for each pair of baths. Compare these standard deviations with the deviations in step 2. The standard deviations are statistical estimates for the population distribution. Based on these estimates and a 95% ($\pm 2\sigma$) criterion for data rejection, would the emf predicted by the least-squares line be rejected? Is a linear fit between emf and ΔT valid for your data? Why?
4. Using the data from step 3 and bath 7, make a plot of emf versus t, where t is the centigrade temperature of the bath (use the ice bath as a reference). For baths below 0°C, use negative emf values. Include vertical bars indicating error limits. Examine this graph and comment on the linearity of emf versus t.
5. Construct a tangent to your curve at 77°K and determine over how large a temperature range a linear approximation would be valid. Such an approximation is used in Section 5.10.

3.3 Plate Characteristics of a Diode and Amplification of a Triode

The purpose of this experiment is to acquaint the student with a few aspects of electronic circuitry through a study of the voltage and current relationships of a vacuum diode and a vacuum triode tube. The experiment is very straightforward and requires only a small amount of time.

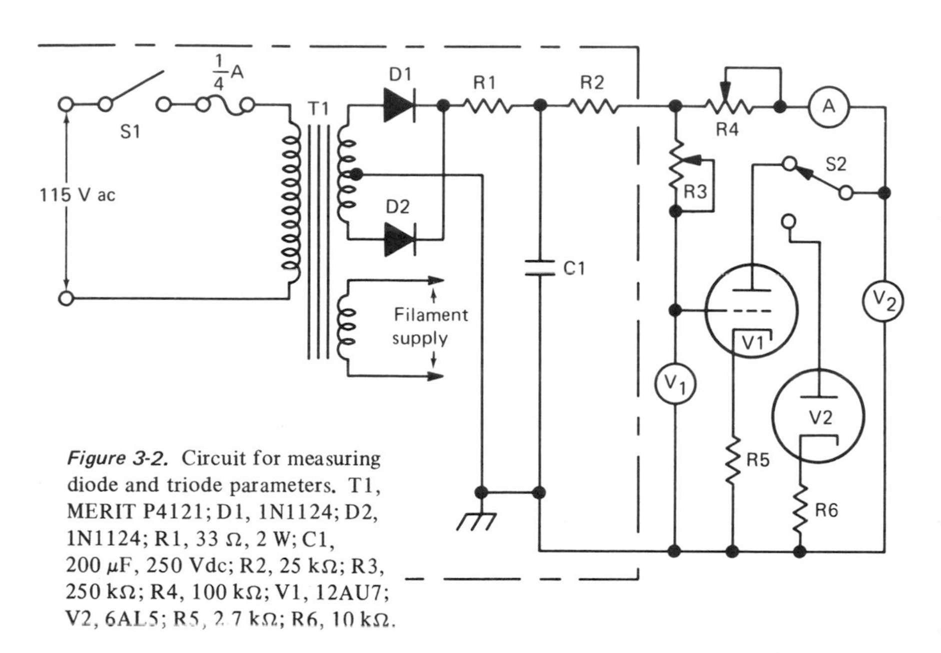

Figure 3-2. Circuit for measuring diode and triode parameters. T1, MERIT P4121; D1, 1N1124; D2, 1N1124; R1, 33 Ω, 2 W; C1, 200 μF, 250 Vdc; R2, 25 kΩ; R3, 250 kΩ; R4, 100 kΩ; V1, 12AU7; V2, 6AL5; R5, 2.7 kΩ; R6, 10 kΩ.

EXPERIMENT.* The circuit of Fig. 3-2 may be used to measure the plate characteristics of a vacuum diode or the amplification of a vacuum triode. Study the circuit and answer the following questions before performing the experiment:

1. What is the function of D1 and D2?
2. Why is T1 center-tapped?
3. What purpose does the R1-C1 combination serve?
4. What kind of circuit is located inside the dashed lines?
5. Comparing with Fig. 2-62, what does the circuit inside the dashed lines replace?
6. Which tube is the diode and which the triode?
7. What is the function of S2?
8. What are the functions of R3 and R4?
9. Why is R5 necessary?
10. What do the meters V_1, V_2, and A measure?

*Refer to Section 2.4 for a discussion of the features of this experiment.

PROCEDURE.

1. Characteristics of the vacuum diode:
   a. With power switch S1 off, plug the circuit into the 115V ac commercial supply.
   b. Turn R3 and R4 completely counterclockwise.
   c. Turn S2 to diode and turn S1 on.
   d. After a few minutes of warm-up, begin turning R4 clockwise, recording plate voltage and plate current at about eight intervals over the plate voltage range 0–80 V.
   e. Repeat step d.
   f. Get six independent readings with the plate at 40 V by turning R4 away from the 40-V setting between readings.
   g. Plot the plate characteristic of the diode. Comment on the linearity of the observed characteristic and on the uncertainty in the plate current.
   h. Suppose that an ac signal whose amplitude is 70V were applied to this diode. Sketch a graph of the input and the output voltage. Comment on the distortion of the output.
2. Amplification of the triode:
   a. With the circuit on as described above, turn R3 and R4 completely counterclockwise.
   b. Switch S2 to triode.
   c. Turn R4 clockwise until the plate current is 2 mA. Record the grid voltage and plate voltage.
   d. Keeping the plate current at 2 mA, find the variation of the grid voltage with the plate voltage. This is achieved by varying the plate voltage and then adjusting the grid voltage to return the plate current to 2 mA.
   e. Repeat steps c and d with constant plate currents of 2.5 and 3 mA. Plot the grid voltage versus the plate voltage for the three fixed plate currents.
   f. Keeping the grid voltage fixed at 3.5 V, measure the variation of the plate current with the plate voltage over the range 10–120 V on the plate. Plot the results.
   g. What is the interpretation of the slope of the graph from step f?
   h. What interpretation may be given to the slopes of the graphs from step e?

3.4 Analysis of Electronic Circuits with an Oscilloscope*

Troubleshooting is one very common application of an oscilloscope. Used in this way the scope becomes a device for investigating how a malfunctioning instrument is responding to various input signals. The scope can be used to trace a

*See Section 2.4 for a fuller discussion of how an oscilloscope operates.

certain input signal through the instrument; in the process the source of the malfunction can be located. In the experiment described here the goal is not troubleshooting but analysis—analysis of the characteristics of an audio amplifier. Figure 3-3 illustrates the system. The signal generator provides either sine or square waves, the latter being somewhat more useful. These ac signals in the audio range (20–20,000 Hz) are applied to the input of an audio amplifier. An elementary public address system amplifier is satisfactory, as is that described in Section 3.5. The real difference in amplifiers can be revealed by comparing the characteristics of such an amplifier with those of a well-designed amplifier in a record or tape player system. The amplifier output should be connected to a load whose input impedance matches the output impedance of the amplifier. This could be a loudspeaker or a resistor of the appropriate size. As shown in Fig. 3-3, the oscilloscope is connected across the input to the load and thus measures the voltage characteristics of the signals delivered to the load. These characteristics are to be compared with those applied to the amplifier input (the scope can be connected across the input for this purpose) and the amplification (gain) and distortion of the amplifier assessed. Because square-wave distortion is easier to recognize and because a square wave of a given frequency can be decomposed into a sine wave of that frequency and sine waves of harmonics of that frequency, the qualitative features of amplifier distortion are generally investigated with square-wave input. Figure 3-4 illustrates good amplification and uniform response over a range of frequencies [part (a)], distortion of low frequencies or poor low-frequency response [part (b)], and high-frequency distortion [part (c)]. In all these figures the leading edge of the wave form is assumed to be on the left.

A quantitative investigation of the gain of an amplifier and its frequency dependence may be carried out using either square- or sine-wave input.

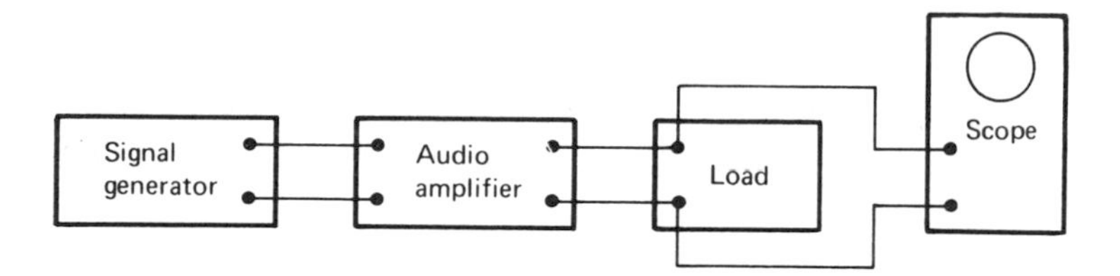

Figure 3-3. Apparatus for investigating amplifier characteristics.

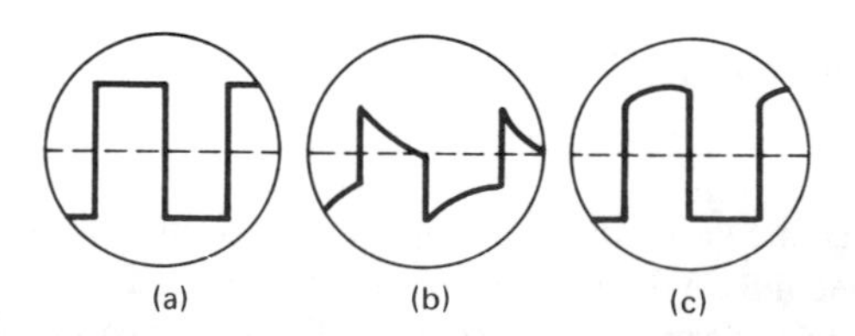

Figure 3.4. Schematic response of an amplifier to a square-wave input. (a) Uniformly good response; (b) poor low-frequency response; (c) poor high-frequency response.

PROCEDURE.

1. If the vertical amplifier of the scope is not calibrated, carry out this operation first so that in the actual experiment you will know how to assign voltages to various observed peak heights. Most scopes have an internal reference voltage (a 1.0- or 0.5-V peak-to-peak sine wave is typical) which can either be switched or connected with wire to the vertical amplifier. With this reference connected, the vertical amplifier gain should be adjusted so that it corresponds to a 1-V/cm vertical deflection on the visual scope output when the vertical attenuator (voltage control) on the control panel is set at 1V. Alternatively and for more reliable measurements over a wider voltage range a variable and accurate external ac voltage supply should be applied to the vertical input and a peak height versus applied voltage calibration curve should be constructed in the range where measurements are to be made. An accurate calibration is not necessary for this experiment.
2. Once the scope is calibrated, connect the apparatus as shown in Fig. 3-3 except connect the scope across the amplifier input. Measure the voltage applied to the amplifier and record in graphical form the applied waveform.
3. Connect the scope across the load and measure the output voltage and waveform.
4. Repeat steps 2 and 3 for several frequencies in the range 20–20,000 Hz.
5. If time permits, repeat steps 2 and 3 for signals of various amplitudes.

ANALYSIS. Report the amplifier gain and distortion characteristics as a function of input signal amplitude and frequency. Comment on methods for improving the frequency response if it is poor.

3.5 Elementary Circuit Construction. An Amplifier and Power Supply

This experiment is in two parts. In the first section a power supply is constructed which is suitable for providing the bias voltage of the following triode amplifier circuit. Although these tube-type circuits are seldom used in audio amplifiers now that transistor circuits are available, these circuits do provide an interesting introduction to circuit construction.

CONSTRUCTION OF A FILTERED B+ POWER SUPPLY*. In this experiment the goal is to construct and test an elementary dc power supply which is suitable for supplying the B+ requirement of an amplifier. In fact this power supply is quite suitable for the amplifier experiment which follows in the next section. The circuit is a standard R-C-filtered full-wave rectifier and is shown in Fig. 3-5. As expected the components include a center-tapped step-up transformer including a 6.3-V winding to the heat tube filaments, a tube containing two diode

*Refer to Section 2.4 for a discussion of the circuit constructed here.

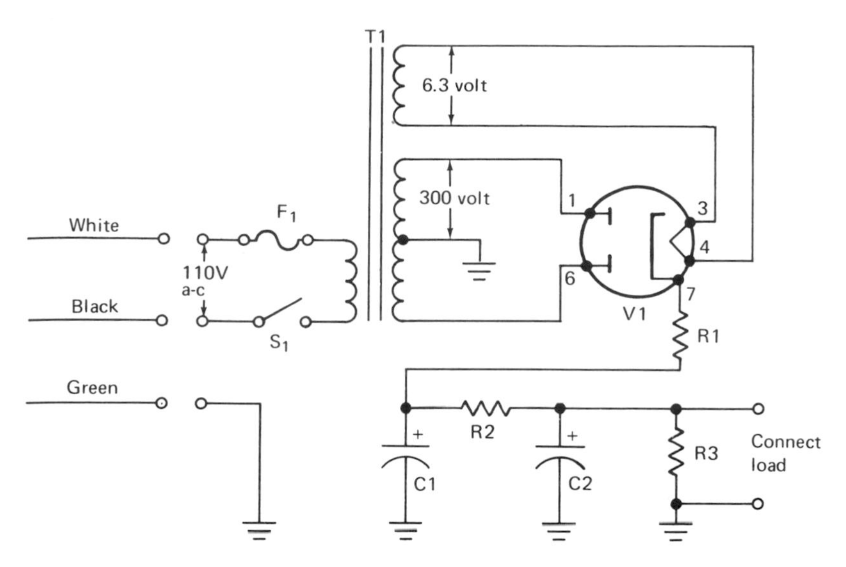

Figure 3-5. R-C-Filtered full-wave power supply. S1, Single-pole toggle switch; F1, fuse (amp); T1, 300-0-300 transformer with 6.3-V winding; V1, 6 X 4 tube; R1, 500 Ω, 5 W; R2, 500 Ω, 3 W; R3, 1 MΩ, 1 W; C1, 40 μF, 450 V dc; C2, 40 μF, 450 V dc.

circuits, and an R-C filter section which is terminated in a bleeder resistor. The purpose of the last is to provide a path for the discharge of the capacitors even in the absence of an intended load (say an amplifier).

In considering this circuit it is obvious that the transformer and diode characteristics must be compatible. The characteristics of the 6X4 are detailed in any standard vacuum tube manual as follows:

heater (filament) voltage:	6.3 V
heater current:	0.6 A
peak inverse plate voltage (max):	1250 V
steady peak plate current (max):	245 mA
ac plate supply voltage:	0–400 V
dc output voltage (max):	350 V
dc output current (max):	45 mA at each plate

In a routine application as a B+ supply an output voltage near 250 V would be a typical requirement and is well within the range of the above parameters provided the transformer is capable of supplying the ac plate voltage and current. The voltage available at the output will be somewhat lower than that applied to the input of the filter because of voltage drops in R1 and R2.

In constructing the supply a 4 X 6 X 3 in. chassis is satisfactory with T1 and V1 mounted on its top and the filter components attached to a grounded four-lug terminal strip beneath. The toggle switch, fuse holder, and three-wire ac line cord should be attached to or through the chassis walls. Two output terminals should be provided in one of the walls—a black insulated banana jack for the ground terminal and a red one for the "hot" line would be standard. Combination binding posts which accept either banana-type or lug-type connectors would be more desirable. In connecting the 110-V ac power cord, the black (standard hot line) should be connected to the switch, the white (standard neutral line) to the fuse holder, and the green to ground. At the chassis a rubber grommet should be used in the hole through which the line cord passes. Grommets should also be used in the holes through which the transformer wire passes. In summary the parts needed (in addition to those noted in Fig. 3-5) are

1. Chassis
2. Line cord
3. Output terminal jacks, two each
4. Fuse holder and 5-A fuse
5. Four-lug grounded terminal strip
6. Seven-pin miniature tube socket
7. Ground lug
8. Rubber grommets, three each
9. A few feet of hookup wire

General procedure.

1. Lay out the chassis so the mounted parts will be conveniently accessible.
2. Drill or punch holes for all the parts which are to be mounted including one grounding lug which is not discussed above.
3. Mount the parts.
4. Connect and solder the components as indicated in Fig. 3-5. Note that the tube pins are marked on the figure. All the ground points in the filter section should be connected to the single grounded terminal of the four-lug strip, while the transformer center-tap and the green line cord should be connected to the other ground lug. The filter section connections should all be made at the four-lug terminal strip.

Note: Use resin core solder for all circuit wiring. Acid core solder is deleterious to circuit components. Making good electrical connections is primarily a matter of experience; two things indicate a good connection: a bright joint and a joint where the solder flowed as opposed to "balling up."

After the wiring is completed the following tests should be made:

1. Using a VOM (volt-ohmmeter), check the voltages at the primary and secondary of the transformer (ac voltages) and at various points along the filter section (dc voltage) and at the output (dc voltage). Compare these voltages and account for their differences.

2. Using a scope,* check for 60-cycle ac ripple in the dc section at various places along the filter section and at the output. *Remember:* In all these measurements the secondary voltage is on the order of 250 V.

CONSTRUCTION OF A SIMPLE SINGLE-STAGE TRIODE AMPLIFIER. The purpose of this enterprise is to construct a simple triode amplifier suitable for processing audio input signals. The circuit will appear like the first stage of Fig. 2-62 in which the grid bias is achieved by connecting a resistor in the cathode-to-ground part of the circuit. The actual circuit makes use of the 6C4 triode or its very common dual triode equivalent, the 12 AU7A. Here only half of the capabilities of the dual triode is used. Assuming the presence of an input signal, the amplifier itself is comprised of a tube (central ingredient), a filament power supply (usually 6.3 V), a bias (B+) supply (on the order of a few hundred volts generally), and appropriate passive elements for coupling the signal in and out. Following Fig. 2-62 and asuming that no signals whose frequencies are less than 100Hz are studied, the following passive element parameters are satisfactory:

C1:	0.032 μF	R1:	0.1 MΩ
C2:	0.032 μF	R2:	1.9 kΩ
C3:	1.9 μF	R3:	0.1 MΩ

For either of the above tubes the filament supply should deliver 6.3 V at 0.15 mA. The plate (B+) supply should be 250 V dc and can be that discussed in the previous section. The necessary filament voltage can be derived from the 6.3 V secondary of the B+ supply transformer.

The bottom views of the pins of the 6C4 and 12AU7A tubes are shown in Fig. 3-6. They are obviously different and so some care needs to be exercised in making the proper connections to either a seven- or nine-pin tube socket. Assuming the 12AU7A, Fig. 2-62 shows the completed diagram.

Follow the general prescription given for the power supply in setting up and constructing the actual circuit. Once it is together, connect the B+ supply and apply an audio signal as outlined in Section 3.4 and finally test the amplification of the triode as a function of the circuit parameters. Use a 1-kΩ resistor as the load for the amplifier output.

*See Section 3.4 for further details on the use of an oscilloscope.

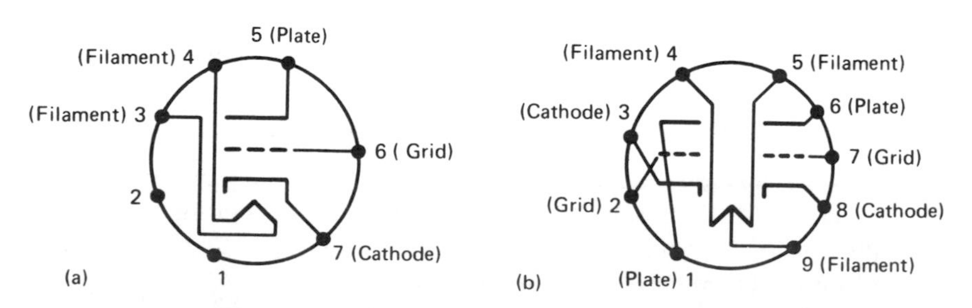

Figure 3-6. Pin connections for (a) 6C4 and (b) 12AU7A.

Chapter 4

Physical Properties of Gases

4.1 Vacuum Practice, Gas Thermometry, and Equations of State

The purpose of this experiment is to acquaint the student with vacuum apparatus, to calibrate a gas thermometer, and to use its properties to investigate the validity of various equations of state. It is assumed that the student is not familiar with vacuum apparatus; therefore, very explicit instructions are given.

THEORY. The thermodynamic and international temperature scales are discussed in Section 2.2, and there the utility of the latter scale is pointed out. The helium gas thermometer has been used to measure accurately the thermodynamic temperatures of the fixed points on the international scale. Assuming the validity of the ideal gas law it can be easily shown that the ideal gas temperature scale and the thermodynamic temperature scale are identical.

Historically, the Celsius or Centigrade temperature scale is based on two fixed points: the ice point ($0^\circ C$) and the steam point ($100^\circ C$) at 1 atm pressure. We shall now consider a method by which an absolute temperature scale may be established using a gas thermometer. In principle, this scale differs from the International Temperature Scale but is identical to the thermodynamic scale. We refer to the latter as absolute temperature.

Consider an experiment performed using a fixed weight and volume of gas. Let p_i and T_i be the pressure and absolute temperature, respectively, at the ice point, and let p_s and T_s be the same quantities at the steam point. If some of the gas is now removed and the above experiment is repeated, new values for p_i

and p_s will be obtained but T_i and T_s will remain the same. Repeating the above experiments with smaller and smaller weights of gas, but keeping the volume constant, results in a set of data which can be plotted in the form $(p_s/p_i)_v$ versus p_i. This will furnish a graph similar to Fig. 4-1. In all these measurements, the pressure above the steam and ice baths is assumed to be 1 atm. Extrapolation (the dotted portion of the curve) to $p_i = 0$ furnishes the limiting ratio of the steam point pressure to the ice point pressure as the total amount of gas goes to zero. Repeating the experiment with a different gas may furnish a curve with a different slope but it will have the same intercept. This common ordinate at the intercept is found to be 1.366. The absolute temperature scale is defined in terms of this value as

$$\frac{T_s}{T_i} \equiv \lim_{p_i \to 0} \left(\frac{p_s}{p_i}\right)_v = 1.366 \tag{4-1}$$

To establish the size of the absolute degree the difference between T_i and T_s is set equal to 100°. That is,

$$T_s - T_i = 100°\mathrm{A} \tag{4-2}$$

Equations (4-1) and (4-2) contain two unknowns and can be solved to furnish

$$T_i = 273.15°\mathrm{A} \tag{4-3}$$

Observe that the definition made in Eq. (4-2) guarantees that the Centigrade and absolute degrees will be of exactly the same size. Hence, 273°A corresponds to 0°C. Also observe that the absolute-temperature-scale-defining equations [Eqs.

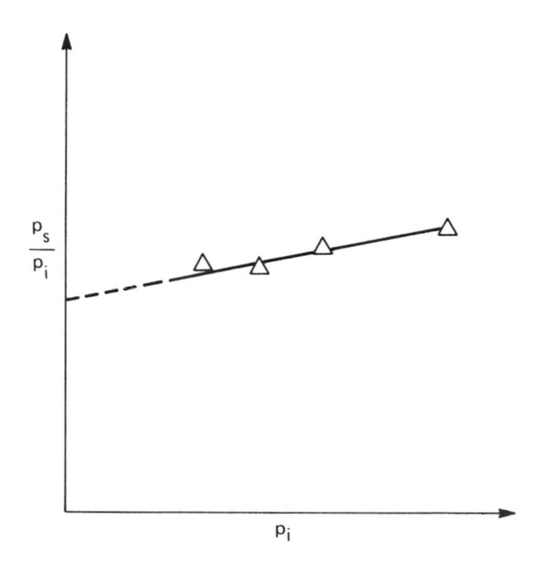

Figure 4-1. Variation of the ratio of steam point pressure to ice point pressure with the ice point pressure (at constant volume).

(4-1) and (4-2)] do not depend on the existence of any particular kind of gas or upon any assumptions about the pressure or volume approaching zero as the temperature approaches absolute zero.

From a macroscopic point of view, and in the absence of externally applied electric and magnetic fields, the complete characterization of a fixed amount of a pure substance is given mathematically by a relationship among the three variables pressure, volume, and temperature. This relation is called the equation of state.

For gases, the simplest equation of state is the ideal gas law:

$$p\bar{V} = RT \tag{4-4}$$

where $\bar{V}$ is the molar volume, p the pressure, R the gas constant in units compatible with those used for p and $\bar{V}$, and T the absolute temperature. The ideal gas law adequately describes gases under conditions where the intermolecular forces can be neglected. This situation occurs when the molecules are, on the average, far apart (i.e., low pressure and high temperature).

When the gas molecules begin to interact strongly with each other, deviations from the ideal gas law begin to be observed. Under these conditions, other equations of state must be used to describe the gas. Some examples are

1. van der Waals' equation:

$$\left(p + \frac{a}{\bar{V}^2}\right)(\bar{V} - b) = RT \tag{4-5}$$

where a and b are constants for a given gas. The parameter b takes into account the finite volume of the gas molecules, and the parameter a takes into account intermolecular forces which are not in balance near the walls of the container.

2. The virial equation of state:

$$\frac{p\bar{V}}{RT} = 1 + B_p(T)p + C_p(T)p^2 + \ldots \tag{4-6}$$

or

$$\frac{p\bar{V}}{RT} = 1 + \frac{B(T)}{\bar{V}} + \frac{C(T)}{\bar{V}^2} + \ldots \tag{4-7}$$

Equation (4-6) is a power series in the pressure p, whereas Eq. (4-7) is a power series in the reciprocal of the molar volume, $1/\bar{V}$. Observe that in the limit of low pressures and high temperature (large $\bar{V}$) both Eqs. (4-6) and (4-7) reduce to the ideal gas law.

Examination of Eqs. (4-4) – (4-7) reveals that the pressure of a fixed amount of gas under conditions of constant volume can be directly related to the absolute temperature. This has been pointed out earlier as being the property which permits one to use a gas as a thermometer after suitable calibration.

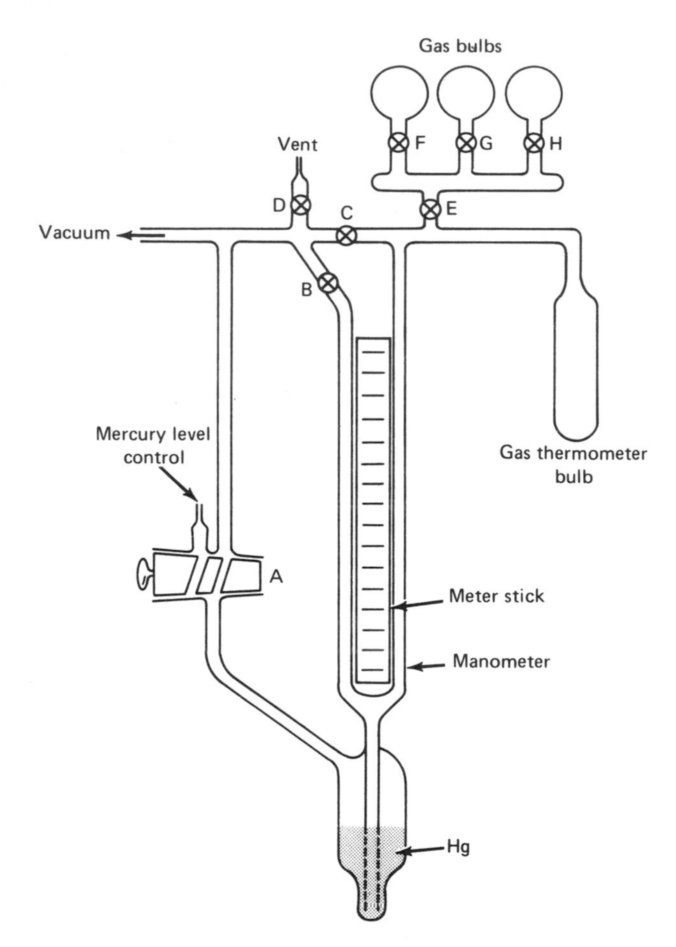

Figure 4-2. Gas thermometry apparatus.

EXPERIMENTAL PROCEDURE. Section 2.1 should be read prior to this experiment. The apparatus is shown in Fig. 4-2. It is to be used with great care in order to avoid breaking the glass and scattering mercury throughout the system. It is suggested that the experimenter study this set of instructions, making certain that the manipulations are understood, before actually performing the experiment. *Caution:* Always turn stopcocks slowly.

1. Evacuate the apparatus. The pump is connected to the tube labeled Vacuum in Fig. 4-2. In carrying out this operation keep stopcocks D, F, G, and H closed and stopcocks A, B, C, and E open. Carefully note that stopcock A has two orientations. When the system is being evacuated it should be turned as shown in Fig. 4-2. Also notice in Fig. 4-2 that the mercury reservoir is constructed so that the small inner tube is isolated from the large reservoir.

2. After evacuation, rotate stopcock A through 180° so as to admit air into the mercury reservoir. This will cause the mercury level to rise in the manometer tube. When the level has been raised to about 20 cm on the meter stick, close stopcock A and measure the heights of the two mercury columns, keeping account of which is the right-hand side and which is the left. This gives you a correction to be applied to all the pressure determinations and furnishes a base reference level for the experiments to be carried out at constant volume. Even though the pressure on both sides of the manometer is the same, the mercury column heights may differ because of adhesion of mercury to the glass.
3. Close stopcock C.
4. Close stopcock E, and then open momentarily and close stopcocks F, G, or H depending on which gas you want to add. Gas is now confined to the volume bounded by valves E, F, G, and H.
5. Open and close stopcock E. Then note roughly the difference in the heights of the mercury columns to see if there is enough gas pressure ("enough" will be specified later for each experiment). If not, reopen and close the stopcock to the desired gas bulb and reopen and close stopcock E. Repeat until sufficient gas pressure is established.
6. When the desired pressure is achieved, be certain stopcock E is closed; then adjust the mercury level to give the same volume in the gas thermometer as originally (i.e., adjust the right-hand column to the same level as in the initial measurement). This will furnish a reading at room temperature.
7. For readings at other temperatures, simply surround the gas thermometer bulb with the desired bath and repeat the above measurements. Remember to correct the readings using the heights initially obtained.
8. When a set of readings is completed, repeat step 1 to evacuate the apparatus.
9. When all the experiments are completed, evacuate the apparatus according to step 1 and vent the system carefully.

TEMPERATURE BATHS. The experiments outlined below call for several temperature baths. The method of their preparation and use is outlined below.

1. Ice point: Mix ice and distilled water in a dewar flask to make a very slushy mixture that can easily be placed around the gas thermometer bulb. Exercise caution here because only a small amount of torque may break the glass tubing.
2. Steam point: Figure 4-3 illustrates the steam generator which is to be used. Steam should be passed through the steam jacket at a rate such that excess steam emerges from the exhaust port but not at a high rate. The rubber tubing connecting the boiling flask with the jacket should be oriented so that no condensed water can remain in it. The exhaust of the steam jacket should be connected to a large beaker or to a drain. When conducting a measurement at the steam point, do not insert the rubber stopper into the neck of the boiling flask until the water is boiling vigorously.

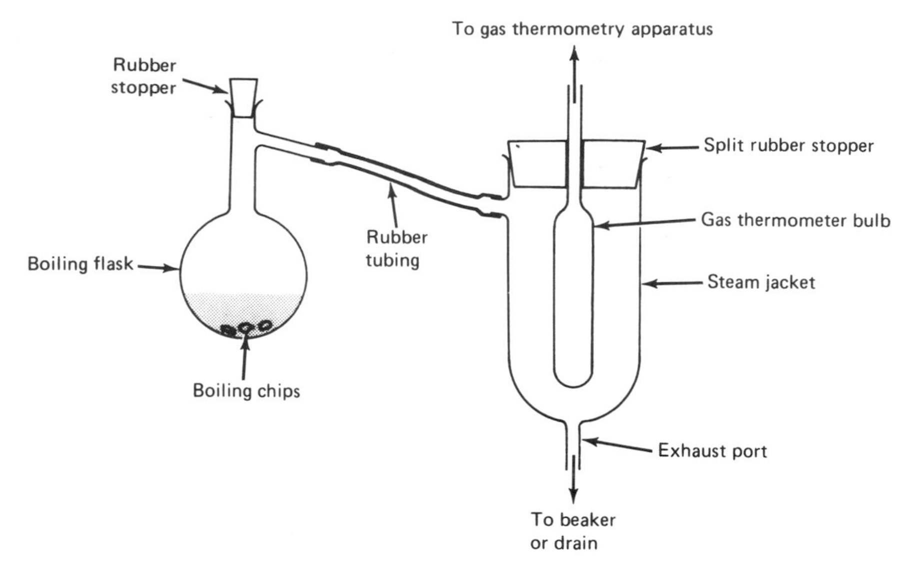

Figure 4-3. Steam generator.

3. Dry ice-acetone or -trichloroethylene: Powdered dry ice should be slowly added to the acetone or trichloroethylene in a dewar flask with stirring. Slow additions are necessary to avoid violent bubbling of the mixture. The dry ice should be added until a small amount remains at the bottom of the dewar. Once the dewar flask is around the gas thermometer bulb, small amounts of dry ice can be added as necessary to maintain the constant temperature.
4. Liquid nitrogen: The liquid nitrogen should be contained in a dewar flask and placed around the gas thermometer bulb very slowly.

EXPERIMENTS

1. Using a pressure (at room temperature) of about 20 cm Hg, run a sample of N_2. Measure the pressure of this sample (at constant volume) at room temperature, the ice point, and the steam point.
2. Repeat step 1 using 10 cm Hg of N_2 and also measure the pressure at the temperature of the dry ice-acetone mixture and at liquid nitrogen temperatures.
3. Using approximately 15 cm Hg of argon, measure pressures at room temperature, dry ice-acetone, and liquid nitrogen points. Repeat these measurements in a different order to check for reproducibility. This will give an idea of the confidence to be placed in the readings.
4. Using about 20 cm Hg of CO_2, measure the pressures at dry ice-acetone, ice point, liquid nitrogen, and room temperatures.

5. Make an estimate of the volume of the gas thermometer bulb and of the tubing connecting the bulb to the manometer. Any part of the system which is not in contact with the constant temperature bath is called *dead volume* and contributes some error to the temperature determinations.
6. Using a mercury-in-glass thermometer, measure the temperature of the room.

CALCULATIONS

1. In a handbook look up values of the constants appearing in van der Waals' equation for CO_2. Construct a graph of the variation of pressure with temperature for both Eqs. (4-4) and (4-5). To obtain $\overline{V}$, use the pressure of CO_2 which was used in Experiment 5 and assume that the ideal gas law is valid at room temperature and this pressure. Extend the graphs until experimentally detectable deviations from the ideal gas law are observed.
2. Using van der Waals' constants from part 1 and using the volume estimate from Experiment 5, how large would the pressure of CO_2 need to be before a 5% deviation from the ideal gas law would be obtained at room temperature?
3. From the results of parts 1 and 2, comment on the feasibility of observing nonideal behavior using the apparatus of this experiment.
4. According to the volume estimates, how large an error does the dead space introduce into the calculations? Assuming that the ideal gas law is valid, derive an expression which could be used to correct the results for the effect of the dead volume.
5. Another source of error is the slight variation with temperature of the volume of the gas thermometer bulb. The coefficient of cubical expansion for Pyrex glass is $0.25 \times 10^{-6}/°C$. The cubical expansion coefficient is defined as the increase in volume per unit volume per degree Centigrade. For example, if the volume at 0°C is 2.00000000 cm^3, volume at 1°C will be 2.00000050 cm^3. Estimate the error introduced in the present experiments by assuming that the volume was constant. Base the calculation on the volume change expected in going from the lowest to the highest temperature used.
6. Using the results of Experiments 1 and 2, plot $(p_s/p_i)_v$ versus p_i and extrapolate to $p_i = 0$ to find the limiting value. How does the value compare with that of Fig. 4-1? What is the slope of the plot? What slope is predicted for a gas obeying the ideal gas law?
7. Using the results of Experiment 1 and assuming the validity of the ideal gas law, derive an expression which will furnish T as a function of p [i.e., use Eqs. (4-1) - (4-4) to calibrate your thermometer]. From the resulting equation and the experimentally measured (Experiment 2) pressures using room, dry ice-acetone, and liquid nitrogen baths, calculate the temperatures of these three baths.

8. From the results of Experiment 3, estimate the overall accuracy of the pressure measurements. Compare this with the sum of the errors predicted in Calculations 4 and 5. Would a correction for the effects of dead volume and volume expansion improve your results? Why?
9. Using the temperatures determined in Calculation 7, plot p versus T for Experiments 3 and 4. Discuss the validity of the ideal gas law in these experiments, explaining any deviations which may occur.

Suggestion: With a little thought prior to this experiment, the taking of the data should be very easy. Plotting the graphs required is very straightforward and can be done while the experiment is in progress, thus shortening the total time required.

4.2 Speed of Sound in an Ideal Gas

The purpose of this experiment is to measure the speed of sound in several different ideal (or nearly ideal) gases at room temperature and ambient pressure. From these data and a simple thermodynamics equation the ratio γ of the heat capacity at constant pressure to the heat capacity at constant volume can be calculated.

THEORY. In physical problems, sound has a technical definition, namely, an alternating stress wave which propagates through a medium. In gases, sound is a pressure wave, implying alternating compressions and rarefactions of the gas which are induced by a mechanically vibrating source and which propagate through the gas with some finite speed. For example, when you speak the oscillations of your vocal cords induce propagating waves in the air. You are heard when these waves interact with and cause vibrations of the eardrum of another individual. Sound waves travel (propagate) relatively slowly, much more slowly than light waves, for example. If you go to a track meet and stand at the finish line of the 100-m dash, you see the smoke from the starter's gun before you hear the discharge. If you hear a fast-flying jet, you detect the sound as coming from far behind where the jet is seen because it moves a significant distance in the time required for the sound to reach you.

We shall now examine some of the thermodynamic aspects of sound-wave propagation. Suppose that a sound wave from some source is propagating through some gas in a given direction x. At any instant of time a snapshot of this wave will appear as alternating regions of varying gas density ρ or pressure p beginning at the source and extending along x. If we find one position of high density and follow it with closely spaced later snapshots, it will appear to move away from the source. The rate of this motion is the speed of sound. It is important to recognize that the individual molecules themselves do not move away from the source at this rate. That is, there is no macroscopic flow of gas here;

rather there are localized and alternating pressure variations and gas motion. From a thermodynamic point of view we assume that these local processes are adiabatic (Q = 0). If the gas is also ideal, then we can show that $p\overline{V}^{\gamma}$ is a constant for an adiabatic process, where $\overline{V}$ is the molar volume and γ is the heat capacity ratio C_p/C_v.

By assumption we have

$$Q = 0 \tag{4-8}$$

and initially

$$p_i\overline{V}_i = RT_i \tag{4-9}$$

Using the first law of thermodynamics and the definition of C_v in terms of the internal energy E in conjunction with Eqs. (4-8) and (4-9), an analysis of the process of compression and rarefaction furnishes

$$p_i\overline{V}_i^{\gamma} = p_f\overline{V}_f^{\gamma}$$

where f refers to the final state.

We shall now turn our attention to finding a relation between the speed of sound, u, and the pressure variations of the propagating wave. To make progress we look schematically at one full wavelength, namely a compression-rarefaction pair as diagrammed in Fig. 4-4. If we require no net gas flow over several wavelengths, we may qualitatively regard as fixed the number of molecules bounded by the imaginary walls at A and B in Fig. 4-4. Within this region, however, the molecules flow back and forth as time passes to give rise to the pressure wave. Given the situation described in Fig. 4-4 we ask how this wave will evolve in time. Focusing attention at position x_0 we see that the higher pressure toward lower x will result in a local net acceleration of the molecules past x_0 toward higher x. If the gas density at x_0 is ρ_0 and the speed at x_0 is u_0, then, since there is no creation or loss of mass, the following conservation equation holds:

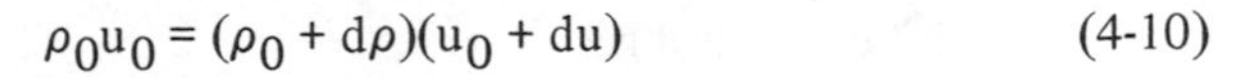

$$\rho_0 u_0 = (\rho_0 + d\rho)(u_0 + du) \tag{4-10}$$

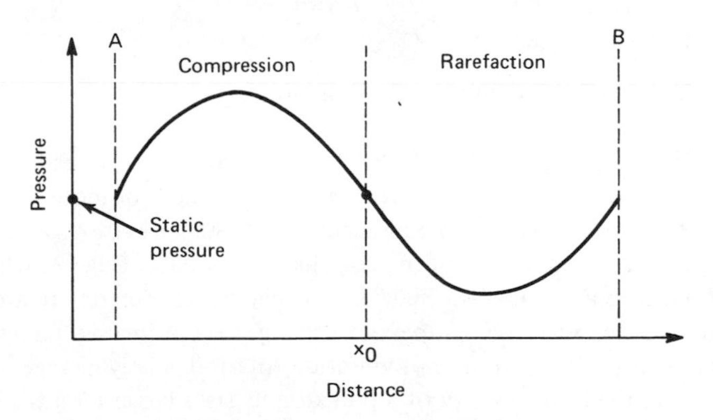

Figure 4-4. Pressure profile over one wavelength of a sound wave.

This equation requires that a local increase in velocity be offset by a local decrease in density to keep the product of the two constant. Furthermore, as the gas passes x_0, its increase in momentum implies an increase in the pressure p. The differential change per unit time in the momentum of the gas passing through a unit area perpendicular to x_0 is equal to

$$(\rho_0 + d\rho)(u_0 + du)^2 - \rho_0 u_0{}^2 = -dp \tag{4-11}$$

Multiplying out Eq. (4-11), keeping only first-order differentials, and substituting from Eq. (4-10) to eliminate du, we obtain

$$u_0{}^2 = \frac{dp}{d\rho} \tag{4-12}$$

This equation furnishes the velocity at any position in terms of the gradient of pressure with respect to density. It is apparent that the fluid is at rest but that the sound wave propagates with speed u_0.

The equation of state of the gas being considered and the kind of process assumed will determine $dp/d\rho$. For an ideal gas undergoing an adiabatic rarefaction-compression cycle we have already shown that $p\bar{V}^\gamma$ is constant. Since the density is $\bar{V}^{-1}$ M, where M is the gram molecular weight, we can write

$$p = A\left(\frac{\rho}{M}\right)^\gamma \tag{4-13}$$

where A is a constant. Differentiating furnishes

$$\frac{dp}{d\rho} = \frac{\gamma A}{M^\gamma}\rho^{\gamma-1} = \gamma\frac{p}{\rho} = \frac{\gamma p\bar{V}}{M} = \frac{\gamma RT}{M} \tag{4-14}$$

Combining Eqs. (4-12) and (4-14) furnishes a thermodynamic expression for the speed of sound in an ideal gas:

$$u_0 = \sqrt{\frac{\gamma RT}{M}} \tag{4-15}$$

If u_0 and T are measured, then γ can be computed.

THE EQUIPARTITION PRINCIPLE. The heat capacity and the heat capacity ratio of a molecule are intimately related to the number of modes of motion available. For example, polyatomic molecules with their vibrations and rotations have more available modes of motion than do monatomic gases where only translational modes are available. Since heat capacity is intimately related to temperature changes and since temperature changes are intimately related to translational energy changes (and not vibration-rotation energy changes), it becomes clear that the heat capacity of a polyatomic gas is larger than the heat capacity of a monatomic gas if some heat is used to excite vibrations and rotations. The equipartition principle assumes that energy put into a gas is distributed in a particular way among all the degrees of freedom. It asserts that each degree of translational freedom carries $\frac{1}{2}$RT cal/mole, that each degree of rotational freedom also carries $\frac{1}{2}$RT, while each degree of vibrational freedom carries RT cal/mole. The total energy of a gas molecule is comprised of its total kinetic energy and its total potential energy, or, viewed another way, it is the sum of the translational, rotational, and vibrational energies. Every gas-phase atom or molecule has 3 degrees of translational freedom. An atom has no rotational and vibrational degrees of freedom, while a diatomic molecule has 2 degrees of rotational freedom and 1 degree of vibrational freedom. Linear polyatomic molecules have 2 degrees of rotational freedom and 3N - 5 degrees of vibrational freedom, where N is the number of atoms in the molecule. Nonlinear polyatomics have 3 degrees of rotational freedom and 3N - 6 degrees of vibrational freedom.

According to the equipartition principle each degree of freedom is fully active. Thus, for a diatomic molecule the total energy is $\frac{7}{2}$ RT; $\frac{3}{2}$ RT from translation, RT from rotation, and RT from vibration. The constant-volume heat capacity is the derivative of E with respect to T and is thus $\frac{7}{2}$R.

The above discussion assumes, of course, that every degree of freedom can absorb energy continuously. This is not true, especially for vibrations and rotations, where the quantized energy levels are separated by energies often greater than a *thermal quantum* of kT. Unless kT is about equal to or exceeds the energy level spacing, that particular channel will be closed and the heat capacity will be lower than predicted by the strict application of the equipartition principle. Since rotational levels are more closely spaced than vibrational levels, we expect the rotational channels to open at lower temperatures than vibrational channels. As a general rule rotations will be open at room temperature except for molecules whose reduced mass is very low (hydrides), whereas vibrations will normally be closed unless they are of very low frequency.

EXPERIMENT. The apparatus is shown in Fig. 4-5 and is comprised essentially of a glass tube (100 cm long × 6 cm in diameter) containing the gas with one end closed by a small radio speaker (2.5 in.) connected to an oscillator operating at a fixed frequency near 1000 Hz. The oscillator frequency is read at various intervals throughout the experiment using the counter. The spherical glass bubble keeps gas off the speaker, seals the end of the tube, and yet transmits the sound waves. A small microphone attached to a movable tube serves as the detector, and its output is amplified and detected with an oscilloscope. A second movable rod carries a Teflon ring, which fits snugly inside the glass tube. The end of the glass tube opposite the speaker is sealed off with a plate. Valves are provided for the admission and removal of the gas. The aluminum end plates can be readily sealed to the glass with epoxy resin. The rod carrying the Teflon ring and the tube carrying the microphone are interfaced with the end plate in a way which allows linear motion and preserves a fairly good vacuum seal. This can be accomplished using snugly fitting Teflon bushings attached to the end plate.

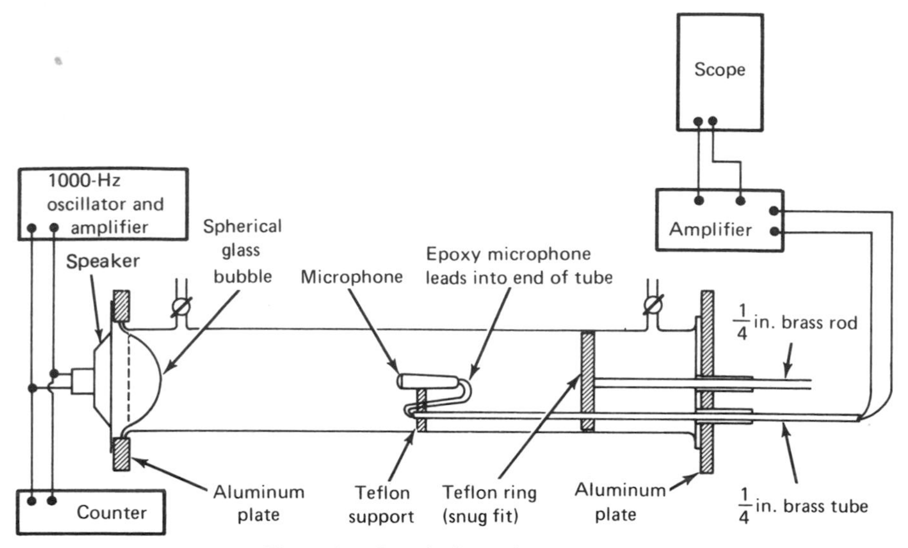

Figure 4-5. Speed of sound apparatus.

In operation the system may be evacuated and then filled with gas through one of the valves, or the system may simply be flushed with both valves open after which both valves are closed to contain the sample for study. All the electronic apparatus should then be turned on and the scope adjusted so it will display a 1000-Hz sine wave. In the experiment a sound-wave cavity is defined by the distance from the speaker to the movable Teflon ring, and this cavity is tuned so that standing waves are formed; that is, pressure waves reflecting off the Teflon ring and moving toward the speaker are in phase with pressure waves moving toward the Teflon ring. Since the speed of sound is uniquely determined by the molecular weight, temperature, and heat capacity ratio of the gas and since the frequency is uniquely determined by the sound source, the wavelength is fixed, and only certain positions of the Teflon ring will tune the cavity. Because both ends of the column are closed, the differential element of gas at these ends has zero velocity and zero displacement but a maximum density and pressure for constructive interference.

To tune the cavity, set the microphone at the midpoint of the cavity and record the amplitude of the oscilloscope output. The cavity midpoint will be a position of maximum pressure or minimum pressure if the cavity is tuned. Why? Slightly lengthen or shorten the cavity by adjusting the position of the Teflon ring. Move the microphone to the new midpoint. If the oscilloscope output increases, a maximum in the standing wave is close, whereas if it drops, a minimum is near. Repeat this procedure until the cavity length is such that the detector indicates a maximum or a minimum pressure at the cavity midpoint.

Leaving the Teflon ring fixed, move the microphone systematically over the length of the cavity, recording oscilloscope amplitude as a function of distance from the source. A graph of the results will locate the maxima (antinodes) and minima (nodes) in the standing pressure wave. The distance between nodes is one-half wavelength. A wide variety of gases can be used; helium, nitrogen, carbon dioxide, methane, and sulfur hexafluoride are suggested.

Measure the temperature of the ambient air in which the glass tube is located.

ANALYSIS

1. For each gas, calculate the speed of sound. Remember, speed equals frequency multiplied by wavelength.
2. Calculate the heat capacity ratio, γ, for each gas.
3. Predict C_p and C_v for each gas using the principle of equipartition of energy. Remember that CO_2 is linear.
4. Compare the results of steps 2 and 3 with literature values. Discuss deviations realistically, considering the uncertainties in the experimental measurement and comparing kT with the vibrational and rotational energy level spacings of the molecules.
5. Estimate the uncertainty in the computed values of the sound speeds. Keep in mind that the wavelengths determined involve the location of several points and that they do not all enter the calculation independently.

REFERENCES

Section 4.1

Many texts deal with the properties of gases and the theoretical models used in their description. Among them are

J. O. Hirschfelder, C. F. Curtiss, and R. B. Bird, *Molecular Theory of Gases and Liquids* (Wiley, New York, 1954).

T. L. Hill, *Lectures on Matter and Equilibrium* (Benjamin, Reading, Mass. 1966).

J. Jeans, *An Introduction to the Kinetic Theory of Gases* (Cambridge University Press, New York, 1960).

F. W. Sears, *An Introduction to Thermodynamics, The Kinetic Theory of Gases, and Statistical Mechanics* (Addison-Wesley, Reading, Mass., 1953).

Section 4.2*

J. H. Jeans, *The Dynamical Theory of Gases* (Dover, New York, 1954) (the first edition of this book appeared in 1916; it is a classic introductory text); H. E. Emmons (ed.), *Fundamentals of Gas Dynamics* (Princeton University Press, Princeton, N.J., 1958); S. Chapman and T. G. Cowling, *The Mathematical Theory of Non-Uniform Gases* (Cambridge University Press, New York, 1939) (this is a classic); J. O. Hirschfelder, C. F. Curtiss, and R. B. Bird, *Molecular Theory of Gases and Liquids* (Wiley, New York, 1954) (a thorough treatise covering theory and comparisons with experimental data). Deal with research continuing to be done in gas dynamics.

Handbook of Chemistry and Physics (Chemical Rubber Publishing Company, Cleveland) (data given are specific heats and require conversion to molar heats); "Selected Values of Chemical Thermodynamic Properties," *National Bureau of Standards Circular 500* (Superintendent of Documents, U.S. Government Printing Office, Washington, D.C.); "JANAF Thermochemical Tables," 2nd ed., *National Bureau of Standards Publications NSRDS-NBS37* (Superintendent of Documents, U.S. Government Printing Office, Washington, D.C.) (the best source). Give heat capacities.

G. Herzberg, *Spectra of Diatomic Molecules* (Van Nostrand Reinhold, New York, 1950); G. Herzberg, *Infrared and Raman Spectra of Polyatomic Molecules* (Van Nostrand Reinhold, New York, 1945); T. Shimanouchi, "Tables of Molecular Vibrational Frequencies," *National Bureau of Standards Publications NSRDS-NBS6, NSRDS-NBS11,* and *NSRDS-NBS17* (Superintendent of Documents, U.S. Government Printing Office, Washington, D.C.). Good sources of energy level spacing data for diatomic and polyatomic molecules.

*Nearly all introductory physics books have a discussion of the propagation of sound waves.

Chapter 5

Thermodynamics

5.1 Heat of Combustion

The purpose of this experiment is to determine the heat of combustion of a molecule using a bomb calorimeter. From this datum and other available thermochemical data, the heat of formation of the molecule is calculated. The apparatus is calibrated using benzoic acid.

THEORY. In thermodynamic studies one frequently holds either the pressure or the volume constant when making a measurement, and thermodynamic quantities are defined which have a rather natural association with one or the other of these experiments. For example, the enthalpy, H, and the Gibbs free energy, G, are most naturally related to constant pressure experiments, while the internal energy, E, and the Helmholtz free energy, A, bear a similar relationship to constant-volume experiments. The mathematical definitions of these quantities permit one to calculate constant-pressure quantities from experimental data obtained at constant volume and vice versa. For example, the relationship between the internal energy E and the enthalpy H is given by the mathematical definition of the enthalpy

$$H = E + pV \tag{5-1}$$

where p is the external pressure and V the volume of the system. The present experiment is carried out in a bomb (constant volume), and a calculation is made of the heat of formation, which is an enthalpy (constant pressure).

The standard heat of combustion at temperature T of a compound containing only carbon, hydrogen, and oxygen atoms is by definition the negative of the enthalpy change attending the complete conversion of 1 mole of the compound to CO_2 (gas) and H_2O (liquid) in the presence of oxygen. The process is presumed to be completed at temperature T and a pressure of 760 Torr = 1 atm. Figure 5-1 illustrates the process schamatically. Since enthalpy is a state function, we may devise any set of processes for passing from reactants at T, 1 atm. to products at T, 1 atm. and by summing appropriately the desired enthalpy can be computed. In Fig. 5-1 we have

$$\Delta H_c^\circ = \Delta H_1 + \Delta H_2 \qquad (5\text{-}2)$$

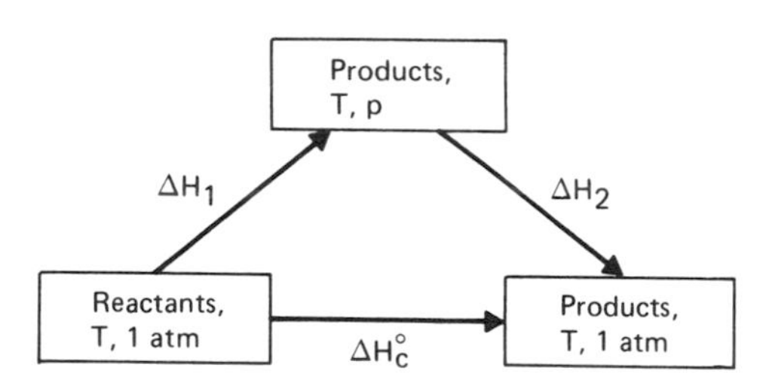

Figure 5-1. Schematic process for obtaining the standard heat of combustion.

Normally, standard enthalpies are tabulated at T = 298°K. For example, the standard heat of combustion of methane at 298°K can be obtained from analysis of the system

$$CH_4(g) + 2O_2(g) \rightarrow CO_2(g) + 2H_2O(l) \qquad (5\text{-}3)$$

where the initial and final states are both characterized by T = 298°K and p = 1 atm. Observe that the reactants and the products include a designation of the physical state. The standard physical state of a substance is defined as the state most stable at 25°C and 1 atm external pressure. The enthalpies of elements in their standard physical state at 25°C and 1 atm are defined to be zero. This definition provides a scale origin from which other enthalpies can be calculated.

In the present experiment, the conditions are those of constant volume rather than constant pressure. Thus, it will be necessary to relate constant-volume conditions to the heat of reaction, ΔH. This is accomplished by relating the internal energy change ΔE to the enthalpy change ΔH.

The natural relationship between internal energy and constant-volume processes may be seen from the relations

$$dE = DQ - DW \qquad (5\text{-}4)$$

where Q is the heat added to the system and W is the work done by the system on the surroundings. The symbol D will serve as a reminder that in general neither Q nor W is a state function and thus DQ and DW are inexact differentials.

On the other hand, E is a state function and dE an exact differential. Assuming only pressure-volume work (i.e., no electrical work and no important electric and magnetic field effects), Eq. (5-4) takes the familiar form

$$dE = DQ - p\,dV \tag{5-5}$$

and under constant-volume conditions

$$dE = DQ \tag{5-6}$$

A schematic diagram of the thermodynamics of the experimental combustion process is shown in Fig. 5-2. The combustion step is imagined to be carried out rapidly and adiabatically so that all the products are formed before any heat transfer occurs to the bomb or heat transfer fluid; thus, $\Delta E_1 = 0$ since $Q_1 = 0$. The second step, ΔE_2, is then imagined to be the cooling of these products (through heat transfer to a heat bath) to the initial temperature T_1. Thus,

$$\Delta E_2 = + \int_{T_3}^{T_1} C_V\,dT \tag{5-7}$$

where C_V is the heat capacity of the products. What we actually measure, as shown in the calorimeter part of Fig. 5-2, is the increase in the heat bath (calorimeter fluid) temperature as a result of the heat transfer in the second step of the above hypothetical process. By using the conservation of energy principle and the total heat capacity, C^{tot}, of that part of the calorimeter which changes temperature we have

$$\Delta E = \int_{T_1}^{T_2} C^{tot}\,dT = -\Delta E_2 = -\Delta E_c^\circ \tag{5-8}$$

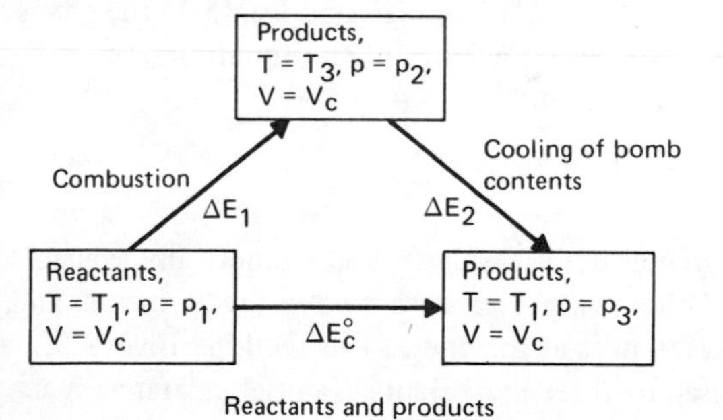

Figure 5-2.

It may be well to recall the origin of Eqs. (5-7) and (5-8). Considering E as a function of T and V, E(T, V), the total differential of E is

$$dE = \left(\frac{\partial E}{\partial T}\right)_V dT + \left(\frac{\partial E}{\partial V}\right)_T dV \tag{5-9}$$

By definition the heat capacity at constant volume, C_V, is $(\delta E/\delta T)_V$. The conditions under which Eq. (5-7) applies are restricted explicitly to constant volume, implying that $dV = 0$ and that

$$dE = C_V\, dT \tag{5-10}$$

The conditions under which Eq. (5-8) is applied are restricted to constant volume of the reactants and products but not necessarily to the heat bath (i.e., there is nothing about the experiment which prevents expansion of the latter). However, the actual conditions are such that the volume change is extremely small so that we lose very little by ignoring the dV term of Eq. (5-9). In this way Eq. (5-8) arises, and C^{tot} is a constant-volume heat capacity. We conclude that a calorimeter measurement of the heat change associated with a constant-volume process will furnish the internal energy change attending that process directly. Further, note that constant-volume conditions render DQ equivalent to an exact differential.

To make use of Eq. (5-8) the total heat capacity as a function of temperature must be available. All parts of the system that change temperature and any phase changes over the temperature interval must be accounted for. In the case where no phase change occurs and where the temperature interval is small (a few degrees Kelvin) the total heat capacity of the system may, to a very good approximation, be taken as constant. Equation (5-8) then becomes

$$\Delta E = C_V^{tot}\, \Delta T = -\Delta E_c^\circ \tag{5-11}$$

Given C_V^{tot}, a measurement of ΔT determines ΔE, which can then be related to ΔH using Eq. (5-1). Observe that to obtain ΔH from ΔE and Eq. (5-1) the change in the pV product is needed. This requires knowledge of the equation of state characterizing the various parts of the system:

$$\Delta H_c^\circ = \Delta E_c^\circ + \Delta(pV) \tag{5-12}$$

EXPERIMENTAL. The calorimeter described here is a commerically available instrument (see Appendix XII). Others have been described in the literature, and a few of these are noted in the References at the end of this section. Figure 5-3 shows a diagram which is comprised of three ingredients essential insofar as specifying the experimental conditions are concerned: (1) an adiabatic shell providing approximate thermal isolation between the inner contents and the universe, (2) a heat transfer fluid (water) whose weight and heat capacity are accurately

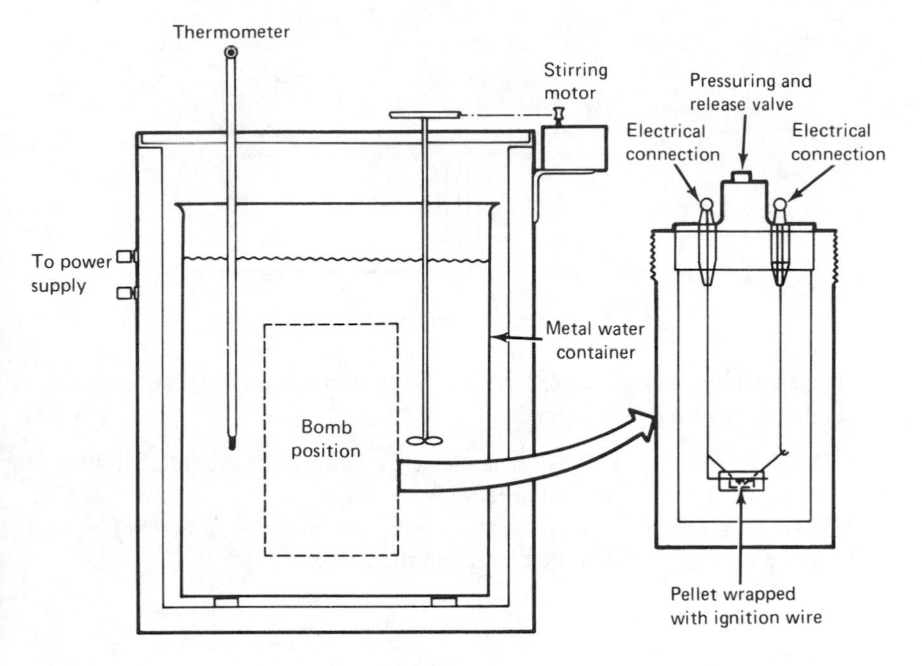

Figure 5-3. Bomb calorimeter.

known, and (3) a constant-volume container (bomb) inside of which the combustion of some material takes place. As mentioned in the previous section the heat capacity of importance includes all parts of the system that undergo a temperature increase. Referring to Fig. 5-3, this includes (1) the bomb, (2) the contents of the bomb, (3) the calorimeter fluid (water), (4) the metal container, (5) the stirrer rod, (6) the thermometer, and (7) the air surrounding the metal container. Generally we omit the latter and frequently (5) and (6) as well.

Some details of the bomb are given in Fig. 5-3. It is comprised of an outer body with a lid assembly and screw ring which positively seals the lid assembly and outer body together with a rubber gasket. The lid assembly itself carries two electrodes and a valve. One of the electrodes is ring-shaped and supports a sample cup, while the other is straight. Ignition wire is used to provide an electrical circuit between these two electrodes and to support a pellet of the sample. Externally the two electrodes are connected to a high-current low-voltage ac transformer capable of delivering 24 V and 5 A at the secondary. Under these conditions the ignition wire heats very rapidly to a high temperature, induces the oxidation of the sample, and is itself oxidized, leaving an electrically open circuit. The primary of the transformer should have a double pole switch in it which can be turned off immediately when the ignition wire burns. A suitable circuit is shown in Fig. 5-4. The thermometer used should be calibrated in hundredths of a degree and cover a small range near room temperature.

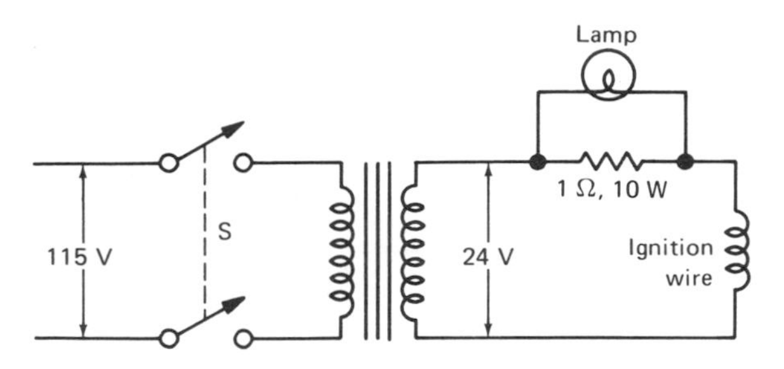

Figure 5-4. Ignition power supply.

The general procedure includes the following steps, assuming that benzoic acid is the standardization material:

1. Weigh accurately a pellet of benzoic acid. The weight should be in the range 0.8–1.0 g and should never exceed 1.1g.
2. Wire the pellet over the cup of the bomb calorimeter as shown in Fig. 5-5. Add a drop of water to the bomb and then close it.

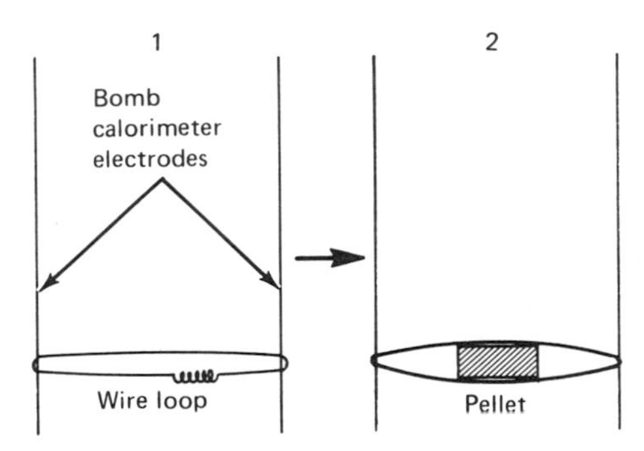

Figure 5-5. Placement of pellet in bomb calorimeter. Step 1: Wrap one loop of wire tightly around the electrodes. Step 2: Insert the pellet between two strands of wire. The tension on the wire should hold the pellet in place.

3. Fill the bomb through its pressuring valve with oxygen to a pressure of approximately 20 atm using the cylinder and connection in the laboratory. Make certain never to exceed 30 atm.
4. Place the bomb in the metal can located inside the adiabatic chamber and make electrical connections to it from the ignition power supply.
5. Add a known amount near 2 liters of water to the metal can at a temperature between 18° and 23°C. The range of the thermometer is small (usually 18°–31°C). How would water at 30°C cause experimental difficulties? Record the temperature of the water as accurately as you can read the thermometer.
6. Cover the calorimeter container and begin stirring the water and recording the temperature.
7. When the temperature is constant, fire the bomb by pressing the ignition switch. As soon as the lamp shown in Fig. 5-5 goes out, release the switch. Record the temperature versus the time until the temperature again becomes constant.

8. Remove the bomb, release the oxygen pressure by depressing the release valve, and weigh the wire which remains.
9. Clean the bomb, empty the water, dry the parts, and repeat the above procedure for a second sample suggested by the instructor. An enormous number of compounds can be used, but special techniques must be employed with some of them. See bomb calorimetry and tabulated data on various compounds in the References. A good beginning list includes naphthalene, anthracene, glucose, sucrose, and starch.

ANALYSIS.

1. Taking the standard heat of combustion of benzoic acid as 771.2 kcal/mole calculate the heat capacity of the system. The heat of combustion of the wire is given on the card to which it is attached or will be provided by the instructor. Assume that any gases formed in the combustion are ideal. From the known heat capacity of water, dissect the total heat capacity into two parts: one for the water and the remainder for the rest of the system.
2. From the naphthalene data and the heat capacity of the bomb, calculate the standard heat of combustion of naphthalene:

$$C_{10}H_8(s) + 12O_2(g) \rightarrow 10CO_2(g) + 4H_2O(l)$$

If you do not use naphthalene, perform step 2 for your compound. Remember that the total heat capacity here will differ from that in step 1 if different amounts of water are used.
3. Comment on sources of error and the magnitude of their influence on the results. Several sources of error are apparent and can be corrected for. First, the heat capacity of the products are not identical in both samples. A correction could be applied for this systematic error by knowing the heat capacity of water and carbon dioxide and by knowing the difference in the amounts of each formed in the reactions actually run. A second source of error arises if any impurities are present. A common impurity is nitrogen in which case some nitric acid and other oxides of nitrogen can be formed. A third source of error is the imperfect adiabaticity of the system. These heat leaks are small but in refined work must be evaluated. Fourth, the stirring process will add some energy to the system.
4.* Using thermochemical data available in various texts and handbooks and the results of step 3, calculate the standard heat of formation of naphthalene, i.e.,

$$10C(s, \text{graphite}) + 4H_2(g) \rightarrow C_{10}H_8(s)$$

Compare your calculated value with the literature value.
5. Calculate the enthalpy change associated with the reaction

$$10C(g) + 8H(g) \rightarrow C_{10}H_8(g) \quad \text{at } 298^\circ K$$

*In steps 4, 5, and 6 work out analogous problems if naphthalene is not used.

6. Using bond energies tabulated in texts and assuming that 145 kcal/mole is required to dissociate carbon atoms, which are double-bonded, calculate the enthalpy change associated with the above reaction. Give a precise definition of the bond energies used. From a thermodynamic point of view, are these bond energies really internal energies or enthalpies? Explain.

PROBLEMS

1. Why is it desirable to add a drop of water to the bomb prior to carrying out the combustion? *Clue:* Examine Eq. (5-3).
2. Suppose that the ideal gas law is not valid under the conditions of these experiments. What effect will this have on the results?

5.2 Binary Liquid-Gas Equilibrium. Boiling Point Elevation.

The purpose of this experiment is to study liquid-gas equilibria in binary solutions and to measure boiling point elevations in dilute systems of volatile solvent and nonvolatile solute. The emphasis is on understanding how these ebullioscopic measurements are made, how they are related through equilibrium thermodynamic theory to the apparent molecular weight of the nonvolatile solute, and how to interpret the calculated molecular weight in terms of the physical-chemical properties of the two participating molecules.

COLLIGATIVE PROPERTIES. In this experiment a study is made of one of the four colligative properties of dilute binary solutions. These binary solutions will contain two components in the liquid phase: a small amount of solute in a large amount of solvent. In the gas phase only solvent molecules are assumed to appear. Hence, we say that the solute is nonvolatile.

A colligative property is a physical property which depends solely on the number of particles of solute present and not on their kind (big, small, ionic, light, heavy, etc.). The colligative property to be investigated in this experiment is boiling point elevation. That is, we measure the increase in the boiling point of some volatile material as small amounts of nonvolatile solute are added. The three other colligative properties are (1) freezing point depression, (2) osmotic pressure, and (3) vapor pressure lowering.

The boiling point elevation and vapor pressure lowering are two intimately related colligative properties. This relationship may be understood from the following considerations. The boiling point of any substance is defined as the temperature at which the vapor pressure of the substance equals the gas pressure above the liquid. Therefore, if we are boiling some substance and then add some solute which lowers the vapor pressure, the substance will stop boiling and the temperature will have to be increased in order to raise the vapor pressure until it is again equal to the gas pressure above the liquid.

The necessary thermodynamic relations related to this experiment are developed in detail in most introductory thermodynamics texts. Only a summary is given here. As usual the Gibbs free energy, G, plays a central role in the description of this two-phase system. In particular the chemical potential μ is of considerable utility. By definition

$$\mu_j^{(ph)} = \left(\frac{\partial G^{(ph)}}{\partial n_j}\right)_{T,p,n_i} \tag{5-13}$$

where the superscript ph labels the phase, j labels the species, n is the concentration variable, T is the absolute temperature, and p is the pressure. Assuming an equilibrium system at constant temperature and pressure in which dn_j moles of j are transferred from the liquid to the gas phase while all other n_is are constant, one can readily show that the chemical potential of species j is the same in both phases at equilibrium. For the sake of clarity in relation to this experiment, let j denote the solvent. By hypothesis, the second component, solute, does not appear in the gas phase. If the gas phase is ideal,

$$\mu_{solvent}^{(l)} = \mu_{solvent}^{(g)} = \mu_0^{(g)} + RT \ln p \tag{5-14}$$

where ℓ denotes liquid and g denotes gas.

If the solution is ideal, then Raoult's law is applicable and relates the equilibrium solvent vapor pressure p to the mole fraction, $X_{solvent}$. That is,

$$p = p^\circ X_{solvent} \tag{5-15}$$

where p° is the vapor pressure of pure solvent at the equilibrium temperature. Combining Eqs. (5-15) and (5-14) furnishes

$$\mu_{solvent}^{(g)} = \mu^{\circ(l)}_{solvent} + RT \ln X_{solvent} \tag{5-16}$$

where

$$\mu^{\circ(l)}_{solvent} = \mu_0^{(g)} + RT \ln p^\circ$$

Applying the Gibbs-Helmholtz equation (see Section 5.7) to Eq. (5-16) results, after considerable algebraic manipulation, in the following approximate expression:

$$\Delta T \cong \frac{RT_0^2 M_0}{1000\Delta\bar{H}_{vap}} m \tag{5-17}$$

where T_0 is the boiling point, M_0 the molecular weight, $\Delta\bar{H}_{vap}$ the molar heat of vaporization, and m the molality. The last quantity refers to solute; all the others refer to solvent. The coefficient of m in Eq. (5-17) is called K_b, the ebullioscopic constant of the particular *solvent.*

Equation (5-17) furnishes a relationship between ΔT and m once K_b is determined. From m, the average molecular weight of solute is calculable if the weight of solute is known. One method of determining K_b for a particular solvent is to measure ΔT with a solute of known gram molecular weight (hence, known molality). This approach will be used in the present experiment.

In discussing Eq. (5-17) it is important to keep in mind that the boiling point change is a colligative property which does not depend on the weight but really only on the mole fraction of solute, which is directly related to the number of solute particles. Also note that the ebullioscopic constant for a given solvent is independent of the solute (i.e., this constant is to be associated with the solvent and not with the solute).

The average molecular weight is calculable only when the weight of solute added is known. The reason for using the term average molecular weight becomes apparent when we consider the following solute equilibrium which could occur in a solution:

$$2Y \rightleftharpoons Y_2$$

Since boiling point elevation is a *particle-counting* phenomenon, the observed ΔT will depend on the position of the equilibrium. For example, if the equilibrium is far to the right, the ΔT observed for a given weight of solute will be smaller than if the equilibrium were far to the left. The experimentally determined molecular weight, M, would then be an average written as

$$\bar{M} = X_Y M_Y + X_{Y_2} M_{Y_2} \tag{5-18}$$

where X_Y is the solute mole fraction of Y and X_{Y_2} the solute mole fraction of Y_2. Solute mole fractions are based on solute only (i.e., neglect solvent). Mathematically, this requires that

$$X_Y + X_{Y_2} = 1 \tag{5-19}$$

Although the boiling point elevation furnishes only an average molecular weight, combining Eqs. (5-18) and (5-19) together with measurement of the weight of solute added permits determination of the equilibrium constant for $2Y \rightleftharpoons Y_2$. This last measurement provides $[Y_0]$, which is a computed concentration of Y assuming that no Y_2 is formed. A conservation equation on the number of Ys, both combined and uncombined, is

$$[Y_0] = [Y] + 2[Y_2] \tag{5-20}$$

How do we compute $[Y_0]$? Just calculate the molar concentration expected on the basis of the weight of solute and volume of solvent used. The equilibrium constant is, as usual,

$$K_{eq} = \frac{[Y_2]_e}{[Y]_e^2} \tag{5-21}$$

where the subscript e on the concentration terms underscores that they are equilibrium values. Using Eqs. (5-18)–(5-20) and remembering the definition of mole fraction in terms of molar concentrations K_{eq} becomes

$$K_{eq} = \frac{(\bar{M} - M_y)(M_{y_2} - M_y)}{(\bar{M} - M_{y_2})^2[Y_0]} \tag{5-22}$$

This expression is in terms of available molecular weights. It may also be written in terms of mole fractions.

EXPERIMENTAL. Two methods are outlined here—the first uses a Beckmann thermometer, while the second makes use of a thermocouple arrangement.

Method 1: Beckmann Thermometry. A typical Cottrell boiling point apparatus is shown in Fig. 5-6. The Beckmann thermometer must be properly calibrated as described in Section 2.2. The apparatus should be clean and dry before each boiling point is determined. The liquid (known weight of solvent and solute) should be added in sufficient quantity to cover the flared end of the inner funnel inside the apparatus. Usually about 100 ml is required. After adding some fresh boiling chips to provide for bubble formation without superheating, the apparatus should be assembled and tap water passed through the condensor. Then the apparatus should be inserted into a beaker of water and the water bath heated until the solution boils evenly and sprays the bulb of the thermometer. The temperature will become constant as the system approaches equilibrium, and the equilibrium temperature is taken as the boiling point of the solution. The boiling points of the following solutions should be determined:

1. Pure CCl_4 (100 ml)
2. 100 ml of CCl_4 + 10 g of naphthalene
3. 100 ml of CCl_4 + 10 g of benzoic acid
4. Pure C_2H_5OH (100 ml)
5. 100 ml of C_2H_5OH + 10 g of naphthalene
6. 100 ml of C_2H_5OH + 10 g of benzoic acid

The weights of the above solutes should be measured to 0.1-g accuracy and the boiling point temperature to 0.01°C.

Some comments about experimental techniques may be helpful:

1. Avoid using any stopcock grease on the joints unless you wish to measure the boiling point of a solvent-stopcock grease solution.
2. Use fresh boiling chips (four or five) and place them under the conical section of the inner funnel. This funnel serves to pump the boiling liquid up through the tubular glass jets onto the thermometer stem.
3. Keep the Beckmann thermometer upright at all times in order to maintain its calibration.

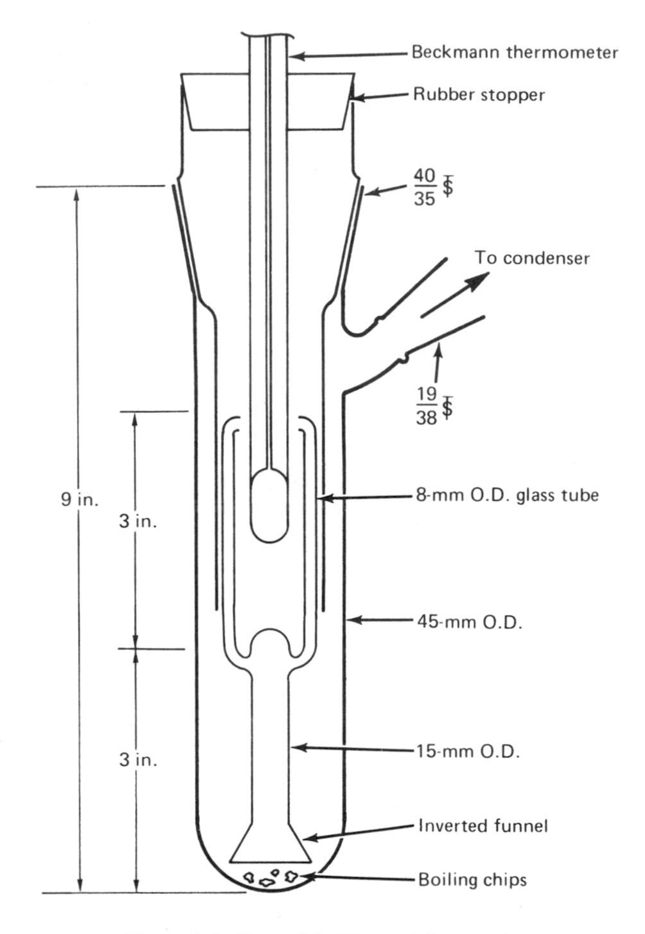

Figure 5-6. Cottrell boiling point apparatus.

4. In the above experiment it is tacitly assumed that the atmospheric pressure is constant. In experiments done over extended periods of time the pressure variation and its influence on boiling temperature must be considered.
5. It is worth remembering that ethanol is inflammable.
6. Swietoslawski and Anderson in the References point out some conceptual difficulties in interpreting the experimental data.

Method 2: Thermocouple Thermometry. A more elaborate but, from the thermometry point of view, sturdier apparatus and one which provides more precise data is a differential ebulliometric device described by Mair in the References which uses thermocouples to measure the temperature difference between the pure boiling solvent and the boiling solution. This method involves two Cottrell-type units like that shown in Fig. 5-6. These units must be modified by replacing the Beckmann thermometer with a thin-walled glass tube containing enough light mineral oil to cover the thermocouple junctions (which are inserted into it) and to make the oil level slightly higher than the level of the glass tube of the inner funnel portion of the Cottrell apparatus. Figure 5-7 gives a block diagram of the apparatus. Copper-constantan thermocouple junctions connected in series are recommended. Five to 10 junctions in each part of the system will multiply by 5 to 10 times (approximately) the voltage developed by one junction. In this way a quite sensitive thermometer can be constructed. Each junction must be fused separately and coated with shellac or some other insulating material. In setting up the apparatus, simply place the junctions in the oil-filled glass tubes described above. Place alternate junctions in the same tube (i.e., first, third, fifth, etc., junctions in one tube and second, fourth, etc., in the other). Each Cottrell apparatus should have its own heating bath.

PROCEDURE. The experiment consists of two basic parts: calibration of the thermocouple system and measurement of boiling point differences of the solutions. In the latter part the procedure is identical to that described in the previous experiment except one of the Cottrell systems always has pure solvent in it. Use the same solution concentrations as described there.

In the calibration the boiling point difference of two pure liquids is measured, and the known temperature difference is related to the measured emf, assuming a linear correlation of the two. Two pairs of solvents are suggested: carbon tetrachloride-benzene and carbon tetrachloride-ethanol. At 760 Torr these pairs have boiling point differences of 3.3° and 1.6°C, respectively. Carry out the calibration by measuring the emf difference with one member of the pair in one apparatus and the second in the other. Make certain to allow the boiling processes to come to equilibrium. If you are not working at 760 Torr pressure, the boiling temperatures of each liquid will need to be adjusted. From the measured emf and the temperature difference given (or adjusted to the proper pressure), calculate

$$k = \frac{\Delta T}{\text{emf}}$$

and use this as the thermocouple calibration constant.

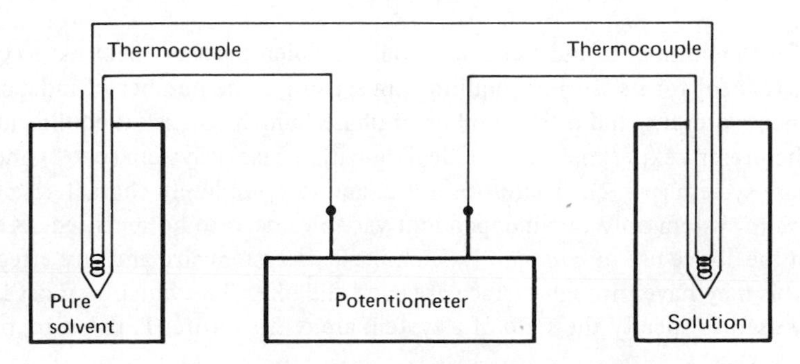

Figure 5-7. Differential ebulliometric apparatus.

Make measurements on the systems described in the previous experiment and treat the data as described there.

Other solvent-solute systems which can be readily employed include benzene-biphenyl, benzene-triphenylmethane, benzene-anthracene, water-ethylene glycol, water-sucrose, and water-sodium chloride. For the last three, the water bath used to heat the solutions must be replaced with an oil bath of light mineral oil. With this kind of bath, electrical heating with a mantel or a directly inserted heating element is preferable to a Bunsen burner. The oil bath should also be constantly stirred.

ANALYSIS. Assume that naphthalene is completely monomeric in solution. Using its molecular weight, calculate K_b in both CCl_4 and C_2H_5OH solutions and compare with the literature values. Using the experimentally determined K_b, determine m, $\bar{M}$, X_Y, and X_{Y_2} for benzoic acid in both the CCl_4 and C_2H_5OH solutions. Explain the differences in the effects of the two solvents. Calculate $\Delta\bar{H}_{vap}$ for CCl_4 and C_2H_5OH and compare with literature values. Prepare a plot of temperature versus time for one of the determinations and comment on the approach of the system to equilibrium.

5.3 Binary Solid-Liquid Equilibria

The purpose of this unit is to study equilibria in two-phase two-component systems where the two phases are solid and liquid. Experimentally, the measurements consist of cooling curves (temperature versus time) which are analyzed to determine the equilibrium freezing point of solutions of variable composition. From the freezing point data a phase diagram is constructed and a freezing point depression constant is calculated. Simple systems with uncomplicated phase diagrams are suggested.

THEORY. A useful tool in the study of equilibria of systems containing more than one component is the phase rule developed by Gibbs:

$$f = c - p + 2 \tag{5-23}$$

where f is the number of independent variables which must be specified to characterize (except for its size) the equilibrium system, c the number of independent components present, and p the number of phases which coexist at equilibrium.

In the present experiment we are dealing with a binary system ($c = 2$) and a two-phase system ($p = 2$). Therefore, $f = 2$, and to completely characterize the equilibrium system only two independent variables need to be specified. Keep in mind the above use of *complete,* which neglects system size and any effects which this may have through surface area and the like. The variables most frequently used to specify the state of a system are temperature, T, pressure, p, and mole fractions, X_i, of each component. For a binary two-phase system the choice of two of these fixes all the others. For example, the pressure and temperature or the pressure and one of the mole fractions in one of the phases would be sufficient. The other variables are then fixed because the chemical potential of each component is the same in every phase, and the mole fractions are related since their sum must equal unity in each phase.

If we choose the pressure to be fixed (usually 1 atm), then temperature and composition must be uniquely related in a binary, two-phase system. We can then plot a temperature versus composition curve for equilibrium mixtures of the two phases. The present experiment involves solid-liquid equilibria. If we start with pure compound A, the temperature at which solid and liquid exist in equilibrium is just the freezing point of pure A. If a small amount of component B is added, the temperature at which solid-liquid equilibrium occurs will be lowered. This is just the colligative property of freezing point depression. A similar argument may be made starting with pure component B. Suppose that we add larger and larger amounts of B to A and measure the freezing point by slowly cooling the liquid until solid appears. At low concentrations of B (small mole fractions) A will freeze out first as the solution is cooled. As more and more B is added we may reach a point at which both A and B freeze out simultaneously. The freezing temperature at this point must necessarily be lower than the freezing points of both pure A and pure B. This point is called the eutectic point. If still more B is added, then A will no longer freeze out first, and the freezing point will increase. These phenomena are illustrated in Fig. 5-8 for the simple benzene-naphthalene system. Phase diagrams can be very complex, especially if the solid can exist in different forms or if compound formation occurs.

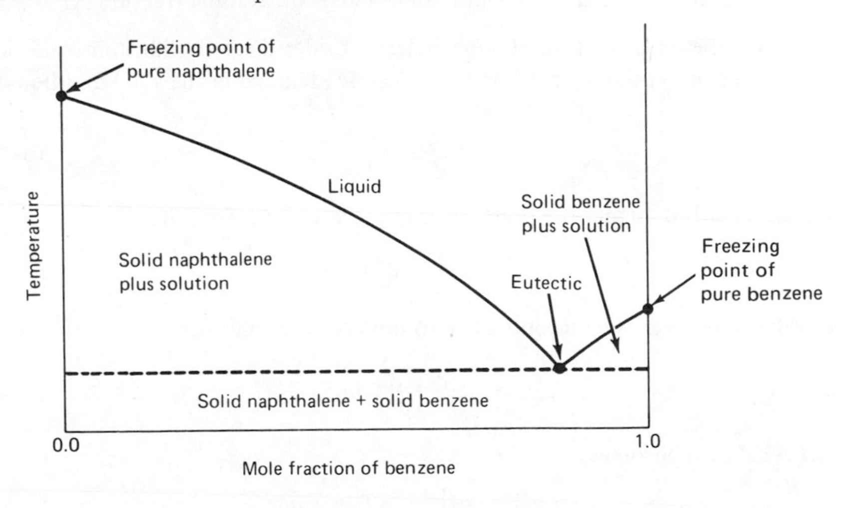

Figure 5-8. Schematic of phase diagram for naphthalene-benzene system.

From thermodynamic considerations we can arrive at the Gibbs-Helmholtz equation (see Section 5.7) which is applicable to the present system when it is at equilibrium:

$$\left(\frac{\partial\, \Delta G^\circ}{\partial T}\right)_p = -\Delta S^\circ = \frac{\Delta G^\circ - \Delta H^\circ}{T} \tag{5-24}$$

or

$$\left[\frac{\partial}{\partial T}\left(\frac{\Delta G^\circ}{T}\right)\right]_p = -\frac{\Delta H_T{}^\circ}{T^2} \tag{5-25}$$

where $\Delta H_T{}^\circ$ is the standard enthalpy change at temperature T.

We also know that the standard Gibbs free energy change ΔG° is related to the equilibrium constant, K_a, by

$$\Delta G^\circ = -RT \ln K_a \tag{5-26}$$

Using (5-25) and (5-26) furnishes

$$\frac{d \ln K_a}{d(1/T)} = -\frac{\Delta H_T^\circ}{R} \tag{5-27}$$

If ΔH_T° is independent of T, a plot of $\ln K_a$ versus 1/T will be linear and furnish ΔH_T° from a determination of the slope.

If we are in a region where only naphthalene freezes out, we can write, for the process occurring,

naphthalene (solid) $\longrightarrow$ naphthalene (solution at mole fraction X)

where X is the mole fraction of naphthalene. Under these equilibrium conditions the chemical potentials of naphthalene must be identical in the solid and liquid phases:

$$\mu_N^{(s)} = \mu_N^{(l)} \tag{5-28}$$

But the solid is pure and the pressure is assumed to be 1 atm; thus,

$$\mu_N^{(s)} = \mu_N^{\circ(s)} \tag{5-29}$$

In addition if we assume Raoult's law to hold for the solution,

$$\mu_N^{(l)} = \mu_N^{\circ(l)} + RT \ln X_N \tag{5-30}$$

Eq. (5-28) thus becomes

$$\mu_N^{\circ(s)} - \mu_N^{\circ(l)} = RT \ln X_N \tag{5-31}$$

$$\begin{aligned} \bar{G}_N^{\circ(l)} - \bar{G}_N^{\circ(s)} &= -RT \ln X_N \\ \Delta\bar{G}^\circ &= -RT \ln X_N \end{aligned} \tag{5-32}$$

Equation (5-32) may be compared directly with Eq. (5-26) from which we conclude that the appropriate K_a to use for the process outlined above is simply X_N, the mole fraction of naphthalene in the solution. Equation (5-27) then becomes

$$\frac{d \ln X_N}{d(1/T)} = -\frac{\Delta H_T^\circ}{R} \tag{5-33}$$

Clearly, a plot of $\ln X_N$ versus $1/T$ furnishes ΔH_T°, the standard enthalpy change associated with the process, in this case the enthalpy of fusion.

Integration of Eq. (5-33) from $X_N = 1$ to $X_N = X_N$, assuming that ΔH_T° is independent of temperature, furnishes

$$\ln X_N = \frac{\Delta H^\circ}{R} \frac{T - T_0}{TT_0} \tag{5-34}$$

where T_0 is the freezing point of pure naphthalene. For this binary system $X_N = 1 - X_B$, where X_B is the mole fraction of benzene (Fig. 5-8). Provided $X_B \ll 1$, we may write

$$\ln (1 - X_B) \cong -X_B \tag{5-35}$$

Substituting (5-35) into (5-34), using $TT_0 \cong T_0^2$, and rearranging yield

$$T_0 - T = \Delta T = \frac{X_B T_0^2 R}{\Delta H^\circ} \tag{5-36}$$

Finally, since $X_B \ll 1$, we can write

$$X_B \cong \frac{n_B}{n} = \frac{M_N m}{1000} \tag{5-37}$$

where M_N is the molecular weight of naphthalene and m is the molality of diphenylamine. Therefore, Eq. (5-36) becomes

$$T_0 - T = \frac{RT_0^2 M_N}{\Delta H^\circ \times 1000} m \tag{5-38}$$

The coefficient of m is known as the freezing point depression constant.

Having now developed experimentally useful relations for ΔH_T° and the freezing point depression constant we turn our attention to the entropy associated with the process of interest. We begin with a general prescription for the Gibbs free energy and require equilibrium conditions. This leads to the following development:

$$\begin{gathered} G = H - TS \\ \Delta G = \Delta H - T\,\Delta S - S\,\Delta T \end{gathered} \tag{5-39}$$

At constant T, the Gibbs free energy change for the process is

$$\Delta G = \Delta H - T\,\Delta S \tag{5-40}$$

At thermal equilibrium where the forward and reverse processes occur at the same rate, $\Delta G = 0$ and

$$\Delta H = T\,\Delta S \tag{5-41}$$

If as we have supposed that ΔH is independent of temperature, then $\Delta H = \Delta H^\circ$ and

$$\Delta S = \frac{\Delta H^\circ}{T} \tag{5-42}$$

Equation (5-42) suggests that as the equilibrium freezing point drops the entropy attending the process must increase.

A cooling curve for a binary mixture is obtained by plotting temperature versus time for a slowly cooling molten mixture of the two components. As heat is transferred from the solution to the surroundings a temperature is reached where at least one component freezes out. Solid material then appears and the cooling rate drops off because the solidification process is exothermic. If we have a composition corresponding to pure materials or to a eutectic mixture, then the phase rule leaves only a single degree of freedom at the freezing point. If the pressure is constant, then no degrees of freedom remain, and when freezing begins the temperature remains fixed until all the material has solidified. Then cooling can occur once more. Figure 5-9(a) illustrates this. For other compositions, the phase rule leaves two degrees of freedom. Again one is taken by fixing the pressure, but the temperature is no longer forced to remain constant. As shown in Fig. 5-9(b) the cooling rate changes when the freezing point is reached and one component begins to solidify, lowering the cooling rate and changing the composition of the mixture toward the eutectic composition. Cooling curves much more complex than those shown in Fig. 5-9 are often observed and indicate the presence of other phenomena such as compound formation and crystallization.

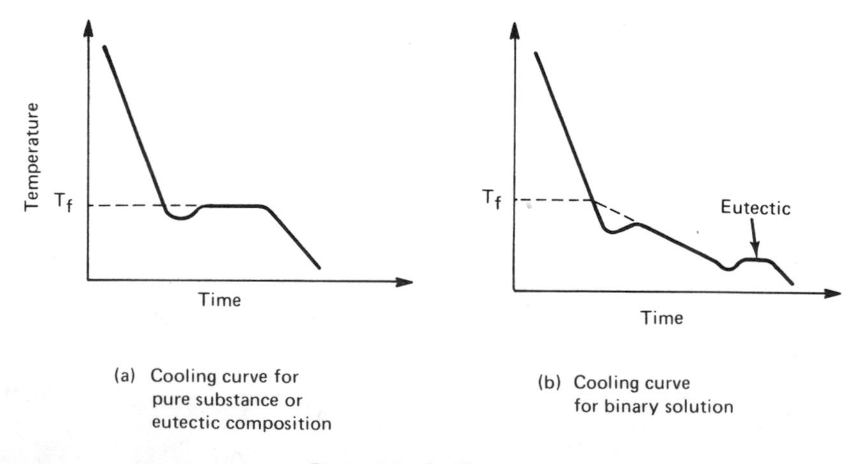

(a) Cooling curve for pure substance or eutectic composition

(b) Cooling curve for binary solution

Figure 5-9. Cooling curves.

EXPERIMENTAL. Considerations of the above theory indicate that some means of preparing several molten solutions of known but variable composition is needed. In addition provision must be made for cooling these solutions and monitoring temperatures. The References enumerate a large number of pairs of compounds which are amenable to study. Those recommended here are convenient because they freeze in readily accessible temperature ranges (–5° to 100°C) and are themselves relatively inexpensive materials. Suggested pairs are

1. Naphthalene–diphenylamine
2. Naphthalene–biphenyl
3. *p*-Dichlorobenzene–biphenyl
4. *p*-Dichlorobenzene–*o*-chlorophenol
5. *p*-Dichlorobenzene–*o*-cresol
6. *p*-Dichlorobenzene–*p*-cresol

One type of apparatus is shown in Fig. 5-10 and should be assembled using glycerin to lubricate the rubber stopper-thermometer joint. Approximate sizes are given in the figure legend. The thermometer in this experiment can be replaced with an iron-constantan thermocouple junction. Figure 2-17 shows the proper electrical circuit. A 0-100-mV potentiometric chart recorder is particularly useful as a voltage-measuring device because cooling curves in the range

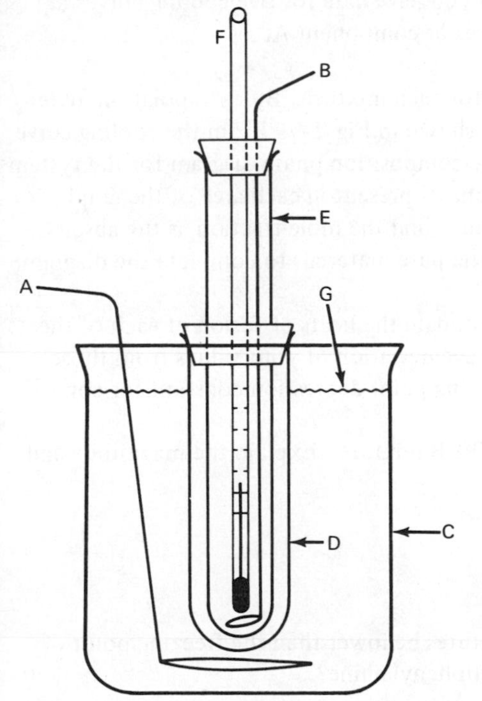

Figure 5-10. Freezing point apparatus. A, Bath stirrer; B, solution stirrer; C, large beaker (3000 ml); D, outer tube (65 mm diameter × 21 cm long); E, inner tube (35 mm diameter × 20 cm long); F, 100-deg thermometer, calibrated in tenths; G, ice bath.

0°-100°C can be plotted directly. If a thermocouple system is used, it should be calibrated over the temperature range of the experiment by using the cooling curves of the pure materials and their known freezing points. Refer to Section 2.2 for additional details.

Record cooling curves for five mixtures of the two components A and B according to the following procedure:

1. Grind the components well, and then place a weighed amount of component A in the test tube. Weigh to an accuracy of 0.01 g. Enough material should be weighed to fill the inner tube of Fig. 5-10 to a depth of about 2 or 3 cm or as much as is necessary to cover the bulb of the thermometer. Now add enough component B so that 80% of the weight of the mixture is A.
2. Melt the solid in a hot water bath and then place it in the freezing point apparatus, stir gently, and record the temperature to one-tenth of a degree every 30 sec. Make certain that there is always some ice present in the ice water bath. Continue to stir. Take only 10 or 12 readings after reaching the initial freezing point of the mixture.
3. Remove the test tube from the apparatus and add enough component B for a 70% mixture in component A and record the temperature versus time exactly as you did in step 2.
4. Continue in the same way until you have data for five cooling curves at 80, 70, 60, 50, and 30% mixtures of component A.

ANALYSIS. Plot a cooling curve for each mixture. By extrapolation, determine the freezing temperature, T_f, as shown in Fig. 5-9. From the cooling curve data, construct a freezing temperature-composition phase diagram for the system and label it appropriately, including phases present in each area of the graph. Plot with the temperature as the ordinate and the mole fraction as the abscissa. You will need the freezing points of the pure materials to complete the diagram. Look these up in a handbook.

From a plot of log X versus 1/T, estimate the heats of fusion of each of the two components. Report the percentage deviation of your values from those given in a handbook. Calculate a freezing point depression constant for component A.

Calculate ΔS for the 90% A and 10% B mixture. Explain the magnitude and sign of this quantity.

PROBLEMS

1. Why should the eutectic temperatures be lower than the freezing point of either pure naphthalene or pure diphenylamine?

2. Using a general expression for the standard enthalpy change, ΔH°, determine under what conditions it will be temperature-independent.
3. What is the difference between ΔG and ΔG° for the equilibrium studied in this experiment?
4. Explain why the mole fraction of benzene does not appear in Eq. (5-33).

5.4 Vapor Pressure of a Pure Liquid

In this experiment, the vapor pressure of a pure liquid is measured as a function of temperature. The data are treated according to equilibrium thermodynamic considerations, and the heat of vaporization is calculated. Variations of the heat of vaporization with temperature are considered.

THEORY. Equilibrium thermodynamic considerations predict a variation of vapor pressure with temperature for pure liquids. The Clausius-Clapeyron equation is generally used to describe this variation. For its usual development, this equation depends on several assumptions:

1. Equilibrium between liquid and gas phases.
2. Vapor is an ideal gas.
3. The molar volume of liquid may be neglected in comparison to the molar volume of vapor.
4. The heat of vaporization is independent of temperature.

Assumption 1 implies that the chemical potential of the gas phase is equal to the chemical potential of the liquid phase. Assumption 2 furnishes an equation of state which is needed to eliminate the volume of gas as a variable. Assumption 3 is needed to eliminate the volume of the liquid from the problem. Assumption 4 permits simple integration of the resulting differential equation.

The starting point for the development of the Clausius-Clapeyron equation is assumption 1, which provides the following mathematical expression connecting differential changes in the chemical potentials of the liquid and vapor phases:

$$d\mu_i^{(v)} = d\mu_i^{(l)} \tag{5-43}$$

where v and l are defined as vapor and liquid, respectively. In the present case, only a single component is being considered and so the subscript i may be dropped.

For a phase containing a single component, μ is simply the Gibbs free energy, G, divided by the number of moles in that phase. Hence,

$$d\mu^{(v)} = -\bar{S}^{(v)}\, dT + \bar{V}^{(v)}\, dp \tag{5-44}$$

where $\bar{S}$ and $\bar{V}$ are molar quantities. The change in molar entropy $\Delta\bar{S}$ attending the process $l \rightarrow v$ may be written as $\Delta\bar{S} \equiv \bar{S}^{(v)} - \bar{S}^{(l)}$. It is a straightforward matter to relate $\Delta\bar{S}$ to the molar enthalpy of vaporization as follows:

$$H = E + pV$$
$$dH = dq - p\,dV + V\,dp + p\,dV \tag{5-45}$$

and for a reversible process $dq = T\,dS$

$$dH = T\,dS + V\,dp \tag{5-46}$$

The vaporization process considered here occurs at constant temperature and pressure; thus,

$$dH = T\,dS$$
$$\int_l^v dH = T \int_l^v dS \tag{5-47}$$
$$\Delta H_{vap} = T\,\Delta S$$

and for 1 mole of material undergoing the process

$$\Delta\bar{H}_{vap} = T\,\Delta\bar{S} \tag{5-48}$$

where $\Delta\bar{H}_{vap}$ is the molar enthalpy of vaporization. Equations (5-43), (5-44), and (5-48) may be combined to furnish a differential equation relating vapor pressure and absolute temperature for a pure liquid. Writing an equation analogous to (5-44) for the liquid phase and combining it with (5-44) according to Eq. (5-43), we obtain, after rearrangement,

$$\frac{dp}{dT} = \frac{\bar{S}^{(v)} - \bar{S}^{(l)}}{\bar{V}^{(v)} - \bar{V}^{(l)}} = \frac{\Delta\bar{S}}{\Delta\bar{V}} \tag{5-49}$$

Combining with Eq. (5-48) we have

$$\frac{dp}{dT} = \frac{\Delta\bar{H}_{vap}}{T\,\Delta\bar{V}} \tag{5-50}$$

Assumption 2 yields

$$\bar{V}^{(v)} = \frac{RT}{p} \tag{5-51}$$

Assumption 3 leads to

$$\Delta\bar{V} = \bar{V}^{(v)} - \bar{V}^{(l)} \cong \bar{V}^{(v)} = \frac{RT}{p} \tag{5-52}$$

Substituting Eq. (5-52) into Eq. (5-50) provides, after rearranging,

$$\frac{d\ln p}{dT} = \frac{\Delta\bar{H}_{vap}}{RT^2} \tag{5-53}$$

Assuming, as in assumption 4, that $\Delta\bar{H}_{vap}$ is independent of temperature we can easily integrate Eq. (5-53) between T_1 and T_2 with corresponding vapor pressures p_1 and p_2. The result is

$$\ln\frac{p_2}{p_1} = -\frac{\Delta\bar{H}_{vap}}{R}\left(\frac{1}{T_2} - \frac{1}{T_1}\right) \tag{5-54}$$

Under conditions satisfying all the above assumptions a plot of ln p versus 1/T will be linear with slope $-\Delta\bar{H}_{vap}/R$.

Frequently we consider gases to be ideal with no real examination of the situation. Discussion of deviations from ideality has been presented in Section 4.1. A different approach will be taken here.

We envisage the following experiment which could be used to test the validity of the ideal gas law. Measure p, $\bar{V}$, and T for a sample of gas over a wide range of conditions and compute $p\bar{V}/RT$ for each set of conditions. If the ideal gas law is valid, $p\bar{V}/RT = 1$; if not, deviations of this quantity from unity will be observed. We define a quantity Z, called the compressibility factor, as

$$Z = \frac{p\bar{V}}{RT} \tag{5-55}$$

Any deviations from ideality are reflected in deviations of Z from unity.

A theoretical estimate of the compressibility is available from the Berthelot equation:

$$Z = 1 + \frac{9}{128}\left(\frac{p}{p_c}\right)\frac{T_c}{T}\left[1 - 6\left(\frac{T_c}{T}\right)^2\right] \tag{5-56}$$

where T_c and p_c are the critical temperature and pressure for a particular substance. Equation (5-56) predicts that Z varies with p and T for a particular substance. At fixed p and T, Z will change from one substance to another because of changes in T_c and p_c. In the problems section of this experiment conditions are examined where deviations from ideality are expected.

EXPERIMENTAL. There is a variety of apparatus for the purpose of measuring vapor pressures. One of the simpler ones, which is straightforward to use and which gives good data, is shown in Fig. 5-11. The experimental data consist essentially of a set of {p, T} values obtained at equilibrium. The pure liquid is placed in the 200-ml reservoir, is vaporized at the bottom of the column, and is recondensed at the top of the column. The temperature of the vapor in equilibrium with the liquid at some fixed pressure is measured at the wick-thermometer interface. The thermometer should range from 0° to 100°C and be calibrated in intervals of 0.2°C. The wick can be made of a piece of cotton cloth (toweling). The small-diameter hole above the liquid and the funnel-shaped tube above it serve to prevent liquid splashing. The ballast volume serves to increase the volume of the system and helps even out pressure fluctuations. In this experiment the pressure above the liquid is adjusted and then the temperature at the wick shown in Fig. 5-11 is recorded until equilibrium is reached.

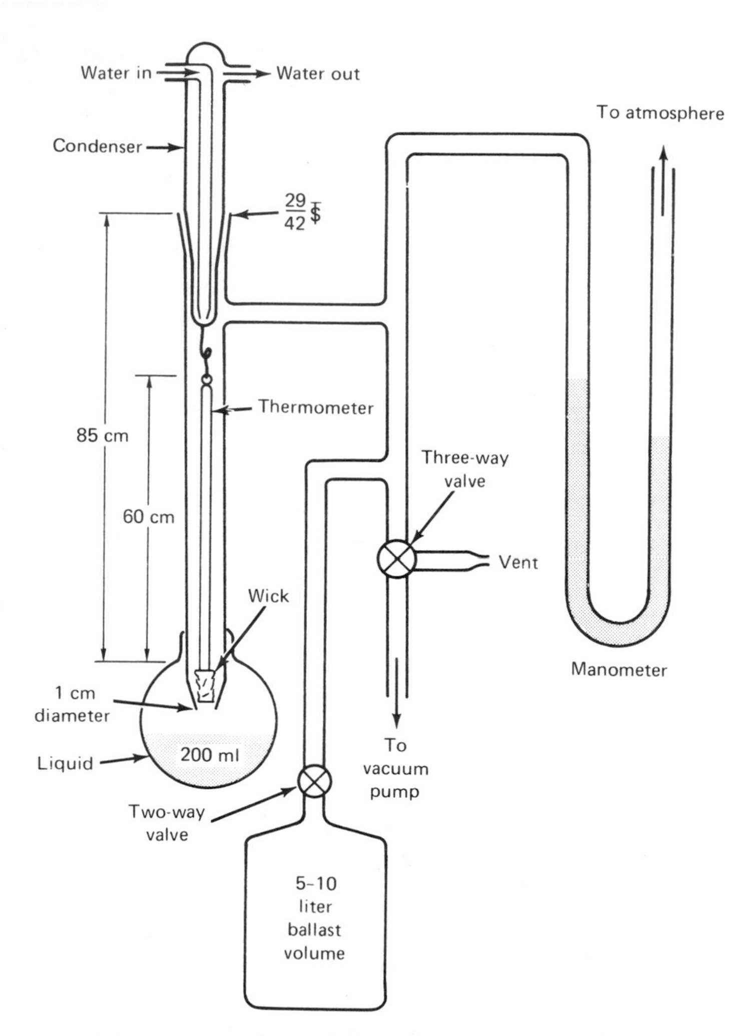

Figure 5-11. Apparatus for determining the vapor pressure of a pure liquid.

The clean and dry system should be first tested for leaks by evacuating the whole system and then isolating it from the pump. If the pressure indicated by the manometer varies with time, leaks are present and the joints need to be cleaned and regreased. Use only a small amount of hydrocarbon grease on the joints. When the evacuated system shows no leaks, the apparatus is ready for use. Close the valve to the ballast volume before venting to minimize evacuation time later in the experiment.

Add about 50 ml of a pure liquid to the reservoir of the tube, reassemble the tube, and evacuate the system as completely as possible with the ballast volume valve open. Then isolate the tube from the pump by closing the three-way valve.

Pass water through the condenser at the upper end of the tube, surround the reservoir with a water bath or heating mantle, and begin heating. Keep the temperature of the water bath about 15°C higher than the temperature of the wick to ensure that the liquid-vapor equilibrium investigated is actually established at the wick.

Record the wick temperature versus time until it becomes constant over a period of about 5 min. A plot of T versus time may be helpful in ascertaining the asymptotic limit of the absolute temperature. Record the pressure difference indicated on the manometer, and with the atmospheric pressure, calculate the pressure inside the tube above the liquid.

Repeat the above measurements for four or five different total pressures by adding increments of air through the vent. The highest pressure measurement should be for atmospheric pressure. Following these measurements for ascending pressures, make two measurements at reduced pressures so that a check can be made of the validity of the approach. There should be no systematic difference in measurements taken by increasing the pressure and those taken by decreasing it. Upon completion of data collection, vent the system to atmosphere and dispose of the remaining liquid. As Fig. 5-11 has it, the manometer is measuring the pressure relative to atmospheric pressure. To extract absolute pressures above the liquid, atmospheric pressure must be determined from a laboratory manometer.

A variety of liquids can be easily studied with this apparatus, including benzene, trichloroethylene, carbon tetrachloride, water, *n*-propyl alcohol, *iso*-propyl alcohol, toluene, acetone, and ethanol. Many of these liquids are inflammable and require commensurate caution around open flames.

ANALYSIS.

1. Plot the experimental data in the form ln p versus 1/T, evaluate the slope, and calculate $\Delta\bar{H}_{vap}$ for the substance studied.
2. Assuming random error in both T and p, derive an *error propagation* expression for $\Delta\bar{H}_{vap}$. Using estimated uncertainties in temperature and pressure, calculate an uncertainty in $\Delta\bar{H}_{vap}$. Refer to Section 1.3.
3. Compare the experimental $\Delta\bar{H}_{vap}$ with a literature value. Keeping in mind the estimated experimental uncertainty, comment on deviations of your experimental value from the literature value.
4. Discuss any curvature of the plot made in step 1. If no curvature was detected, set an upper bound on the variation of $\Delta\bar{H}_{vap}$ with T over the temperature range of your experiments.
5. The use of Eq. (5-54) implies that the assumptions are satisfied. One of the assumptions has been examined above. We shall now examine the others. What experimental criteria were applied to ensure minimal deviations from equilibrium?

6. Evaluate the compressibility factor for the conditions of your experiment corresponding to the highest and lowest temperature. Critical pressures and temperatures are available in many handbooks. Comment on the validity of the ideal gas assumption.
7. Making use of gas and liquid densities, evaluate the error introduced through the neglect of the volume of the liquid in the development of Eq. (5-54).

5.5 Liquid-Vapor Equilibria in Multicomponent Systems

The purpose of this experiment is to study a binary liquid-vapor equilibrium system. Measurements of liquid and vapor compositions will be made by refractometry after suitable calibration. The data will be treated according to equilibrium thermodynamic considerations, which are developed in the theory section.

THEORY. Consider a liquid-gas equilibrium involving more than one species. By definition, an ideal solution is one in which the vapor pressure of a particular component is proportional to the mole fraction of that component in the liquid phase over the entire range of mole fractions. Note that no distinction is made between solute and solvent. The proportionality constant is the vapor pressure of the pure material.

Empirically it has been found that in very dilute solutions the vapor pressure of solvent (major component) is proportional to the mole fraction X of the solvent. The proportionality constant is the vapor pressure, p°, of the pure solvent. This rule is called Raoult's law:

$$p_{solvent} = p^\circ_{solvent} X_{solvent} \quad \text{for } X_{solvent} \cong 1 \tag{5-57}$$

For a truly ideal solution, this law should apply over the entire range of compositions. However, as $X_{solvent}$ decreases, a point will generally be reached where the vapor pressure no longer follows the ideal relationship.

Similarly, if we consider the solute in an ideal solution, then Eq. (5-57) should be valid. Experimentally, it is generally found that for dilute real solutions the following relationship is obeyed:

$$p_{solute} = K X_{solute} \quad \text{for } X_{solute} \ll 1 \tag{5-58}$$

where K is a constant but not equal to the vapor pressure of pure solute. Equation (5-58) is called Henry's law.

The implication of the foregoing discussion is that ideal solutions are not likely to be found. Only very dilute solutions could be considered ideal. Furthermore, there is no theory available which places limits on just how concentrated a solution must be if deviations from ideality are to be observed. Such limits are set by experimentation and vary, of course, from system to system.

A useful way of treating data relating the vapor pressure of any component to its mole fraction in solution is to plot p_i/X_i versus X_i. Variations of the ratio with X_i indicate deviations from ideality. From the discussion of the previous paragraphs a plot such as that shown in Fig. 5-12 might be expected for a binary solution. At high concentrations of component i, Raoult's law should be valid and the ordinate should be constant and equal to the vapor pressure of pure component i. At some lower mole fraction of i (more concentrated in solute) deviations from ideality may be observed and p_i/X_i may increase or decrease. At very low concentrations, the solvent becomes the solute and Henry's law becomes valid. Thus, p_i/X_i becomes constant again and takes on the value of Henry's law constant.

Ideality and deviations from it may be qualitatively understood by considering molecular interactions. For the purposes of this discussion, we view the gas-liquid interface at the molecular level. A crude representation of a dilute binary system is shown in Fig. 5-13. Component i is the more concentrated (solvent).

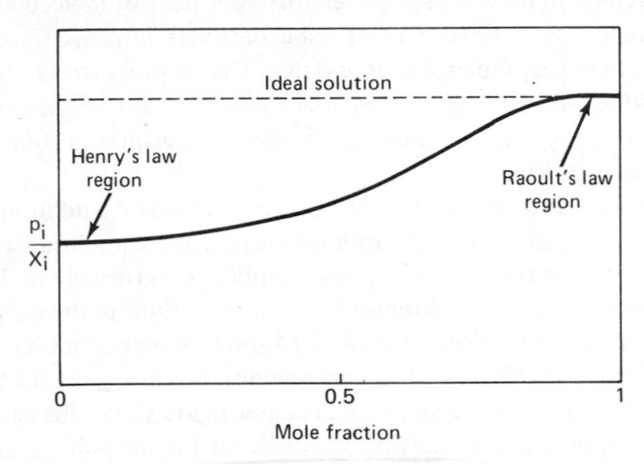

Figure 5-12. Variation of p_i/X_i with X_i.

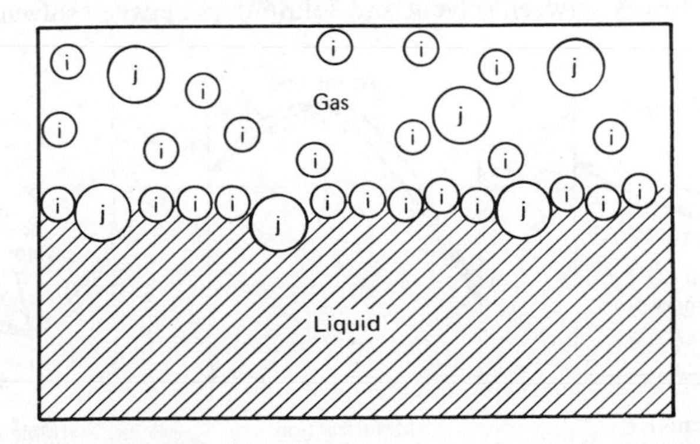

Figure 5-13. Molecular representation of a binary gas-liquid system.

If the system were pure i, then all the j molecules in the surface layer would become i molecules and the observed vapor pressure would be that of the pure liquid. Addition of j molecules to the liquid so that the mole fraction of i is less than unity reduces the concentration of i molecules in the surface layer just as it does in the body of the liquid. Energy is required to transfer a molecule from the liquid to the gas phase, implying that intermolecular potentials are important in determining partial vapor pressures. If molecule i "looks about" and "sees" the same potential field as molecule j regardless of what its neighbors are, then molecule i will require the same energy as molecule j to transfer from the liquid phase to the gas phase. The partial vapor pressures will then be directly proportional to the concentration of each component in the solution and will extrapolate to the vapor pressure of each component when pure i or pure j is used. Thus, a sufficient condition for an ideal solution is to require the interaction potential between (i, j) pairs to be identical to that between (i, i) pairs and (j, j) pairs.

Experimentally observed deviations from ideality are then attributed to differences in interaction potentials between different pairs of molecules. From this point of view deviations from ideality become very intuitive since there is no a priori reason to expect molecules of widely differing properties to have the same interactions. At very high or very low concentrations of one component the constancy of p_i/X_i is easily understood since the partners in these extremes are mostly of a single kind.

From an experimental point of view a binary, two-phase equilibrium system at constant pressure can be characterized by plotting the equilibrium boiling point as a function of the liquid-and vapor-phase compositions (usually mole fractions). These two compositions are different because of variations in the escaping tendency of the molecules involved. Figure 5-14 shows temperature versus composition diagrams for hypothetical ideal and nonideal mixtures. At the left is a typical ideal (or nearly ideal) system. The center figure shows the case of a strong negative deviation from ideality in which the boiling point of some mixtures rises above that of the highest boiling pure component. At the molecular level stronger forces between solvent and solute than between solvent-solvent are

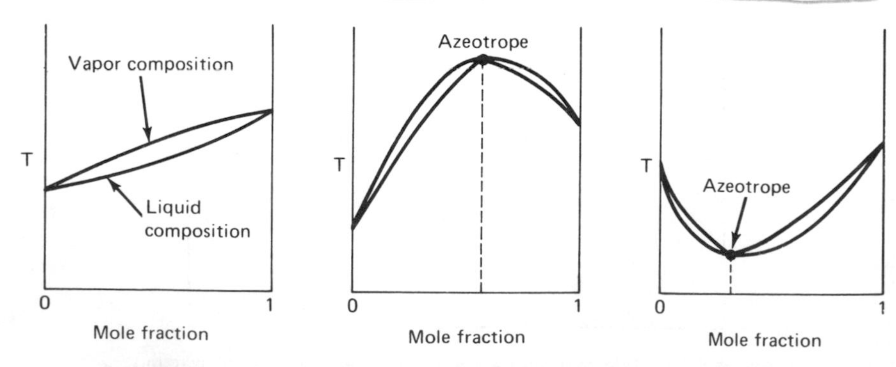

Figure 5-14. Temperature versus composition for a binary two-component system.

implicated. At some intermediate composition the liquid- and vapor-phase compositions come together at the so-called azeotropic composition. Separation of one component from another by fractional distillation is impossible at this composition because of the identicalness of the two phases. The figure at the right in Fig. 5-14 is also a nonideal mixture in which a positive deviation from ideality is noted.

As usual, the equilibrium properties of these solutions are developed by considering the Gibbs free energy or the chemical potential. For the present case of a binary two-phase system at equilibrium, we require the chemical potential, μ_i, of each component to have the same value in every phase:

$$\mu_i^{(l)} = \mu_i^{(g)} \tag{5-59}$$

The chemical potential is related to molar volume $\bar{V}_i$ and pressure p_i by the relation

$$d\mu_i = \bar{V}_i \, dp_i \tag{5-60}$$

If the gas is ideal, substituting for $\bar{V}_i^{(g)}$ in terms of the ideal gas law furnishes

$$d\mu_i^{(g)} = RT \frac{dp_i}{p_i} \tag{5-61}$$

which upon integration leads to

$$\mu_i^{(g)} = RT \ln p_i + A \tag{5-62}$$

where A is a constant of integration. Thermodynamic quantities having dimensions of energy are generally measured with respect to some standard state, the value of the thermodynamic variable at the standard state being agreed upon by convention. For gases the standard state is usually defined as a pressure of 1 atm and the particular temperature at which Eq. (5-62) is applied. Definition of the standard state permits evaluation of A (i.e., set p = 1 atm):

$$\begin{aligned} A &= \mu_i^{\circ(g)} \\ \mu_i^{(g)} &= \mu_i^{\circ(g)} + RT \ln p_i \end{aligned} \tag{5-63}$$

where $\mu_i^{\circ(g)}$ is the standard-state chemical potential. Substituting Raoult's law (an ideal solution law) into Eq. (5-63) leads to

$$\mu_i^{(g)} = \mu_i^{\circ(l)} + RT \ln X_i \tag{5-64}$$

where $\mu_i^{\circ(l)} \neq \mu_i^{\circ(g)}$, which serves to emphasize an important point. If you are making calculations involving standard states, make certain the standard state used is consistent with the concentration units being used (i.e., just as $p_i \neq X_i$, so $\mu_i^{\circ(l)} \neq \mu_i^{\circ(g)}$).

Equation (5-64) depends for its development on both the ideal gas law and the ideal solution law. Use of these laws permitted the connection between $\bar{V}_i^{(g)}$ and p_i and then X_i. Deviations from both these laws do occur. To examine how these deviations are formally treated, we return to Eq. (5-60) which is stripped of all ideal approximations and is, therefore, valid for any component in any phase in an equilibrium system at constant temperature.

To make progress toward integrating Eq. (5-60) under nonideal conditions an equation of state is needed [i.e., $\bar{V}_i = \bar{V}_i(T, p, n_j)$]. Generally speaking, valid equations of state are not readily available in analytical form. Rather, the experimental PVT data have simply been tabulated for those systems that have been studied. With these data, Eq. (5-60) can then, in principle, be integrated numerically to furnish the change in chemical potential between two pressures.

An alternative, introduced by G. N. Lewis, preserves the form of Eq. (5-63) or (5-64) but does not require ideality. In this formulation the pressure is replaced by a quantity f_i, called the fugacity of component i, and Eq. (5-61) becomes

$$d\mu_i^{(\alpha)} = RT \frac{df_i^{(\alpha)}}{f_i^{(\alpha)}} \tag{5-65}$$

which upon integration between some arbitrarily chosen standard state and the actual state of the equilibrium system yields

$$\mu_i^{(\alpha)} = \mu_i^\circ + RT \ln \frac{f_i}{f_i^\circ} \tag{5-66}$$

where μ_i° and f_i° refer to the standard state. In this formulation all deviations from ideality are contained in the fugacities. Generally they are not amenable to quantitative theoretical interpretation and as such are simply empirical experimental quantities. The form of Eq. (5-66) however, is quite convenient because it preserves the form of all equilibrium constant expressions.

Usually Eq. (5-66) is written in terms of an activity a_i, defined as

$$a_i = \frac{f_i}{f_i^\circ} \tag{5-67}$$

so that

$$\mu_i^{(\alpha)} = \mu_i^\circ + RT \ln a_i \tag{5-68}$$

Equation (5-68) is quite general, and the activity a_i may be measured with respect to any definable standard state. In whatever way the standard state is defined it is clear from Eq. (5-68) that when $a_i = 1$ the standard-state chemical potential is reached. Compare Eqs. (5-63) and (5-64) for two ideal cases. In these terms, all thermodynamic quantities become functions of activities.

Conclusion: We have exchanged one difficulty for another. The form of thermodynamic relations has been preserved through the introduction of the fugacity. However, the problem of understanding nonideality has not been advanced at all. From one point of view the fugacity is simply a *fudge factor* which brings experimental data into the form of an ideal equation. From a different point of view, the use of Eq. (5-68) requires a method of obtaining activities from experimental measurements of concentrations. In either case, the right-hand side of Eq. (5-68) is not written in terms of directly measurable experimental quantities.

The activity a_i appearing in Eq. (5-68) is related to a concentration variable through the use of an activity coefficient γ_i. This coefficient depends on both the species i and the concentration scale. For the present experiment the activity coefficient of interest is that of a species present in a solution at mole fraction X_i and $a_i = \gamma_i X_i$. If the vapor phase of this species is ideal, the activity of species i at equilibrium is also given by $a_i = p_i/p_i^\circ$, where p_i° is the vapor pressure of pure species i and p_i the vapor pressure measured at concentration X_i. The result for γ_i is

$$\gamma_i = \frac{p_i}{p_i^\circ} X_i \tag{5-69}$$

EXPERIMENTAL. A refractometer is used in this experiment to measure mole fractions. To obtain mole fraction information from refractive index measurements, the refractometer must be calibrated (i.e., construct a graph of refractive index versus composition) using solutions of known composition. This calibration curve can then be used "in reverse" to find the composition of unknown mixtures from an experimental measurement of the refractive index.

Refer to Section 2.3 for a description of the refractometers and operating instructions. The boiling apparatus is shown in Fig. 5-15. An integral part is a reservoir heated by a resistive filament, as shown, or alternatively by a heating mantle. The filament should be silver-soldered to the ends of a pair of $\frac{3}{32}$ in.-diameter brass rods. These rods in turn are connected to the double pole-double throw switch in the secondary of a low-voltage, high-current transformer (6 V at 8 A is satisfactory) whose primary is operated at 115 V ac. With a solution in this reservoir and boiling, the vapor phase is condensed and trapped in the tube below the condenser. Under equilibrium conditions in the boiling solution–vapor-phase system, the trapped condensate represents the vapor phase, while the liquid remaining in the reservoir represents the solution phase.

In what follows chloroform-acetone mixtures are assumed. The same procedure of sample treatment and analysis works very well for benzene-ethanol mixtures.

Measure the refractive index of chloroform-acetone mixtures of the following compositions and use the data as a calibration of the refractometer system:

Chloroform (%)	Acetone (%)
1. 100	–
2. –	100
3. 25	75
4. 50	50
5. 75	25

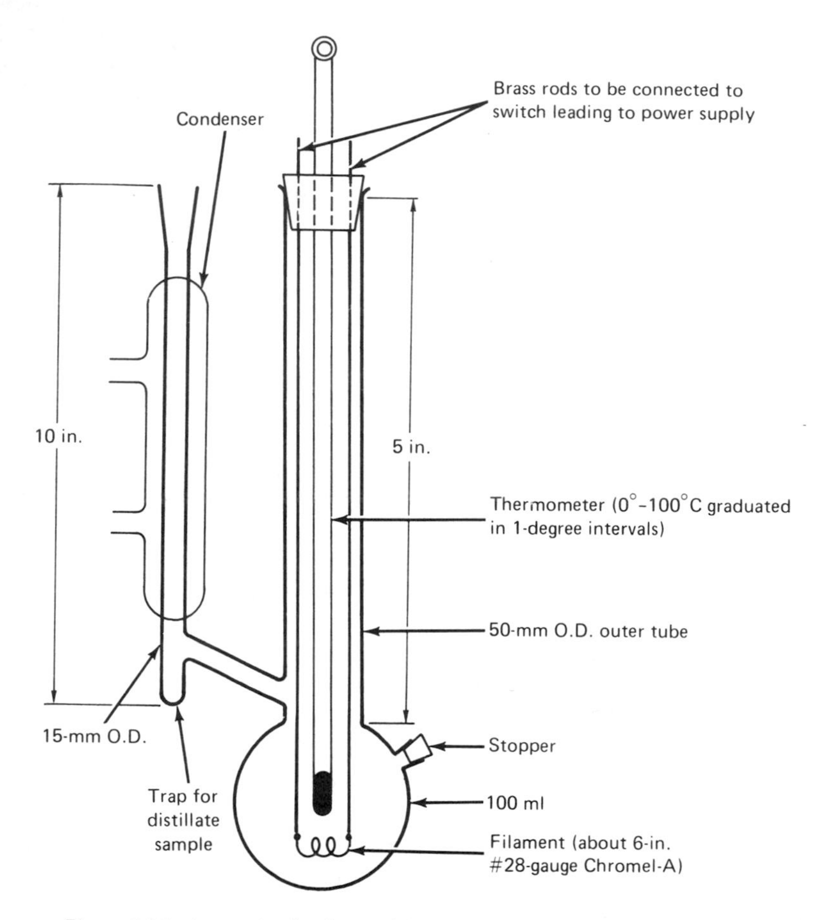

Figure 5-15. Apparatus for determining liquid and vapor compositions as a function of temperature.

The system studied will be chloroform-acetone. Samples of the following solutions should be run starting set A with 25 ml of acetone and set B with 25 ml of chloroform:

Increments added to existing solutions	A (ml of acetone added to 25 ml of chloroform)	B (ml of chloroform added to 25 ml of acetone)
First	4.0	5.0
Second	4.0	5.0
Third	3.0	5.0
Fourth	3.0	5.0
Fifth	3.0	5.0
Sixth	3.0	5.0

Notice that you will begin set A with pure chloroform and add six consecutive portions of acetone to the boiling liquid. If in the course of the run the flask becomes too full, pour out about 10 ml—make certain the heater is off—and continue the addition of acetone. Set B begins with pure acetone.

The following procedure should be followed:

1. Place 25 ml of pure chloroform in the flask through the side arm. Stopper the flask and start water flowing through the condenser. Record the barometric pressure.
2. Close the switch to the heater to start the liquid boiling. *Caution:* Never operate the heater unless liquid covers it. If the heater is operated dry, the filament will burn out.
3. After the distillate trap is filled, empty by tilting the flask so that the distillate will flow back into the body of the liquid. *Remember:* If the heated filament protrudes above the solution, it may burn out. Therefore, open the switch before tilting the flask. The distillate trap should be emptied at least twice before the sample of any solution is withdrawn.
4. After the distillate trap has been filled a third time, allow 2 min for overflow to establish steady-state conditions and then stop heating. Remove the distillate sample with the long-stemmed eyedropper and immediately place into one of the sample holders. Several small test tubes will be needed for samples. Place the sample holders into an ice water bath. The distillate sample represents the vapor-phase composition.
5. Take a sample from the flask as representing the liquid phase at the same time the distillate sample is taken.
6. Record the equilibrium temperature corresponding to these two equilibrium compositions.
7. Measure the refractive index of both samples shortly after taking the sample to avoid sample loss via evaporation. Repeat the refractive index measurements after a few minutes to check for evaporative losses and reproducibility.
8. Record the barometric pressure.

ANALYSIS.

1. Plot a calibration curve which relates mole percent of one of the components to the refractive index of the mixture.
2. From the calibration, determine the composition of the samples taken from the vapor and liquid phases in each experiment.
3. Construct a phase diagram using temperature versus mole percent composition. See Section 5.3 for a brief discussion of phase diagrams. In the present case, there will be two compositions plotted at each equilibrium boiling temperature—one for the liquid phase and one for the vapor phase.

4. In liquid-vapor equilibrium systems, there sometimes exists a particular composition at which the liquid phase and the vapor phase both have the same composition. This is called an azeotropic mixture. What is the azeotropic composition in the chloroform-acetone system and what is its boiling point?
5. Construct a plot of p_i/X_i versus X_i and comment on the ideality or non-ideality of chloroform-acetone mixtures. Discuss any observations from a molecular viewpoint.
6. Calculate and plot the activity coefficient γ_a for acetone as a function of the mole fraction of acetone. What inferences may be made from this plot?

PROBLEMS

1. Consider the qualitative molecular interaction model discussed in the text of this experiment. If molecule A is more strongly attracted to itself than it is to molecule B, make a qualitative plot of the vapor pressure of A versus the mole fraction of B, clearly showing any deviations from ideality.
2. In Eqs. (5-63) and (5-64), define $\mu_i^{\circ(l)}$ and $\mu_i^{\circ(g)}$.
3. Derive an expression for the equilibrium constant K_a in terms of mole fractions and coefficients of activity.
4. Consider an azeotropic mixture of two components. Would fractional distillation be an appropriate way to separate the two components? Why?
5. According to the phase diagram what can be said about the attraction between acetone molecules as compared to the attraction between chloroform and acetone molecules? What might be the origin of this attraction?
6. Is the vapor pressure at the azeotropic composition more or less than that predicted by Raoult's law? Explain.

5.6 Partial Molal Volume

The purpose of this experiment is to determine the partial molal volume of an aqueous solution of a strong electrolyte as a function of its molality. In the theoretical section partial molal quantities and the mathematics of homogeneous functions are developed. A description of two methods of handling the data is also presented.

THEORY. At the outset we shall present a quite general description of homogeneous extensive thermodynamic variables. Thermodynamic variables fall into two classes: Extensive variables are directly proportional to the number of molecules of each component; intensive variables are independent of the number of molecules present. Examples of extensive thermodynamic variables are volume, internal energy, enthalpy, entropy, Gibbs free energy, and Helmholtz free energy. Examples of intensive thermodynamic variables are temperature, pressure, refractive index, and partial molal quantities.

Consider a certain volume V of an aqueous solution of sodium chloride. We can express this volume as

$$V = n_{NaCl}V^{\circ}_{NaCl}k_{NaCl} + n_{H_2O}V^{\circ}_{H_2O}k_{H_2O} \tag{5-70}$$

where V°_i is the volume of 1 mole of pure species i, n_i the number of moles of species i, and k_i a proportionality factor which depends only on the ratio of n_{NaCl} to n_{H_2O}. Suppose that we make a different volume of solution at the same temperature, pressure, and n_{NaCl}/n_{H_2O} ratio. This solution is obtained by multiplying the amounts of NaCl and H_2O by a constant, λ. The resulting volume V' is then

$$V' = \lambda V = \lambda n_{NaCl}V^{\circ}_{NaCl}k_{NaCl} + \lambda n_{H_2O}V^{\circ}_{H_2O}k_{H_2O} \tag{5-71}$$

Since Eq. (5-71) is Eq. (5-70) multiplied by λ raised to the first power, we say that the total volume is a homogeneous function of degree 1 in the number of molecules (or moles) of each component.

In general a function, f, of the independent variables u, v, and w is said to be a homogeneous function of degree n if

$$f(\lambda u, \lambda v, \lambda w) = \lambda^n f(u, v, w) \tag{5-72}$$

Since the intensive thermodynamic variables are, by definition, those which are independent of the number of molecules of each component, they must be homogeneous functions of zero degree in the number of moles of each component. By contrast all the extensive variables are homogeneous functions of first degree.

A very useful mathematical property of homogeneous functions is given by Euler's theorem, which is stated as follows: If f(u, v, w) is a homogeneous function of degree n, then

$$nf(u, v, w) = u\left(\frac{\partial f}{\partial u}\right)_{v,w} + v\left(\frac{\partial f}{\partial v}\right)_{v,w} + w\left(\frac{\partial f}{\partial w}\right)_{v,u} \tag{5-73}$$

Applying this theorem to the total volume of an aqueous solution of sodium chloride we have

$$V = V(n_1, n_2, T, p) \tag{5-74}$$

where n_1 denotes NaCl and n_2 denotes H_2O. Now V is homogeneous of degree 1 in n_i. Hence, we can also write

$$V = n_1\left(\frac{\partial V}{\partial n_1}\right)_{n_2,T,p} + n_2\left(\frac{\partial V}{\partial n_2}\right)_{n_1,T,p} \tag{5-75}$$

Comparing Eq. (5-75) with Eq. (5-70) furnishes

$$\left(\frac{\partial V}{\partial n_i}\right)_{n_j,T,p} = V_i^\circ k_i \tag{5-76}$$

Generally we define the partial molal volume, $\bar{V}_i$, as

$$\bar{V}_i = \left(\frac{\partial V}{\partial n_i}\right)_{n_j,T,p} \tag{5-77}$$

In words, the partial molal volume of component i is given by the rate of change of the total volume with the number of moles of component i holding constant the number of moles of all other species, the temperature, and the pressure.

Rewriting Eq. (5-75) furnishes

$$V = n_1\bar{V}_1 + n_2\bar{V}_2 \tag{5-78}$$

We may look into the physical content of Eq. (5-78) by examining it under certain conditions. Suppose that two pure substances A and B have molar volumes $\bar{V}_A^\circ$ and $\bar{V}_B^\circ$, respectively. Imagine mixing together 1 mole of each. What is the new total volume, V_T? The answer depends on how molecules A and B interact. If the interaction were just right, we could get $V_T = \bar{V}_A^\circ + \bar{V}_B^\circ$. This is an idealized case, and normally we expect some change in total volume on mixing because of changes in the magnitudes of various forces. We conclude then that $\bar{V}_1$ differs from V_1°, the molar volume of pure component 1, because $\bar{V}_1$ involves forces that are absent in $\bar{V}_1^\circ$, namely, solvent-solute interactions. It is not easy to describe these interactions quantitatively and in most instances we simply use a comparison of $\bar{V}_1$ and $\bar{V}_1^\circ$ as a qualitative guide.

For a two-component system we can arrive at another interesting equation as follows. Taking total derivatives of Eqs. (5-74) and (5-78) and setting them equal, we obtain

$$n_1\, d\bar{V}_1 + n_2 d\bar{V}_2 - \left(\frac{\partial V}{\partial p}\right)_{n_i,T} dp - \left(\frac{\partial V}{\partial T}\right)_{n_i,p} dT = 0 \tag{5-79}$$

Under conditions of constant temperature and pressure Eq. (5-79) reduces to the Gibbs-Duhem equation

$$\frac{d\bar{V}_1}{d\bar{V}_2} = -\frac{n_2}{n_1} = -\frac{X_2}{X_1} \tag{5-80}$$

where X_i is the mole fraction of component i which is defined as

$$X_i = \frac{n_i}{\sum_j n_j} \tag{5-81}$$

Observe that Eq. (5-80) furnishes a functional relationship between $\bar{V}_1$ and $\bar{V}_2$. This means that $\bar{V}_1$ and $\bar{V}_2$ are not both independent variables (i.e., $\bar{V}_1$ cannot be changed while holding $\bar{V}_2$ constant).

Thus far, our attention has been restricted to the volume. However, since all other extensive thermodynamic variables are also homogeneous functions of degree 1 in the amounts of each component, we can obtain relations analogous to Eqs. (5-74), (5-78), (5-79), and (5-80) for each of these variables provided the system contains only two components.

In the remainder of this development we shall search for relations between $\bar{V}_1$ or $\bar{V}_2$ and experimentally available quantities such as the total volume V, the molality m, and the solution density d. Two general methods of handling the data will be presented. Both methods use Eq. (5-78) but in different ways.

First Method. Equation (5-78) can be arbitrarily rewritten in the form

$$V = n_1 Q + n_2 \bar{V}_2^\circ \tag{5-82}$$

where $\bar{V}_2^\circ$ is the partial molal volume of pure water (i.e., the volume occupied by 1 mole of water at the experimental temperature and pressure), and Q is defined as the apparent partial molal volume of NaCl. V, $\bar{V}_2^\circ$, n_1, and n_2 can be obtained from experimental measurement of density, weight of solvent, weight of solute, temperature, and pressure. Hence, from Eq. (5-82), Q can be determined. Thus, the problem of determining $\bar{V}_1$ and $\bar{V}_2$ is one of relating these quantities to Q since the right-hand side of Eq. (5-82) must be regarded as an assumed form. Note that Q is an "invented" variable which is related in a well-defined way to certain experimentally measurable quantities.

To relate Q to $\bar{V}_1$ and $\bar{V}_2$, we arbitrarily choose to make n_2 correspond to 1000 g of solvent, which for water becomes 55.51 moles. Then n_1 becomes the number of moles of solute in 1000 g of solvent or the molality, m. Hence,

$$\begin{aligned} V &= m\bar{V}_1 + 55.51\bar{V}_2 \\ &= mQ \ + 55.51\bar{V}_2^\circ \end{aligned} \tag{5-83}$$

where V is the volume of solution containing 55.51 moles of water. From Eq. (5-83) Q can be calculated for any known V, m, and $\bar{V}_2^\circ$.

By definition

$$\bar{V}_1 = \left(\frac{\partial V}{\partial n_i}\right)_{n_2,T,p} \tag{5-84}$$

and from Eq. (5-83), $\bar{V}_1 = (\partial V/\partial m)_{T,p}$. Differentiating Eq. (5-83) furnishes

$$\bar{V}_1 = Q + m\frac{dQ}{dm} \tag{5-85}$$

and

$$\bar{V}_2 = \bar{V}_2^\circ - \frac{m^2(dQ/dm)}{55.51} \tag{5-86}$$

From Eqs. (5-85) and (5-86) it is clear that experimental determination of $\bar{V}_2^\circ$, m, and dQ/dm suffice to determine $\bar{V}_2$ and $\bar{V}_1$.

Thus far in our discussion we have not particularized the development to a class of solutions. Thus, Eqs. (5-85) and (5-86) can be applied to any kind of one-phase, two-component system whether it be solid, liquid, or gas. Generally the application is to liquid solutions, which can be divided into two classes: ionic and nonionic. For dilute aqueous solutions of strong electrolytes such as NaCl, the Debye-Hückel treatment may be applied as described in Section 9.1 to show that the apparent partial molal volume Q should vary linearly with the $\sqrt{m}$. Since Eqs. (5-85) and (5-86) require dQ/dm, it will be useful to plot Q versus $\sqrt{m}$, to find $dQ/d\sqrt{m}$, and to use the relation

$$\frac{dQ}{dm} = \frac{d\sqrt{m}}{dm}\frac{dQ}{d\sqrt{m}} = \frac{1}{2\sqrt{m}}\frac{dQ}{d\sqrt{m}} \tag{5-87}$$

Using Eq. (5-87) in Eqs. (5-85) and (5-86) furnishes

$$\bar{V}_2 = \bar{V}_2^\circ - \frac{m}{55.51}\frac{\sqrt{m}}{2}\frac{dQ}{d\sqrt{m}} \tag{5-88}$$

$$\bar{V}_1 = Q + \frac{\sqrt{m}}{2}\frac{dQ}{d\sqrt{m}} \tag{5-89}$$

From the experimental data on Q, m, $\bar{V}_2^\circ$, and the slope of a plot of Q versus $\sqrt{m}$, one can determine $\bar{V}_2$ and $\bar{V}_1$.

The molality of the NaCl solution is related to the molarity by

$$m = \left(\frac{d}{M} - \frac{M_1}{1000}\right)^{-1} \tag{5-90}$$

where M is the molarity, d the density, and M_1 the molecular weight of NaCl. It should be noted carefully that V in Eq. (5-83) is not determined directly by experiment but is the volume of solution containing 55.51 moles of water.

Second Method. This alternative method, called the intercept method, makes use of Eq. (5-78) in a different way. Dividing Eq. (5-78) by $n_1 + n_2$ yields an expression in terms of the mole fractions of the two components:

$$\bar{V} = X_1\bar{V}_1 + X_2\bar{V}_2 = X_1(\bar{V}_1 - \bar{V}_2) + \bar{V}_2 \tag{5-91}$$

where

$$\bar{V} \equiv \frac{V}{n_1 + n_2} \tag{5-92}$$

Note that $\bar{V}$ is experimentally available.

Plotting $\bar{V}$ versus X_1 furnishes a graph (generally nonlinear) whose tangent measured at X_1 has a slope of $\bar{V}_1 - \bar{V}_2$ and an intercept of $\bar{V}_2$. Since the labels 1 and 2 can be interchanged, the same argument obviously applies to the second component of the binary mixture.

There are alternative sets of concentration variables which one can use in the intercept method. For example, a plot of solution density versus molar concentration (grams per liter) of one of the components can be used. Returning to Eq. (5-78) and dividing both sides by V, we obtain

$$\frac{n_1}{V}\bar{V}_1 + \frac{n_2}{V}\bar{V}_2 = 1 \tag{5-93}$$

but $n_i/V = C_i/M_i$, the concentration in grams per liter of species i divided by the molecular weight of species i. The density ρ of the solution is related to C_i by

$$\rho = \frac{C_i}{X_i^w} \tag{5-94}$$

where X_i^w is the weight fraction of i and is expressed as

$$X_i^w = \frac{w_i}{\sum_j w_j} \tag{5-95}$$

where w_j is the weight of species j in the solution. Employing Eqs. (5-94) and (5-95) in Eq. (5-93) gives

$$\rho X_1^w \frac{\bar{V}_1}{M_1} + \rho X_2^w \frac{\bar{V}_2}{M_2} = 1 \tag{5-96}$$

This equation can be simplified by noting that for a two-component system $X_1^w + X_2^w = 1$. Inserting this in Eq. (5-96) to eliminate X_1^w and then replacing X_2^w by C_2/ρ, we obtain the desired expression:

$$\rho = \frac{\bar{V}_1}{M_1} + \left(1 - \frac{\bar{V}_1/M_1}{\bar{V}_2/M_2}\right)C_2 \tag{5-97}$$

Clearly the slope of ρ versus C_2 at a given value of C_2 is $\{1 - [(\bar{V}_1/M_1)/(\bar{V}_2/M_2)]\}$, and the tangent to the curve at this C_2 has an intercept $\bar{V}_1/M$. Thus, measurement of the slope and intercept suffice to determine both $\bar{V}_1$ and $\bar{V}_2$ at the selected value of C_2.

EXPERIMENTAL. As described above binary liquid solutions can be of two types: electrolyte and nonelectrolyte. From a physical point of view we expect the nature of the forces to be quite different in these and it is therefore of some interest to study and compare both types.

In any case the measurements involve the use of a pycnometer, one type of which is shown in Fig. 5-16. This Ostwald-Sprengel type of device is used here to determine the density of solutions. To compute the density of a solution the volume of the pycnometer is determined by weighing it full and empty of some material whose density is known at the temperature of the measurement. Here distilled water at room temperature is used. Reiterating, the first stage of this

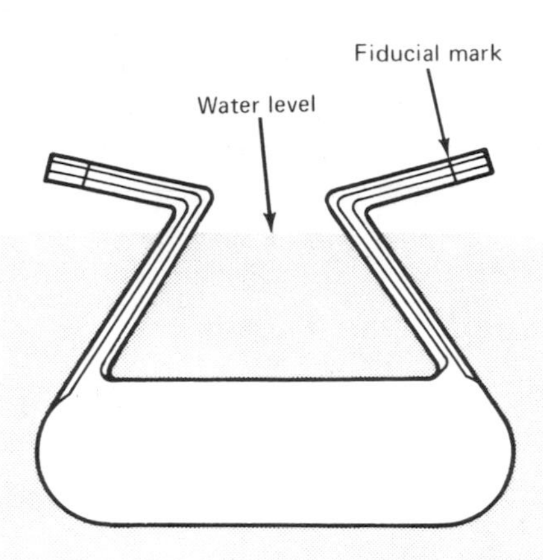

Figure 5-16. Pycnometer.

experiment consists of weighing the dry, clean, and empty pycnometer; then weighing it filled with water at a known temperature (measure it); and finally calculating the pycnometer volume from the weight of water and its density, the latter being available in any handbook of chemistry. It is important to make these weighings as carefully and accurately as possible. The balance should furnish the weight to at least 1 mg and preferably 0.1 mg. The pycnometer should have a capacity compatible with the upper limit of the weight the balance can comfortably handle and at the same time be large enough to permit the determinations to have some precision. Assuming a balance capable of handling 100g, a pycnometer of 25–30 cm is convenient. The fiducial marks in Fig. 5-16 are used as filling marks; the menisci of the solutions should be located at these marks. In filling the pycnometer, air bubbles are to be avoided for obvious reasons. To fill, dip one arm of the pycnometer into the solution and apply suction with an aspirator or a rubber bulb. To avoid bubbles, keep the aspirator tilted so that it fills without trapping any air. When the pycnometer is nearly full, reduce the rate of filling and bring the solution past the fiducial marks so that it is overfilled. Remove the suction device carefully so that no solution is lost. If the pycnometer ends up slightly underfilled, then use an eyedropper full of solution to fill to the fiducial marks. To do this without trapping air it is necessary to tilt the pycnometer until the solution completely fills one capillary arm. Then apply the eyedropper to this arm, and with one motion level the pycnometer and force solution from the eyedropper into the arm. For an overfilled pycnometer, just tilt until a small amount of solution runs out of one arm, relevel, and wipe away what ran out. Adjust the level on each side to the fiducial marks, wipe the outside dry, record the solution temperature, and weigh the filled pycnometer. Repeat this procedure for all the solutions. Between measurements rinse the pycnometer with small amounts of the solution to be measured next.

Strong Electrolyte Solutions. Aqueous strong electrolyte solutions can be easily prepared by dissolving a readily soluble salt in water. Sodium chloride, potassium iodide, and sodium acetate are excellent candidates. Two procedures for making the solutions may be employed: (1) dissolving a known weight of salt in water to make a known volume of solution using a volumetric flask and (2) dissolving a known weight of salt in a known weight of water. The latter procedure is described here. For any of the salts described above the concentration range 0–3 M should be covered using four or five solutions of different concentration. It is necessary to make at least enough solution to fill the pycnometer, and it is actually desirable to have a considerable excess. From the molecular weight of the salt and the known amount of solution needed, calculate the approximate weights of the two components needed to make up the four or five solutions mentioned above. Weigh a clean and dry container (beaker or Erlenmeyer flask), add about the amount of water needed, and weigh again. Add a known weight of salt and weigh again. One of the latter weighings is clearly redundant, but it will serve as an easily obtained check of internal consistency. After preparing the solutions, fill and weigh the pycnometer as described above.

Nonelectrolyte Solutions. Just as in the case of electrolyte solutions, there is a variety of possibilities here. Water-*n*-propanol, water-*iso*-propanol, and benzene-carbon tetrachloride are good candidates. The constituents of these binary solutions are all liquids at room temperature so that measurements can be made from both ends of the mole fraction or weight fraction scale. In what follows attention is focused on the *n*-propanol–water system. Other systems would be studied in a similar way.

Prepare mixtures of 10, 15, 20, 50, and 75% by weight of *n*-propanol. Make enough of each to more than fill the pycnometer. Use an accurate balance capable of measuring 1 mg and preferably 0.1 mg. Weigh the empty, clean, and dry mixing vessel; add water and weigh again; and finally add *n*-propanol and weigh the mixture. The density of water at 25°C is near 1 g/cm^3, and the density of *n*-propanol at 25°C is near 0.8 g/cm^3. With these two figures, the volumes of each required to give approximately the weight percentage desired can be calculated.

At a given temperature (room temperature), make pycnometer measurements as described above on these five solutions and of pure *n*-propanol. Make measurements quickly after filling to avoid evaporative loss problems. There may be some problems with bubble formation in the propanol-water mixtures. It is advisable to overfill the pycnometer and let it stand for a few minutes before adjusting the levels to the fiducial marks. If small bubbles persist, estimate their volume and forge ahead.

ANALYSIS.

Strong Electrolyte Solutions.

1. From the average empty weight of the pycnometer and the average weight of the pycnometer filled with distilled water, calculate the volume of the pycnometer by making use of the density of water at the experimental temperature. The density of water is tabulated in the *Handbook of Chem-*

istry and Physics. Estimate the error incurred as a result of the buoyancy of air and make this correction if necessary (i.e., what weight of air does the pycnometer hold?). These data furnish $\bar{V}_2^\circ$.

2. Calculate an accurate value for the molality of each of the salt solutions using Eq. (5-90), and then using Eq. (5-83), calculate Q for each solution. Keep in mind that Eq. (5-83) assumes that 55.51 moles of water were used in making up the solution. Hence, V of Eq. (5-83) may need to be an adjusted experimental V. Estimate the uncertainty in each value of Q taking into account errors in V, m, and $\bar{V}_2^\circ$. Be explicit (i.e., mathematics, not guesswork).
3. Plot Q versus $\sqrt{m}$. Using a least-squares procedure, fit the data to a straight line; this fit will furnish the slope and intercept. Plot this line on the graph.
4. Using the data now at hand, calculate $\bar{V}_1$ and $\bar{V}_2$ for each molality.
5. Determine $\bar{V}_1$ and $\bar{V}_2$ using the intercept method and compare the two methods.
6. Comment on the variation of the partial molal volumes with molality by considering the intermolecular or interatomic forces which are present.
7. How well do the data obey the Gibbs-Duhem equation [Eq. (5-80)]? A plot of X_2/X_1 versus $\Delta\bar{V}_1/\Delta\bar{V}_2$ may be useful.

Nonelectrolyte Solutions.

1. Determine the pycnometer volume as described above.
2. Plot the experimental data according to Eq. (5-97) with the *n*-propanol concentration along the abscissa.
3. Determine $\bar{V}_1$ and $\bar{V}_2$ at concentrations of 15 and 75% by weight of *n*-propanol. Use the intercept method.
4. At 50% by weight *n*-propanol, is the volume of the solution greater than, less than, or equal to the volume calculated by adding the volumes each pure component would occupy? What sort of forces could account for this result?

PROBLEMS

1. Using the property of homogeneous functions given in Eq. (5-72), prove Euler's theorem given that f(u, v, w) is a homogeneous function of degree n.

2. Derive Eq. (5-79) using Eqs. (5-74) and (5-78).

3. Write an expression analogous to Eq. (5-80) for the Gibbs free energy. $\bar{G}_i$ is of special importance in chemical thermodynamics and is called the chemical potential and usually given the special symbol μ_i.

4. Derive Eq. (5-86) using the definition of $\bar{V}_2$ and the equations given in the text of this write-up.

5. Derive Eq. (5-90) using definitions of M and m.

5.7 Variation of Equilibrium Constant with Temperature

The purpose of this experiment is to collect equilibrium pressure-temperature data from which the equilibrium constant and its variation with temperature can be calculated. The basic relations connecting the experimental data with the thermodynamic quantities are developed. The system under study is $N_2O_4 \rightleftharpoons 2\,NO_2$. Both molecules are gases at the temperatures used in these experiments.

THEORY. In its most general form the equilibrium constant describing a process is given in terms of the activities a_i of the species participating in the process. In this experiment we shall consider the gas-phase equilibrium

$$N_2O_4 \rightleftharpoons 2NO_2$$

for which the equilibrium constant is

$$K_a = \frac{(a_{NO_2})^2}{(a_{N_2O_4})} = \frac{\gamma_{NO_2}^2 p_{NO_2}^2}{\gamma_{N_2O_4} p_{N_2O_4}} = \frac{\gamma_{NO_2}^2}{\gamma_{N_2O_4}} K_p \tag{5-98}$$

where γ_i is the activity coefficient of species i. If the gases are ideal, then by definition $\gamma_i = 1$ and we have

$$K_a = K_p = \frac{p_{NO_2}^2}{p_{N_2O_4}} \tag{5-99}$$

In this experiment we assume the validity of Eq. (5-99) and proceed to relate K_p to the experimentally available quantities, total pressure p, temperature T, total volume V, and total weight of gas w. The pressures are normally expressed in atmospheres.

The equation of state for both gases is, by hypothesis, the ideal gas law. Thus, we have

$$pV = nRT \tag{5-100}$$

where n is the total number of moles of gas and is given by

$$n = n_{NO_2} + n_{N_2O_4} \tag{5-101}$$

We define α, the fractional dissociation of N_2O_4, by

$$\alpha \equiv \frac{n_{NO_2}}{2n_{N_2O_4}^{\circ}} \tag{5-102}$$

where $n^{\circ}_{N_2O_4}$ is the number of moles of N_2O_4 present if no dissociation takes place. A numerical value of $n^{\circ}_{N_2O_4}$ is calculable from the weight w of sample used by dividing by the molecular weight of N_2O_4, M.

In terms of these quantities

$$n_{N_2O_4} = \frac{w}{M}(1 - \alpha) \tag{5-103}$$

$$n_{NO_2} = 2\frac{w}{M}\alpha \tag{5-104}$$

$$p_{N_2O_4} = \frac{(1 - \alpha)}{(1 + \alpha)}p \tag{5-105}$$

$$p_{NO_2} = \frac{2\alpha}{1 + \alpha}p \tag{5-106}$$

Direct substitution of Eqs. (5-105) and (5-106) into (5-99) furnishes

$$K_p = \frac{4\alpha^2}{1 - \alpha^2}p \tag{5-107}$$

Examining this expression makes clear the necessity of stating the units of pressure when reporting K_p values. It remains to relate α to experimentally measured quantities. Progress toward this end begins by combining Eqs. (5-99) and (5-100):

$$pV = (n_{NO_2} + n_{N_2O_4})RT \tag{5-108}$$

Substituting Eqs. (5-103) and (5-104) into (5-108) yields

$$pV = \left(\frac{w}{M}\right)(1 + \alpha)RT \tag{5-109}$$

Solving for α and substituting into Eq. (5-98) we find

$$K_p = \frac{4[(pVM/RTw) - 1]^2}{2 - (pVM/RTw)}\left[\frac{VM}{RTw}\right] \tag{5-110}$$

This equation or its equivalent provides the basis for the experimental determination or K_p. In one way or another we must gain access to p, V, T, and w.

The variation of the equilibrium constant with temperature is obtained from the relationship between the standard Gibbs free energy change of the process and the equilibrium constant. Again assuming that the ideal gas equation of state is valid, the desired relationship is

$$-\Delta G^{\circ} = RT \ln K_p \tag{5-111}$$

Working with the left-hand side of Eq. (5-111) and beginning with

$$G = E + PV - TS \tag{5-112}$$

one can derive, by differentiation,

$$\left(\frac{\partial G^\circ}{\partial T}\right)_p = -S^\circ = \frac{G^\circ - H^\circ}{T}$$

or

$$\left(\frac{\partial\, \Delta G^\circ}{\partial T}\right)_p = -\Delta S^\circ = \frac{\Delta G^\circ - \Delta H^\circ}{T} \tag{5-113}$$

which is called the Gibbs-Helmholtz equation. Multiplying Eq. (5-113) by 1/T furnishes

$$\frac{1}{T}\left(\frac{\partial\, \Delta G^\circ}{\partial T}\right)_p - \frac{\Delta G^\circ}{T^2} = -\frac{\Delta H^\circ}{T^2} \tag{5-114}$$

The left-hand side of Eq. (5-114) is identical to

$$\left[\frac{\partial}{\partial T}\left(\frac{\Delta G^\circ}{T}\right)\right]_p = -\frac{\Delta H^\circ}{T^2} \tag{5-115}$$

Equation (5-115) furnishes the temperature dependence of the Gibbs free energy [i.e., the temperature dependence of the left-hand side of Eq. (5-111)]:

$$-\left[\frac{\partial}{\partial T}\left(\frac{\partial\, \Delta G^\circ}{\partial T}\right)\right]_p = \frac{\Delta H^\circ}{T} = \frac{\partial}{\partial T}(R \ln K_p) \tag{5-116}$$

This can be rearranged to furnish

$$\frac{d \ln K_p}{d(1/T)} = -\frac{\Delta H^\circ}{R} \tag{5-117}$$

Hence, a plot of $\ln K_p$ versus $1/T$ is predicted to have slope $-\Delta H^\circ/R$. If ΔH° is not dependent on T, then $\Delta H^\circ/R$ is a constant with respect to this variable and the plot will be linear. ΔH° is the standard enthalpy change associated with the forward reaction. In the present case it is the heat of dissociation of N_2O_4.

EXPERIMENTAL. Several different experimental methods may be used to determine the variables on the right-hand side of Eq. (5-110). Three methods are presented here. In the first, a fixed weight of material in a known volume is used, and the total pressure is measured as a function of temperature. In the second method the pressure and volume are fixed, and the sample weight is measured as a function of temperature. The third method also keeps the total pressure and volume fixed and measures the partial pressure of NO_2 by a spectroscopic technique. In this method K_p is calculated using Eq. (5-99) once the optical absorption data have been converted to pressures.

At the outset it is important to appreciate the corrosive and toxic nature of NO_2 and to take commensurate precautions. Breathing oxides of nitrogen will lead to formation of nitric and nitrous acid in the respiratory system and in severe cases causes edema of the lungs. Prolonged breathing of low concentrations can lead to pulmonary edema.

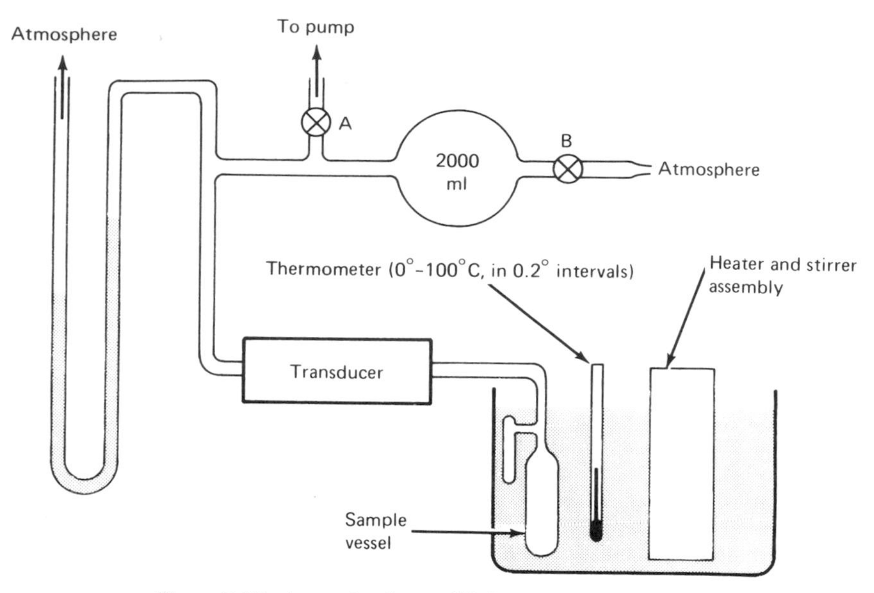

Figure 5-17. Apparatus for equilibrium constant experiment.

First Method. The apparatus for the first method is shown in Fig. 5-17. The N_2O_4-NO_2 vessel should have a volume of about 200 ml of which 2 ml or less is not submerged in the heating bath. This vessel is permanently attached to the pressure transducer. The volume of this vessel can be determined in a variety of ways with varying degrees of accuracy. One very simple way which gives results of sufficient reliability is to fill the vessel with water prior to connecting it to the transducer and measuring the volume or the weight of this water. Since some glass blowing is done during assembly, a correction will be necessary for the reduction in volume.

A known weight of NO_2-N_2O_4 is sealed into a thin-walled glass tube and is inserted into the 12-mm-diameter side arm of the vessel. A glass-encased magnet is inserted, and the container tube is closed (see Figure 5-18). The vessel is then evacuated through the constricted side arm and is sealed off under vacuum by using a glass-blowing torch to completely close the constriction. By dropping the magnet on the ampule the sample is released into the vacuum system.

A good procedure for preparing the ampules is outlined in Fig. 5-19. A vacuum line is used to which a storage cylinder of nitrogen dioxide and a standard taper-capillary tube device are attached. The stopcocks shown should be lubricated with Kel-F which is relatively inert toward NO_2. The capillary should be sealed to a $\frac{5}{20}$ $ joint and should be connected, after weighing, to the vacuum line with sealing wax (for example, picein cement). The vacuum line is then pumped out, valve A is closed, and a small amount of NO_2 is admitted. After standing a couple minutes, valve A is reopened and the NO_2 is trapped in the dry ice-acetone trap as the system is evacuated. Close valve A again and admit enough N_2O_4-NO_2 mixture to correspond to about 0.3 g. The boiling point of NO_2 is about 21°C.

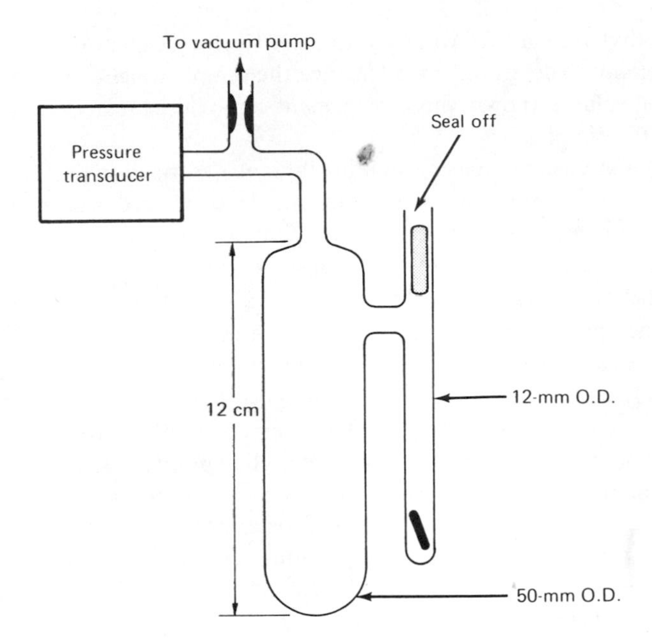

Figure 5-18. Detail of a sample vessel.

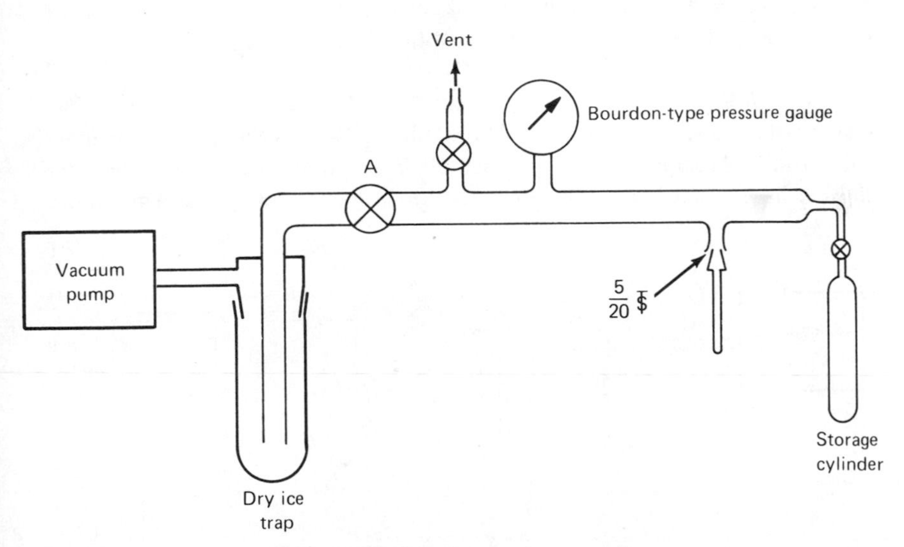

Figure 5-19. Sample ampule preparation.

Using a liquid nitrogen bath, freeze this NO_2–N_2O_4 sample into the lower tip of one of the capillaries, and while frozen seal off the capillary as short as possible. Make sure to keep both parts of the resulting capillary assembly together since a second weighing will be needed to determine the sample weight. Pump out the system, turn off the pump, vent through stopcock B, and quickly remove the trap to a hood. Warm up the $\frac{5}{20}$ joint, remove the standard taper, clean it with trichloroethylene and then with alcohol, and finally weigh both parts of the capillary assembly to determine by difference the sample weight. With the sample weight and volume known, this arrangement can be used many times without changing the sample.

Pressures are measured using a null technique described in Section 2.1. The null point may be determined with either an electronic pressure transducer or a glass spoon gauge, which are both described in Section 2.1. In either case the nulling device isolates the corrosive N_2O_4–NO_2 mixture from the mercury manometer. The temperature range should be 5°–60°C, and the experiment should begin at the upper end of this range. Using the nulling device according to instructions, take pressure versus temperature readings while the sample tube is being slowly cooled by adding cool water to the water bath. The rate of cooling should not exceed 1°C every 2 min. This technique will prove just as reliable as trying to hold the temperature constant at several points in the range. If an open-ended manometer like the one shown in Fig. 5-17 is used, it will be necessary to determine the atmospheric pressure before the sample pressure can be determined. Valves A and B are used to adjust the pressure—A to decrease and B to increase it. The 2000-ml flask is used to increase the volume of the "nulling" side of the transducer arrangement so that small changes in the amount of air through addition or removal do not cause large changes in the pressure. This gives the experimenter better control. The data for this method consist of pressure versus temperature points, which are used in Eq. (5-110) to obtain values for K_p.

Second Method. The apparatus for the second method is shown in Fig. 5-20 and consists of a vacuum line suitable for filling glass sample bulbs in a manner similar to that described above. The sample bulbs should be about 250 ml in volume and have either Teflon plug high-vacuum stopcocks or standard glass bore

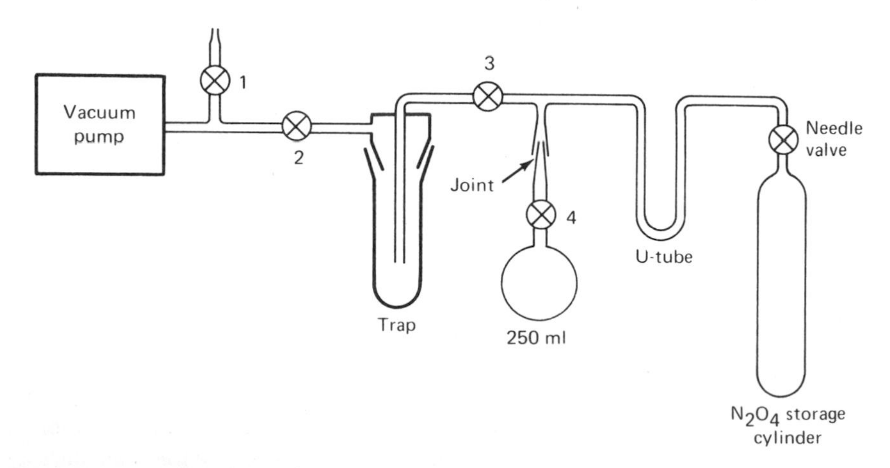

Figure 5-20. Apparatus for filling sample bulbs.

high-vacuum stopcocks greased with Kel-F grease. Both the volume and evacuated weight of the sample bulb should be determined. The volume may be determined at the end of the experiment, but the evacuated weight should be determined both at the start and at the end. The joint shown in Fig. 5-20 can be either a standard taper sealed with picein wax or an ungreased O-ring joint. If the former is used, the joint must be carefully cleaned with trichoroethylene followed by ethanol before it is weighed. In operation, it is important that no N_2O_4-NO_2 gas reach the vacuum pump. This is achieved by using a liquid nitrogen or dry ice-acetone trap as shown. Furthermore, it is desirable to place the entire apparatus in a well-ventilated hood.

Since N_2O_4 boils at 21°C, it is not necessary to use a manometer to measure the pressure of a sample; the vapor pressure above a liquid sample at a temperature slightly greater than 21°C may be used. The vacuum line should be evacuated and the liquid nitrogen placed around the cold trap. After closing stopcock 3 the needle valve at the N_2O_4 cylinder should be opened slowly and a pressure equal to the vapor pressure allowed to fill the sample bulb through stopcock 4. Stopcocks 4 and 2 should then be closed and 3 opened to freeze out the N_2O_4 in the sample line. It may turn out that the material from the cylinder needs to be degassed. If so, the U-tube shown can be used to degas the sample frozen at liquid nitrogen temperature.

Once the sample bulb is filled it is inserted into a heat bath (water or air) at temperatures slightly above room temperature (say around 30°C) and allowed to come to thermal equilibrium for about 5 min. Refer to Section 2.2. The stopcock is then momentarily opened to allow pressure equilibration with the atmosphere. After a few minutes it is opened again. Repeating this process the system should reach a condition where no further NO_2 fumes escape on opening the valve. The dried bulb is then weighed to an accuracy of 0.1 mg using a double pan balance. This procedure is repeated with the same bulb of gas up to temperatures of about 60°C with measurements every 10°C. After the highest temperature measurement has been made the sample should be removed by evacuation through the liquid nitrogen trap and the evacuated bulb should be weighed. Then by filling with distilled water (open the stopcock while the tube is beneath water and the bulb evacuated) and weighing, the volume of the bulb can be calculated from the density of water and the weight of water in the bulb. At the close of the experiment, evacuate the sample bulb and line, close valve 3, turn off the vacuum pump, vent through valve 1, remove the trap to a hood, and open valve 3.

The data for this method then consist essentially of gas densities as a function of temperature. These data are used in Eq. (5-110) with density $\rho = w/V$.

Third Method. The third method of measuring the equilibrium properties of this system employs (1) optical absorption and Beer's law to determine the concentration of NO_2, (2) a solid-vapor equilibrium at –30.6°C to fix the total pressure, and (3) a thermostat to vary the temperature of an optical absorption cell.

A schematic diagram of the apparatus is shown in Fig. 5-21. The vacuum system construction and valve lubrication should be carried out with the same considerations in mind as the weighing method described above. Two new features appear in this vacuum system: a bromobenzene trap and an optical absorption cell. The former is simply a cold finger which during the experiment is maintained at –30.6°C (see Section 2.2 for instructions on making slush baths). The latter should be a Pyrex tube about 40 mm in diameter and 10 cm long with flat Pyrex windows on the ends. These can be sealed on with epoxy resin or, preferably, glass-blown on.

Once the vacuum system is together, evacuated, and leak-checked, valve 2 should be closed and about $\frac{1}{2}$ atm of N_2O_4 admitted from the storage cylinder. After closing the needle valve, this N_2O_4 should be frozen into the 1000-ml flask by placing liquid nitrogen around its lower tip. After the pressure has been reduced, a dry ice bath should be placed around the trap and valve 2 opened to degas the sample. Valve 4 can then be closed and the 1000-ml vessel allowed to warm up to room temperature. This material can serve as a reserve for a large number of experiments.

To carry out an experiment the sample should be degassed by pumping on it while frozen under liquid nitrogen as above. Then the mercury source (medium pressure mercury lamp such as Hanovia HA-100) should be activated and a photocurrent reading should be taken with the cell empty (I_0). The phototube can be powered by a pair of 45-V B batteries or a 45-V dc power supply. The radiation from the lamp passes through a convex lens, an interference filter (5460Å), the cell, a second convex lens, and finally the phototube. Distances are determined primarily by the focal lengths of the lenses. The source and phototube should be placed near the focal point of the lens nearest them. All these optical components must be aligned and distances adjusted to give the maximum signal at the phototube. It is necessary to make this alignment permanent (at least over the course of one experiment).

After aligning and measuring I_0, the 1000-ml storage vessel should be warmed to room temperature with valve 4, open to provide a sample. After warming, valve 4 should be closed and the bromobenzene slush bath placed around the cold finger. The temperature of the slush bath should be recorded using a calibrated iron-constantan thermocouple. This temperature should be monitored throughout the course of a complete experiment. It will be used to calculate the vapor pressure above the solid N_2O_4. Starting at room temperature the absorption cell should be slowly heated (about 1°/min), and the variables slush bath temperature, absorption cell temperature, and photocell current should be measured at regular time intervals. It is not essential that all three be measured simultaneously so long as the time is recorded with each measurement. Plots of each versus time then permits the proper correlation. The slow heating schedule is sufficient to maintain the cell at conditions very near equilibrium. It is important to keep the bromobenzene slush bath at nearly a constant temperature; this can be accomplished by regularly adding small amounts of dry ice or liquid nitrogen while stirring the slush.

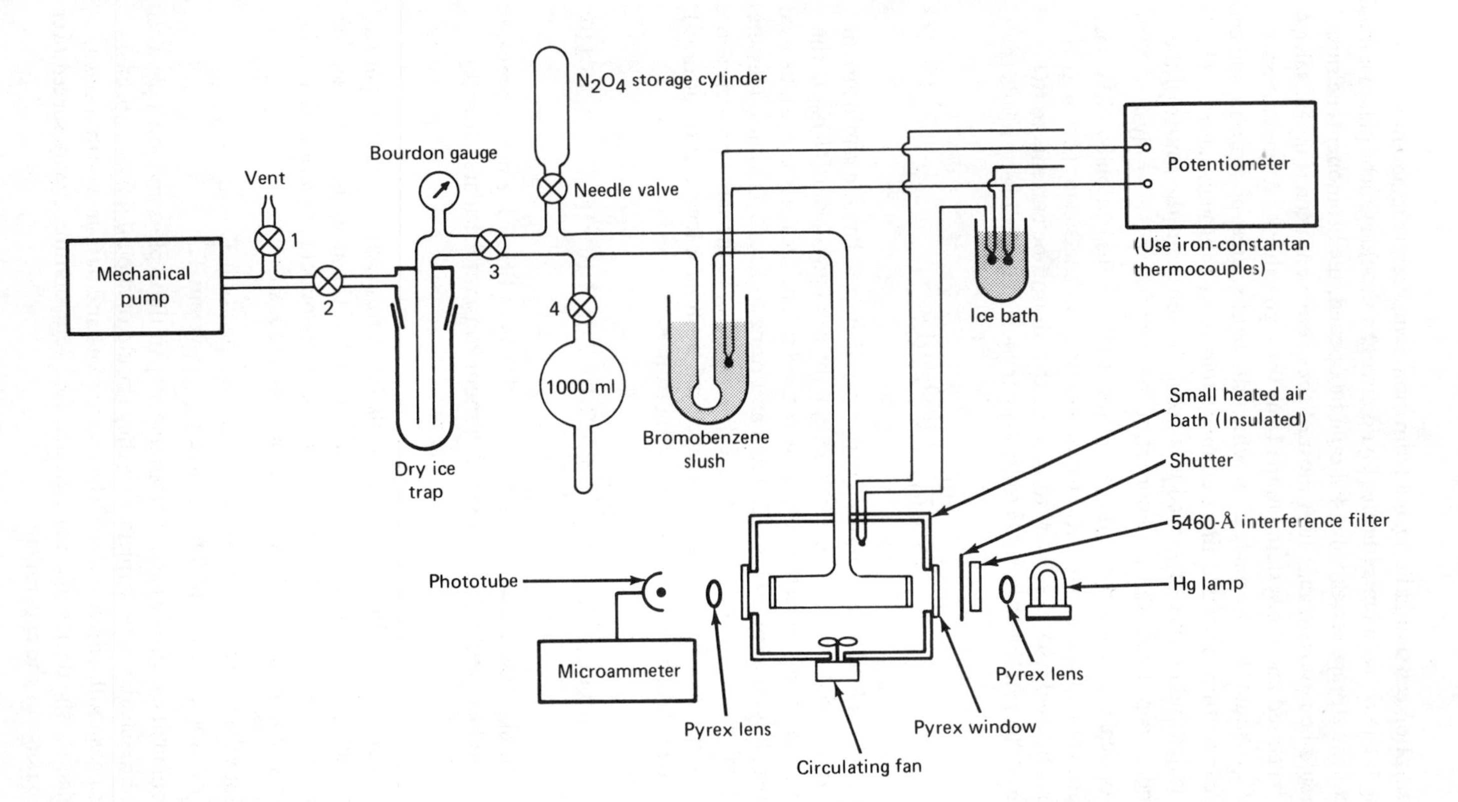

Figure 5-21. Optical method of studying $N_2O_4 \rightleftharpoons 2NO_2$. See section regarding the insulated air bath.

The absorption cell should be heated from room temperature to about 70°C. After reaching the maximum temperature, liquid nitrogen should again be placed around the tip of the storage vessel, valve 4 should be opened, the bromobenzene slush bath should be removed, and, after most of the sample has been trapped, valve 4 should be closed and the system pumped out through valve 3. A final measurement of I_0 should then be made after which the light source, phototube, and cell heaters can be turned off and the vacuum system vented through valve 1. The dry ice trap should then be removed to a hood. Throughout the course of the experiment keep the shutter closed most of the time to avoid photolysis of NO_2.

The optical data collected in this experiment reflect absorption by NO_2 since N_2O_4 does not absorb at 5460 Å. A molar extinction coefficient ($\epsilon = 40.0\ M^{-1}\ cm^{-1}$) can be used with Beer's law to convert the phototube currents to NO_2 pressures. The symbol M^{-1} stands for liters per mole. According to Beer's law

$$\log_{10} \frac{I_0}{I(t)} = \epsilon [NO_2(t)]\, \ell \tag{5-118}$$

where I_0 is the photocurrent reading with the cell empty, $I(t)$ the photocurrent at some time t, ϵ the extinction coefficient, ℓ the cell length, and $[NO_2(t)]$ the concentration of NO_2 which was present at time t. The ideal gas law can be used to convert NO_2 concentrations calculated according to Eq. (5-118) into pressures.

The total pressure in the absorption cell is identical to the vapor pressure above the solid in the cold finger. From the measured slush bath temperature at time t, the pressure p(t) can be computed in cm Hg from

$$\log_{10} p(t) = -\frac{2460.0}{T(t)} + 9.581 + 7.62 \times 10^{-3} T(t) \tag{5-119}$$

From the total pressure and partial pressure of NO_2 at time t the partial pressure of N_2O_4 can be determined and the equilibrium constant calculated from Eq. (5-99).

ANALYSIS. While the detailed manipulations of the numbers vary depending on which method of investigation is used, the goal of the analysis, of course, does not. What we wish to calculate are (1) degree of dissociation as a function of T, (2) K_p as a function of T, (3) $\Delta H°$, (4) $\Delta G°$, and (5) $\Delta S°$.

1. Plot w/M or the total pressure p versus temperature T.
2. If method 3 is used, plot p_{NO_2} and $p_{N_2O_4}$ versus T.
3. Comment on any deviations from linearity in these plots and, on a physical and molecular basis, explain why they should or should not be expected.
4. Using smooth curves through the above plots and data taken from them, calculate the degree of dissociation at four temperatures evenly spaced over the range of the experiment.

5. Calculate K_p at each of the above four temperatures using appropriate data. Use pressures in atmospheres.
6. Determine ΔH°.
7. Determine ΔG° and ΔS° at 30° and 60°C. Rationalize the algebraic sign and magnitudes of both these quantities in terms of the process involved.
8. Compare K_p, ΔH°, and ΔG° with literature values.

PROBLEMS

1. Beginning with Eq. (5-112), develop Eq. (5-117) explicitly including all the mathematical details and assumptions which are implicit in the text.
2. Develop Eq. (5-110).

5.8 Heterogeneous Equilibria

In this experiment the equilibrium between solid ammonium carbamate and its gaseous dissociation products will be studied as a function of temperature by measuring the vapor pressure at various temperatures. The data will be used to calculate ΔH° of the process. Other thermodynamic quantities will also be considered.

THEORY. The process being studied in this experiment is

$$NH_2CO_2NH_4\,(\text{solid}) \rightarrow 2NH_3\,(\text{gas}) + CO_2\,(\text{gas}) \tag{5-120}$$

We assume equilibrium conditions at the temperature of operation. Hence, many considerations discussed in previous experiments are applicable. It will prove useful to review the theory section of Section 5.7 since many details are omitted here.

The general form of the equilibrium constant of reaction (1) is

$$K_a = \frac{\left[a_{NH_3}\right]^2 \left[a_{CO_2}\right]}{\left[a_{NH_2CO_2NH_4}\right]} \tag{5-121}$$

where the a's are activities. The activity of pure solids are defined to be unity for 1 atm pressure and the temperature of the experiment. Since ammonium is a solid and presumably pure, its activity is unity if the pressure is 1 atm. The thermodynamic properties of condensed phases are not sensitive functions of the pressure (i.e., thermodynamic properties do not change markedly as the pressure is increased or decreased from 1 atm). Therefore, the activity of solid ammonium carbonate may be accurately approximated by unity provided the pressure is not several orders of magnitude lower or higher than 1 atm. If we consider the gases of Eq. (5-120) to be ideal, $a_{NH_3} = p_{NH_3}$ and $a_{CO_2} = p_{CO_2}$.

Assuming the validity of the above approximations, the equilibrium constant expression becomes

$$K_p = p_{NH_3}^2 p_{CO_2} \tag{5-122}$$

Since p_{NH_3} and p_{CO_2} are related to the total pressure, experimental measurement of the total pressure is sufficient to determine K_p.

Thermodynamic analysis presented in Section 5.7 shows that

$$\frac{d(\ln K_p)}{d(1/T)} = -\frac{\Delta H^\circ}{R} \tag{5-123}$$

where ΔH° is the standard enthalpy change associated with process (5-120). Plotting ln K_p versus 1/T then permits determination of ΔH° at any temperature.

Equation (5-123) arises from the relationship between ΔG° and K_p:

$$-\Delta G^\circ = RT \ln K_p \tag{5-124}$$

Thus, knowledge of K_p permits determination of ΔG° at any temperature.

The Gibbs-Helmholtz equation can then be used to relate ΔH° and ΔG° to ΔS° at any temperature:

$$-\Delta S^\circ = \frac{\Delta G^\circ - \Delta H^\circ}{T} \tag{5-125}$$

EXPERIMENTAL. The apparatus for this experiment is diagrammed in Fig. 5-22 and consists of a sample vessel, a pressure-measuring device, a temperature-measuring device, a vacuum system, and a thermostat. The range of values covered

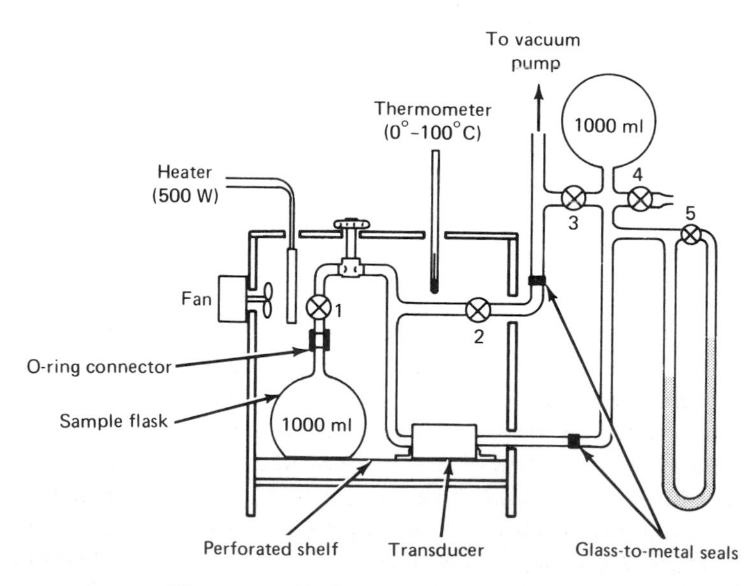

Figure 5-22. Equilibrium constant apparatus.

in temperature and pressure are 25°–60°C and a few torr to 1 atm, respectively. Any set of components commensurate with these ranges is satisfactory. We have selected one convenient set in which a thermometer is used to record temperature and a magnetic reluctance transducer is used to record pressure. The oven should enclose a volume of about 1 ft^3 and should be constructed and heated in a manner described in Section 2.2. The pressure transducer and its operation are described in Section 2.1. The tubing inside the oven should all be $\frac{1}{4}$-in. copper or brass. Valves 1 and 2 should be bellows-sealed metal valves (Hoke 4111 series, for example) which are silver-soldered to the tubing. The sample vessel is a 1000-ml flask with a flattened base which rests on a perforated metal shelf. This flask is connected to the metal part of the system with an O-ring connector (Cajon Ultra-torr fitting, for example). Outside the vacuum wall two metal-to-glass seals (preferably stainless steel-to-Pyrex seals) are used to connect onto the manometer. These could be replaced with O-ring-type connectors like that described above.

The data to be collected are pairs of variables $\{p, T\}$ corresponding to the equilibrium pressure p above the solid sample at some temperature T. To proceed, first place 4 or 5 g of finely divided sample in the flask and spread it out more or less evenly over the flat bottom. Such a sample has a relatively large surface area readily accessible to molecules from the gas phase above it. These features promote a rapid approach to equilibrium. Ammonium carbamate will pick up water to form ammonium carbonate and so it is important to keep the sample dry.

After placing the sample in the flask, attach it to the vacuum system with the O-ring connector, making certain that it is tightly sealed. The next step is to evacuate the system and "zero" the transducer circuitry. With all the valves open except 4 and 1, begin pumping on the system. After the pump quiets, zero the transducer circuitry. This setting will be checked later to ascertain the significance of electronic drift in the transducer circuitry. Valve 3 should now be closed and valve 1 opened very slowly—slowly in order to avoid blowing the sample around the vacuum system. After pumping for a few minutes the pump should again become relatively quiet. Remember that valves 1 and 2 have rather small orifices and that the interconnecting tubing is relatively small. Both tend to reduce pumping speeds. After pumping several minutes, close valve 1, continue pumping, and recheck the transducer zero, noting the amount of drift if any. Heat the oven to 60°C as quickly as possible and hold it there for several minutes. Again check the transducer zero. Next reopen valve 1 and immediately close valve 2. To do this without a large heat loss it is advisable to cut two holes in the front face of the oven directly in line with the valves. With an appropriately designed tool one can reach through these holes to manipulate the valves. When not in use the holes can be stoppered. After closing valve 2, the transducer should indicate an out-of-null condition reflecting a pressure differential. Null the transducer by admitting air through valve 4. Record the pressure versus time in this way while holding the oven at 60°C. This serves to check on equilibrium conditions. Once equilibrium is reached, begin slowly lowering the temperature. Strive for an *average* rate of 1°C/min by lowering the heater voltage slightly and letting the temperature stabilize. Take pressure and temperature readings every few minutes, building up a set of at least 10 points distributed over the temperature range. When a temperature near ambient is reached, take a final pressure reading, close valve 1, open valve 2, and note whether or not the transducer zeros. If it does not, note how much adjustment is necessary to rezero and use this in correcting the pressure readings.

Finally shut off the pump; open valves 1, 3, and 5; and slowly vent the system through valve 4.

ANALYSIS.

1. Plot pressure versus temperature and construct a smooth curve representing the data. From the smoothed data, compute and plot ln p versus 1/T for five temperatures in the range of the data. Compare the results with ln p (torr) = –2741.9/T + 11.1448 taken from Egan *et al.* in the References.
2. Calculate K_p at 30°, 40°, 50°, and 60°C.
3. Plot ln K_p versus 1/T and calculate $\Delta H°$ for the process being considered. Egan etal. in the References report the heat of dissociation of ammonium carbamate as 376 kcal/mole.
4. Calculate $\Delta G°$ and $\Delta S°$ at 30° and 60°C. Explain the sign and magnitude of these thermodynamic quantities on a molecular basis.
5. Using the Berthelot expression for the compressibility, Z, compute Z for the conditions of these experiments. Refer Section 5.4. For NH_3, T_c = 132.4°C and p_c = 111.5 atm. For CO_2, T_c = 31.1°C and p_c = 73.0 atm. Comment on expected deviations from gas-phase ideality in this experiment.
6. Estimate the error in the value of $\Delta H°$ based on estimated uncertainties in T and p.

PROBLEMS

1. For what amount of substance is the $\Delta H°$ computed? Explain.

5.9 Gas Chromatography and Thermodynamics

The purpose of this experiment is to calculate thermodynamic properties of solutions based on gas-liquid chromatographic measurements. The mathematical relations connecting experimentally measured quantities with the thermodynamic properties of solutions are developed in the theory section.

THEORY. We shall restrict ourselves at the outset to elution in gas-liquid chromatography. Elution implies that the sample being analyzed is swept (eluted) through the material, which causes the separation of the sample components. In all chromatography there are two phases: one fixed and the other mobile. The term gas-liquid chromatography implies that the mobile phase is a gas, while a liquid is the fixed phase. The liquid phase is held fixed by adsorbing it on some kind of solid support. The solid material is packed inside a tube to form a column.

In the above-described technique the physical separation between two components in the sample relies upon variation with components of the molecular interaction between the gas and liquid phase. If there are two gaseous components, A and B, in the sample and if A interacts with the fixed liquid phase more strongly than B, then the carrier gas sweeps species B through the supported liquid phase more rapidly than it does A. As a result B is eluted more rapidly than A. By using a suitable detector after the column (thermal conductivity, flame ionization, etc.) it is possible to measure the time, t_i, required to elute species i. Other measurable parameters are column carrier gas flow rate, column temperature, and pressure drop across the column.

We shall now turn to a more quantitative description of gas-liquid chromatography. A typical chromatogram of a sample containing a small amount of air and a pure sample of interest is shown in Fig. 5-23.

The true retention time t_i' is given by

$$t_i' = t_s - t_0 \tag{5-126}$$

where t_0 is the time at which the sample is injected and t_s the time of sample elution. A second time, the net retention time t_i, is used, and it is defined by the expression

$$t_i = t_s - t_a \tag{5-127}$$

where t_a is the time at which the air peak appears. The difference between t_i' and t_i is called the dead time of the column under the conditions of the experiment. Of greater theoretical utility than t_i is V_i, the retention volume of carrier gas, which is given by the expression

$$V_i = t_i\bar{F} \tag{5-128}$$

where $\bar{F}$ is the average flow rate through the column in milliliters per unit time and at the column temperature.

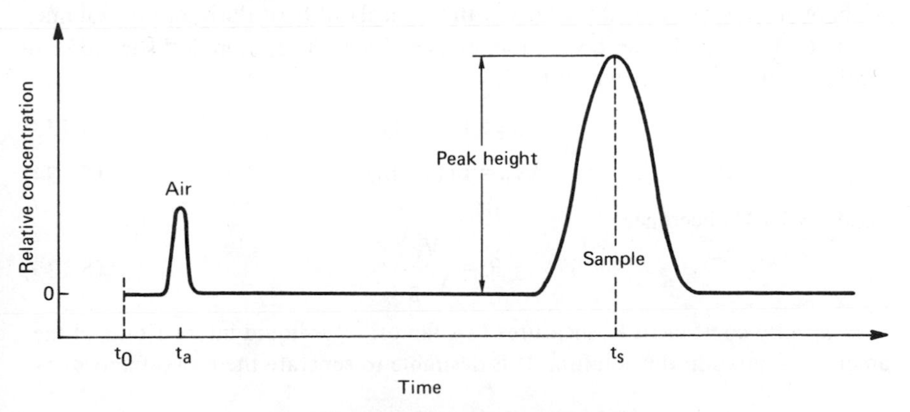

Figure 5-23. Schematic chromatogram.

Assuming that the air in the sample has, at most, a negligible interaction with the liquid phase of the column, we may say that the speed, S_{cg}, of the carrier gas through the column is proportional to $1/(t_i' - t_i)$, while the speed, S_i, of the sample peak through the column is proportional in the same way to $1/t_i'$. Therefore, the ratio of the two speeds is

$$\frac{S_i}{S_{cg}} = \frac{t_i' - t_i}{t_i'} = \left[\frac{t_s - t_a}{t_a - t_0} + 1\right]^{-1} \tag{5-129}$$

This ratio of speed is also given by the fraction of sample molecules in the gas phase at any instant. This fraction depends on the interaction of species i with the stationary phase and we may write

$$\frac{S_i}{S_{cg}} = \frac{\text{amount of sample in gas phase at any instant}}{\text{total amount of sample}} \tag{5-130}$$

Inverting Eq. (5-130) furnishes, upon dissecting its denominator into two parts,

$$\begin{aligned}\frac{S_{cg}}{S_i} &= \frac{\text{sample in gas phase + sample in liquid phase}}{\text{sample in gas phase}}\\ &= 1 + \frac{\text{sample in liquid phase}}{\text{sample in gas phase}}\\ &= 1 + R_i\end{aligned} \tag{5-131}$$

where R_i is the capacity coefficient, which is a measure of both the average interaction of the sample species i for the liquid phase and the amount of liquid phase in the column. Combining Eqs. (5-129) and (5-131) furnishes

$$\frac{t_s - t_a}{t_a - t_0} = R_i \tag{5-132}$$

The derivation of Eq. (5-132) implicitly assumes that the carrier gas flow rate and other column parameters do not change during the course of a single experiment. We have already mentioned the greater theoretical utility of the retention volumes and to each of the time interval parameters of Eq. (5-132) there is associated a retention volume. That is, we write

$$V_i = \bar{F}(t_s - t_a) \tag{5-133}$$

$$V_{cg} = \bar{F}(t_a - t_0) \tag{5-134}$$

and Eq. (5-132) becomes

$$R_i = \frac{V_i}{V_{cg}} \tag{5-135}$$

The capacity coefficient R_i measures two factors: gas-liquid interaction and the amount of liquid in the column. It is desirable to separate these two factors because we are especially interested in the first one. To do this we define a partition coefficient K_i as the ratio of concentrations of species i in the liquid phase and gas phase:

$$K_i \equiv \frac{C_i^{(l)}}{C_i^{(g)}} = R_i \frac{V_{cg}}{V_l} = \frac{V_i}{V_l} \tag{5-136}$$

where V_l is the volume of liquid phase in the column. The ratio defining K_i is also expressible in terms of assumed equations of state. If we take the ideal gas law for the gas phase, then

$$C_i^{(g)} = \frac{p_i}{RT} \tag{5-137}$$

while for the liquid phase we write

$$C_i^{(l)} = \frac{\rho_l X_i}{M_l} \tag{5-138}$$

where ρ_l is the density of the pure liquid phase, M_l its molecular weight, and X_i the mole fraction of the eluting species i. At equilibrium between the liquid and gas phases p_i and X_i are related by

$$p_i = p_i^\circ \gamma_i X_i \tag{5-139}$$

where p_i° is the vapor pressure of pure i and γ_i is the mole fraction activity coefficient of species i. Substituting Eqs. (5-137)–(5-139) into Eq. (5-136) yields, after rearrangement,

$$V_i = \frac{V_l \rho_l RT}{M_l p_i^\circ \gamma_i} = \frac{w_l RT}{M_l p_i^\circ \gamma_i} \tag{5-140}$$

where w_l is the weight of liquid phase used to prepare the column. The parameters appearing on the right-hand side of Eq. (5-140), except for γ_i, are readily available from experimental data. We may therefore regard Eq. (5-140) as furnishing the activity coefficient if V_i can be calculated from experimental data. To accomplish this the following parameters must be determined:

1. Flow rate F of gas at output of column
2. Net retention time, t_i
3. Pressure p_0 at output of column (normally atmospheric pressure)
4. Pressure p_i at input of column
5. Column temperature, T_c
6. Ambient temperature, T_a (normally room temperature)
7. Vapor pressure p_w of water at T_a

The average flow rate is related to the flow rate at the output of the column by

$$\bar{F} = \frac{3}{2}\left[\frac{(p_i/p_0)^2 - 1}{(P_i/P_0)^3 - 1}\right] F_{cg} \tag{5-141}$$

where F_{cg} is the carrier gas flow rate at the column output port. The total flow rate F is generally measured with a soap film flow meter. The gas flow through this burette is composed of carrier gas and water vapor. To find F_{cg} the total flow F must be reduced according to the relation

$$F_{cg} = \frac{p_0 - p_w}{p_0} F \tag{5-142}$$

Thus,

$$\bar{F} = \frac{3}{2}\left[\frac{(p_i/p_0)^2 - 1}{(p_i/p_0)^3 - 1}\right]\left[\frac{p_0 - p_w}{p_0}\frac{T_c}{T_a} F\right] \tag{5-143}$$

where the factor T_c/T_a corrects the measured flow to the column temperature. Using Eq. (5-143) in Eq. (5-138) and substituting into Eq. (5-140) yield

$$\gamma_i = \frac{w_1 RT}{M_1 p_i^\circ \bar{F} t_i} \tag{5-144}$$

from which activity coefficients may be calculated.

Similarly the partition coefficient is calculable from

$$K_i = \frac{\bar{F} t_i \rho_1}{w_1} \tag{5-145}$$

The net retention volume V_i divided by the weight of liquid in the column w_1 is called the specific retention volume V_i^*. Frequently the standard specific retention volume V_i° is used and is calculated from

$$V_i^\circ = V_i^* \frac{273}{T_c} \tag{5-146}$$

From Eq. (5-140)

$$V_i^* = \frac{RT_c}{M_1 p_i^\circ \gamma_i} \tag{5-147}$$

and

$$V_i^\circ = \frac{273R}{M_1 p_i^\circ \gamma_i} \tag{5-148}$$

$$\ln V_i^\circ = \ln \frac{273R}{M_1} - \ln p_i^\circ - \ln \gamma_i \tag{5-149}$$

Differentiating Eq. (5-149) with respect to temperature furnishes an equation whose right-hand side is directly relatable to the thermodynamic properties of the liquid-vapor equilibrium:

$$\frac{d \ln V_i^\circ}{d(1/T)} = -\frac{d \ln p_i^\circ}{d(1/T)} - \frac{d \ln \gamma_i}{d(1/T)} \tag{5-150}$$

The first term on the right-hand side is just the negative of the molar heat of vaporization at temperature T divided by the gas constant. The second term is just the heat associated with the mixing with the solvent divided by the gas constant. Hence,

$$\frac{d \ln V_i^\circ}{d(1/T)} = -\frac{\Delta H_{vap}}{R} + \frac{\Delta H_{mix}}{R} \tag{5-151}$$

If the solution is ideal, $\gamma_i = 1$ and a plot of ln V_i° versus 1/T is linear and furnishes the molar heat of vaporization. In cases where the system is not ideal but where ΔH_{vap} and ΔH_{mix} vary only slowly with temperature, a plot of ln V_i° versus 1/T furnishes the algebraic sum of the two enthalpies ΔH_s the molar heat of solution.

EXPERIMENTAL. The apparatus* shown in Fig. 5-24 is suitable and consists of a gas chromatograph using thermal conductivity detection which has been altered slightly so the pressure at the input to the columns can be measured and

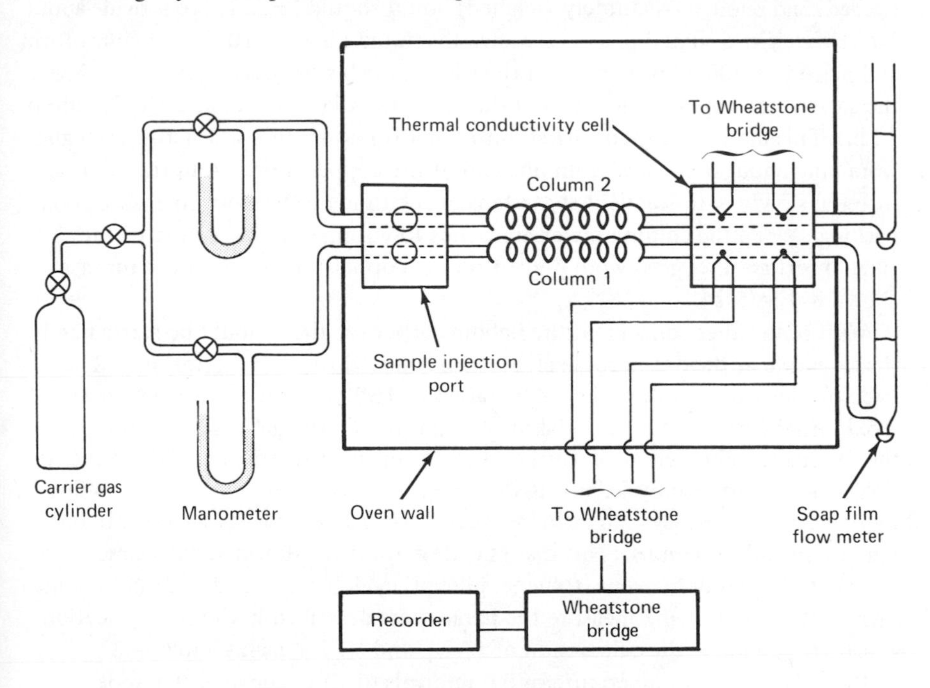

Figure 5-24. Gas chromatography apparatus.

*A suitable gas chromatograph kit is available from Gow-Mac Instrument Co., Madison, N.J. See Appendix XIII for detailed addresses of suppliers.

so that injection can be made into either column. Either a mercury manometer or a commercial Bourdon-type gauge may be used to measure the input pressure. The injection is accomplished with a syringe containing the liquid sample and a small amount of air.

In general terms the chromatograph operates as follows. Injection of sample into the moving carrier gas provides the means of achieving liquid-gas equilibrium in the column where the sample gas is retained for a time period depending on its interaction with the liquid phase. When the sample is eluted the composition of the carrier gas arriving at the thermal conductivity cell changes, thereby changing the thermal conductivity of the gas surrounding the heated filaments, which are part of a balanced Wheatstone bridge circuit. As the thermal conductivity changes, the filament temperature and resistance changes, thus causing imbalance in the Wheatstone bridge circuit. The imbalance voltage is detected by the strip chart recorder.

Since the calculations clearly depend on the amount of liquid phase in the column, it is important to prepare the columns carefully. One of the columns is prepared by mixing dinonylphthalate (M = 418.6 g/mole) on dried 80–100 mesh Chromosorb. The other is prepared by mixing tricresylphosphate (M = 368.4 g/mole) on the same solid support. Four feet of $\frac{1}{4}$-in. copper tubing is satisfactory for column length. For columns of this length about 10 g of Chromosorb is needed, and enough (accurately weighed) liquid should be added to provide about 20% by weight of liquid phase. Dissolve the liquid phase in 100 ml of chloroform and place in a 600-ml beaker. Add the Chromosorb slowly with stirring. Allow the solvent to evaporate slowly, and then heat the Chromosorb at 50°C for about 12 hr. This material is then packed into the 4-ft copper tubes. First insert a glass wool plug about 6 in. back from one end of the copper tubes. Add the column material slowly with frequent sharp taps of the tube on the floor to ensure good packing. When the column material reaches to within 6 in. of the top of the tube, insert another glass wool plug. Coil the copper tube to fit in the furnace shown in Fig. 5-24.

With both tubes connected, the helium carrier gas flow should be started and the flow rate at the output of each column adjusted to about 75 ml/min. The injector block temperature should be raised to 160°C and the oven temperature fixed initially at about 50°C. Making certain that carrier gas is passing through the thermal conductivity cell, turn on the power to the filaments. Adjust the recorder output to zero by balancing the Wheatstone bridge circuit of the conductivity cell. When the system becomes stable, use a 10-μl syringe to inject 0.5-μl samples into the columns. This makes certain that air is injected each time.

Inject samples of benzene, toluene, phenol, cyclohexane, and cyclohexene into each of the columns and measure the parameters described in the theory section, which are needed in the calculation of γ_i, K_i, and V_i° for each sample.

Raise the column temperature in 10° intervals to 80°C, and at each temperature inject a sample of benzene into each column, making the same measurements as those above.

ANALYSIS. Calculate γ_i, K_i, and V_i° for each sample. Compare the activity coefficients with accepted literature values. Calculate for benzene, by preparing and analyzing appropriate graphs, the molar heat of solution on both columns.

Explain in terms of intermolecular forces, molecular geometries, etc., the variations from system to system of the quantities calculated above.

5.10 Physical Adsorption at the Gas-Solid Interface

The purpose of this experiment is to calculate the surface area of a solid by determining the amount of argon adsorbed as a function of the argon gas-phase pressure over the solid. The data are fitted to the B.E.T. theoretical equations. This furnishes parameters characteristic of the material from which one can calculate the surface area.

THEORY. When atoms or molecules in the gas phase at pressures less than their vapor pressure become attached to a solid material they are said to be adsorbed. The process by which this event occurs is called adsorption. Adsorptions are of two general types: chemisorption in which a chemical bond is formed between the adsorbent and adsorbate and physisorption in which the adsorbent-adsorbate interaction is a result of van der Waals forces which arise from induced dipole-induced dipole effects. The potential energy of the latter interactions is approximately

$$V = -\frac{C}{r^6} \tag{5-152}$$

where C is a constant depending on the species involved and r the distance of the atom or molecule from the solid surface adsorption site. Physical adsorption is of concern in this experiment.

Note that we have introduced the notion of a site. If we took an atomic level view of the surface, we would find a nonhomogeneous situation; the potential energy depends on the position over the surface. Sites are those positions which show strong attraction for species and are the positions at which adsorption occurs.

The extent of adsorption is usually characterized by computing, at a particular gas pressure, the number of moles of gas adsorbed and converting it to the corresponding gas-phase volume at STP. A graph of STP volume adsorbed versus p/p_0 is then constructed, where p is the measured gas pressure above the sample and p_0 is the vapor pressure of the adsorbate at the adsorbent temperature. The resulting plot is called an isotherm. The form of the isotherm depends on the area of the adsorbent, the temperature, and the nature of the species involved.

Langmuir has developed a theory for the adsorption isotherm which predicts that the volume adsorbed will rise to finite maximum value as p/p_0 is increased. This theory is applicable to situations in which only a single layer (monolayer) of adsorbate is formed. The finite maximum corresponds to the formation of a complete monolayer.

In most physical adsorption processes, however, the adsorption does not cease when a monolayer is completed; rather, multiple layers of adsorbate are formed. Brunauer, Emmett, and Teller have developed a theory of multilayer adsorption, the B.E.T. theory, which predicts a rapid increase of the volume adsorbed as p/p_0 approaches unity. (What happens when p/p_0 equals unity?) The validity of the B.E.T. theory rests upon five assumptions:

1. All sites are equally suitable for adsorption.
2. Adsorbed species in the first layer are localized; they cannot move over the surface.
3. Each adsorbed species furnishes a site for further adsorption.
4. There is no lateral interaction between species in a given layer.
5. All species in the second, third, and higher layers are assumed to behave like those of the liquid. The first layer is unique because of direct adsorbent-adsorbate interactions.

In the B.E.T. theory, the surface of the adsorbent is divided into subareas labeled i, where i denotes the number of layers of adsorbate covering that particular subarea. The label S_i is used to denote the total subarea covered with i adsorbed layers. For example, S_0 is the area of exposed surface and S_{10} is the area covered with 10 layers.

We assume (1) the existence of equilibrium and (2) that the adsorption-desorption phenomenon occurs only at the gas-surface interface. The adsorption of species from the gas into layer $i - 1$ increases S_i, while desorption of species from layer i decreases S_i. At equilibrium, the principle of microscopic reversibility requires the rate of desorption to be equal to the rate of adsorption into layer i. The rate of adsorption into layer i is governed by the number of collisions of gaseous species with the surface per square centimeter per second and by the surface of subarea $i - 1$. Applying the kinetic theory of gases, the number of collisions per square centimeter per second is $\frac{1}{4}n\bar{v}$, where n is the number of species per cubic centimeter is the gas phase and $\bar{v}$ is the average molecular velocity. Making substitutions for n and $\bar{v}$ in terms of pressure p, mass, and temperature T furnishes:

$$\frac{1}{4}n\bar{v} = \frac{N_0 p}{(2\pi RTM)^{1/2}} \tag{5-153}$$

where N_0 is Avagadro's number, R the gas constant, and M the molecular weight.

With this result, for the rate of adsorption into layer i we can write

$$\text{rate of adsorption} = \frac{f_{i-1} S_{i-1} p N_0}{(2\pi RTM)^{1/2}} \tag{5-154}$$

where f_{i-1} is the fraction of collisions with layer $i - 1$ which results in adsorption.

The rate of desorption from layer i is governed by the concentration of species in this layer and the size of the energy barrier that must be overcome in order to escape. The concentration term is just S_i. We write the rate constant for desorption in Arrhenius form:

$$k_i = A_i \exp\left\{-\frac{E_i}{RT}\right\} \tag{5-155}$$

where A_i is the preexponential factor and E_i a measure of the energy barrier that must be overcome. The rate of desorption from layer i at equilibrium is then

$$\text{rate of desorption} = S_i A_i \exp\left\{-\frac{E_i}{RT}\right\} \tag{5-156}$$

Requiring equilibrium provides

$$\frac{f_{i-1} S_{i-1} p N_0}{(2\pi RTM)^{1/2}} = S_i A_i \exp\left\{-\frac{E_i}{RT}\right\} \tag{5-157}$$

Making use of assumption 5 we can say that

$$\begin{aligned} f_i &= f_{i-1} = f, & i &= 3, 4, \dots \\ A_i &= A_{i-1} = A, & i &= 3, 4, \dots \\ E_i &= E_{i-1} = E_L, & i &= 3, 4, \dots \end{aligned} \tag{5-158}$$

where E_L is the heat of liquefaction (generally written ΔH_L). Solving Eq. (5-157) for S_i furnishes

$$S_1 = \frac{f_0}{A_1} \frac{pN_0}{(2\pi RTM)^{1/2}} \exp\left\{+\frac{E_1}{RT}\right\} S_0 \tag{5-159}$$

$$S_2 = \frac{f_1}{A_2} \frac{pN_0}{(2\pi RTM)^{1/2}} \exp\left\{+\frac{E_2}{RT}\right\} S_1 \tag{5-160}$$

$$S_i = \frac{f_{i-1}}{A_i} \frac{pN_0}{(2\pi RTM)^{1/2}} \exp\left\{+\frac{E_i}{RT}\right\} S_{i-1} \tag{5-161}$$

Making use of relations (5-158) and (5-159), Eq. (5-161) can be rewritten as

$$S_i = \frac{f}{A}\left[\frac{pN_0}{(2\pi RTM)^{1/2}} \exp\left\{\frac{E_L}{RT}\right\}\right]^{i-1} S_1 \tag{5-162}$$

for $i > 1$.

Further,

$$S_i = x^{i-1} S_1 = c x^i S_0 \quad \text{for } i > 1$$

where x is defined by Eq. (5-162) and c is given by

$$C = \frac{Af_0}{A_1 f} \exp\left\{\frac{E_1 - E_L}{RT}\right\} \tag{5-163}$$

If we desire to compute the total surface area A of the adsorbent, a sum over all subareas is required:

$$\begin{aligned} A &= \sum_{i=0}^{\infty} S_i \\ &= \sum_{i=1}^{\infty} cS_0 x^i + S_0 \end{aligned} \tag{5-164}$$

The property that can be measured is the STP adsorbed volume. We desire a relationship between this volume and the adsorbent area. The total volume of gas adsorbed is given by

$$V = V_0 \sum_{i=0}^{\infty} icS_0 x^i \tag{5-165}$$

where v_0 is the volume adsorbed per square centimeter at monolayer coverage. Note that $Av_0 \equiv v_m$ is the volume adsorbed at complete monolayer coverage. Taking the ratio of Eq. (5-165) to Eq. (5-164) furnishes, after rearranging,

$$\frac{V}{V_m} = \frac{c \sum_{i=0}^{\infty} ix^i}{c \sum_{i=1}^{\infty} x^i + 1} \tag{5-166}$$

Using the following series properties

$$\sum_{i=0}^{\infty} ix^i = \frac{x}{(1-x)^2} \tag{5-167}$$

$$\sum_{i=1}^{\infty} x^i = \frac{x}{1-x} \tag{5-168}$$

substituting into Eq. (5-166) furnishes

$$\frac{V}{V_m} = \frac{cx}{(1-x)(1-x+cx)} \tag{5-169}$$

Making use of assumptions 1 and 3 we note that an infinite number of layers can be adsorbed provided the saturation vapor pressure, p_0, is maintained. This requires that $V \rightarrow \infty$ and from Eq. (5-169) $x \rightarrow 1$. Examining the behavior of x we find $x = p/p_0$ and from Eq. (5-169), after some rearrangement,

$$\frac{p}{V(p_0 - p)} = \frac{1}{V_m C} + \frac{C - 1}{V_m C}\frac{p}{p_0} \tag{5-170}$$

Plotting $p/[V(p_0 - p)]$ versus p/p_0 furnishes a curve from which both V_m and C can be obtained. Assuming a value for V_0 (the volume adsorbed per square centimeter at monolayer coverage) one can then compute the surface area, A.

EXPERIMENTAL. All the apparatus needed for this experiment are described in detail below except for the vacuum pumping system. Two mechanical vacuum pumps are required: one for the high-vacuum line containing the sample under study and a second for controlling the mercury reservoirs, shown in Fig. 5-25. Referring to Fig. 5-25 the high-vacuum system connects to the pumping system at the constant-volume manometer. From the constant-volume manometer the line should go to a liquid nitrogen trap, then to a small glass-type oil diffusion pump, and finally to a mechanical vacuum pump. The mercury reservoir control pump is connected directly to the bottom vacuum line shown in Fig. 5-25. Refer to Section 2.1 for a description of the necessary components and especially Fig. 2-1 for a block diagram of the type of high-vacuum system needed in this experiment.

The volumetric adsorption apparatus is shown in Fig. 5-25. It consists of a sample bulb containing a sample whose total geometric surface area is between 100 and 300m^2. This large area is required so that adsorption of gas-phase species will be detectable. The sample container is connected to a series of calibrated volumes (volumetric apparatus). These volumes are measured before the apparatus is assembled using the density-weight-volume relations for mercury or distilled water and the instructor will provide them.

The volumetric apparatus typically has six or seven bulbs connected with small-bore capillary tubes which are file-marked. The total volume should be in the neighborhood of 150 ml with individual volumes on the order of 75, 40, 20, 10, 5, 2, and 1 ml. This assembly is surrounded with a large glass tube to provide strength and the possibility of surrounding the volumetric apparatus with a liquid heat bath medium. Two other volumes will be provided: (1) the free volume defined as that of the sample container and connecting tube which terminates in stopcock E and (2) the dead volume defined as that bounded by stopcock E, stopcock B, mark 8 of the volumetric apparatus, and the mercury level in the left-hand manometer tube. The argon gas used in the adsorption studies is stored in a flask, and gas-phase argon pressures are measured with the connected manometer. The left-hand tube is marked in this experiment with a file mark or, as shown in the figure, with a metal wire contact point sealed in the vacuum wall. As shown, this

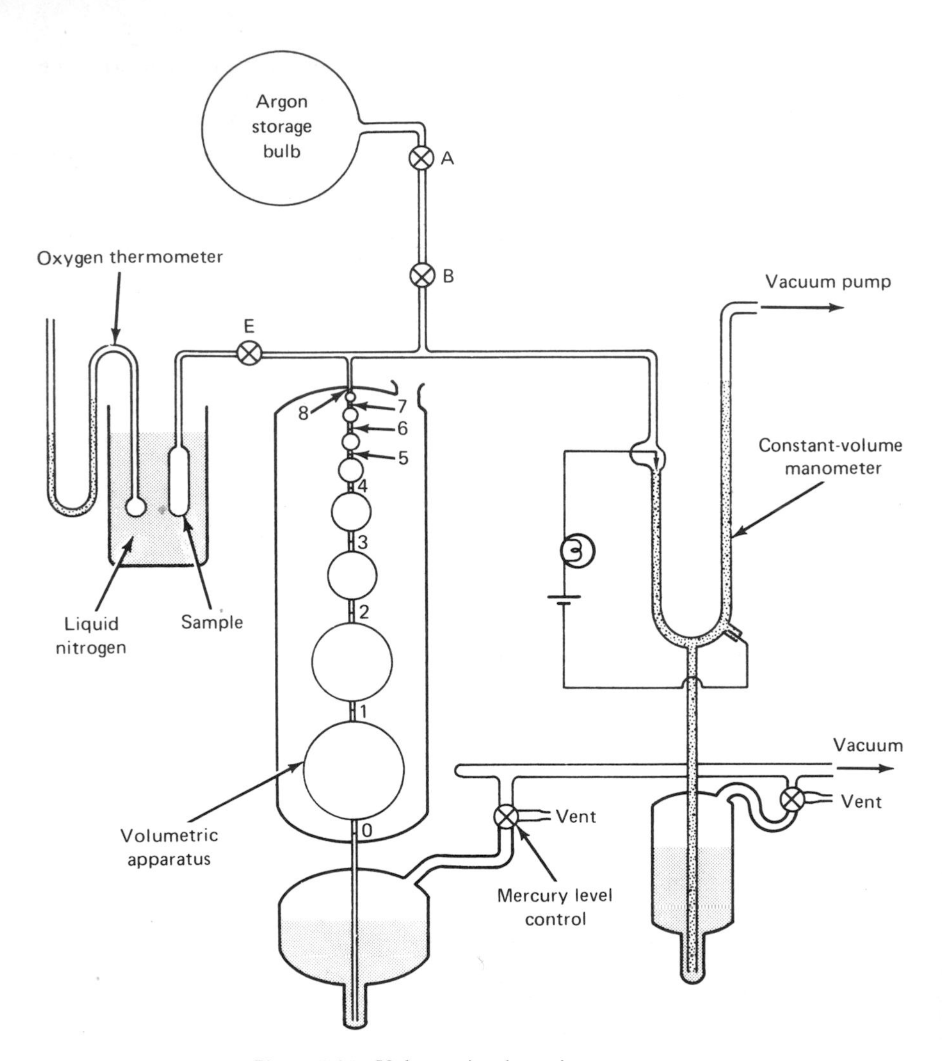

Figure 5-25. Volumetric adsorption apparatus.

point is connected through a lamp and flashlight battery to the mercury column of the manometer. When the mercury is raised in the manometer, the upper contact is reached and the lamp is on. Below this point the mercury-wire contact is broken and the lamp is off. Temperatures in the low-temperature bath are measured using an oxygen thermometer. This device is described in Section 2.2.

To begin an experiment, close stopcock E, open stopcock B, and lower the mercury levels in the volumetric apparatus and the manometer to their lowest points by turning the mercury level control stopcocks so that the mercury-containing flasks are evacuated. Do this very slowly if other parts of the vacuum system are not already evacuated. Evacuate (slowly down to about 10^{-5} Torr) the sample portion of the system by slowly opening valve E. Remember, this container has a dusty sample in it which will scatter throughout the apparatus if the pressure is lowered suddenly. Once evacuated, it is a good idea to go through the motions of adjusting the mercury levels in the absence of an argon sample. Raise the level in the volumetric apparatus to point 8 and then practice getting the level adjusted to various points on the way back down to zero. Then practice adjusting the U-tube mercury level. Note that any time this level falls below the base of the U, the sample will be pumped away. Return the volumetric to zero, adjust the U-tube level to near the contact, and close stopcock E.

Argon is now admitted to a pressure between 15 and 20 Torr (1 Torr = 1 mm Hg). To admit the argon, close stopcock B and then open and close stopcock A. Reopen stopcock B to expand the argon into the system. Measure the pressure with the U-tube manometer; repeat the above steps as necessary until the required pressure is achieved. Adjust the mercury level of the volumetric apparatus to position zero and that of the manometer to just contact the metal lead wire. Measure the pressure accurately (use a cathetometer if available) and note the room temperature. These measurements along with the volumes permit calculation of the total number of moles of gas used in the experiment or the STP volume V_T.

A dewar of liquid nitrogen is placed around the sample, and the sample stopcock E is now opened. It is noted that the pressure drops; this is due to a volume expansion in the system and loss of gas by adsorption. The pressure should be read with the manometer every 5 min until equilibrium is attained (20 min). In making these measurements, keep the mercury levels constant at zero in the volumetric apparatus and at the metal contact in the manometer. When equilibrium is reached, the room temperature is noted and the temperature of the liquid nitrogen bath is obtained by reading the vapor pressure of the oxygen thermometer and using a graph of log vapor pressure versus T^{-1}. The volumes, pressures, and temperatures on both sides of the sample stopcock are known. Thus, the respective STP volumes can be calculated and are called V_0 and V_0'. These are the volumes corresponding to room and liquid nitrogen temperature, respectively. The difference

$$V_{ads} = V_T - (V_0 + V_0') \tag{5-171}$$

is the volume of gas adsorbed at position zero on the volumetric bulbs.

The mercury in the volumetric bulbs is raised to succeeding positions (X = 1, 2, ..., 8), and after pressure equilibrium is attained at each point, the pressure, room temperature, and liquid N_2 bath temperature are all measured. V_X and V_X' are then calculated, and values for

$$V_{ads} = V_T - (V_X + V_X') \tag{5-172}$$

are obtained as the volumes of gas adsorbed at the various pressures. The vapor pressure of argon at the bath temperature enters into the analysis and is obtained from a graph of ln p versus 1/T. The vapor pressure data are available in standard handbooks.

ANALYSIS. Calculate and plot the appropriate quantities in Eq. (5-170) and determine v_m and C. Assuming one argon atom covers 16×10^{-16} cm^2, calculate the surface area of the sample. Plot the adsorption isotherm and comment on deviations from Langmuir theory.

PROBLEMS

1. Calculate the error in volume adsorbed at position 2 caused by a ±1°C error in room temperature.
2. Calculate the total moles of argon introduced into the system by both the ideal gas and van der Waals, gas equations of state.
3. Rearrange Eq. (5-170) into a form suitable for construction of an isotherm. Assuming that $V_m = 10$ cm^3 and $C = 1$, construct a B.E.T. isotherm for $0 \leqslant p/p_0 \leqslant 1$. At about what value would you predict monolayer coverage?
4. If Langmuir's theory is valid up to monolayer coverage, sketch in what Langmuir's theory would predict for $0 \leqslant p/p_0 \leqslant 1$.
5. Beginning with Eq. (5-166) develop Eq. (5-170) including all steps and assumptions.

5.11 Lattice Energy of Solid Argon

The purpose of this experiment is to study the variation with temperature of the vapor pressure of solid argon. From the data the heat of sublimation and the lattice energy of solid argon are calculated. In the theory section the thermodynamic and potential-energy relations are developed.

THEORY. In this experiment the following phase transition is studied as a function of temperature:

$$Ar(solid, T, p) \longrightarrow Ar(gas, T, p) \tag{5-173}$$

It is assumed that the phase change is carried out under equilibrium conditions, rendering the differential Gibbs free energy change identically equal to zero. Furthermore, the temperature, pressure, and chemical potential for the gas phase must be identically equal to the values of these parameters in the solid phase. The differential Gibbs free energy change may be written as

$$dG = V\,dp - S\,dT + \sum_i \mu_i\,dn_i = 0 \tag{5-174}$$

It can also be expressed in terms of the chemical potential of each component in each phase and the number of moles of each component in each phase:

$$dG = \sum_i \mu_i\,dn_i + \sum_i n_i\,d\mu_i = 0 \tag{5-175}$$

Setting Eq. (5-174) equal to Eq. (5-175) and dividing both sides by n, the total number of moles in one phase, the following relation is obtained:

$$\sum_i X_i\, d\mu_i + \overline{S}\, dT - \overline{V}\, dp = 0 \tag{5-176}$$

where $X_i = n_i/n$, $\overline{S} = S/n$, and $\overline{V} = V/n$. Equation (5-176) is a general relationship for a single phase containing any number of components. In the present experiment there is a single component and two phases. Hence, i = 1 and $X_i = 1$ in Eq. (5-176). Two equations like Eq. (5-176) may be immediately written down, one for the solid phase and one for the gas phase. Labeling the solid phase s and the gas phase g, the two equations are

$$d\mu^s + \overline{S}^s\, dT - \overline{V}^s\, dp = 0 \tag{5-177}$$

$$d\mu^g + \overline{S}^g\, dT - \overline{V}^g\, dp = 0 \tag{5-178}$$

Since it is assumed that the phase transition is carried out at equilibrium, $\mu^s = \mu^g$. Hence,

$$d\mu^s = d\mu^g \tag{5-179}$$

Using Eq. (5-179), Eqs. (5-177) and (5-178) can be combined to furnish

$$(\overline{S}^g - \overline{S}^s)dT = (\overline{V}^g - \overline{V}^s)dp$$

or

$$\frac{dp}{dT} = \frac{\overline{S}^g - \overline{S}^s}{\overline{V}^g - \overline{V}^s} = \frac{\Delta\overline{S}}{\Delta\overline{V}} = \frac{\Delta\overline{H}}{T\,\Delta\overline{V}} \tag{5-180}$$

The last equality in Eq. (5-180) follows from the fact that the phase transition is carried out at equilibrium which implies that $\Delta\overline{G} = 0$. Therefore, the following series of equations can be written:

$$\overline{G} = \overline{H} - T\overline{S}$$

$$\Delta\overline{G} = \Delta\overline{H} - T\,\Delta\overline{S} - \overline{S}\,\Delta T = 0 \tag{5-181}$$

but T = constant, and therefore $\Delta T = 0$. Hence, $\Delta\overline{S} = \Delta\overline{H}/T$.

If use is made of the inequality, $\overline{V}^g >> \overline{V}^s$, Eq. (5-180) can be approximated by

$$\frac{dp}{dT} \cong \frac{\Delta\overline{H}}{\overline{V}^g T} \tag{5-182}$$

Observing that $d \ln p = dp/p$ and $d(1/T) = -dT/T^2$, Eq. (5-182) may be reexpressed as

$$\frac{d \ln p}{d(1/T)} = \frac{-\Delta\overline{H}T}{p\overline{V}^g} \tag{5-183}$$

This is an equation of the same form derived in Section 5.4. In this case the process is solid → gas instead of liquid → gas.

If gas behaves ideally, then $p\overline{V} = RT$. However, if the gas deviates from ideality, some other equation of state must be used. Frequently, deviations from ideality are measured in terms of a dimensionless parameter Z called the compressibility factor. This parameter is defined by the following equation:

$$Z = \frac{p\bar{V}^g}{RT} \tag{5-184}$$

Equation (5-184) indicates that $Z = 1$ for an ideal gas, $Z < 1$ indicates negative deviations from ideality, and $Z > 1$ indicates positive deviations from ideality. Observe that Z varies with p, V, and T. However, over small temperature ranges, Z is approximately independent of T. Using Eq. (5-184) in Eq. (5-183) and assuming that Z is approximately constant over the temperature range of interest, one observes that a plot of ln p versus 1/T will furnish a linear plot with slope = $-\Delta\bar{H}/RZ$. In the present experiment, data will be obtained which will furnish a plot of ln p versus 1/T and a method, to be described below, will be given which will allow corrections for nonideality.

The compressibility factor approaches the ideal gas value of unity as p approaches zero or $\bar{V}$ approaches infinity. Consequently, the variation of $p\bar{V}/RT$ with p or $\bar{V}$ can be represented as a power series in p or $1/\bar{V}$. Observe that a power series in p will approach the ideal gas equation of state as p approaches zero and that a power series in $1/\bar{V}$ will approach the ideal gas equation of state as $\bar{V}$ approaches infinity. The resulting equations of state are called the virial equations. When the power series expansion is in terms of p, the virial equation becomes

$$\frac{p\bar{V}}{RT} = 1 + Bp(T)p + Cp(T)p^2 + \ldots \tag{5-185}$$

In Eq. (5-185), $B_p(T)$ is called the second virial coefficient and $C_p(T)$ is called the third virial coefficient. Notice that the virial coefficients depend on the temperature.

Methods are available from statistical mechanics for determining the virial coefficients if the intermolecular potential is given. One potential which is frequently used in describing the intermolecular forces between two molecules is the Lennard-Jones potential. This particular potential is useful when considering the interaction of two molecules or atoms which have no net charge or permanent dipole moment. In the present experiment the interaction between argon atoms is being considered; hence, the Lennard-Jones potential is expected to be applicable. This potential gives the variation of intermolecular or interatomic potential with distance between the centers of the two interacting particles according to the following equation:

$$V(r) = 4\epsilon\left\{\left(\frac{\sigma}{r}\right)^{12} - \left(\frac{\sigma}{r}\right)^{6}\right\} \tag{5-186}$$

In this equation, observe that V(r) approaches zero as r approaches infinity. Hence, when the two interacting particles are far apart their potential is defined to be zero. As r is decreased (meaning that the particles are being brought closer together) the potential becomes negative and at some particular value of r reaches a minimum. For smaller values of r, the Lennard-Jones potential rises very rapidly because of the $1/r^{12}$ term. The so-called *well depth* is given by ϵ, and σ is the value of r for which the potential is zero near the origin. A schematic plot of the Lennard-Jones potential is shown in Fig. 5-26. Physically, the $1/r^6$ term arises from induced dipole-induced dipole interactions and is attractive, whereas the $1/r^{12}$ term arises from the interaction of the electron densities surrounding the two nuclei and this interaction is strongly repulsive at small separations. The Lennard-Jones parameters for argon are $\epsilon/k = 121°K$ and $\sigma = 3.40Å$, where k is Boltzmann's constant.

The relationship between the Lennard-Jones parameters and the virial coefficients is obtained from statistical mechanics, as has already been mentioned. The results are tabulated in terms of so-called *reduced* variables, T* and B*, which are defined by the following relations:

$$T^* = \frac{kT}{\epsilon} \tag{5-187}$$

$$B^* = \frac{B(T)}{\frac{2}{3}N\pi\sigma^3} \tag{5-188}$$

In Eq. (5-188), σ should be expressed in centimeters to furnish the molar virial coefficient. Values of T* and B* are given in Table 5-1. Note that T* and B* are dimensionless. $N = 6.023 \times 10^{23}$ molecules/mole.

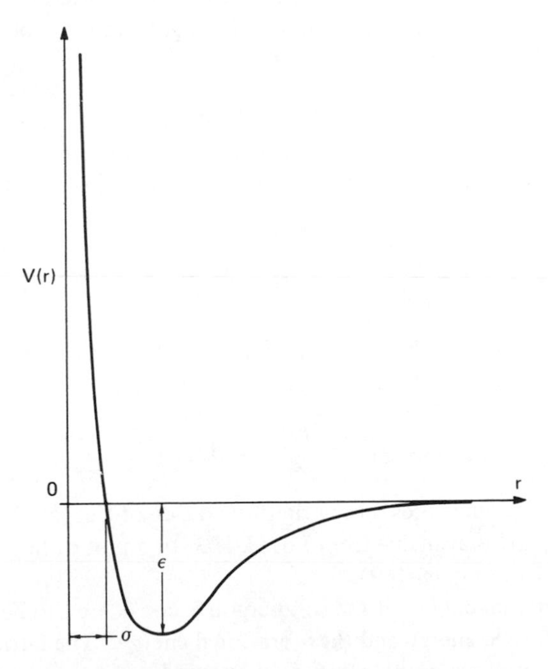

Figure 5-26. Lennard-Jones potential.

Table 5-1. VIRIAL COEFFICIENT FOR THE LENNARD-JONES 6-12 POTENTIAL*

T*	B*	T*	B*	T*	B*
0.30	-27.88	0.55	-7.27	0.80	-3.73
0.35	-18.75	0.60	-6.20	0.85	-3.36
0.40	-13.30	0.65	-5.37	0.90	-3.05
0.45	-10.75	0.70	-4.71	0.95	-2.77
0.50	-8.72	0.75	-4.18	1.00	-2.54

*J. O. Hirschfelder, C. F. Curtiss, and R. B. Bird, *Molecular Theory of Gases and Liquids* (Wiley, New York, 1954).

Observe that B(T) is not equivalent to $B_p(T)$. B(T) is the second virial coefficient when the equation of state is expanded in terms of $1/\bar{V}$. Hence, B(T) can be used in the equation of state:

$$\frac{p\bar{V}}{RT} = 1 + \frac{B(T)}{\bar{V}} + \ldots \tag{5-189}$$

The molar internal energy change associated with the sublimation of argon is related to the molar change in enthalpy by

$$\Delta\bar{H} = \Delta\bar{E} + \Delta(p\bar{V}) = \Delta\bar{E} + p\bar{V}^g \tag{5-190}$$

In Eq. (5-190) use has been made of the approximation that $\bar{V}^g >> \bar{V}^s$. The change in the internal energy per mole $\Delta\bar{E}$ for the sublimation is simply the difference between the molar internal energy of the gas and the molar internal energy of the solid. Therefore, one can write

$$\Delta\bar{E} = \bar{E}_{gas} - \bar{E}_{solid} \tag{5-191}$$

If the internal energy of the gas is defined to be identically zero when the absolute temperature, T, is zero, then the molar internal energy of argon (monatomic) will be

$$\bar{E}_{gas} = \frac{3RT}{2} \tag{5-192}$$

Making use of Eqs. (5-190), (5-191), and (5-192), the following expression for the molar internal energy of the solid argon is obtained:

$$\bar{E}_{solid} = \frac{3RT}{2} - \Delta\bar{H} + p\bar{V}^g \tag{5-193}$$

Observe that all the quantities on the right-hand side of Eq. (5-193) are available; T is measurable, $\Delta\bar{H}$ is available from Eq. (5-183) or a plot of ln p versus 1/T, and $p\bar{V}^g$ is available from Eq. (5-189).

The molar internal energy of the solid argon is considered to be made up of two parts: the lattice energy and the vibrational energy. The lattice energy is defined as the potential energy an argon atom would have if it was completely at rest in the solid lattice. The potential energy is defined to be zero when the argon atom is an infinite distance from the lattice. The vibrational energy is the energy involved in the thermal motion of the argon atom away from its equilibrium (rest) position in the solid lattice.

The lattice energy is the quantity which we desire to calculate in this experiment and it is given from Eq. (5-193) by

$$\bar{E}_{lat} = \frac{3RT}{2} - \Delta\bar{H} + p\bar{V}^g - \bar{E}_{vib} \tag{5-194}$$

The molar vibrational energy for argon at low temperatures is small and can be calculated from the Debye theory of lattice vibrations. The Debye theory is a statistical mechanical theory which assumes that the argon atoms in the lattice can vibrate with a wide variety of frequencies which are distributed in the same way as black-body radiation. According to the Debye theory the molar vibrational energy is given by

$$\bar{E}_{vib} = \frac{9}{8}R\Theta + 3RTD\frac{\Theta}{T} \tag{5-195}$$

In Eq. (5-195), Θ is called the Debye temperature and $D(\Theta/T)$ is called the Debye function. For argon at 70°K, $\Theta = 85°K$ and $D(\Theta/T) = 0.617$. Using these values, Eq. (5-195) furnishes the molar vibrational energy of solid argon. Using this value in Eq. (5-194) furnishes the molar lattice energy of solid argon at 70°K.

The lattice energy of a molecular crystal can be calculated theoretically if data are available on the potential energy between atoms in the crystal. The Lennard-Jones potential is useful for calculating the potential between two interacting particles. The same forces are present in the molecular crystal of argon, but there are many more than two interacting atoms. For example, pick a given argon atom and examine the surrounding environment. There are several other argon atoms very close, and each of them influences the potential of this single atom. If it is assumed that the total potential energy is the sum of all the potential energies taking the atoms in pairs, then for a face-centered cubic crystal lattice such as argon the result is

$$\bar{E}_{lat} = 2N_0\epsilon\left\{12.13\left(\frac{\sigma}{A}\right)^{12} - 14.45\left(\frac{\sigma}{A}\right)^{6}\right\} \tag{5-196}$$

where A is the distance between nearest neighbors, N_0 is Avogadro's number, and σ and ϵ are the Lennard-Jones parameters. Notice that the major portion of the repulsion comes from nearest-neighbor interactions (summing the pair potentials over nearest neighbors alone would furnish 12.00 for both coefficients). Since $1/r^6$ falls off much more slowly than $1/r^{12}$, the coefficient of the second term in Eq. (5-196) is greater than 12.00. The above equation permits a calculation of the lattice energy of solid argon which can be compared with the experimental result obtained using Eq. (5-194).

EXPERIMENTAL. In this experiment argon and liquid nitrogen are used because both are readily available commercially and argon solidifies with a readily measured vapor pressure at liquid nitrogen temperatures. As suggested in the theoretical background section the goal is to measure the vapor pressure in equilibrium with solid argon at several temperatures around that of liquid nitrogen. At 1 atm pressure, liquid nitrogen has a temperature of 77°K. If the pressure is lowered with the liquid nitrogen in a container which allows no energy to come into the system (a dewar or Thermos bottle), the equilibrium liquid temperature drops. In outline form the experiment thus requires a dewar, vacuum pump, sample container, sample, temperature-measuring device, and pressure-measuring device. These components are shown schematically in Fig. 5-27 along with valves to provide control of the system. The numbers refer to valves which will be assumed here to be stopcocks. The crucial component is the sample container-thermometer combination which is shown in more detail in Fig. 5-28. Clearly the sample container must be compatible with the liquid nitrogen container, and for best results a wide-mouth stainless steel dewar holding about 2 liters of liquid is recommended. However, Pyrex dewars and larger liquid nitrogen storage vessels can be used. Considerable caution should be used with Pyrex dewars so as to avoid flying glass in case of breakage. Figure 5-28 assumes a wide-mouth stainless steel dewar. In the upper portion of Fig. 5-28 some detail of the dewar assembly is given. This assembly consists of a metal lid which seals to the lip of the dewar by a flat rubber gasket. The heating wire wrapped around the rim of the lid is for the purpose of keeping the lid near room temperature. Stainless steel is preferable for the lid, although brass can be used. The sample container and thermocouple well should be machined out of a single piece of copper and then silver-soldered onto the thin-walled stainless steel support tube. The thermocouple should be held in its well with a clamping screw. In making the thermocouple connections at the vacuum wall, be certain to connect the same type of wire to both sides of

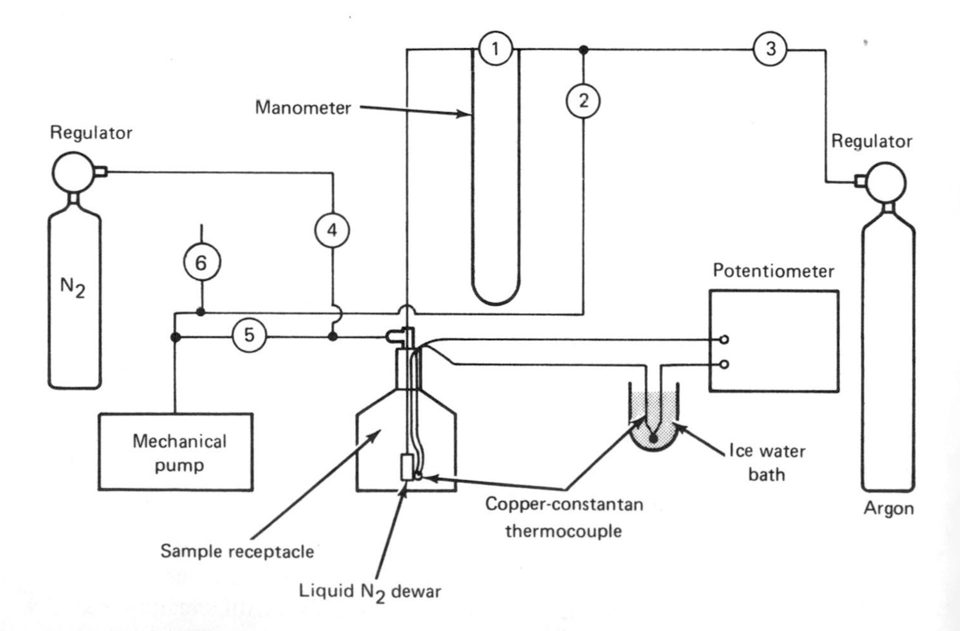

Figure 5-27. Apparatus for measurement of argon lattice energy.

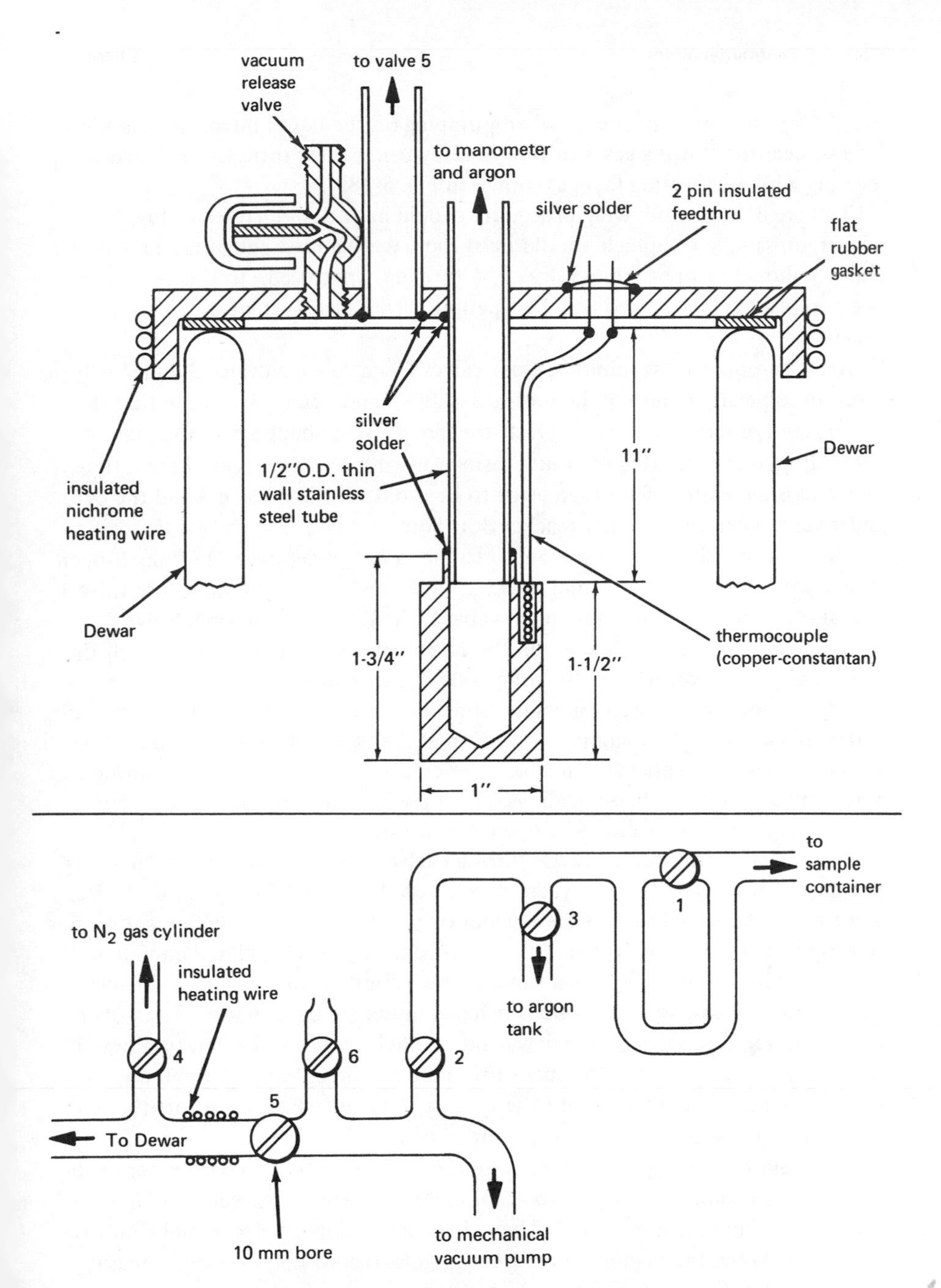

Figure 5-28. Details of argon receptacle and valve arrangement.

the two-pin feedthrough. The vacuum release valve can be any sort of needle valve which can be screwed into the lid. In the lower part of Fig. 5-28 the valving arrangement for evacuation and sample handling is shown. All these valves can be standard Pyrex stopcocks with 4-mm bores except valve 5, which should have a 10-mm bore to allow large pumping speeds during evacuation of the dewar. To avoid "freezing up" stopcock 5 when pumping on the liquid nitrogen, it is advisable to heat the flowing gas near this valve by using a few turns of insulated heating wire or a heating tape, as shown in Fig. 5-28.

To take data the following procedure should be followed (refer to Fig. 5-27): The argon sample receptacle should be at room temperature and outside the dewar. Valve 1 should be open while valves 3, 4, 5, and 6 are closed. In this condition the pump should be started and, by opening valve 2, the system should be evacuated.

After pumping a few minutes, close valves 2 and 1. Open valve 3 and slowly bleed in some argon, noting the pressure at the manometer. Also note that the argon is not entering the receptacle at this point. What happens to the manometer if argon is admitted without closing valve 1? Nothing, even though the pressure is increasing. This is an error to be avoided. Close valve 3 and the cylinder valve when the pressure reaches 800 Torr.

Open valve 1 slowly to admit argon to the sample receptacle. Leaving it open, pump out the system by opening valve 2. This flushes the system. Close valve 1, add another 800 Torr of argon, open valve 1, then close it, and open valve 2. This leaves the system with argon in the receptacle connected to one side of the manometer and a vacuum on the other side of the manometer.

After recording the argon pressure, slowly immerse the receptacle in the liquid nitrogen and make the vacuum connection at the top of the dewar. Add ice water to the copper-constantan thermocouple reference dewar and begin measuring emf versus time. Continue these readings until they become constant. Keep the vacuum release valve of Fig. 5-28 open during this time.

By evacuating the dewar slowly through valve 5 the temperature of the nitrogen may be lowered. Keep the vacuum release valve closed during this time. Record emf, pressure, and time as the evacuation takes place. When the pressure ceases to drop over a period of 10 min, the experiment may be considered complete.

Close valve 5 and slowly open valve 1; then admit gaseous nitrogen through valve 4 until atmospheric pressure is reached in the nitrogen dewar. Then open the vacuum release valve on the dewar lid assembly, remove the dewar, allow the sample to warm up to room temperature, shut off the pump, and finally open valves 6 and 2 to completely vent the system. *Caution:* Never shut off the pump when the argon receptacle is at low temperatures.

Ascertain the atmospheric pressure in the laboratory using a barometer or the above experimental apparatus. To accomplish the latter, close valves 3, 5, and 6, turn on the pump, and evacuate. Then close valve 1, open valve 6, and shut off the pump. When the manometer stops changing, record the pressure. Finally open valve 1 to complete the experiment.

ANALYSIS. Translate the measured emfs to temperature using data for the region near 77°K, which is available in various handbooks. By constructing a graph of emf versus temperature from the handbook data, interpolation may be carried out to obtain the temperatures corresponding to the measured emfs. It may be necessary to apply a small correction to the handbook values if the thermocouple is imperfect.

To make an approximate correction, look up in a handbook the nitrogen vapor pressure as a function of temperature and plot it as ln p versus 1/T. From the logarithm of the atmospheric pressure, determine the temperature at which the nitrogen would boil using the above graph. Remember that boiling occurs when vapor pressure equals atmospheric pressure. The resulting temperature should agree with the liquid nitrogen temperature measured with the thermocouple in your experiment prior to the time you began pumping on it. If the two do not agree, assume that a correction should be applied to all the temperatures—a correction equal to the difference in the measured (systematically incorrect) and expected temperatures.

ANALYSIS. Determine the molar heat of vaporization of argon by constructing and analyzing the appropriate graph. Calculate $\Delta\bar{S}$ for the process at 77°K and explain the sign and magnitude of this quantity. Give reasons the process Ar (liquid) → Ar (gas) might have a different $\Delta\bar{S}$.

Assume that solid argon forms a face-centered cubic crystal and that the unit cell has sides of length 5.43Å. Calculate both a theoretical and an experimental value for the lattice energy of solid argon and compare the two.

Calculate the expected error in $\Delta\bar{H}$ arising from the experimental measurement which appears to be most uncertain.

PROBLEM

1. Define the potential energy of a pair of argon atoms as being zero when they are separated by an infinite distance. Construct a potential-energy curve and an energy level diagram showing the lattice energy and the vibrational energy of the pair.

REFERENCES

Section 5.1*

J. M. Sturtevant in A. Weissberger (ed)., *Physical Methods of Organic Chemistry*, Vol. I, (Wiley-Interscience, New York, 1959), Part I, Ch. X. Contains a full discussion of bomb calorimetry.

E. J. Prosen and F. D. Rossini, *J. Res. Natl. Bur. Standards* **27**, 289 (1941); H. M. Huffman and E. L. Ellis, *J. Am. Chem. Soc.* **57**, 41 (1935). Contain a full description of the type of bomb calorimeter used here.

*General information regarding heats of reaction can be found in the References listed in Appendix XI.

R. S. Jessup, *J. Res. Natl. Bur. Standards* **29**, 247 (1942), gives the accepted heat of combustion. Benzoic acid standard samples are available from the National Bureau of Standards, Washington, D.C.

E. W. Washburn, *J. Res. Natl. Bur. Standards* **10**, 525 (1933), and Sturtevant, *op. cit.* Describe the so-called *Washburn corrections* that should be made in very accurate work.

R. C. Wilhoit, *J. Chem. Educ.* **44A**, 685 (1967). A review of bomb calorimetry and available instrumentation. This reference also contains a useful tabulation of error limits associated with different kinds of measurements.

F. D. Rossini (ed.), *Experimental Thermochemistry* (Wiley-Interscience, New York, 1956). An excellent general reference on the measurement of heats of reaction.

F. D. Rossini, D. D. Wagman, W. H. Evans, S. Levine, and I. Jaffe, *Selected Values of Chemical Thermodynamic Properties* (U.S. Government Printing Office, Washington, D.C., 1952). Tabulated data on various compounds. Additions to this publication have appeared from time to time in the form of technical notes from the National Bureau of Standards. These notes are numbered 270-1, 270-2, 270-3, 270-4, etc.

Section 5.2 (See the references in Appendix XI.)

F. G. Cottrell, *J. Am. Chem. Soc.* **41**, 721 (1919); E. Washburn and J. W. Read, *J. Am. Chem. Soc.* **41**, 729 (1919). The original papers describing the ebulliometer used in this book. It is interesting to note how Cottrell's name has "stuck," while Washburn's has not.

W. Swietoslawski, *Ebulliometric Measurements* (Van Nostrand Reinhold, New York, 1945). A very useful monograph on ebulliometric measurements.

W. Swietoslawski and J. R. Anderson in A. Weissberger (ed.), *Technique of Organic Chemistry,* Vol. I (Wiley-Interscience, New York, 1959), Part I, Ch. VIII. A second useful general source of ebulliometric information.

B. J. Mair, *J. Res. Natl. Bur. Standards* **14**, 345 (1935). An interesting article describing the use of a differential Cottrell-type apparatus using thermocouple temperature measurements.

Section 5.3 (See the references in Appendix XI.)

J. Timmermans, *Physico-Chemical Constants of Pure Organic Compounds,* Vols. I and II (Elsevier, Amsterdam, 1950 and 1965). An excellent source of data for the physical properties of pure organic compounds.

J. Timmermans, *Physico-Chemical Constants of Binary Systems* (Wiley-Interscience, New York, 1959–1960). A second source of data on an enormous number of binary systems; a four-volume set.

H. M. Glass and W. M. Madgin, *J. Chem. Soc. (London)* 193 (1933); H. M. Glass and W. M. Madgin, *ibid.* 1431 (1933); H. M. Glass, W. M. Madgin. and F. Hunter, *ibid.* 260 (1934); H. M. Glass and W. M. Madgin, *ibid.* 1292 (1934). A series of four papers describing work on some of the systems mentioned. These papers also show how to determine activities from the type of data collected in this experiment.

C. M. Mason, B. W. Rosen, and R. M. Swift, *J. Chem. Educ.* **18**, 473 (1941). Includes phase diagrams for the systems biphenyl-*p*-dichlorobenzene, naphthalenebiphenyl, and *m*-dinitrobenzene-*p*-nitrolotuene.

Section 5.4 (See the references in Appendix XI.)

S. W. Tobey, *J. Chem. Educ.* **35**, 352 (1958). Contains the type of apparatus recommended for this experiment.

W. Ramsay and S. Young, *J. Chem. Soc. (London)* **47**, 42 (1885). Describes the Ramsay-Young apparatus, a similar type of vapor pressure apparatus but one which is more difficult to use.

G. W. Thomson in A. Weissberger (ed.), *Physical Methods of Organic Chemistry,* Vol. I (Wiley-Interscience, New York, 1959), Part I, Ch. IX. Contains a full description of vapor pressure methods.

Handbook of Chemistry and Physics (Chemical Rubber Publishing Company, Cleveland). Any edition has an adequate list of vapor pressures as a function of temperature.

Section 5.5 (See the references in Appendix XI.)

K. Denbigh, *The Principles of Chemical Equilibrium* (Cambridge University Press, New York, 1971), Ch. 8 and 9.

G. N. Lewis and M. Randall (rev. by K. S. Pitzer and L. Brewer), *Thermodynamics* (McGraw-Hill, New York, 1961).

J. Timmermans, *Physico-Chemical Constants of Binary Systems* (Wiley-Interscience, New York, 1959–1960). An excellent source of data for a large number of binary systems; a four-volume set.

N. A. Lange, *Handbook of Chemistry,* 10th ed. (Handbook Publishers, Inc., Sandusky, Ohio, 1967), pp. 1496–1505. Contains a table of azeotropic mixtures.

Section 5.6 (See the references in Appendix XI.)

J. Timmermans, *The Physico-chemical Constants of Binary Systems in Concentrated Solutions* (Wiley-Interscience, New York, 1960). An excellent compendium of data on the physical properties of binary systems; a four-volume set.

A. R. Fowler and H. Hunt, *Ind. Eng. Chem.* **33**, 90 (1941). The *n*-propanol–water system.

S. E. Wood and J. P. Brusie, *J. Am. Chem. Soc.* **65**, 1891 (1943). The benzene-carbon tetrachloride system.

J. S. Rowlinson, *Liquids and Liquid Mixtures,* 2nd ed. (Butterworths, London, 1969). A general treatise on solutions.

N. Bauer and S. Z. Levin in A. Weissberger and B. W. Rossiter (eds.), *Physical Methods of Chemistry,* Vol. I (Wiley-Interscience, New York, 1972), Part IV, Ch. II. Contains a description of density measurements.

J. F. Coetzee, *Solute-Solvent Interactions* (Dekker, New York, 1969).

Section 5.7 (See the references in Appendix XI.)

N. I. Sax, *Dangerous Properties of Industrial Materials* (Van Nostrand Reinhold, New York, 1957); M. N. Gleason, R. E. Gosselin, and H. C. Hodge, *Clinical Toxicology of Commerical Products* (Williams & Wilkins, Baltimore, 1957). Good sources of information on the dangerous properties of compounds.

F. H. Verhoek and F. Daniels, *J. Am. Chem. Soc.* **53**, 1250 (1931). An early reference using method 1 with a spoon gauge null detector.

L. Harris and K. L. Churney, *J. Chem. Phys.* **47**, 1703 (1967); F. S. Wettack, *J. Chem. Educ.* **49**, 557 (1972). Describe method 3. The first contains thermodynamic data.

W. F. Giauque and J. D. Kemp, *J. Chem. Phys.* **6**, 40 (1938). Studies the vapor pressure above solid N_2O_4. This reference also contains thermodynamic quantities of interest.

T. C. Hall, Jr., and F. E. Blacet, *J. Chem Phys.* **20**, 1745 (1952). Reports the optical absorption spectra of NO_2 and N_2O_4.

JANAF Thermochemical Tables, *Volume NBS 37,* 2nd ed. (National Bureau of Standards, U.S. Government Printing Office, Washington, D.C., 1971). A good source for thermochemical data.

Section 5.8 (See the references in Appendix XI.)

Handbook of Chemistry and Physics, any ed. (Chemical Rubber Publishing Company, Cleveland). The equilibrium data measured here are tabulated as vapor pressure data for inorganic compounds.

E. P. Egan, Jr., J. E. Potts, Jr., and Georgette D. Potts, *Ind. Eng. Chem.* **38**, 454 (1946); T. R. Briggs and V. Migrdichian, *J. Phys. Chem.* **28**, 1121 (1924). Two journal references on the equilibrium studied here.

M. J. Astle, *Industrial Organic Nitrogen Compounds* (Van Nostrand Reinhold, New York, 1961). Contains a discussion of the properties of many organic nitrogen compounds.

Section 5.9

S. Dal Nogare and R. S. Juvet, *Gas Chromatography* (Wiley-Interscience, New York, 1962); J. H. Knox, *Gas Chromatography* (Wiley, New York, 1962); J. H. Purnell, *Gas Chromatography* (Wiley, New York, 1962). A few of the large number of texts discussing gas chromatography.

S. Kenworthy, J. Miller, and D. E. Martin, *J. Chem. Educ.* **40**, 541 (1963). Describes the methods used here.

J. M. Miller, *Experimental Gas Chromatography* (Gow-Mac Instrument Co., Madison, N.J., 1965). An excellent short description of the workings of a gas chromatograph using thermal conductivity.

Section 5.10 (See the references in Appendix XI.)

S. Ross and J. P. Olivier, *On Physical Adsorption* (Wiley-Interscience, New York, 1964); A. W. Adamson, *Physical Chemistry of Surfaces* (Wiley-Interscience, New York, 1960); G. A. Somorjai, *Principles of Surface Chemistry* (Prentice-Hall, Englewood Cliffs, N.J., 1972). A few of the many textbook references on physical adsorption.

S. Brunauer, P. H. Emmett, and E. Teller, *J. Am. Chem. Soc.* **60**, 309 (1938). An early BET paper.

T. L. Hill, *J. Chem. Phys.* **17**, 762 (1949); T. L. Hill, *ibid.* **17**, 520 (1949); T. L. Hill, *ibid.* **16**, 181 (1948); T. L. Hill, *ibid.* **15**, 767 (1947); T. L. Hill, *ibid.* **14**, 263 (1946); T. L. Hill, *ibid.* **14**, 441 (1946). Treat the statistical mechanical theory of adsorption.

Handbook of Chemistry and Physics, any ed. (Chemical Rubber Publishing Company, Cleveland). Vapor pressure data for argon and oxygen is tabulated.

Section 5.11 (See the references in Appendix XI.)

L. D. Landau and E. M. Lifshitz, *Statistical Physics* (Addison-Wesley, Reading, Mass., 1958). Gives a good description of the statistical mechanics involved here.

J. O. Hirschfelder, C. F. Curtiss, and R. B. Bird, *Molecular Theory of Gases and Liquids* (Wiley, New York, 1954). An enormous volume treating in detail many of the problems of interest in this experiment.

H. J. M. Hanley, J. A. Barker, J. M. Person, Y. T. Lee, and M. Klein, *Molec. Phys.* **24**, 11 (1972). A recent contribution to the extensive literature on the Ar-Ar interatomic potential; a good place to begin to appreciate the enormous interest that continues in this field.

J. E. Lennard-Jones, *Proc. Roy. Soc. (London)* **A106**, 441, 463 (1924). An interesting early account of an attempt to determine a molecular force field from the equation of state of a gas or the temperature dependence of the viscosity of a gas. These papers record attempts by Lennard-Jones to write the force field between two atoms as the superposition of an attractive and repulsive force.

Chapter 6

Kinetics

6.1 Ionic Reactions in Solution

The purpose of this experiment is to determine the rate of the iodide ion (I^-) reaction with persulfate ion ($S_2O_8{}^{2-}$) under various experimental conditions. In the theory section equations suitable for rationalizing the data are developed.

THEORY. The area of physical chemistry which attempts to answer the question "How rapidly do chemical reactions take place?" is called kinetics. Empirically the rate of a chemical reaction can often be formulated in terms of a rate equation which furnishes a relationship between the differential change of the concentration of a particular species per unit time and the concentrations of various species present in the system. This relationship is not a theoretical one but arises from an empirical analysis of experimental data.

As an example of a rate equation, consider the reaction of ethylene bromide with potassium iodide in the liquid phase using methanol as the solvent. The overall stoichiometry is given by

$$C_2H_4Br_2 + 3KI \rightarrow C_2H_4 + 2KBr + KI_3$$

Empirically, this reaction is found to obey the following rate equation:

$$\frac{d[KI_3]}{dt} = k[C_2H_4Br_2]\,[KI] \tag{6-1}$$

The left-hand side of Eq. (6-1) is explicitly the differential change of the concentration of KI_3 per unit time. The right-hand side tells one how this rate of change depends on the concentrations of $C_2H_4Br_2$ and KI. The proportionality constant, k, is called the rate constant. The order of a reaction is given by the sum of the exponents of the concentration terms appearing in the rate equation. Since the sum of the exponents on the concentrations is 2, the above reaction is said to be second order. Observe that the stoichiometry does not furnish the form of the rate law. Rather the rate law is furnished in principle by the mechanism. A mechanism consists of a set of molecular scale rate processes which when considered together can explain the observations. Part of this experiment is oriented toward determining the mechanism.

In the present experiment the reaction of persulfate ion ($S_2O_8^{2-}$) with iodide ion (I^-) will be investigated. The stoichiometry of this reaction in aqueous solution is

$$S_2O_8^{2-} + 2I^- \rightarrow I_2 + 2SO_4^{2-} \tag{6-2}$$

The following form of the rate equation will be used:

$$-\frac{d[S_2O_8^{2-}]}{dt} = k[S_2O_8^{2-}]^n[I^-]^m \tag{6-3}$$

Reaction (6-2) and its rate equation (6-3) can be studied using small amounts of thiosulfate as a detector for iodine molecules. Thiosulfate ions react rapidly with I_2 in the following manner:

$$2S_2O_3^{2-} + I_2 \rightarrow 2I^- + S_4O_6^{2-} \tag{6-4}$$

Reaction (6-4) is extremely rapid compared to reaction (6-2), and the equilibrium strongly favors the formation of $2I^-$ and $S_4O_6^{2-}$. Hence, the thiosulfate reacts rapidly and quantitatively with iodine molecules.

Now consider a situation in which thiosulfate, iodide, and persulfate ions are mixed so that the concentration of thiosulfate is much less than the concentration of persulfate. In this situation the persulfate ion reacts with iodide ion to form I_2. The I_2 in turn reacts rapidly with thiosulfate ions, regenerating the iodide ions and using up the thiosulfate ions. Notice that for every $S_2O_8^{2-}$ ion that reacts two $S_2O_3^{2-}$ ions are used up but there is no change in the iodide ion concentration. Therefore, as the reaction proceeds the overall effect is to use up $S_2O_8^{2-}$ and $S_2O_3^{2-}$ while the I^- concentration remains constant. However, the $S_2O_3^{2-}$ concentration is small compared to the $S_2O_8^{2-}$ concentration; thus, after a certain period of time the $S_2O_3^{2-}$ is completely used up. At this point I_2 begins to be formed, and its concentration increases continuously since reaction (6-4) is no longer occurring. The time at which I_2 begins to be formed is indicated by the appearance of the blue starch-iodine complex.

In summary, use is made of a known small amount of $S_2O_3^{2-}$ to follow the reaction of $S_2O_8^{2-}$. The time required for complete removal of the $S_2O_3^{2-}$ is determined visually by the observation of a blue starch-iodine complex.

Let a, b, and c be the initial concentrations of $S_2O_8^{2-}$, I^-, and $S_2O_3^{2-}$, respectively. Let x be the amount of $S_2O_8^{2-}$ that has reacted at time t. Then at time t the concentrations are

$$[S_2O_8^{2-}] = a - x \tag{6-5a}$$

$$[S_2O_3^{2-}] = c - 2x \tag{6-5b}$$

$$[I^-] = b \qquad \text{if } [S_2O_3^{2-}] > 0 \tag{6-5c}$$

$$[I^-] = b + (c - 2x) \qquad \text{if } [S_2O_3^{2-}] = 0 \tag{6-5d}$$

Substitution into Eq. (6-3) furnishes

$$\frac{dx}{dt} = k(a - x)^n b^m \qquad \text{when } [S_2O_3^{2-}] > 0 \tag{6-6}$$

$$\frac{dx}{dt} = k(a - x)^n (b - 2x + c)^m \qquad \text{when } [S_2O_3^{2-}] = 0 \tag{6-7}$$

Equation (6-6) furnishes the rate when $[S_2O_3^{2-}]$ is present and Eq. (6-7) furnishes the rate after $[S_2O_3^{2-}]$ is depleted. From Eq. (6-5b) the value of x when the $[S_2O_3^{=}]$ is depleted is

$$x = \frac{c}{2} \tag{6-8}$$

Experimentally, the time required to reach the point of $[S_2O_3^{2-}]$ depletion is measured. This time will be denoted as $\bar{t}$ in the following discussion. From Eqs. (6-6) and (6-8) and the experimental times, the values of m and n can be estimated. Consider the case where $a \gg x$. In this case Eq. (6-6) can be approximated as

$$\frac{dx}{dt} \cong ka^n b^m \tag{6-9}$$

Another consequence of $x \ll a$ is

$$\frac{dx}{dt} = \frac{\Delta x}{\Delta t} = \frac{x}{t} = \frac{c}{2\bar{t}} \qquad \text{when } [S_2O_3^{2-}] \to 0 \tag{6-10}$$

The procedure for evaluating m and n is as follows. Consider two experiments run under the following conditions of time and concentrations where p is a constant:

Experiment	a	b	c	$\bar{t}$
1	A	B	C	$\bar{t}_1$
2	pA	B	C	$\bar{t}_2$

From the combination of Eqs. (6-9) and (6-10) using Experiment 1,

$$kA^nB^m = \frac{c}{2\bar{t}_1} \tag{6-11}$$

Using Experiment 2,

$$k(pA)^nB^m = \frac{c}{2\bar{t}_2} \tag{6-12}$$

The ratio of Eq. (6-12) to Eq. (6-11) is

$$\frac{\bar{t}_1}{\bar{t}_2} = \frac{(pA)^n}{A^n} = p^n \tag{6-13}$$

Equation (6-13) can be solved for n. A similar set of experiments in which a and c are maintained fixed and b is varied will furnish m. Having determined m and n, Eq. (6-6) can be integrated to furnish x as a function of t. Applying the boundary conditions at $t = 0$ and $t = \bar{t}$ permits determination of both the integration constant and the rate constant, k.

The investigation of chemical reactions has revealed that the rate of a reaction depends on the temperature. Generally speaking, the rate is found to increase with T, the absolute temperature. Examining Eq. (6-6) reveals that, for the rate to be temperature-dependent, the rate constant must be temperature-dependent. In a large number of cases, the variation of k with T is found to obey the Arrhenius equation

$$k = A \exp\left\{-\frac{E_a}{RT}\right\} \tag{6-14}$$

where A and E are constants referred to as the frequency factor or preexponential factor and Arrhenius activation energy, respectively. R is the gas constant in calories per mole. Notice that a plot of ln k versus 1/T will be a straight line of slope $= -E_a/R$ if the reaction being studied obeys the Arrhenius law. The intercept is ln A.

A theory referred to as the absolute rate theory or transition state theory is frequently used to describe chemical reactions. This theory predicts that for the elementary bimolecular reaction

$$A + B \rightarrow \text{products}$$

the following occurs:

$$A + B \rightarrow AB^{\neq} \rightarrow \text{products}$$

This equation suggests that the interaction of A and B forms an activated complex denoted by $AB^{\neq}$. This complex may be thought of as being a pseudomolecule which is formed when A and B come together with sufficient energy to react (i.e., form products). This transient complex, which has seldom been observed directly in any experiment, is assumed to be in equilibrium with the reactants, and the equilibrium constant is calculated by the methods of statistical thermodynamics. For present purposes we shall write the equilibrium constant as

$$K_{eq}^{\neq} = \frac{[AB]^{\neq}}{[A]\,[B]} \tag{6-15}$$

According to the transition state theory the rate is

$$-\frac{d[A]}{dt} = [AB]^{\neq}\,\frac{kT}{h} \tag{6-16}$$

where k is Boltzmann's constant, h Planck's constant, and T the absolute temperature.

Making use of (6-15) furnishes

$$-\frac{d[A]}{dt} = K_{eq}^{\neq}\,[A]\,[B]\,\frac{kT}{h} \tag{6-17a}$$

and

$$k(T) = K_{eq}^{\neq}\,\frac{kT}{h} \tag{6-17b}$$

where k(T) is the temperature-dependent rate constant.

From elementary thermodynamics the variation of an equilibrium constant with absolute temperature is

$$\Delta G^{\circ\neq} = -RT \ln K_{eq}^{\neq} \tag{6-18}$$

where $\Delta G^{\circ\neq}$ is the free energy change accompanying the formation of 1 mole of $AB^{\neq}$ in its standard state from the reactants A and B in their standard states at temperature T. Furthermore,

$$\Delta G^{\circ\neq} = \Delta H^{\circ\neq} - T\,\Delta S^{\circ\neq} \tag{6-19}$$

Using Eqs. (6-19) and (6-18) in Eq. (6-17) furnishes a thermodynamic form for the rate coefficient:

$$\begin{aligned} k(T) &= \frac{kT}{h} \exp\left\{-\frac{\Delta G^{\circ\neq}}{RT}\right\} \\ &= \frac{kT}{h} \exp\left\{\frac{\Delta S^{\circ\neq}}{R}\right\} \exp\left\{-\frac{\Delta H^{\circ\neq}}{RT}\right\} \end{aligned} \tag{6-20}$$

The quantities $\Delta G^{\circ\neq}$, $\Delta H^{\circ\neq}$, and $\Delta S^{\circ\neq}$ are called the free energy of activation, the enthalpy of activation, and the entropy of activation, respectively.

Differentiating the logarithm of Eq. (6-20) with respect to T yields

$$\begin{aligned} \frac{d \ln k(T)}{dT} &= \frac{1}{T} + \frac{d \ln K_{eq}^{\neq}}{dT} \\ &= \frac{RT + \Delta H^{\circ\neq}}{RT^2} \end{aligned} \tag{6-21}$$

Comparison with the derivative of the logarithm of Eq. (6-14) leads to

$$E_a = RT + \Delta H^{\circ\neq} \tag{6-22}$$

Using this result in Eq. (6-14) and comparing the results with Eq. (6-20) provides a relationship between $\Delta S^{\circ\neq}$ and A. If the resulting $\Delta S^{\circ\neq}$ is positive, this is taken to indicate an increase in entropy in going from reactants to activated complex. Most often $\Delta S^{\circ\neq}$ is negative, indicating a decrease in entropy for the process.

Up to this point it has been tacitly assumed that the system being dealt with was ideal so that the equilibrium constant was given by Eq. (6-15). Actually, rate constants are found to be functions of the solvent concentration and of the ionic strength in addition to the temperature and reactant concentration mentioned earlier.

The equilibrium constant $K^{\neq}_{eq}$ in such a nonideal situation becomes a function of the activity coefficient, γ_i, of each species. This coefficient takes into account the effects of the solvent and ionic strength while preserving the form of the thermodynamic equations derived using the ideal gas or ideal solution models. Stated another way, we require that nonideal systems obey equations of the same form as ideal systems. To achieve this requirement use is made of activities, fugacities, and activity coefficients.

In the present experiment, ionic strength effects must be accounted for. The ionic strength, I, is defined by the relation

$$I = \frac{1}{2} \sum_i m_i Z_i^2 \tag{6-23}$$

where m_i and Z_i are the molality and charge on the ith kind of ion. One theory frequently used in treating ionic effects is the Debye-Hückel theory described in Section 9.1. It is based on the assumption of complete dissociation of the electrolyte, and all deviations from ideal behavior are assumed to arise from electrical interactions caused by ions. When the ions are considered to be point charges the so-called Debye-Hückel limiting law, which relates the activity coefficient to the ionic strength, arises. Its form is

$$-\log \gamma_i = AZ_i^2\sqrt{I} \tag{6-24}$$

where γ_i is the activity coefficient, Z_i the charge, I the total ionic strength, and A a constant. An activity coefficient γ_i may be defined for any concentration scale. Here we use molarity, and the γs are molar activity coefficients defined by the relation

$$a_i = M_i\gamma_i \tag{6-25}$$

where a_i is the activity and M_i is the molarity of species i. Notice that the activity coefficient measures deviations from ideality and that by using activities in the form of Eq. (6-25) the thermodynamic equations for nonideal solutions will have the same form as those for ideal solutions. Furthermore, as the value of M_i

approaches zero, a_i approaches M_i and, hence, γ_i must approach unity. Also notice that according to Eq. (6-24) the activity of an ion depends on the total ionic strength and not solely on its own concentration. The Debye-Hückel theory does not take into account the finite size of the ions and, as a result, tends to overestimate the effects of interionic attraction and repulsion. This becomes a serious defect when the ionic concentration is large.

In terms of activities the equilibrium constant expression for the reaction becomes

$$K_{eq}^{\neq} = \frac{\gamma^{\neq}[AB^{\neq}]}{\gamma_A \gamma_B [A][B]} \tag{6-26}$$

and the rate expression becomes

$$-\frac{d[A]}{dt} = k'K_{eq}^{\neq} \frac{\gamma_A \gamma_B}{\gamma^{\neq}} [A][B] \tag{6-27}$$

Hence,

$$k(T) = k'K_{eq}^{\neq} \frac{\gamma_A \gamma_B}{\gamma^{\neq}} \tag{6-28}$$

In the limit as $\gamma_i \longrightarrow 1$ (i.e., at infinite dilution)

$$k(T) \longrightarrow k_0(T) = k'K_{eq}^{\neq} \tag{6-29}$$

Hence

$$\log k(T) = \log k_0(T) + \log \frac{\gamma_A \gamma_B}{\gamma^{\neq}}$$

$$= \log k_0(T) + \log \gamma_A + \log \gamma_B - \log \gamma^{\neq} \tag{6-30}$$

Equation (6-24) furnishes γ_i as a function of I. Using this relationship in Eq. (6-30) gives:

$$\log k(T) = \log k_0(T) - A\sqrt{I}\,(Z_A^2 + Z_B^2 - Z_{\neq}^2) \tag{6-31a}$$

$$Z_A + Z_B = Z_{\neq} \tag{6-31b}$$

$$\log k(T) = \log k_0(T) - A\sqrt{I}\,[Z_A^2 + Z_B^2 - (Z_A + Z_B)^2] \tag{6-31c}$$

$$\log k(T) = \log k_0(T) + 2AZ_A Z_B \sqrt{I} \tag{6-31d}$$

From Eq. (6-31d) it is clear that a plot of log k(T) versus $\sqrt{I}$ should furnish a straight line having an intercept log $k_0(T)$ if the Debye-Hückel theory is valid.

EXPERIMENTAL. The following solutions are to be used:

$K_2S_2O_8$: 0.01 M

KI: 0.10 M

$Na_2S_2O_3$: 0.001 M

KNO_3: 0.4 M

starch: 0.2%

These solutions should be freshly made using deionized or distilled water. The thiosulfate should be standardized, using potassium iodate or potassium dichromate as a primary standard. The procedure is described in most quantitative analysis texts. The starch solution is made by mixing 1 g of soluble starch with a little cold water to make a paste and then adding it to 100 cc of boiling water. After 2 min of boiling the solution can be cooled to room temperature for use.

The reaction mixtures should be prepared by pipetting known amounts of KI, KNO_3, starch, and $Na_2S_2O_3$ into one clean container and the $K_2S_2O_8$ solution into another; to start the reaction, these two mixtures should be poured together quickly and mixed thoroughly. To evaluate m and n, suggested reaction mixtures are

ml of 0.1-M KI	ml of 0.001-M $Na_2S_2O_2$	ml of 0.4-M KNO_3	ml of 0.2% starch	ml of 0.10-M $K_2S_2O_8$	ml of H_2O
5	5	10	5	5	20
5	5	10	5	15	10
10	5	10	5	5	15

The range of times on these three reaction mixtures should vary from 5 to 20 min (if carried out at room temperature). Repeat these three runs to check for reproducibility. After finding m and n the student can make up two more reaction mixtures to give two additional values for $\bar{t}$, so that a graph of the integrated form of Eq. (6-6) versus t with about five points may be obtained. (The ionic strengths, controlled by the amount of KNO_3 solution, and the amount of starch solution should be kept the same for these extra points.)

Actually, once m and n are known we need only one measurement of $\bar{t}$ to find k using Eq. (6-6). The plot of the integrated form of Eq. (6-6) versus t is made to verify the values for m and n and to obtain a more accurate value for k(T).

The student should also determine k(T) at two other temperatures (40° and 60°C) so as to make a plot of log k versus 1/T. Use the three concentrations listed above. To reach these temperatures, use the constant-temperature bath in the laboratory. The ionic strength of the reaction mixtures should still be kept the same as for the determination of m and n.

To study the variation of k with ionic strength, make a series of four or more measurements at different ionic strengths, holding the temperature constant. The ionic strengths should range from that used for the determination of m and n to near zero. Vary the ionic strength by changing the concentration of the *inert* salt KNO_3 while keeping the reactant concentrations constant.

ANALYSIS. Using the data, find m and n to the nearest integer value and estimate the experimental uncertainties in these values. The mechanism of the reaction is thought to consist of the following two bimolecular steps:

$$S_2O_8^{2-} + I^- \rightarrow (IS_2O_8)^{3-} \quad (1)$$

$$(IS_2O_8)^{3-} + I^- \rightarrow I_2 + 2SO_4^{2-} \quad (2)$$

$$(IS_2O_8)^{3-} \rightarrow I^- + S_2O_8^{2-} \quad (3)$$

where $k_2 \gg k_1$ and $k_2 \gg k_3$. Ascertain whether or not this mechanism is compatible with your results by developing an expression for $d[S_2O_8^{2-}]/dt$. The steady-state approximation for $[IS_2O_8^{3-}]$ will be useful; namely, $d[IS_2O_8^{3-}]/dt = 0$. Under the above-described conditions the overall rate is expressed with high accuracy in terms of a single elementary reaction. Thus, the observed order becomes the molecularity. Postulate a structure for the activated complex neglecting solvent molecules. Why are solvent molecule effects especially significant in ionic reactions?

Using the integrated form of Eq. (6-6), make a plot of x versus $\bar{t}$ and determine k(T) for each temperature. Plot the appropriate form of the Arrhenius equation from which A and E_A are easily determined. Calculate the thermodynamic parameters $\Delta S^{\circ\neq}$, $\Delta H^{\circ\neq}$, $\Delta G^{\circ\neq}$, and $K^{\neq}_{eq}$. Comment on the significance of the magnitude and sign of each of these quantities.

From the data taken with varying ionic strength and constant T, plot $\log k(T)$ versus $\sqrt{I}$ and determine $k_0(T)$ and A of Eq. (6-31d). At I = 0.25, calculate, using Debye-Hückel theory, the values of $\gamma^{\neq}$, γ_{I^-}, and $\gamma_{S_2O_8^{2-}}$. From these results and $K^{\neq}_{eq}$, calculate $[AB^{\neq}]$ assuming that $[I^-] = 0.01M$ and that $[S_2O_8^{2-}] = 0.001$ M.

6.2 Dissociation Rates of Transition Metal Complex Ions

The purpose of this experiment is to measure the rates of dissociation of some transition metal complex ions and to correlate the results with predictions based on the crystal field theory of the stability of these complexes. Spectrophotometry is used as an analytical tool to obtain the rate of disappearance of the complex. In the theory section introductory crystal field theory and inorganic reaction rate theory are discussed. Dissociation rate measurements of Co(II), Ni(II) complexes with 1,10-phenanthroline are recommended.

THEORY. In this experiment we shall focus our attention on six-coordinated transition metal ion complexes of the form $(ML_6)^{2+}$, where M is the central metal ion bearing a positive charge (+2*e*) and L represents the ligand (neutral molecule or ion) which is chemically bound to the central metal ion. The Co^{2+}, Ni^{2+}, and Cu^{2+} complexes used all react with Hg^{2+} and release the original metal ion. The rates of these reactions vary significantly with the central metal ion, as will be illustrated in the experiment. We shall seek a rationalization of these differences in terms of the electronic and geometric structures of the metal ions and complexes involved.

First we set down a classification scheme for transition metal complex substitution reactions. There are two possible substitutions: the metal ion or the ligand. In accord with the traditions of organic chemistry these are differentiated on the basis of the nucleophilic or electrophilic character of the species attacking the complex. Since transition metal ions are electrophilic (electron-accepting) while ligands are nucleophilic (electron-donating), this scheme neatly divides the two types of substitution reactions into S_N and S_E, where S indicates substitution, N indicates nucleophilic, and E indicates electrophilic. These two categories are further dissected on the basis of the reaction mechanism. Two mechanisms denoted 1 and 2 are employed. The dissociation mechanism (1) goes in two steps and for an S_N process involving two different ligands, L and L′, is

$$ML_6{}^{2+} \rightleftharpoons ML_5{}^{2+} + L \tag{6-32}$$

$$L' + ML_5{}^{2+} \rightarrow (ML_5L')^{2+} \tag{6-33}$$

It is denoted as an S_N1 mechanism. The other type of mechanism proceeds in a single concerted step as

$$L' + ML_6{}^{2+} \rightarrow (ML_5L')^{2+} + L \tag{6-34}$$

and is denoted S_N2. The S_E counterparts of these are

$$(ML_5L')^{2+} \rightleftharpoons ML_5{}^{2+} + L' \tag{6-32a}$$

$$M^{2+} + L \rightarrow ML^{2+} \tag{6-33a}$$

denoted S_E1 and

$$M'^{2+} + ML_5L^{2+} \rightarrow (M'L_5L)^{2+} + M^{2+} \tag{6-35}$$

denoted S_E2, where M^{2+} and M'^{2+} are two different metal ions.

In these mechanisms we have anticipated in some respects the kind of complexes that will occur in this experiment. In particular we shall be interested in reactions involving complexes of the form $(ML_4L'_2)^{2+}$, where $L = H_2O$ and L'_2 is a bidentate ligand (bidentate means that it acts as two normal ligands and binds to the central metal atom at two points). Complexes of this type are still six-coordinate, and assuming an S_E1-type mechanism we are interested in the sequence

$$(ML_4L'_2)^{2+} \rightleftharpoons (ML_4L' - L')^{2+} \tag{6-36}$$

$$M'^{2+} + (ML_4L' - L')^{2+} \rightarrow (M'L'_2)^{2+} + ML_4{}^{2+} \tag{6-37}$$

The L′–L′ symbol denotes the bidentate ligand and we assume in reaction (6-36) that one of the coordinating positions of the ligand is dissociated. Reaction

(6-37) follows in some way, not necessarily in one step, to allow L′–L′ to bind to the new metal ion M′. It should be noted that M′ in Eqs. (6-34), (6-35), and (3-37) has, in aqueous solution, H_2O bound to it. The distinction in these mechanisms is made, of course, at the molecular encounter level where different sequences of encounters lead to different overall results—the first mechanism (S_N1 or S_E1) requiring dissociation prior to forming the new complex, whereas the second (S_N2 or S_E2) does not.

Turning now to the features which control the rates of these reactions we expect the usual variations with temperature, ionic strength, and solvent that are discussed in Sections 6.1 and 6.3. Of more interest in this experiment are the relative rates of $ML'_2{}^{2+}$ complexes for the same L′–L′ but different M^{2+}. The source of differences lies buried in the electronic structure of the metal ion and the way in which these electrons behave in the presence of ligands.

If, for the purposes of simplification, we ignore the coordinated water molecules in every case, then the reaction mechanism of importance in this experiment is exemplified by the Co^{2+} case as

(6-38)

where the ligand is 1,10-phenanthroline. This is one form of an S_E1 mechanism. One question which arises in connection with this sequence of steps is identification of the rate-controlling step. In all the cases investigated in this experiment step 3 is fast. We assume step 2 to be faster than step 1, and therefore step 1 becomes rate controlling. The rate of appearance of the mercury complex or the rate of disappearance of the cobalt complex measures the rate of step 1 directly. If we inquire about the energetics of step 1, we are led to an investigation of the stability of the initial complex as one parameter to use in correlating kinetic data. A second important parameter is the stability of the product of step 1.

The stabilities of these complexes can be discussed in terms of size, charge, and d-electron structure of the central metal ion. As a general rule greater charge and/or smaller ionic radii lead to slower dissociation rates because the ligands are more tightly bound. The ions of interest, Cu^{2+}, Ni^{2+}, and Co^{2+}, all have the same charge. Their ionic radii are 0.72Å, 0.78Å, and 0.82Å, respectively. Using the general size-charge rule, one would predict the rates of dissociation to be ordered $Cu^{2+} < Ni^{2+} < Co^{2+}$. In actual fact the experimental order is $Cu^{2+} \gg Ni^{2+} < Co^{2+}$. Thus, other factors must intervene and we turn to the d-electron structure as one possibility. Remember that Eq. (6-38) has tacitly ignored the coordinated water.

For the three ions employed here the ground-state electron configurations are

$$Co^{2+}:\quad (1s)^2(2s)^2(2p)^6(3s)^2(3p)^6(4s)^0(3d)^7$$

$$Ni^{2+}:\quad (1s)^2(2s)^2(2p)^6(3s)^2(3p)^6(4s)^0(3d)^8$$

$$Cu^{2+}:\quad (1s)^2(2s)^2(2p)^6(3s)^2(3p)^6(4s)^0(3d)^9$$

These ions typically form octahedral complexes when they are fully coordinated. The geometry is illustrated in Fig. 6-1, where the metal ion defines the origin of the coordinate system and the four monodentate ligands (H_2O) and one bidentate ligand (1,10-phenanthroline) are located at opposing sides along the x, y, and z axes, each the same distance from the origin. Figure 6-2 illustrates the typical d-orbital geometries for the isolated metal ion. These orbitals are energetically degenerate in the free metal ion, but the formation of the metal-ligand bonds not only lowers the potential energy of the composite ligand-metal system but also causes, according to crystal field theory, a splitting (removal of degeneracy) of the metal ion d-orbitals. The argument is based essentially on

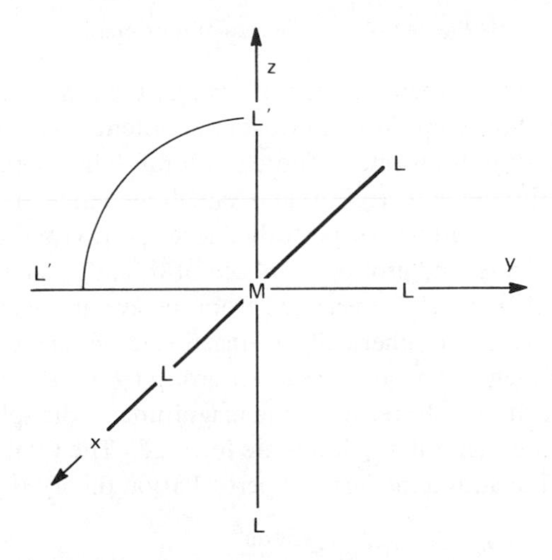

Figure 6-1. Geometry of an octahedral complex.

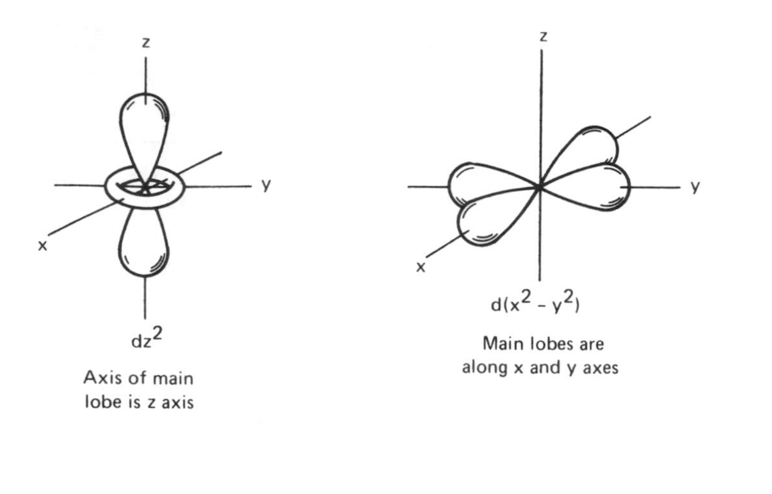

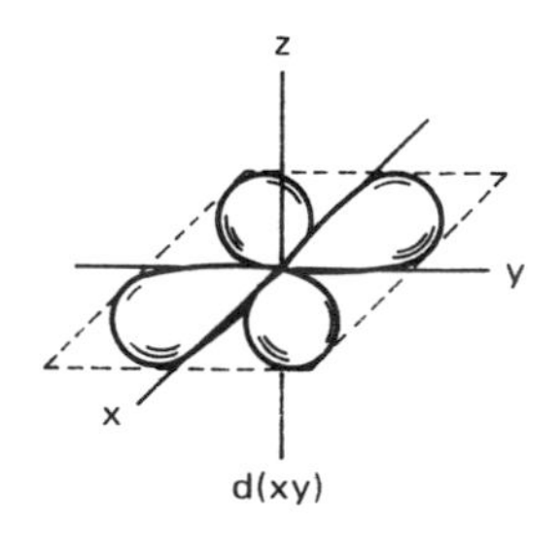

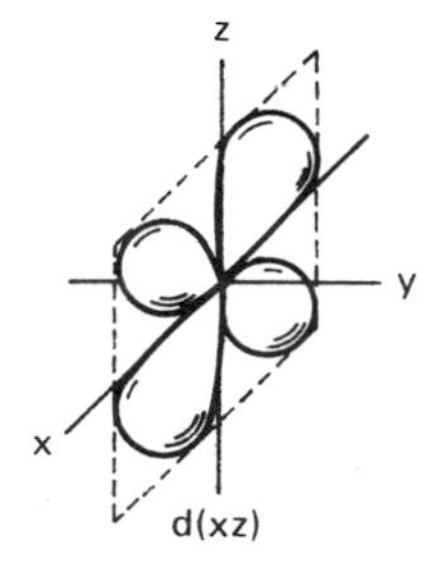

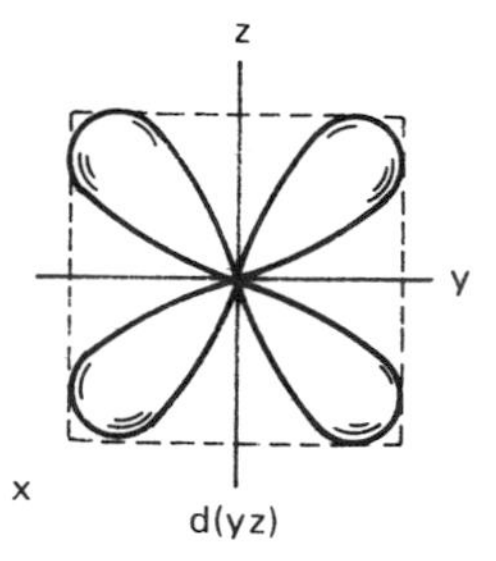

Lobes are in xy plane; axes at 45° with respect to x and y axes

Lobes are in xz plane; axes at 45° with respect to x and z axes

Lobes are in yz plane; axes at 45° with respect to y and z axes

Figure 6-2. Geometries of d-orbitals.

electrostatic repulsion between electron orbitals of the ligand and the d-orbitals of the metal atom. Since repulsion increases the potential energy and the increase is higher for shorter distances, those d-orbitals lying closest to the ligands have their potentials raised most. For an octahedral complex the crystal field theory thus predicts two groups of perturbed levels, one group comprised of d_{z^2} and $d_{x^2-y^2}$ and a second group comprised of d_{xy}, d_{yz}, and d_{xz}. Figure 6-3 illustrates the splitting, which is measured from the average energy which these d-orbitals would have in the spherically averaged electric field of the ligands. The upper group is denoted e_g and the lower group t_{2g} on the basis of group theoretical considerations. In terms of the magnitude of the splitting the e_g levels are raised more than the t_{2g} levels are lowered. The total splitting is 10Dq and on the basis of quantum mechanical perturbation theory is given by

$$10Dq \cong \frac{5e\mu a^4}{r^6} \tag{6-39}$$

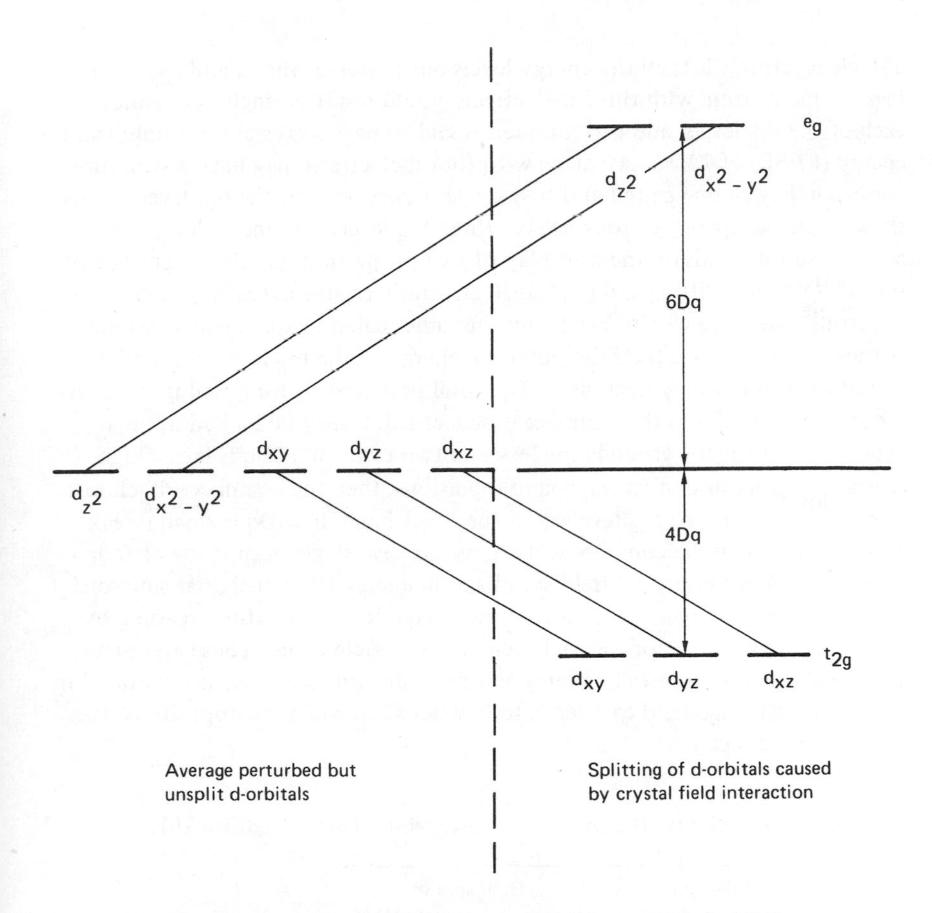

Figure 6-3. **Splitting of metal ion d-orbitals in an octahedral field.**

where e is the electron charge, μ the ligand dipole moment, a^4 the average value of the fourth power of the d-electron distance from the nucleus, and r the distance of the ligand to the metal. A theoretical evaluation of the right-hand side of Eq. (6-39) is stymied by the complexity of calculating a^4. If one of the e_g orbitals is not filled, one can, however, obtain 10Dq experimentally from the absorption spectrum of the complex ion. Excitation of an electron from t_{2g} to e_g corresponds to absorption of a photon of energy 10Dq. Equation (6-39) clearly depends on molecular properties of the metal atom and the ligands; thus, 10Dq is expected to vary from complex to complex. The relative stability of a given complex is, according to this theory, intimately related to the numbers of electrons actually occupying the split levels of Fig. 6-3. The stabilization energy is computed by assigning +4Dq to each electron in the t_{2g} levels and -6Dq to each electron in the e_g levels. It is worth remembering that the diagram of Fig. 6-3 is an approximation for the behavior of single electrons. Occupation by more than a single electron shifts all the energy levels but preserves the e_g and t_{2g} groupings. A metal atom with three d-electrons would result in single occupancy of each of the t_{2g} levels, and the complex is said to have a crystal field stabilization energy (CFSE) of 12Dq. An atom with four d-electrons may have a structure corresponding to one paired and two single occupancies of the t_{2g} levels or may show single occupancy of four levels: three t_{2g} levels and one e_g level. The actual result depends on the interplay of two properties: (1) the magnitude of the crystal field splitting and (2) the electrostatic repulsion (pairing energy) occurring when two electrons occupy the same region of space (same orbital). Because of the latter effect, the potential energy of the t_{2g} levels is raised if any one of them is multiply occupied. The total potential is also raised if an e_g level is occupied. Insofar as the complex is concerned, it may be looked upon as choosing on energetic grounds the lesser of two evils. If 10Dq is large (high crystal field) because of strong ligand repulsions, then the complex will choose to pair electrons in the t_{2g} levels. On the other hand, if 10Dq is small (weak crystal field), then the complex will choose to have single occupancy of four levels. The effective crystal field stabilization energy (that energy arising solely on the basis of occupancy and not on the energetic considerations leading to that occupancy) varies for the high-field and low-field cases. These arguments apply equally well to cases involving any number of d-electrons, as illustrated in Table 6-1. The high-field case leads to low net spin, while the opposite is true for the low-field case.

Table 6-1. CRYSTAL FIELD STABILIZATION ENERGIES (OCTAHEDRAL)

Configuration	Low spin or high field (Dq)	High spin or low field (Dq)
d^0	0	0
d^1	4	4
d^2	8	8
d^3	12	12
d^4	16	6
d^5	20	0
d^6	24	4
d^7	18	8
d^8	12	12
d^9	6	6
d^{10}	0	0

Knowledge of the stability of an octahedral transition metal ion complex is not sufficient to determine a lower bound for the activation energy of reaction (6-38). In addition the crystal field stabilization energy of the $(ML_4L' - L')^{2+}$ product of step 1 is needed. By difference the lower bound for the activation energy can be set. In estimating the CFSE for $(ML_4L' - L')^{2+}$ we need to keep two things in mind: (1) L (water) and L′ are not identical and so the tacit assumption of a constant Dq in Table 6-1 is suspect and (2) the geometries assumed for $(ML_4L'_2)^{2+}$ and $(ML_4L' - L')^{2+}$ are not exact. We assume a low-field square-pyramid $(ML_4L' - L')^{2+}$ complex. Accounting for the d-orbital crystal field splitting in such a complex leads to the diagram of Fig. 6-4, which also illustrates the relevant geometry. Calculating the CFSE as before leads to Table 6-2.

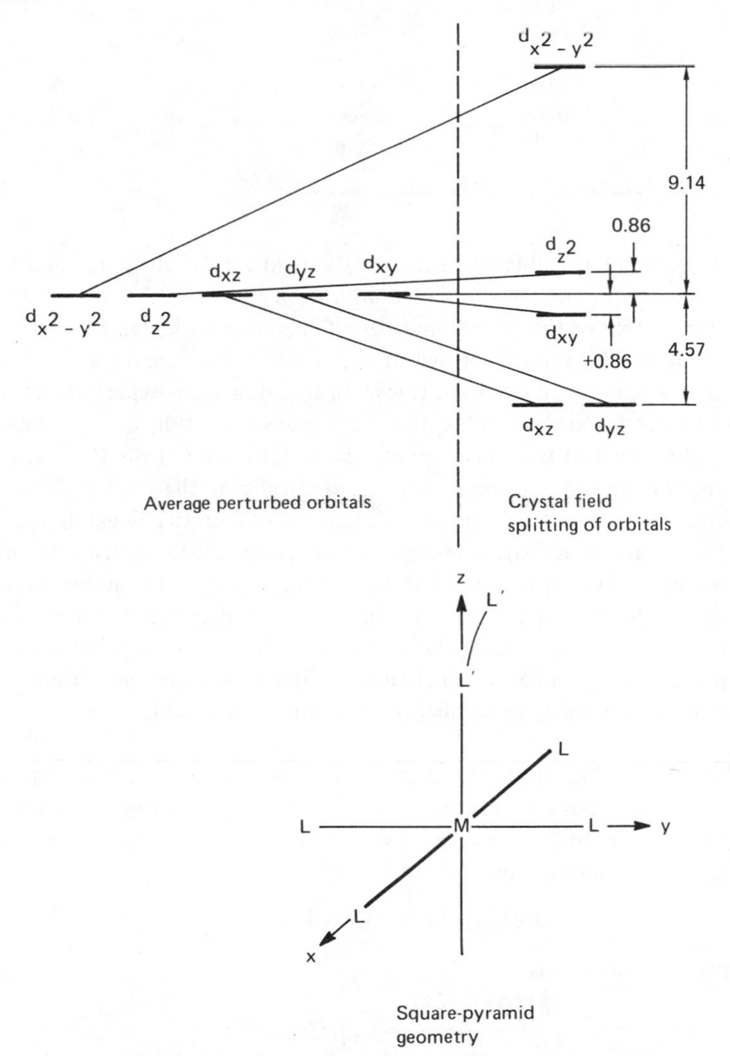

Figure 6-4. Crystal field splitting for square-pyramid complex and square-pyramid geometry.

Table 6-2. CRYSTAL FIELD STABILIZATION AND ACTIVATION ENERGIES FOR SQUARE-PYRAMID COMPLEX FORMED FROM OCTAHEDRAL COMPLEX (LOW FIELD)

Configuration	CFSE (Dq)	CFAE (Dq)
d^0	0	0
d^1	4.57	-0.57
d^2	9.14	-1.14
d^3	10.00	2.00
d^4	14.57	1.43
d^5	19.14	0.86
d^6	20.00	4.00
d^7	19.14	-1.14
d^8	10.00	2.00
d^9	9.14	-3.14
d^{10}	0	0

The last column in Table 6-2 is the crystal field activation energy and is calculated by subtracting the crystal field stabilization energy of the square-pyramid complex from that of the corresponding octahedral complex of Table 6-1. On this basis the d^7, d^8, and d^9 configurations of Co^{2+}, Ni^{2+}, and Cu^{2+} are expected to dissociate with different rates. In actual fact the experimental trend is borne out with the Ni^{2+} complex dissociating slowly and the Co^{2+} somewhat more rapidly and the Cu^{2+} very rapidly. In fact, the rate of the Cu^{2+} case is anomalously large and is accounted for theoretically on the basis of Jahn-Teller distortion. The crystal field activation energy is a theoretically estimated lower bound for the actual activation energy. In fact, any negative activation energy would be called zero. It is worth noting that the source of the activation energy in this simple theory is the shift in d-orbital energies that occurs when an octahedral complex stretches one of the metal-ligand bonds to the point of breakage. In the process there is a continual change in the geometry of the system from, as we have modeled it here, an octahedron to a square pyramid.

EXPERIMENTAL. In this experiment 1,10-phenanthroline is used as the bidentate ligand in aqueous solution, and the optical properties of the transition metal complex are used as a means of establishing the rate of the dissociation reaction, which can be denoted

$$(ML'_2)^{2+} \xrightarrow{k_1} (ML' - L')^{2+} \tag{6-40}$$

The rate of this process is

$$-\frac{d[(ML'_2)^{2+}]}{dt} = k_1[(ML'_2)^{2+}] \tag{6-41}$$

Integrating Eq. (6-41) from $t = t_0$ to $t = t$ furnishes

$$\log_{10} \frac{[(ML'_2)^{2+}]_{t_0}}{[(ML'_2)^{2+}]_t} = \frac{k_1(t - t_0)}{2.303} \tag{6-42}$$

Spectrophotometry is used to determine the concentrations. Many commercial ultraviolet spectrophotometers provide optical density (OD), defined as

$$OD = \log_{10} \frac{I_0}{I_{abs}} = \sum_i (\epsilon_i C_i) l \tag{6-43}$$

where I_0 is the transmitted intensity in the absence of the sample, I_{abs} the intensity absorbed by the sample, ϵ_i the optical extinction coefficient of species i at a particular wavelength, C_i the concentration of species i, and l the path length of the light rays through the sample. To be useful the concentration terms on the right-hand side of Eq. (6-43) must be related to the concentration terms appearing in Eq. (6-42). To accomplish this one must identify all the species which are absorbing and connect their concentrations by means of the chemical equations that apply. In the present experiment the optical absorption involves electrons essentially isolated on the ligand, 1,10-phenanthroline. When the ligand is bound to a metal ion the electron energy levels are perturbed and the absorption spectra are shifted. Different metal ions shift the spectra by various amounts. Conditions in this experiment are chosen so that the free ligand concentration is very small compared to the complex concentration. The optical density is thus determined by the concentrations of two species $(ML'_2)^{2+}$ and $(HgL'_2)^{2+}$, where M is Co, Ni, or Cu. By stoichiometry, these two concentrations are related by the expression

$$[(HgL'_2)^{2+}]_t = [(ML'_2)^{2+}]_0 - [(ML'_2)^{2+}]_t \tag{6-44}$$

where $[(ML'_2)^{2+}]_0$ is the initial concentration of the metal-ligand complex. Substituting into Eq. (6-43) furnishes at any time t the expression

$$(OD)_t = \left\{\epsilon_1 [(ML'_2)^{2+}]_t - \epsilon_2 [(ML'_2)^{2+}]_t + \epsilon_2 [(ML'_2)^{2+}]_0\right\} l \tag{6-45}$$

where ϵ_1 and ϵ_2 are the extinction coefficients for the metal-ligand and mercury-ligand complexes, respectively. If the reaction is assumed to go to completion, then at long times ($t \to \infty$) Eq. (6-44) becomes

$$[(HgL'_2)^{2+}]_\infty = [(ML'_2)^{2+}]_0 \tag{6-46}$$

since

$$[(ML'_2)^{2+}]_\infty = 0 \tag{6-47}$$

Using Eqs. (6-46) and (6-47) in (6-45) and solving for $[(ML'_2)^{2+}]_t$, we obtain

$$[(ML'_2)^{2+}]_t = \frac{(OD)_t - (OD)_\infty}{(\epsilon_1 - \epsilon_2)l} \tag{6-48}$$

which when substituted into Eq. (6-42) furnishes

$$\log_{10} \left\{ \frac{(OD)_{t_0} - (OD)_\infty}{(OD)_t - (OD)_\infty} \right\} = \frac{k_1(t - t_0)}{2.303} \tag{6-49}$$

Hidden in Eqs. (6-48) and (6-49) are some algebraic sign features which may be confusing. For example, it may turn out that $\epsilon_2 > \epsilon_1$ in Eq. (6-48) in which case $\epsilon_1 - \epsilon_2$ is negative. This is compensated for by the fact that the numerator of Eq. (6-48) would also be negative in such a case. Thus, it is irrelevant for this experiment whether the mercury complex product absorbs more or less strongly than the metal complex reactant. It is, however, crucial that $\epsilon_1 \neq \epsilon_2$, since if the two are equal, the right-hand side of Eq. (6-48) becomes indeterminate and the spectroscopic method would be inapplicable.

The absorption spectra of phenanthroline and its metal ion complexes show three maxima in the ultraviolet region, and as mentioned above the positions of these maxima vary with the metal ion. For the solutions used here these maxima appear as shown in Table 6-3. The data will be taken at wavelengths slightly off the maxima in the 2700-Å region in order to maximize the difference between ϵ_1 and ϵ_2 and thus obtain the best sensitivity.

Table 6-3. ABSORPTION MAXIMA IN 1,10-PHENANTHROLINE COMPLEXES*

Complexing ion	λ_{max} (Å)	ϵ_{max} (M^{-1} cm^{-1})	λ_{max} (Å)	ϵ_{max} (M^{-1} cm^{-1})	λ_{max} (Å)	ϵ_{max} (M^{-1} cm^{-1})
None	2900	9,000	2650	29,500	2260	42,000
Co^{2+}	2910	11,000	2695	34,000	2260	36,000
Cu^{2+}	2940	10,300	2720	34,500		
Ni^{2+}	2920	11,000	2700	36,500	2280	36,500
Hg^{2+}	2930	11,000	2715	33,500		

*K. Sone, P. Krumholz, and H. Stammreich, *J. Am. Chem. Soc.* 77, 777 (1955).

In all the experiments undertaken, it is important to achieve rapid and thorough mixing since in some cases the reaction goes to completion in about 100 sec. To accomplish this the two reactants are added simultaneously, rapidly, and in known amounts to a 1-cm silica absorption cell using two syringes. This is illustrated in Fig. 6-5. One recommended procedure is (assuming an instrument which furnishes optical density directly)

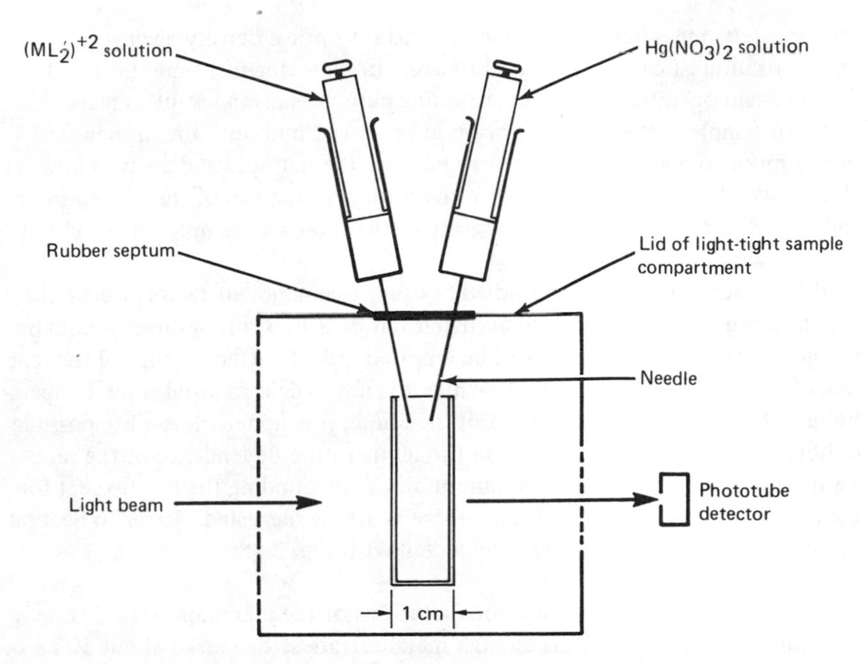

Figure 6-5. Mixing apparatus for measuring complex ion dissociation rates.

1. Properly position the empty cell in the sample compartment.
2. Fill 2-cc syringes with reactants and insert them through the septum in the sample compartment lid so that they will line up with absorption cell.
3. Position the lid on the sample compartment.
4. Set the spectrometer balance control so that the instrument indicates an optical density of zero for the empty cell when the wavelength is set at the position of interest.
5. Add the contents of the syringes rapidly and simultaneously. If the output is on chart paper, the chart should be moving at a known and fairly rapid speed during this operation to establish an adequate time scale. If the output must be read from a meter, two people are essential: one to add the reagents and another to keep time and make readings. The starting time of the experiment should be taken as the instant when half the reactants have been added.
6. Record the optical density versus time until no further change is detected. The samples are chosen so that the optical densities are about 0.2–0.6. In some cases the change in optical density is small so that care must be exercised to achieve useful results.

Frequently in this kind of spectroscopy a blank sample is prepared and its optical density is measured several times during the course of one experiment. This implies a two-compartment cell holder which can be shifted in and out of the beam. Its measurement provides a constant optical density against which the sample of interest can be compared to assist in removing instrument drift. In double-beam operation, as we are assuming here, it is often useful to place a standard sample in the reference beam in order to "null out" the influence of one or more components. This works because the net optical density is now representative of the difference between the optical properties of the absorption cell and the reference cell. For this experiment the reference sample is helpful but not necessary.

If these reactions are to be studied at other than ambient temperatures, the solutions must be thermostatted at higher temperatures, the syringes should be thermostatted, the syringes should be emptied quickly after filling, and the reaction cell should be thermostatted. While it is impossible to avoid some temperature fluctuations in the procedure outlined here, it is nonetheless quite possible to obtain reasonably reliable data on the temperature dependence of the rate. To thermostat the cell a large aluminum block surrounding the cell (except for the light path) and heated with a cartridge heater is suggested. Refer to Section 2.2 and especially the discussion connected with Fig. 2-18.

EXPERIMENTS. Prepare aqueous solutions of the metal ions (Hg^{2+}, Co^{2+}, Ni^{2+}, and Cu^{2+}) by dissolving enough metal nitrate salt to make about 50 ml of 2×10^{-3} M solution except in the case of Hg^{2+} where the concentration should be about 2×10^{-2} M. The Hg^{2+} solution should be stabilized by adding one or two drops of dilute nitric acid. The ligand solution should be prepared by dissolving enough 1,10-phenanthroline monohydrate (MW = 198) in water to make 50 ml of 0.2×10^{-3} M solution. It may be necessary to heat the water up to about 60° or 70°C to speed up the solution process. In some experiments outlined below concentrations other than these may be called for. They can be prepared by dilution of the stock mixtures.

Experiment 1: Spectra of Complex and Ligand Solutions

1. Co^{2+}: Mix equal amounts (5 ml) of Co^{2+} and ligand solutions. After standing for 5 min, fill the 1-cm absorption cell and scan the spectrum downward from 3600 Å. Set the optical density of the cell to zero at 3600 Å.
2. Hg^{2+}: Mix equal amounts (5 ml) of Hg^{2+} and ligand solutions. Scan the spectrum in the same way as for step 1.
3. Scan the spectrum of the ligand solution over the same range.

Experiment 2: Dissociation Rates of Co^{2+}, Ni^{2+}, and Cu^{2+} 1:1 Complexes With 1,10-Phenanthroline

1. Co^{2+}: Use 2 ml of the $(CoL'_2)^{2+}$ complex solution prepared in Experiment 1. Mix in the absorption cell with 2 ml of Hg^{2+} solution at 1×10^{-2} M made by dilution of stock solution. Follow the kinetics at 2660 Å. For a blank, use a similar sample that has been allowed to stand for 5 min.

2. Co^{2+}: Repeat Experiment 1 but use 2 ml of Hg^{2+} solution at 0.05 × 10^{-2} M.
3. Cu^{2+}: Prepare 10 ml of $(CuL'_2)^{2+}$ solution by mixing 5 ml each of the stock Cu^{2+} and ligand solutions. Mix 2 ml of this 1:1 complex with 2 ml of stock Hg^{2+} solution. Follow the kinetics at 2860 Å. For a blank, use a similar sample that has stood for 5 min.
4. Ni^{2+}: Prepare 20 ml of 1:1 complex with Ni^{2+} by mixing 10 ml each of stock solutions. Using 2 ml of 1:1 complex and 2 ml of Hg^{2+} stock solution, follow the kinetics at 2860 Å. For a blank, use a similar sample that has stood for 2 or 3 days. This reaction is very slow at 25°C and should not be run if time is short.

Experiment 3: Temperature Dependence of the Dissociation Rate of 1:1 Ni^{2+}, 1,10-Phenanthroline Complexes At 25°C the dissociation rate of the Ni^{2+} complexes is quite slow and easily followed if one is patient enough to wait. An interesting experiment is to follow the kinetics at several temperatures in the range 35°–60°C and try to extract an activation energy from the data. Four separate experiments are suggested commencing at 60°C and working downward. Use solutions of the concentrations employed in Experiment 2.

A wide variety of other complexes can be investigated in much the same manner as described above. Substituted phenanthrolines and other transition metals have been studied. Among them are 5-methylphenanthroline, 5-nitrophenanthroline, 5-chlorophenanthroline, Fe^{2+}, Mn^{2+}, Ag^{1+}, Cd^{2+}, and Zn^{2+}.

ANALYSIS. For each experiment, plot the appropriate quantitites in Eq. (6-49) and determine k_1. Comment on the quality of the fit of your data to a first-order rate expression. Determine the half-life, $t_{1/2}$, of each sample. By definition the half-life is the time required for the complex concentration to fall to one-half its value at any time t.

For the Ni^{2+} complexes, estimate, from the temperature dependence of k_1, the activation energy for the dissociation step and compare it with that based on crystal field theory. Use a Dq value of 4.5 kcal/mole.

Comment on the dependence of k_1 on metal and ligand concentrations, on the qualitative applicability of the crystal field theory method of estimating activation energies for dissociation, and on how the tacit elimination of complexed water from any considerations might influence the results.

6.3 Gas-Phase Unimolecular Decomposition

The purpose of this experiment is to study as a function of temperature the unimolecular decomposition of molecules in the gas phase. Theoretical expressions are developed which provide physical interpretations of some of the experimentally determined parameters.

THEORY. The molecularity of a reaction is defined only for elementary steps and is taken as the number of species which must come together if a certain event takes place. Unimolecular reactions are those in which the species present in the interaction region of the relevant potential-energy surface are derived from a single stable molecular species. In the parlance of activated complex theory we would say that the activated complex is derived from a single molecule. By definition then, the rate of unimolecular loss of molecule A is given by

$$-\frac{d[A]}{dt} = k[A] \tag{6-50}$$

where k is the rate coefficient. Among the common types of unimolecular processes are isomerization, decomposition, fluorescence, and phosphorescence. In the present experiment decomposition will be studied because it provides a change in total pressure and thus a convenient means of monitoring the course of the reaction. In early studies of what turned out to be unimolecular decompositions, pressure-dependent rate coefficients were obtained when the overall rates were analyzed using the form of Eq. (6-50). At high pressures k was constant, but at intermediate pressures it began to decrease significantly. This observation suggested the importance in the mechanism of steps other than unimolecular decomposition. The so-called Lindemann mechanism provided a very intuitive qualitative explanation of both the high-pressure and falloff regions. Consider a unimolecularly decomposing gas A diluted in a large excess of inert diluent M, where M is any body in the system that can serve as an energy transfer agent. The Lindemann mechanism for this system is

$$A + M \xrightarrow{k_a} A^* + M \tag{1}$$

$$A^* + M \xrightarrow{k_b} A + M \tag{2}$$

$$A^* \xrightarrow{k_c} B + C \tag{3}$$

The essence of this mechanism is the dissection of the reactant molecules into two categories, A and A*. The distinction between A and A* is on the basis of their energy, A* being defined as reactant molecules having sufficient internal energy to decompose and A as all other reactant molecules. In a qualitative way this mechanism permits a microscopic look at how molecules behave in the system. A bimolecular process is responsible for putting A molecules into the energetic category capable of reacting. This implies the need in any quantitative theory for consideration of intermolecular energy transfer. Once in category A* two possibilities arise: (1) unimolecular decomposition independent of the pressure of M and (2) deactivation in a bimolecular encounter which depends linearly on the pressure of M.

Turning now to a more quantitative description we search for an expression which gives the observed rate of loss of A in terms of k_a, k_b, and k_c. From the mechanism,

$$-\frac{d[A]}{dt} = k_a[A][M] - k_b[A^*][M] \tag{6-51}$$

The concentration of A^* is generally not directly measurable; it is normally, however, very small compared to A, and its change with time is also small compared to the rate of change of A. If these conditions are met, we may then make use of the steady-state approximation which sets $d[A^*]/dt = 0$:

$$\frac{d[A^*]}{dt} = k_a[A][M] - k_b[A^*][M] - k_c[A^*] = 0 \tag{6-52}$$

Substituting (6-52) into (6-51) to eliminate A^* furnishes

$$-\frac{d[A]}{dt} = \frac{k_c k_a[M]}{k_c + k_b[M]}[A] \tag{6-53}$$

Comparing this with Eq. (6-50) we see immediately that for a fixed M the overall rate would indeed be first order but that changing M will change the rate coefficient, k. Explicitly we write the M- and T-dependent rate coefficient as

$$k(M, T) = \frac{k_c k_a[M]}{k_c + k_b[M]} \tag{6-54}$$

For large M (high pressure) $k_c \ll k_b[M]$ and

$$k(M, T) \sim \frac{k_c k_a}{k_b} \equiv k_\infty \tag{6-55}$$

The coefficient k_∞ is valid in the high-pressure limit. Note that it is composed of essentially two factors, k_c the unimolecular rate coefficient, and the ratio k_a/k_b, which is just the equilibrium constant for the activation-deactivation sequence. Thus, at high pressures the rate is controlled by the unimolecular step, while the other two steps are of sufficient rapidity to maintain an equilibrium population of energetic A^* molecules. On the other end of the pressure scale $k_b[M] \ll k_c$ and

$$k(M, T) \sim k_a[M] \tag{6-56}$$

Under these circumstances an equilibrium population of A^* is not maintained because bimolecular collisions are too infrequent. As a result the rate is controlled by the activation step and every activated molecule undergoes unimolecular decomposition. At intermediate pressures the full expression, Eq. (6-54), must be used. In summary the Lindemann mechanism predicts that the rate will be first order at high pressures, second order at low pressures, and intermediate order between the two extremes of pressure.

While Eq. (6-53) is very helpful in rationalizing the observed results, it does not provide any detailed quantitative explanation of the magnitudes of k_a, k_b, and k_c. The approach to a quantitative theory is through a model. Many models of varying degrees of complexity have been proposed, and we shall present a quite useful model labeled RRK theory after its inventors: Rice, Ramsperger, and Kassel. This theory provides the overall rate coefficient k(M, T) of Eq. (6-54) in terms of two measured kinetic properties and other independently determined molecular properties. The postulated mechanism is the Lindemann mechanism.

With a mechanism at hand we now set down the basic postulates of RRK theory regarding the disposition of the energy in the system and the time dependence of some of the molecular events. First we assume that vibrational excitation and vibrational energy transfer are most significant; thus, the reactant molecule is presumed to be describable as a collection of s coupled oscillators, where s is some number less than or equal to the total number of vibrational degrees of freedom. In principle, we may regard s as those vibrational degrees of freedom which can and do contribute their energy toward achievement of the activated complex configuration. Contrariwise the reactant may possess some degrees of vibrational freedom which are not at all effective in contributing to the energy (activation energy) needed to get the molecule into the activated complex configuration. Such ineffective degrees of vibrational freedom will not be included in the collection s. The activated complex is to be distinguished from an energetic A* on the basis of internuclear configuration. Accessible regions of the potential-energy surface are determined by the total energy in the molecule, and access to the activated complex region requires that the so-called critical oscillator (or critical coordinate) possess at least a certain minimum energy. In unimolecular decomposition reactions we intuitively expect the critical coordinate to be the length of the bond being broken. While this is approximately true it is not rigorously so because other internuclear distances and angles change as the decomposition occurs and the critical coordinate turns out to be some linear combination of bond lengths and angles, with the major contribution from the bond being broken. A simple example is illustrative; consider the gas-phase unimolecular decomposition of ethane for which reaction (3) of the Lindemann mechanism is

$$C_2H_6^* \rightarrow 2CH_3 \tag{3}$$

The species $C_2H_6^*$ is regarded as a set of oscillators which are excited. One of these, the critical oscillator, is assumed to possess an energy of at least E_0 if reaction (3) occurs. Intuitively we expect, and correctly so, that the C–C distance must correspond closely to the critical oscillator. Taking account of simultaneous changes in $-CH_3$ geometry we also expect minor contributions to the critical coordinate from this part of the molecule. If, hypothetically, we begin with a $C_2H_6^*$ in which none of the excitation appears in the C–C bond, we conclude immediately that the activated complex region is inaccessible. Because the oscillators are coupled, however, energy can migrate into the C–C bond and result in the activated complex configuration.

At this point we shall state further postulates and then proceed to develop the mathematical relations which incorporate them:

1. The thermal energy is assumed to be distributed randomly among the oscillators on every collision of A with M.

2. The vibrational energy of the s oscillators describing the molecule is rearranged randomly by means of intramolecular energy transfer with a frequency ω that is characteristic of the molecule.
3. Classical mechanics provides a suitable description of the nuclear dynamics.

The first two postulates permit the use of statistical mechanical methods in treating the partitioning of the energy among the oscillators. The third postulate appears very questionable because we know oscillators are quantized. However, many of the features of unimolecular decomposition can be explained using classical mechanical methods. A prominent exception is the effect of isotopic substitution, which can be accounted for only on the basis of quantum statistical mechanics. The latter treatment is beyond the scope of this discussion.

The random disposal postulates are attractive because they permit an analysis to be carried out without consideration of individual collisions or individual molecules. Instead we shall deal only with probabilities and averages. To begin we shall consider the activation step of the Lindemann mechanism:

$$A + M \xrightarrow{k_a} A^* + M \tag{1}$$

The rate of formation of A^* in this process is $k_a[A][M]$, and we are searching for a theoretical expression for k_a derivable from the above postulates. When A and M come close enough together for energy exchange to occur, by definition, a collision has occurred. The rate coefficient for this collision process is simply the number of collisions per second which would occur if both A and M were at unit concentration. This coefficient, the collision frequency, is usually denoted Z and is calculated on the basis of relations developed in kinetic theory of gases:

$$Z = \sigma^2 \left(\frac{8\pi kT}{\mu}\right)^{1/2} \tag{6-57}$$

where σ^2 is the effective cross-sectional area which the two species have for the collision process, μ the reduced mass of the pair $\{A, M\}$, k Boltzmann's constant, and T the absolute temperature. The square-root term accounts for the average relative velocity between A and M and its variation with temperature in a thermally equilibrated system. Because it is normalized to unit concentrations, the collision frequency Z has dimensions $cm^3\ sec^{-1}$. The rate coefficient k_a is less than Z because not every collision leaves A with sufficient total internal energy to react. Out of Z we need to pick those collisions which leave A^* behind. On the basis of the oscillator model we simply ask for the *probability* that a collection of s oscillators will possess energy $E \geqslant E_0$, where E_0 is the minimum energy required for reaction. For later use it will also be useful to ask for the probability that a collection of s oscillators has energy in the range $E \rightarrow E + dE$.

The development of mathematical relations for these probabilities is based on Boltzmann statistics and classical mechanics. The fundamental statistical relation is

$$P_1(E)dE \propto \exp\{-\frac{E}{kT}\}\,dE \tag{6-58}$$

$P_1(E)dE$ is the probability that a single oscillator has a total energy in the range $E \rightarrow E + dE$. Equation (6-58) simply states that this probability is proportional to a Boltzmann factor. Equation (6-58) is valid because of the random collisional energy disposal postulate. The proportionality constant need not be remembered since it is easily calculable by the normalization condition on $P_1(E)$:

$$\int_0^\infty P_1(E)\,dE = 1 \tag{6-59}$$

Writing the proportionality constant of Eq. (6-58) as A we have

$$\int_0^\infty A\exp\{-\frac{E}{kT}\}\,dE = 1 \tag{6-60}$$

Performing the quadrature and solving for A furnishes $A = 1/kT$ and

$$P_1(E)\,dE = \frac{1}{kT}\exp\{-\frac{E}{kT}\}\,dE \tag{6-61}$$

From Eq. (6-61) the probability that a single oscillator has energy greater than or equal to E_0 is simply

$$P_1(E \geqslant E_0) = \int_{E_0}^\infty P_1(E)\,dE = \exp\{-\frac{E_0}{kT}\} \tag{6-62}$$

With these expressions for a single oscillator at hand we shall now proceed in an attempt to generalize to s oscillators by explicitly considering three oscillators. The probability that three oscillators have energies which sum to be in the range $E \rightarrow E + dE$ is given by

$$P_3(E)\,dE = \int_{E_1=0}^{E}\int_{E_2=0}^{E-E_1}\left[\frac{1}{kT}e\{-\frac{(E-E_2-E_1)}{kT}\}\right] \times\left[\frac{1}{kT}e\{-\frac{E_2}{kT}\}\right]\left[\frac{1}{kT}e\{-\frac{E_1}{kT}\}\right]dE_2\,dE_1 \tag{6-63}$$

The three factors in the integrand of this equation are the probabilities for oscillators 3, 2, and 1 having energies in the ranges $E_3 \rightarrow E_3 + dE_3$, $E_2 \rightarrow E_2 + dE_2$, and $E_1 \rightarrow E_1 + dE_1$, respectively. The following boundary conditions apply:

$$E_3 = E - E_1 - E_2$$
$$E_1 \leqslant E \qquad (6\text{-}64)$$
$$E_2 \leqslant E - E_1$$

Performing the quadrature indicated in Eq. (6-63) provides

$$P_3(E)\, dE = \frac{1}{2} \frac{E^2\, e^{-E/kT}}{(kT)^3}\, dE \qquad (6\text{-}65)$$

Generalized to s oscillators the corresponding relation is

$$P_s(E)\, dE = \frac{1}{(s-1)!} \frac{E^{s-1}}{(kT)^s}\, e^{-E/kT}\, dE \qquad (6\text{-}66)$$

The probability that s oscillators have energies summing to some value $E \geqslant E_0$ is

$$P_s(E \geqslant E_0) = \int_{E_0}^{\infty} P_s(E)\, dE = e^{-E_0/kT} \left[\left(\frac{E_0}{kT}\right)^{s-1} \frac{1}{(s-1)!} + \left(\frac{E_0}{kT}\right)^{s-2} \frac{1}{(s-2)!} + \dots + 1 \right] \qquad (6\text{-}67)$$

The integration indicated in Eq. (6-67) is performed by the repeated application of the integration-by-parts formula. If $E_0 \gg kT$, as it often is, then the following approximation is quite accurate:

$$P_s(E \geqslant E_0) \cong e^{-E_0/kT} \left[\left(\frac{E_0}{kT}\right)^{s-1} \frac{1}{(s-1)!} \right] \qquad (6\text{-}68)$$

Equation (6-68) multiplied by Z [Eq. (6-57)] furnishes the rate coefficient for the activation of A (i.e., the collision frequency times the probability that A is left with energy greater than or equal to the critical energy):

$$k_a = Ze^{-E_0/kT} \left[\left(\frac{E_0}{kT}\right)^{s-1} \frac{1}{(s-1)!} \right] \qquad (6\text{-}69)$$

Using Eq. (6-66) we can write an energy-dependent excitation rate coefficient as

$$k_a(E)\, dE = \frac{Z}{(s-1)!} \frac{E^{s-1}}{(kT)^s}\, e^{-E_0/kT}\, dE \qquad (6\text{-}70)$$

$k_a(E)\, dE$ is the rate coefficient for the excitation of A into the differential energy range $E \rightarrow E + dE$.

With Eq. (6-68) or (6-69) we have one of the rate coefficients needed in evaluating Eq. (6-54). Restricting ourselves to energy-dependent rate coefficients Eq. (6-54) becomes

$$k(M, T, E) = \frac{k_c(E)k_a(E)M}{k_c(E) + k_b(E)M} \tag{6-71}$$

We have $k_a(E)$ but need $k_c(E)$ and $k_b(E)$. The latter coefficient corresponds to deactivation of those A* molecules whose energy is in the range $E \rightarrow E + dE$, where $E > E_0$. Since E_0 is normally much larger than kT and by hypothesis the total energy of A* and M is partitioned randomly on every collision, it is clear that the probability that deactivation occurs on one collision is very close to unity. We therefore take $k_b(E) = Z$ independent of energy, assuming that σ^2 for A-M collisions is identical with that for A*-M collisions.

To evaluate $k_c(E)$ we make use of the second postulate, which requires the vibrational energy to exchange intramolecularly and randomly with frequency ω. Process (3) itself requires that energy greater than or equal to E_0 be in the critical oscillator. We thus search for the probability that the critical oscillator has $E_c \geqslant E_0$ given that the collection s has total energy $E \rightarrow E + dE$. If the total energy is E and the critical oscillator has energy E_c, then the $s - 1$ remaining oscillators have energy $E - E_c$. The probability we desire can be written as

$$P_s(E_c \geqslant E_0, E)$$

$$= \frac{\displaystyle\int_{E_0}^{E} \left[[1/(s-2)!] / [E - E_c)^{s-2}/(kT)^{s-1}] \; e^{-(E-E_c)/kT} \right] \left[(1/kT) e^{-E_c/kT} \right] dE_c}{[1/s - 1)!] \, [E^{s-1}/(kT)^s] \, e^{-E/kT}} \tag{6-72}$$

The first factor in the integrand is the probability that $s - 1$ oscillators have a total energy in the range $E - E_c \rightarrow E - E_c + dE$. The second factor is the probability that the critical oscillator has energy in the range $E_c \rightarrow E_c + dE_c$. The denominator provides the appropriate normalization to a set of s oscillators whose total energy is fixed at E. Performing the quadrature furnishes

$$P_s(E_c \geqslant E_0, E) = \left(\frac{E - E_0}{E} \right)^{s-1} \tag{6-73}$$

The energy-dependent unimolecular decomposition rate is then the product of this probability with the frequency of energy exchange, ω:

$$k_c(E) = \omega \left(\frac{E - E_0}{E} \right)^{s-1} \tag{6-74}$$

Equation (6-71) thus becomes

$$k(M, T, E)\, dE = \frac{Z[M]\,\omega[(E - E_0)/E]^{s-1}\{E^{s-1}/[(s-1)!(kT)^s]\}e^{-E/kt}\, dE}{\omega[(E - E_0)/E]^{s-1} + Z[M]} \tag{6-75}$$

Integrating over E furnishes the experimentally measurable k(M, T):

$$k(M, T) = \int_{E_0}^{\infty} \frac{\omega(E - E_0)^{s-1}}{(kT)^s(s-1)!} \exp\left\{-\frac{E}{kT}\right\}\left[\frac{1}{1 + (\{\omega[(E - E_0)^{s-1}/E]\}/Z[M])}\right] dE \tag{6-76}$$

The evaluation of Eq. (6-76) must be done by numerical means.

In the limit of high pressures $k(M, T) \to k_\infty$ and Eq. (6-76) becomes

$$k_\infty = \int_{E_0}^{\infty} \frac{\omega(E - E_0)^{s-1}}{(kT)^s(s-1)!} e^{-E/kT}\, dE \tag{6-77}$$

Substituting $y = (E - E_0)/kT$, Eq. (6-77) becomes

$$k_\infty = \frac{\omega e^{-E_0/kT}}{(s-1)!} \int_0^{\infty} y^{s-1} e^{-y}\, dy$$

$$= \omega e^{-E_0/kT} \tag{6-78}$$

The form of Eq. (6-78) is identical to the Arrhenius form of the temperature dependence of the rate coefficient. Clearly, if we work at sufficiently high pressure both ω and E_0 are determinable experimentally from the standard ln k versus 1/T plot. With this at hand the parameters in Eq. (6-76) are all determined except s, which can be adjusted arbitrarily to give a good fit to the data. With a little reflection it becomes evident that we have developed a formal theory of unimolecular reactions in which none of the parameters are directly calculable (i.e., ω, s, and E_0 are not calculated theoretically). It is possible using the model introduced by Marcus to calculate ω directly (RRKM theory after Rice, Ramsperger, Kassel, and Marcus) using quantum statistical mechanical methods, which are beyond the scope of this text. In RRKM theory E_0 and s remain as experimentally determined parameters.

EXPERIMENTAL. A variety of methods is available for following the rate of a gas-phase reaction; one of the simplest, variation of total pressure with time, is used here. Obviously this method suffices only for reactions in which there is a change in the number of moles of gas as the reaction proceeds. In turn, the interpretation of observed pressure changes requires knowledge of the stoichiometry of the reaction. Assuming that the overall stoichiometry is $A \rightarrow B + C$, we shall proceed to develop relations between the total pressure and the rate of loss of A. Beginning with a sample of pure A, let the initial total pressure be p°. Note that in this case $[M] = [A]$. At any subsequent time, t, the total pressure p(t) is given by

$$p(t) = p_A + p_B + p_C \tag{6-79}$$

while the pressure of B and C are related by

$$p_B = p_C = p^\circ - p_A \tag{6-80}$$

Substituting Eq. (6-80) into Eq. (6-79) for p_B and p_C furnishes

$$p_A = 2p^\circ - p(t) \tag{6-81}$$

In the high-pressure region

$$-\frac{dp_A}{dt} = k_\infty p_A \tag{6-82}$$

which on integration furnishes

$$k_\infty t = \ln \frac{p^\circ}{p_A} = \ln \left[\frac{p^\circ}{2p^\circ - p(t)}\right] \tag{6-83}$$

From Eq. (6-83) it is apparent that a plot of $\ln (2p^\circ - p)$ versus t should result in a straight line of slope $-k_\infty$. If the reaction is moderately fast, it may be difficult to accurately measure p°. Indirect measurements of p° can be obtained in two ways. The early pressure data can be extrapolated to $t = 0$, and thus p° is obtained. Another method is to allow the reaction to go to completion, which gives a measure of p_∞, the maximum value of p. Since $p_A = 0$ for complete reaction, we see from Eq. (6-81) that

$$p^\circ = \frac{1}{2} p_\infty \tag{6-84}$$

For a first-order reaction, specific rate constants can be evaluated from the time required for a specified fraction of the reaction to occur. Thus, the half-time and third-time are given by

$$t\left(\frac{1}{2}\right) = \frac{\ln 2}{k} = \frac{0.691}{k}$$
$$t\left(\frac{1}{3}\right) = \frac{\ln 1.5}{k} = \frac{0.406}{k} \tag{6-85}$$

For the reaction $A \rightarrow B + C$, the total pressure at $t(\frac{1}{2})$ and $t(\frac{1}{3})$ can be easily determined. Furthermore, a check on the reaction order may be obtained by calculating $t(\frac{1}{2})/t(\frac{1}{3})$. According to Eq. (6-85) this ratio should be 1.70. If the reaction were zero order, the ratio would be 1.5, while if it were second order, the ratio would be 2.0.

There are several possible molecules which may be used including cyclopentene, di-*t*-butyl peroxide, 1,1-dichloroethane, norbornylene, and 2,5-dihydrofuran. The 2,5-dihydrofuran system is described here. This molecule has been shown by Wellington and Walters (see the References) to decompose according to a pressure-independent first-order rate coefficient over the range 5–270 Torr. Below 5 Torr, the first-order rate becomes pressure-dependent [see Eq. (6-54)]. The only products appear to be furan and hydrogen; hence, the stoichiometry fits the above analysis for A → B + C:

$$\text{O} \longrightarrow \text{O} + H_2$$

As mentioned above, we follow this reaction by measuring total pressure as a function of time after adding a sample to a reaction vessel at a fixed temperature. Except for a timing device, the necessary apparatus is diagrammed in Fig. 6-6. Of crucial importance is the construction of the furnace, which should be capable of achieving steady temperatures in the range 350°–500°C. To achieve this requirement we recommend a cylindrical Pyrex reaction vessel of at least 500-cc volume surrounded by a massive block of aluminum bored out to fit the reaction vessel. After covering this block with a thin layer of insulation (asbestos or mica) heating element wire can be wound around it. Use enough wire to give at least 30-Ω total resistance. It may be advisable to wind the wire closer together near the end of the core at which the reaction vessel is connected to the vacuum

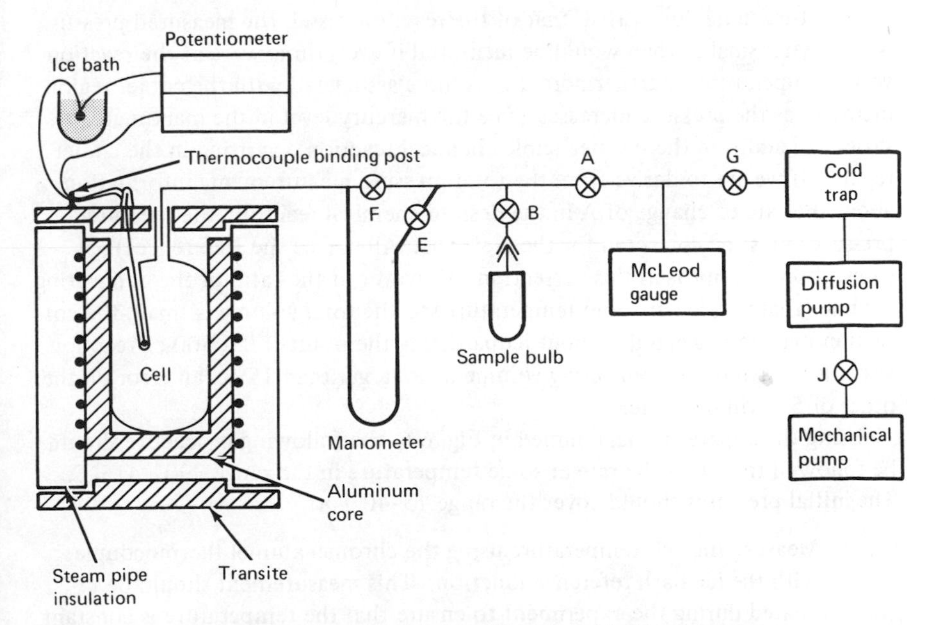

Figure 6-6. Schematic drawing of gas-phase kinetics apparatus.

line. After the heating wire is in place, surround the core with at least 2 in. of asbestos or commercially available steam pipe insulation. To cover the cylinder ends, use Transite and fill the space between the Transite and the cell with particulate insulation. Provide means for inserting a thermocouple into the well in the reaction vessel and for making electrical heating connections to a variable transformer. See Calvert and Pitts (1966) and Maccol (1961) in the References for some additional construction details.

It is sometimes a problem to keep the thermocouple in the well. One suggestion is to press a small amount of asbestos into the well after the thermocouple is inserted in order to form a wedge. A second suggestion is to bind the thermocouple leads permanently to the Transite end plate in order to prevent manipulation of the leads at the ice bath or the potentiometer from disturbing the thermocouple position in the well.

With a furnace designed in this way stable temperatures can be achieved provided the power source is voltage-stabilized. It should be noted that any alterations in the reaction vessel temperature take several hours to stabilize. Therefore, it is not possible during the course of a single lab period to make runs at different temperatures. In the instructions and analysis below, only one temperature is assumed. These instructions may easily be extended to other temperatures by running the experiment at different temperatures during consecutive lab periods.

Another important consideration in the experimental design is the volume of tubing that connects the reaction vessel to the pressure-measuring device, in this case the manometer. Since this volume contains reactant (and product) but at a temperature much lower than that of the reaction vessel, the measured pressure is somewhat smaller than would be measured if everything were at the reaction vessel temperature. Furthermore, the volume associated with the cooler region increases as the pressure increases since the mercury level in the manometer varies. In addition there is negligible chemical reaction occurring in the cooler region. If we are to derive from the total pressure measurements information about the rate of change of A in the desired chemical reaction, the measured pressures must be corrected for these effects. Allen (see the References) has shown how to approach this correction. However, if the ratio of the connecting volume (that volume at room temperature) to the total volume is small, the correction may be neglected without jeopardizing the results. If a 500-cc reaction vessel is used and the connecting volume is no larger than 15 cc, an error on the order of 5% will be made.

Using the apparatus diagrammed in Fig. 6-6, the following procedure should be followed to obtain the rate at some temperature in the range 390°–415°C. The initial pressures should cover the range 15–40 Torr.

1. Measure the cell temperature using the chromel-alumel thermocouple with the ice bath reference junction. This measurement should be repeated during the experiment to ensure that the temperature is constant to within 0.5°C/hr.

2. While the temperature is being measured, evacuate the system (all stopcocks in Fig. 6-6 should be open except B). The McLeod gauge should be used to determine if the pressure is less than 10^{-3} Torr.
3. Close stopcocks E and F. Freeze the sample with liquid nitrogen and then open stopcock B for 1 or 2 min in order to pump off any air in the sample bulb. The nitrogen should remain over the sample for 60 sec before B is opened. This ensures that the sample has come to liquid nitrogen temperature before the pumping is begun.
4. Close B, remove the liquid nitrogen, and allow the solid to warm up to room temperature so that any dissolved gases will vaporize into the evacuated space above the sample. Warm water (60°–80°C) should be used to warm the sample, to temperatures even above room temperature.
5. Again freeze the sample with liquid nitrogen and then open stopcock B to pump off additional gases.
6. Repeat steps 4 and 5, close B, and then allow the sample to warm up to room temperature. Do not remove the liquid nitrogen from around the sample while it is being evacuated. (*Note:* Steps 3, 4, and 5 are commonly referred to as *outgassing* or *degassing* the sample.)
7. Close stopcocks A, E, and F. With B open, obtain the desired vapor pressure in the sample bulb by employing an appropriate temperature bath (ice water or the like). Open F for 2 or 3 sec to get the sample into the reaction vessel. Start the timer immediately and record the pressure.
8. Return the vapor outside cell system to the sample bulb by condensing with liquid nitrogen and then close B.
9. Record the pressure reading from the manometer as quickly as possible. Continue to record the pressure every 20 or 30 sec until t equals $t(\frac{1}{2})$ or slightly longer [the experimentalist should calculate $p(\frac{1}{2})$ from the approximate initial pressure].
10. When a run is complete, evacuate the cell by opening A and F. Repeat the experiment using a different pressure of reactant.

ANALYSIS. For each run, find p° by extrapolation to zero time and then plot total pressure versus time and draw a smooth curve through the data. From the resulting plots, find $t(\frac{1}{2})$ and $t(\frac{1}{3})$ for each run and ascertain their first-order character.

Using the smoothed pressure versus time plots, construct plots of $\ln (2p^\circ - p)$ versus time for each run. Weighting the early points most heavily, use a least-squares procedure to determine k_∞. Determine ω and E_0 from the temperature dependence of k_∞ if data at more than one temperature are available.

PROBLEMS

1. Explain how a unimolecular decomposition can appear to be second order.

2. In samples like those used in the above experiments, there is no inert diluent M. Explain how A can play the role of M in the Lindemann mechanism and how as a result the equations developed in the theory section retain their validity.

3. It has been reported that the first-order rate coefficient for the decomposition of 2,5-dihydrofuran is independent of pressure down to 5 Torr. At 1 Torr the rate coefficient drops to 90% of its high-pressure value. Using an average temperature for the experiments performed and an effective cross-sectional area for collisions of 10 $Å^2$, calculate, using Eq. (6-76), the expected variation of k(M, T) with M using different values of s. The numerical quadrature program developed in Section 1.7 will be quite useful in the requisite numerical analysis. Examination of the variation of the magnitude of the integrand with E should provide a clue about handling the infinite upper limit of Eq. (6-76). From the results of this calculation, comment on how sensitive the k(M, T) versus M curves are to the choice of s.

4. In Section 6.1 a thermodynamic formulation of a rate coefficient is given as

$$k(T) = \frac{kT}{h} e^{\Delta S^{\neq}/R} e^{-\Delta H^{\neq}/RT} \tag{6-86}$$

The preexponential factor of the Arrhenius equation is set equal to $(ekT/h)e^{\Delta S^{\neq}/R}$. Explain why this relation can or cannot be applied to the interpretation of Eq. (6-78). If such an application can be made, determine the entropy of activation and comment on both its sign and magnitude.

6.4 Non-Boltzmann Gas-Phase Kinetics

The purpose of this experiment is to illustrate non-Boltzmann effects in a simple photochemically induced gas-phase reaction. Rate coefficient ratios in both nonthermalized and partially thermalized systems are determined.

THEORY. The normal approach to chemical kinetics at the undergraduate level assumes implicitly that all the reactants are describable by Boltzmann energy distributions whose temperatures are all given by the macroscopic temperature of the heat bath in which the reacting system is immersed. Caution must be exercised in employing this assumption, and in this experiment we shall illustrate a system in which one of the reactants has a distinctly non-Boltzmann distribution of energies.

The preparation of this system makes use of the photodissociation of one of the hydrogen halides, HX, or the deuterated counterparts, DX. The electronic absorption spectra of these molecules, observable in the ultraviolet and vacuum ultraviolet regions of the electromagnetic spectrum, are continuous, have long wavelength cutoffs, and possess maxima. Such spectra arise as a result of transitions from the ground state of HX or DX to a repulsive electronically excited state. Schematically the potential energy curves are shown in Fig. 6-7. Some complexity is introduced because of several possible electronically excited states. The main effect of this multiplicity is loss of knowledge, for some photon energies, of the electronic state of the halogen product, i.e., whether it is $^2P_{3/2}$ or $^2P_{1/2}$. The result of this loss of knowledge is illuminated in Problem 1 at the end of this section. Assuming for the moment that only one excited state is important and that it leads to ground-state atomic products, the relevant potential energy curves and associated energies are shown in Fig. 6-8. Assuming the photolysis of HX is accomplished with monochromatic light (one wavelength) the photons all have the same energy $h\nu$. Of this a certain amount, D_0, is required to rupture the HX bond, while the remainder must be transformed from electronic energy to nuclear translational energy as the nuclei move

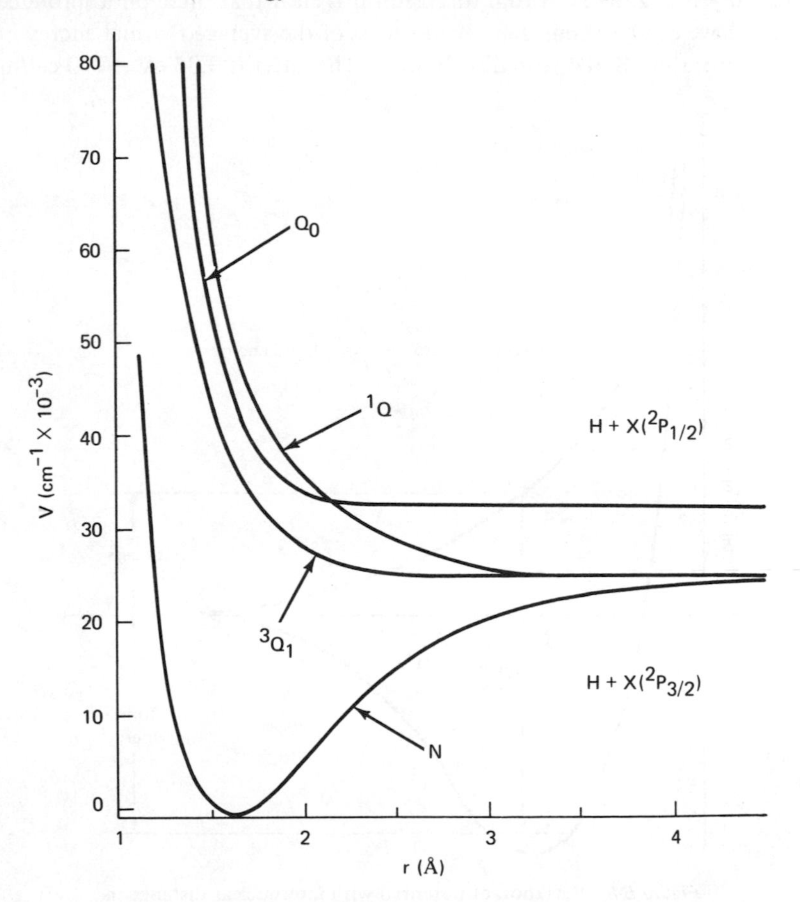

Figure 6-7. Potential-energy curves for HX. Scales on the ordinate and the abscissa are for HI.

toward infinite separation. Because the halogens are very massive compared to the hydrogens, most of the excess energy appears as translational energy of the latter (refer to Problem 1). To a very good approximation, the initial total excess energy is given by

$$E_{XS} = h\nu - D_0 \tag{6-87}$$

and the energy of hydrogen atom E_H in the laboratory is very closely approximated by

$$E_H = \frac{m_X}{m_{HX}} E_{XS} \tag{6-88}$$

where m_{HX} and m_X are masses of the hydrogen halide and halogen atom, respectively. Table 6-4 furnishes the energies of the bonds in some of these molecules along with the excess energy and initial laboratory hydrogen atom energy for photolysis at 2537 Å. From this result it is clear that these photoproduced H atoms have energies considerably in excess of the average thermal energy of a room-temperature Boltzmann distribution. The latter is 0.03 eV = 770 cal/mole.

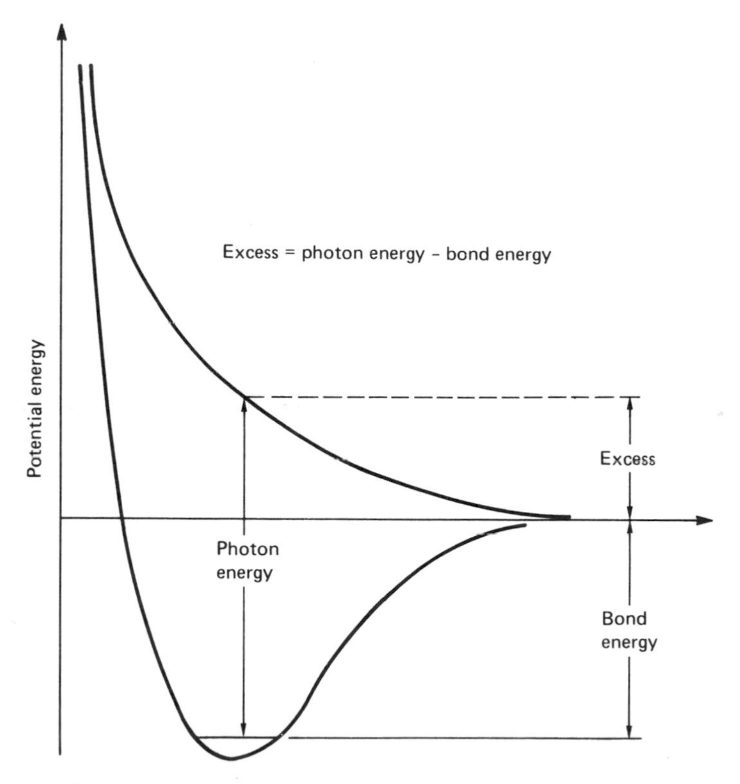

Figure 6-8. Variation of potential with internuclear distance in photodissociation of HX.

Table 6-4. HOT HYDROGEN ATOMS PRODUCED AT 2537 Å

Molecule	D_0 (eV)	E_{XS} (eV)	E_H (eV)
HCl	4.430	0.455	0.443
DCl	4.481	0.404	0.393
HBr	3.754	1.131	1.117
DBr	3.801	1.084	1.071
HI	3.056	1.829	1.815
DI	3.097	1.788	1.774

Once formed in the midst of other molecules, these translationally excited hydrogen atoms may react or simply lose a bit of energy on their first collision, similarly on their second collision, etc. The net result is a steady-state distribution of H atom energies which has a hot atom part and a thermal part. The distinction between a hot atom and a thermal atom is somewhat arbitrary. In this discussion we shall choose to define the hot atom distribution as containing all the H atoms which cannot be accounted for by a Boltzmann distribution whose temperature is the macroscopic photolysis cell temperature. Qualitatively this implies that the low-energy H atoms (~ 0.03 eV) are described as thermal, whereas the high-energy atoms are described as being hot. In passing it is important to note that the hot atoms do not necessarily possess the initial translational energy they started with. Looked at another way this means that an atom may still be hot after several small energy loss collisions.

A question naturally arises regarding the effects of these hot atoms on the subsequent kinetics in any given system. Since experimentally the simplest example is the HBr case, we choose it for illustrative purposes. Consider a mixture of HBr and Br_2 photolyzed at 2537 Å. At this wavelength Br_2 does not absorb and HBr has an extinction coefficient of 1.03 liters $mole^{-1}$ cm^{-1}. The following mechanism will account for the processes which occur:

$$\mathrm{HBr} + h\nu \xrightarrow{k_a} \mathrm{H} + \mathrm{Br} \tag{1}$$

$$\mathrm{H} + \mathrm{HBr} \xrightarrow{k_b} \mathrm{H_2} + \mathrm{Br} \tag{2}$$

$$\mathrm{H} + \mathrm{Br_2} \xrightarrow{k_c} \mathrm{HBr} + \mathrm{Br} \tag{3}$$

$$2\mathrm{Br} + \mathrm{M} \xrightarrow{k_d} \mathrm{Br_2} + \mathrm{M} \tag{4}$$

A steady-state analysis of this mechanism, using I_a to denote the rate of formation of H and Br in step (1), furnishes the following equation for the rate of Br_2 formation:

$$\frac{d[\mathrm{Br_2}]}{dt} = \frac{I_a}{1 + (k_c/k_b)[\mathrm{Br_2}]/[\mathrm{HBr}]} \tag{6-89}$$

The quantum yield of Br_2, Φ_{Br_2}, defined as the number of Br_2 molecules formed per photon absorbed by HBr, is given by dividing (6-89) by I_a. Inversion of the resulting equation gives

$$\Phi^{-1}_{Br_2} = 1 + \frac{k_c}{k_b}\frac{[Br_2]}{[HBr]} \tag{6-90}$$

It is of paramount importance to understand that k_c and k_b are bimolecular rate coefficients but for nonthermal energy distributions. There is no *apriori* reason to suspect that the ratio k_c/k_b should be equal to its thermal counterpart; rather we intuitively expect the ratio to show an energy dependence. This point may be clarified by considering the Arrhenius expression for the ratio of the thermal rate coefficients:

$$\frac{k_c(T)}{k_b(T)} = \frac{A_c}{A_b} \exp\left\{-\frac{E_c - E_b}{RT}\right\} \tag{6-91}$$

While this expression is definitely not valid for the hot atom case, we may nevertheless suggest that, crudely speaking, (6-91) shows an energy dependence because it has a temperature dependence.

The energy dependence, if any, in k_c/k_b can be shown qualitatively by adding some inert gas to the reaction mixture. These added molecules will accept some energy from the hot hydrogen atoms when a collision occurs. In this fashion the energy distribution of the reacting H atoms is displaced toward lower energies and the ratio k_c/k_b will change. Suppose we assume the moderating inert gas to consist of hard spheres. This may be a good approximation if He, Ar, Xe, or Kr are used. Further assume that the motion of the inert species prior to collision may be neglected in comparison to the velocity of the hot H atom. Regarding the hydrogen atom as a hard sphere, we can apply the laws of conservation of energy and momentum to such a collision to find the average fractional energy loss per collision to be

$$f = \frac{2m_A m_B}{(m_A + m_B)^2} \tag{6-92}$$

A picture of the collision is given in Fig. 6-9, where Θ_A and Θ_B are the scattering angles. In Problem 2 at the end of this section a head-on collision is examined

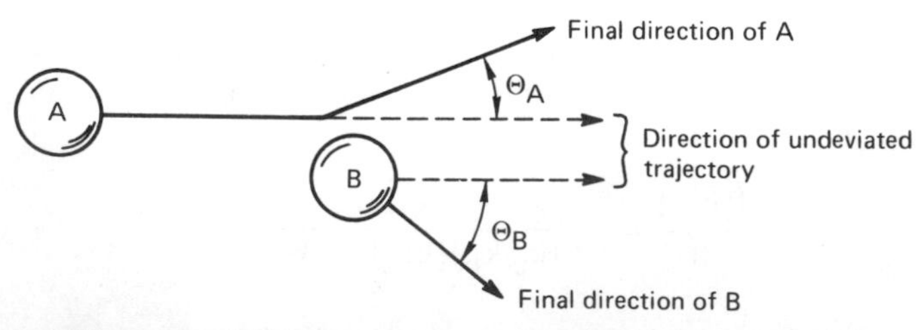

Figure 6-9. Collision of particles and attending scattering angles.

and the results of the calculation show that maximum energy transfer occurs when the masses m_A and m_B are equal. The most efficient of the rare gas moderators for hydrogen atoms is predicted on this basis to be helium, and this is confirmed experimentally. Using thermal rate coefficients measured for k_b and k_c we find that $E_b > E_c$. On this basis we expect thermalization effects to result in an increase in k_c/k_b, as shown qualitatively in Fig. 6-10. This figure predicts that large amounts of added inert gas completely thermalize the hot atoms prior to their entering into reactions. Under these circumstances $k_c/k_b = k_c(T)/k_b(T)$. The amount of inert gas required for complete transformation of the hot into thermal atoms depends on the efficiency of the moderator (thermalizing species). In any case the approach to the asymptotic value is quite slow.

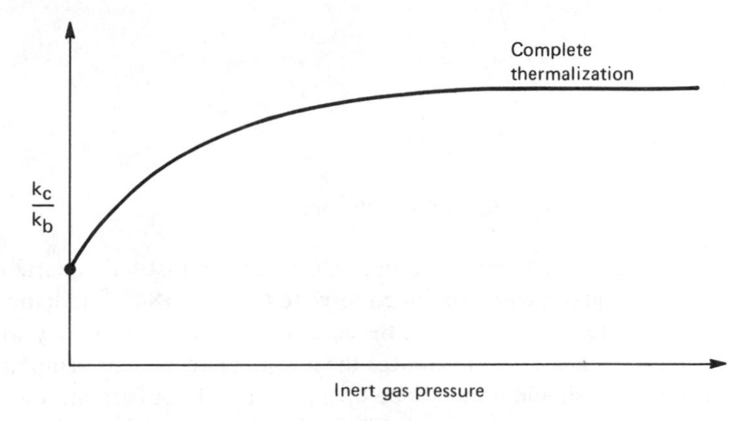

Figure 6-10. Thermalization effects on k_c/k_b.

EXPERIMENTAL. Two experiments are to be performed; in the first HBr is photolyzed at 1849 Å and in the second HBr mixed with He is photolyzed. In both these experiments Br_2 is generated by the photolysis to furnish the reactant mixture described in the theory section. Absorption spectroscopy is used to determine the concentrations of Br_2 and HBr at various times during the course of the photolysis. On the basis of elapsed time and concentration measurements values of k_3/k_2 are determined.

The apparatus consists of the following components: (1) a vacuum line for preparing reactant mixtures, (2) a photolysis cell, (3) a low-pressure mercury arc lamp, and (4) a spectrophotometer. Figure 6-11 illustrates the apparatus, which consists of a conventional oil-pumped vacuum system separated from a mercury manometer by a pressure transducer or spoon gauge. This is necessary because HBr reacts slowly with mercury vapor and, more importantly, mercury must be rigorously excluded from the reaction vessel to avoid photosensitized reactions. These reactions result when mercury atoms absorb the resonance lines from the lamp and transfer the resulting electronic excitation to other molecules in the reaction vessel.

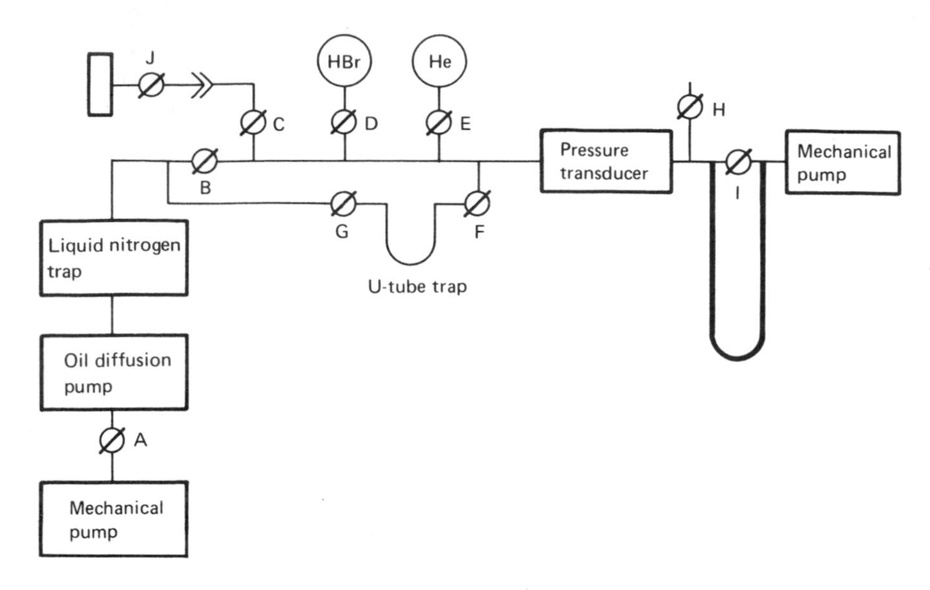

Figure 6-11. Vacuum system.

The photolysis cell is diagrammed in Fig. 6-12 and consists of a quartz tube with high-quality quartz windows which serve to transmit 1849-Å radiation. The stopcock on this cell is greaseless because HBr and Br_2 react slowly with stopcock greases. Figure 6-13 illustrates the low-pressure mercury lamp* and its electrical circuit. It should normally be operated at 120-mA current. Caution must be exercised in working around this source because the ultraviolet radiation can cause severe eye sunburn. The output of the lamp consists mainly of 2537- and 1849-Å radiation. If it is desirable, the 2537-Å line can be removed with a γ-irradiated LiF disc, whereas the 1849-Å line can be filtered with a Vycor filter.

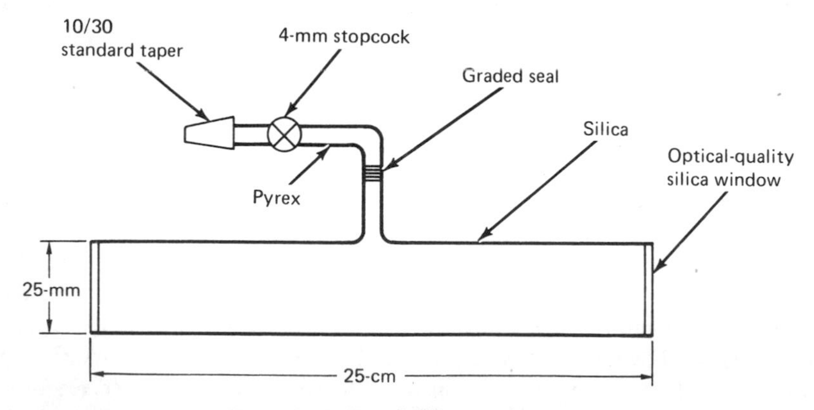

Figure 6-12. Fused silica reaction vessel.

*A lamp of this type may be purchased from Englehard-Hanovia Inc. Refer to Appendix XIII for further details.

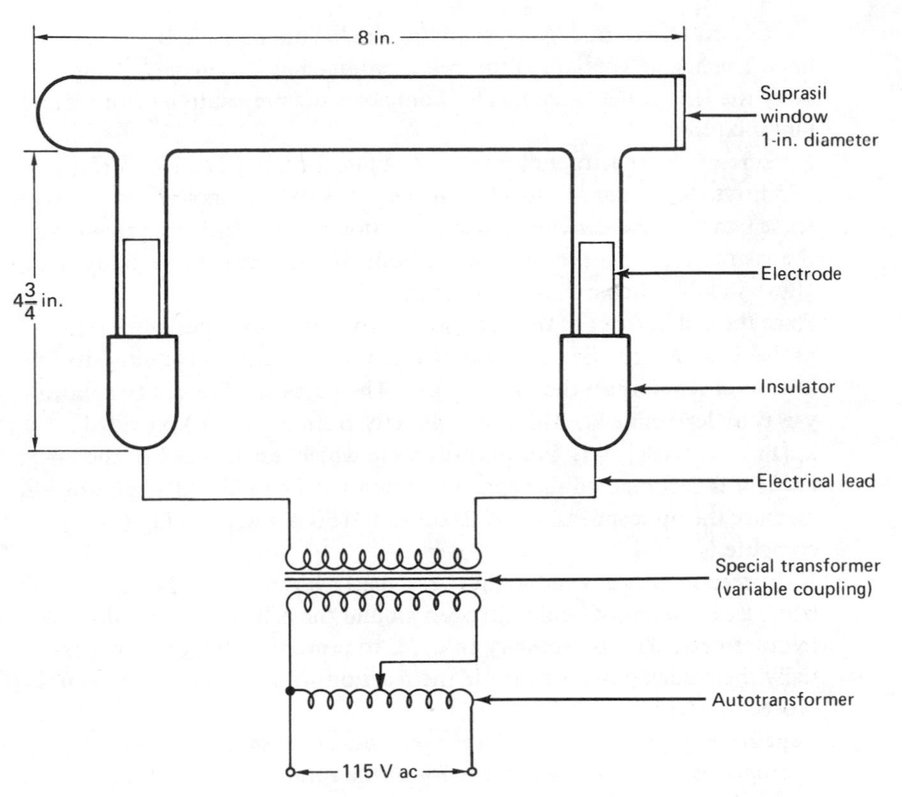

Figure 6-13. Low-pressure mercury arc.

PROCEDURE.

1. After evacuating the reaction vessel and connecting lines, admit a pressure of about 30 Torr of HBr into the volume bounded by stopcocks B, D, E, and G of Fig. 6-11. Place a dewar flask of liquid nitrogen around the U-tube trap and then open valve G slowly and pump all the HBr into the trap. Continue pumping for 2 or 3 min. This procedure will degas the HBr, removing traces of air and hydrogen. Close valve G.
2. Allow the U-tube trap to warm slowly and permit the HBr pressure in the photolysis cell to rise, monitoring it with the pressure transducer. Refer to Section 2.1 for further details. When the pressure reaches roughly 15 Torr, close valve F and replace the liquid nitrogen dewar around the U-tube. Measure the pressure accurately using the pressure transducer, close valve J, and then pump the HBr in the connecting line back into the U-tube trap.
3. If the experiment calls for added helium, close stopcock F, and admit about 250 Torr of helium into the volume bounded by stopcocks B, J, D, E, and F. Open stopcock J momentarily; this operation permits the pressure in the photolysis cell and connecting line to equilibrate but allows very little of the HBr to diffuse out of the reaction cell. Measure the total

photolysis cell pressure by determining the helium pressure in the vacuum line. The helium pressure in the cell is obtained by difference. Pump away the HBr in the U-tube. This completes the preparation of the reactant mixture.

4. Measure on the spectrophotometer the optical density at 2100 Å (ϵ_{HBr} = 155 liters mole^{-1} cm^{-1}) and 4160 Å (ϵ_{Br_2} = 170 liters mole^{-1} cm^{-1}). With these numbers and the empty cell extinction values which are provided by the instructor, the concentrations of both HBr and Br_2 can be computed. $[Br_2]$ should, of course, be close to zero.
5. Place the cell in front of the light source for a measured period of time, probably about 20 min. Because of variations in light source intensity, the instructor will furnish the time period. The purpose of this initial photolysis is to determine I_a. This arises directly from Eq. (6-89) provided $k_c[Br_2]/k_b[HBr] \ll 1$. For photolyses in which less than 3% of the initial HBr is decomposed this approximation will be valid. After photolysis, measure the optical densities at 2100 and 4160 Å and from Eq. (6-89) compute I_a.
6. If the reaction began with no inert gas, reattach the cell to the vacuum line, place a dewar of liquid nitrogen around the cell, and pump off the hydrogen gas. This is necessary in order to prevent hydrogen from partially thermalizing the system. If the reaction started with added helium, proceed directly to step 7.
7. Repetitiously photolyze and determine optical densities, pumping off hydrogen as in step 6 if it exceeds 3% of the total pressure and if no helium is present. From the data, compute $\Delta[Br_2]/\Delta t$ and $[Br_2]/[HBr]$.

ANALYSIS. Plot the appropriate quantities from a rearranged form of Eq. (6-89) so that a linear relation is obtained. One plot should be made for the experiment conducted without inert gas. Another plot should be made for the experiment with added helium. Least-squares analysis would be helpful. From the results, compare k_c/k_b for the two systems.

PROBLEMS

1. The energy of the $^2P_{1/2}$ state of bromine lies 3860 cm^{-1} above the ground state $^2P_{3/2}$ level. If HBr is dissociated, what range of photon energies assure the production of $^2P_{3/2}$ bromine atoms? If the photolysis is performed at 1849 Å, what are the possible hot hydrogen atom energies? Spectroscopy and theoretical considerations must be used to decide what fraction of each are formed.

2. Consider the head-on collision of a hard sphere of mass m_A and speed V_A with a hard sphere of mass m_B initially at rest. On collision, how much energy is transferred to m_B? For maximum energy transfer from m_A to m_B, what choice should be made for the mass of B? Prove it.

3. Compute the fractional energy loss per collision for hydrogen atoms produced by photolysis of HBr at 2537 and 1849 Å. If we write the rate of thermalization as

$$H^* + M \rightarrow H + M \qquad k_M \text{ (many collisions)}$$

how may k_M vary with initial hydrogen atom energy? Assume that thermalization is achieved when the energy is reduced below 0.1 eV.

4. The addition of helium in this experiment serves to change the energy of reacting hydrogen atoms. Suggest another experimental technique of changing the hydrogen atom energy. Comment on the equivalence or lack of it between these two techniques.

6.5 Trajectory Calculations in Chemical Kinetics

The detailed theoretical description of a chemical reaction is essentially a dynamical problem—a problem in which the motion (dynamics) of the electrons and nuclei involved must be considered. Since 1958, trajectory calculations frequently have been used in theoretical studies of bimolecular gas reactions. The significance of these calculations is that they permit examination of the detailed dynamics of individual reactive collisions, whereas the earlier equilibrium theories, such as transition state theory, circumvent with various hypotheses the necessity of examining individual encounters. The major limitation of trajectory calculations is their classical nature, which renders them ineffective in describing any kinetic system where quantum mechanical features are important. At present the seriousness of this limitation cannot be determined because neither definitive experiments nor accurate quantum mechanical theoretical calculations are available for comparison. The available experimental data for bimolecular reactions can, in most cases, be described adequately in classical terms; however, experimental uncertainties may be large enough in most kinetics experiments to render any small quantum effects unobservable. It is possible to introduce a few quantum mechanical features into a classical trajectory calculation; for example, the initial conditions can be specified so that the reactant molecules are in vibrational-rotational eigenstates. However, this does not really alter the nature of the calculation because the dynamics are still assumed to be classical.

The development of a rate coefficient by means of trajectory calculations involves four essential ingredients:

1. Potential-energy surface
2. Equations of motion
3. Reaction cross section
4. Energy distribution function

Given the potential-energy surface and the classical equations of motion, the reaction cross section can be calculated. With it and the experimental energy distribution function for collisions, a rate coefficient can be calculated. Figure 6-14 is a block diagram illustrating the connections.

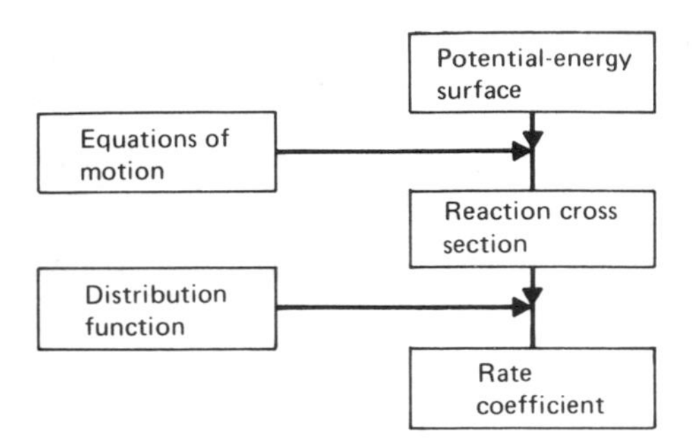

Figure 6-14. Schematic development of rate coefficient.

The above points are discussed below in relation to a three-particle system and the prototype of all bimolecular three-center reactions:

$$A + BC \rightarrow AB + C$$

Special emphasis is given to the hydrogen isotope exchange reaction:

$$H + H_2 \rightarrow H_2 + H$$

THE POTENTIAL-ENERGY SURFACE. The first ingredient in a trajectory calculation of a rate coefficient is the potential-energy surface, which specifies the potential field in which the nuclei move. For the prototype reaction,

$$A + BC \rightarrow AB + C$$

the potential energy is a function of the three interparticle distances, r_{AB}, r_{BC}, and r_{AC}, as defined in Fig. 6-15.

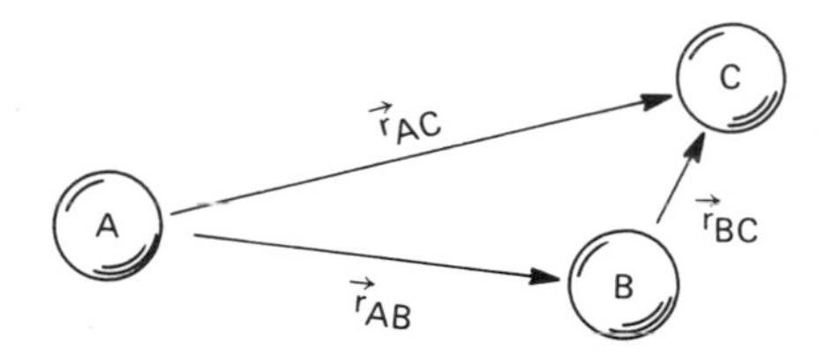

Figure 6-15. Coordinates of three-particle system.

The potential-energy function and its variation with internuclear separation arises as a result of the kinetic and potential energies of the electrons as well as the internuclear Coulombic potential. To understand how this arises we shall consider the total quantum mechanical Hamiltonian operator for the ABC system and its associated energy eigenvalue equation. The total Hamiltonian operator is composed of kinetic-energy terms for both the nuclei and electrons together with potential-energy terms to account for electron-electron repulsions, electron-nucleus attractions, and nucleus-nucleus repulsions. Formally, we write these terms in the order given above as

$$\mathcal{H} = -\frac{\hbar^2}{2m_A}\nabla_A^2 - \frac{\hbar^2}{2m_B}\nabla_B^2 - \frac{\hbar^2}{2m_C}\nabla_C^2 - \frac{\hbar^2}{2m_e}\sum_i \nabla_i^2$$

$$+\sum_{\substack{i>j \\ \text{all } j}} \frac{e^2}{r_{ij}} - \sum_{i,I} \frac{Z_I e^2}{r_{iI}} + \sum_{\substack{I>J \\ \text{all } J}} \frac{Z_I Z_J e^2}{R_{IJ}} \tag{6-93}$$

where lowercase subscripts refer to electrons and uppercase to nuclei. In Eq. (6-93) m refers to mass; r and R refer to interparticle distances; Z refers to nuclear charge; e is the electronic charge; A, B, C, I, and J refer to nuclei; and i and j refer to electrons. Note that in the potential terms describing electron-electron repulsion and nucleus-nucleus repulsion the sums are taken for $i > j$ and $I > J$ to avoid including each pair-wise contribution twice.

The energy eigenvalue equation is

$$\mathcal{H}\Psi = E\Psi \tag{6-94}$$

As is often done, we shall seek to solve the differential equation (6-94) by the method of separation of variables–in this case the separation of nucleus and electron coordinates. Equation (6-93) indicates that this cannot be rigorously done because the term containing r_{iI} prevents the partitioning of $\mathcal{H}$ into two parts, one containing solely electron coordinates and the other solely nuclear. To proceed, an approximation, known as the Born-Oppenheimer approximation, is used. It achieves a separation of variables by requiring the nuclear velocities to be very small compared to electron velocities. Translated into mathematical terms the Born-Oppenheimer approximation permits the electronic motion problem to be solved with the nuclei fixed. The energy eigenvalue of the resulting differential equation is a function of the internuclear distances. These energy eigenvalues at different internuclear distances furnish the potential field in which the nuclei move.

Explicitly, let

$$\Psi \cong \psi_n \psi_e \tag{6-95}$$

where ψ_e is the solution of the fixed-nuclei equation; namely,

$$\left(-\frac{\hbar^2}{2m_e}\sum_i \nabla_i^2 + \sum_{\substack{i>j \\ \text{all } j}} \frac{e^2}{r_{iI}} + \sum_{\substack{I>J \\ \text{all } J}} \frac{e^2}{R_{IJ}}\right)\psi_e = E_{el}\,\psi_e \tag{6-96}$$

and ψ_n is the solution of

$$\left(-\frac{\hbar^2}{2m_A}\nabla_A^2 - \frac{\hbar^2}{2m_B}\nabla_B^2 - \frac{\hbar^2}{2m_C}\nabla_C^2 + E_{el}\right)\psi_n = E\psi_n \tag{6-97}$$

Notice that the eigenvalues of Eq. (6-96) serve as the potential for Eq. (6-97), the nuclear motion equation.

The separation of Eq. (6-95) requires, for validity, that the variation of ψ_e with internuclear distance be very slow in comparison to the variation of ψ_e with electronic coordinates. This can be demonstrated by substituting $\psi_n\psi_e$ into Eq. (6-94) and remembering that ψ_e is a function of both electron and nucleus coordinates. As long as the velocities of the nuclei are small compared to electron velocities the above approximation of Born and Oppenheimer will be valid. The nuclei may then be regarded as moving in the average field set up by the electrons.

In summary, the description of nuclear dynamics as motion on a potential-energy surface is possible because the Born-Oppenheimer approximation is adequate. The potential-energy surface itself is constructed by solving Eq. (6-96) (usually only approximately) for each possible set of values of the internuclear distances with the resulting eigenvalues $E_{el}(R_{IJ})$ serving as the potential for the nuclear motion.

The solution of Eq. (6-96) has never been obtained in analytical form for any real molecular ABC system. As is well known, the only eigenvalue problems solvable, at present, in analytical form are one-electron problems. Thus, approximate methods of one kind or another must be used. One of the oldest methods is that of London, which we shall apply here to a three-electron system. The eigenvalues of Eq. (6-96) are given approximately by the London equation as

$$E_{el} = Q \pm \{\tfrac{1}{2}[(a-b)^2 + (b-c)^2 + (c-a)^2]\}^{1/2} \tag{6-98}$$

where Q is known as a Coulombic integral and a, b, and c are exchange integrals. Development of Eq. (6-98) depends on the valence bond method of treating molecular problems. A resumé of this method is given in the Appendix to this experiment. From there we shall learn that the valence bond approach leads to an expression of the above form if the basis set orbitals are orthogonal and multiple exchange integrals are neglected. The London equation for the ground state of H_3 has the form

$$\begin{aligned} E_- = \langle ABC|\mathcal{H}|ABC\rangle - [&\tfrac{1}{2}(\langle ABC|\mathcal{H}|BAC\rangle - \langle ABC|\mathcal{H}|ACB\rangle)^2 \\ &+ \tfrac{1}{2}(\langle ABC|\mathcal{H}|ACB\rangle - \langle ABC|\mathcal{H}|CBA\rangle)^2 \\ &+ \tfrac{1}{2}(\langle ABC|\mathcal{H}|CBA\rangle - \langle ABC|\mathcal{H}|BAC\rangle)^2]^{1/2} \end{aligned} \tag{6-99}$$

Using the London equation, an estimate of the electronic energy E_- can be obtained provided the integrals on the right-hand side of (6-99) can be evaluated. If one inserts the fixed-nucleus Hamiltonian into (6-99), it becomes immediately apparent that evaluation of these integrals is not simple. It is, however, possible to estimate them by a semiempirical approach which makes use of experimental spectroscopic data for the diatomic molecules involved in the valence bond formulation. To understand why this is a reasonable approach, consider the elements of (6-99). For example, $\langle ABC|\mathcal{H}|BAC\rangle$ is an exchange integral in which electrons on nuclei A and B are allowed to exchange but the one on C does not. Physically, we expect this to occur when A and B are close together and C is a large distance from both A and B. We conclude that the energy involved in $\langle ABC|\mathcal{H}|BAC\rangle$ is closely related to the properties of the diatomic molecule AB. Similar arguments obtain for the other elements of (6-99). The Coulombic integral $Q = \langle ABC|\mathcal{H}|ABC\rangle$ represents the total energy of the system of fixed nuclei where no exchange of electrons is permitted. This energy arises because of the Coulomb interaction of the electrons and nuclei in the system. The Coulomb integral can be also divided into *diatomic* parts corresponding to the diatomic molecules that can be constructed from three atoms (i.e., AB, BC, AC). Hence, we write Q as

$$Q = Q_{AB} + Q_{BC} + Q_{AC}$$

and using J for the exchange integrals we write

$$\langle ABC|\mathcal{H}|BAC\rangle = J_{AB}$$

$$\langle ABC|\mathcal{H}|ACB\rangle = J_{BC}$$

$$\langle ABC|\mathcal{H}|CBA\rangle = J_{AC}$$

Having now dissected the London equation into diatomic parts, we shall proceed to introduce experimental spectroscopic information (making the result semiempirical) as a means of estimating the energy eigenvalues of the fixed-nucleus Hamiltonian, Eq. (6-96). This introduction is accomplished by recognizing that the potential energy for nuclear motion in a diatomic molecule is the sum of a Coulomb integral and an exchange integral and can be written as

$$V_{AB}(r_{AB}) = Q_{AB}(r_{AB}) + J_{AB}(r_{AB}) \tag{6-100}$$

with similar expressions for V_{BC} and V_{AC}. This expression is a diatomic London equation which assumes that the atomic orbitals describing the molecule AB are orthogonal, just as was done in Eq. (6-99) for the three-atom system. The form of V_{AB} is assumed known from spectroscopic work. For the H_3 case, the diatomics are all H_2, and the spectroscopic properties necessary to construct V_{AB} are well known. One widely used method utilizes the Morse equation which furnishes V_{AB} as a function of r_{AB}, the internuclear separation, in the form

$$V_{AB}(r_{AB}) = D_{AB}[\exp\{-2a_{AB}(r_{AB} - r_{0AB})\} - 2\exp\{-a_{AB}(r_{AB} - r_{0AB})\}] \tag{6-101}$$

where D_{AB} is the dissociation energy of the diatomic molecule, r_{0AB} the equilibrium internuclear separation, and a_{AB} the curvature of the potential well evaluated at the equilibrium internuclear distance. All these quantities are assumed known. Figure 6-16 shows V_{AB} schematically. Note that V_{AB} is taken to be zero for infinitely separated atoms. With V_{AB} the London equation can be written as a function of either the Q_{ij}s or the J_{ij}s. To complete the solution another assumption is needed which relates the Q_{ij}s and the V_{ij}s. One assumption introduced by Eyring and Polanyi (see the References) is that Coulombic energy is a constant fraction of the potential energy regardless of internuclear separation. Mathematically this is written as

$$Q_{AB}(r_{AB}) = \lambda_{AB} V_{AB}(r_{AB}) \tag{6-102}$$

where λ_{AB} is independent of r_{AB}. Although this assumption cannot be justified for every internuclear separation, especially small separations, it may be approximately valid for the internuclear separations near the equilibrium configuration of a diatomic molecule and these are usually the most significant in the chemical reaction. Generally speaking, however, this approximation is a poor one. We use it here because it is simple. More useful approximations have been developed by Sato and Raff et al. (see the References) among others.

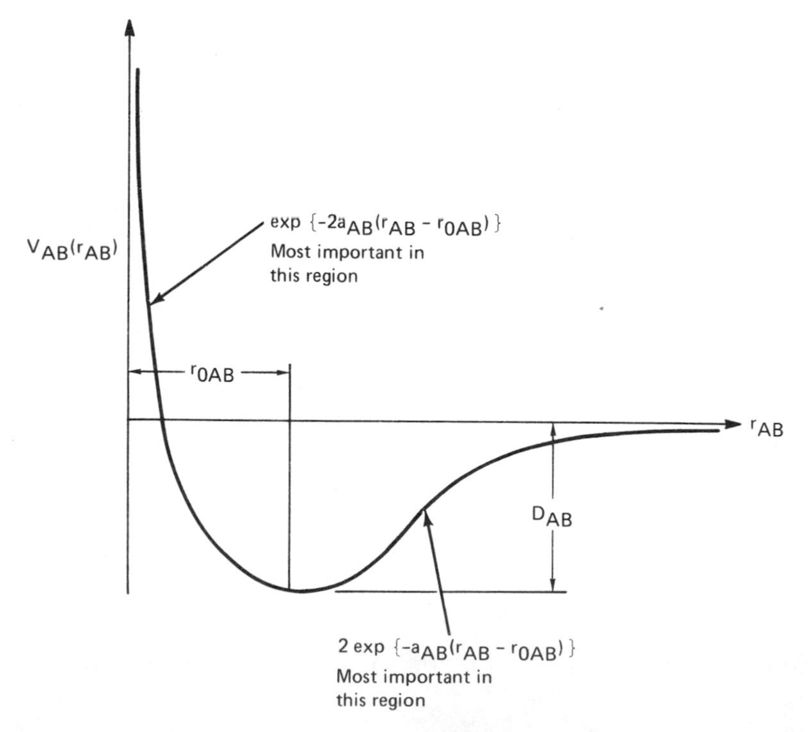

Figure 6-16. Morse potential-energy function.

Substituting (6-101) and (6-102) into (6-99) furnishes an equation in V_{ij} and λ_{ij}:

$$E_- = \lambda_{AB}V_{AB} + \lambda_{BC}V_{BC} + \lambda_{AC}V_{AC} - \{[(1-\lambda_{AB})V_{AB}]^2 + [(1-\lambda_{BC})V_{BC}]^2 + [(1-\lambda_{AC})V_{AC}]^2 - (1-\lambda_{AB})(1-\lambda_{BC})V_{AB}V_{BC} - (1-\lambda_{AB})(1-\lambda_{AC})V_{AB}V_{AC} - (1-\lambda_{BC})(1-\lambda_{AC})V_{BC}V_{AC}\}^{1/2} \quad (6\text{-}103)$$

Applied to H_3 where the diatomic potentials all have the same form and $\lambda_{AB} = \lambda_{BC} = \lambda_{AC} = \lambda$, Eq. (6-103) simplifies to an easily evaluated analytical form for the potential-energy surface, namely

$$E_- = \lambda[V(r_{AB}) + V(r_{BC}) + V(r_{AC})] - (1-\lambda)\{V^2(r_{AB}) + V^2(r_{BC}) + V^2(r_{AC}) - V(r_{AB})V(r_{BC}) - V(r_{AB})V(r_{AC}) - V(r_{BC})V(r_{AC})\}^{1/2} \quad (6\text{-}104)$$

To find E_- at any particular configuration (r_{AB}, r_{BC}, r_{AC}), λ must be chosen and the Morse Eq. (6-101) for H_2 evaluated and substituted into (6-104). By evaluating (6-104) for a large number of sets of values of the three internuclear distances the potential-energy surface can be generated. This surface is four-dimensional (E_-, r_{AB}, r_{BC}, and r_{AC}) and thus not directly amenable to graphical representation. By constraining the three H atoms to lie along a line at all times, one of the dimensions can be eliminated and the potential energy can be represented as a topological map like that shown in Fig. 6-17 for the linear H_3

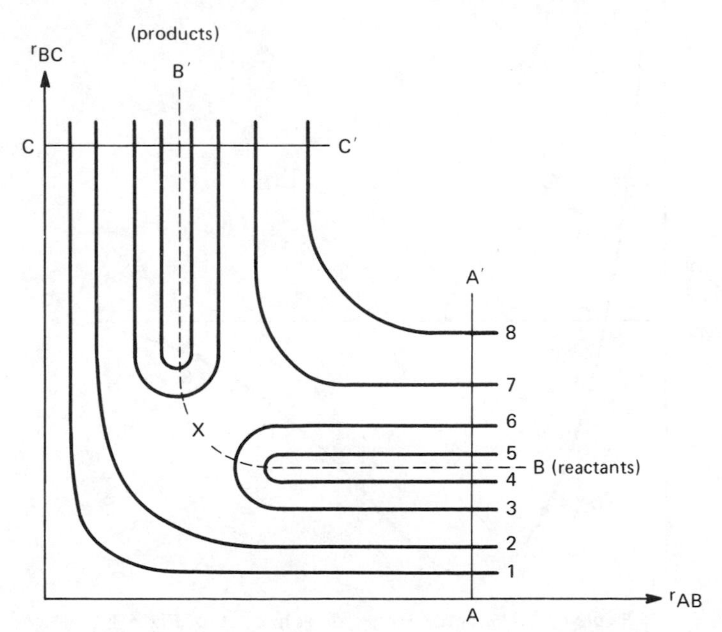

Figure 6-17. Schematic representation of the potential-energy surface for the linear H_3 system.

system $H_A - H_B - H_C$. The lines in this figure are constant potential-energy contours (i.e., E_ has the same value at every point along one of these lines). To identify qualitatively whether a given contour corresponds to a large or small potential-energy contour, consider the line AA′ in Fig. 6-17. If the three atoms are at configuration 1, the r_{AB} distance is large and the r_{BC} distance is small, corresponding to atom A being far removed from molecule BC, which is itself compressed. At configuration 8 atom A is still far from molecule BC but BC is now expanded. Motion along the line AA′ from configuration 1 to configuration 8 therefore corresponds to stretching the diatomic molecule BC, and the potential energy of the system along this line is given approximately by the Morse equation for BC. Figure 6-18 shows the potential along AA′ with the corresponding configurations labeled. We conclude that the lowest energies (most stable configuration) are associated with contours 4 and 5 and the highest with 1 and 8.

Turning now to a qualitative description of a reaction for a linear system of three hydrogen atoms, let the reaction be written as

$$H_A + H_B - H_C \rightarrow H_A - H_B + H_C$$

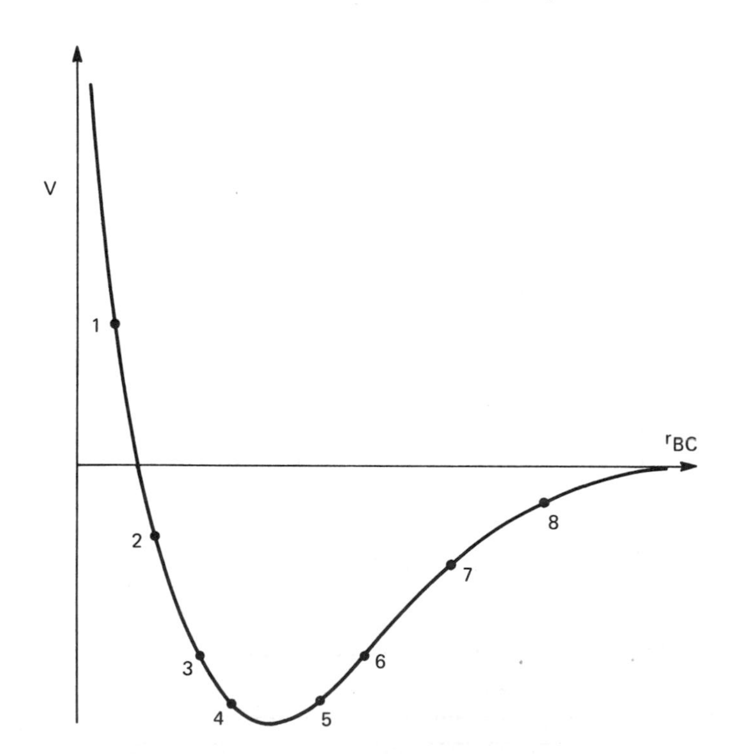

Figure 6-18. Potential energy along line AA′ of Fig. 6-17.

On the potential-energy surface this corresponds to changing the configuration from region AA′ to region CC′ (reactant configuration to product configuration). The lowest-energy path which passes from reactants to products is the dotted line BB′ in Fig. 6-17 and is called the reaction coordinate. Figure 6-19 shows the variation of the potential energy along this coordinate from B to B′. Clearly a potential-energy barrier must be overcome in passing from reactants to products. Its height is called the classical activation energy and is related to the Arrhenius activation energy determined experimentally, but the relation is not a simple one. The maximum of Fig. 6-19 corresponds to the position denoted by X in Fig. 6-17 and corresponds physically to bringing all three H atoms close together in what is called the activated state. The region around X is known as the interaction region, and the details of the dynamics in this region determine whether a given collision is reactive or nonreactive. Given a potential energy surface [Eq. (6-104) or some other approximation] the problem of calculating the rate coefficient for the hydrogen isotope exchange reaction is basically one of studying the dynamics (molecular motion) of a point on the surface (the point represents the system) under a wide variety of initial conditions. Such a study furnishes the reaction cross section, which is intimately related to the rate coefficient. We shall turn now to a brief description of the equations of motion describing the dynamics of the three-particle linear H_3 system.

EQUATIONS OF MOTION. To study dynamics on a potential-energy surface one must choose between classical and quantum mechanics. Although quantum mechanics likely will give a more accurate description of the nuclear motion on the potential-energy surface, the mathematical formulation of the problem is very complex, and it has only very recently become possible to do a reasonably

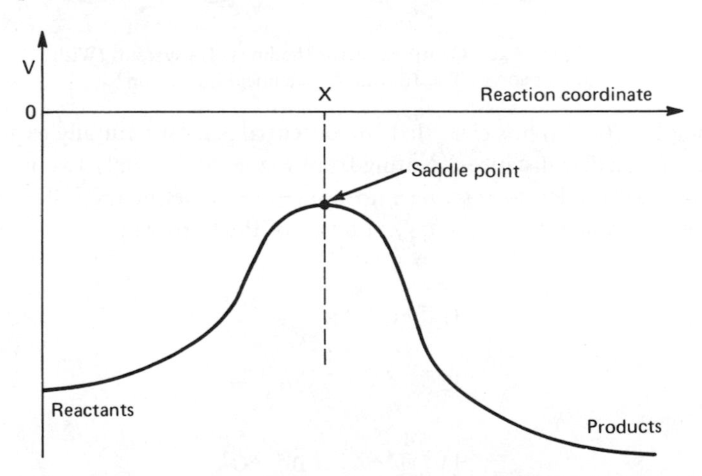

Figure 6-19. Potential energy along the reaction coordinate BB′ of Fig. 6.17. Zero of potential taken to be the infinite separation of all three atoms.

accurate calculation of the quantum dynamics of the hydrogen isotope exchange reactions. (See McCullough and Wyatt, Truhlar and Kuppermann, and Karplus and Tang in the References.) As new numerical procedures are developed and computer systems become faster, progress in quantum mechanical studies can be made. Classical mechanical studies of the dynamics are themselves not simple, but they are considerably easier than quantum mechanical ones. Furthermore, because the de Broglie wavelengths of nuclei involved in chemical reactions are generally small compared to distances over which the potential energy has steep slopes, it is expected that a classical description will be reasonably accurate.

Limiting the discussion to classical mechanics we shall now proceed to set up the equations of motion (refer to Fig. 6-14) for a linear system of three H atoms in Hamilton's form. Figure 6-20 shows the coordinate system and particle location. The Hamiltonian H is given by the sum of the kinetic T and potential V energies at any interparticle configuration. In terms of generalized coordinates and Cartesian momenta

$$H = T + V = T_A + T_B + T_C + V$$

$$= \frac{P_A{}^2}{2m} + \frac{P_B{}^2}{2m} + \frac{P_C{}^2}{2m} + V(q_1, q_2, q_3) \qquad (6\text{-}105)$$

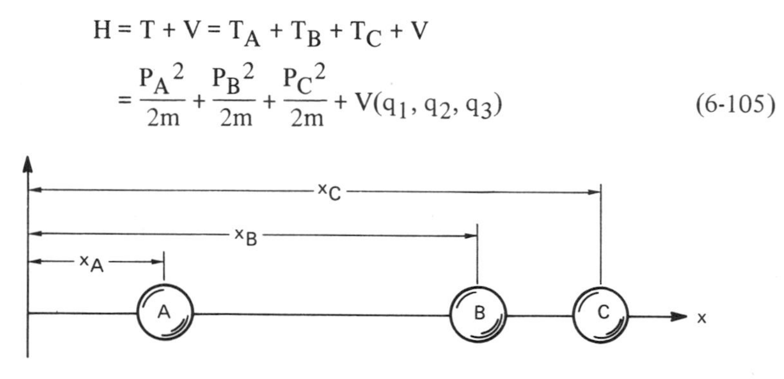

Figure 6-20. Coordinates for the linear H_3 system. (With permission of The Journal of Chemical Education.)

Examining Eq. (6-104) it is clear that the potential is most naturally expressed in terms of interparticle distances. Noting from Fig. 6-20 that only two interparticle distances are needed to describe a linear system we define the following generalized coordinates $\{q_1, q_2, q_3\}$ in terms of the Cartesian coordinates $\{x_A, x_B, x_C\}$:

$$q_1 = x_C - x_B$$

$$q_2 = -x_A + \frac{1}{2}(x_B + x_C)$$

$$q_3 = \frac{1}{3}(x_A + x_B + x_C)$$

where q_1 is the position of B with respect to C, q_2 the position of A with respect to the center of mass of BC, and q_3 the position of the center of mass of the whole system. The inverses of these transformations are needed to evaluate (6-105) and they are given by

$$x_A = -\frac{2}{3} q_2 + q_3$$

$$x_B = q_3 - \frac{1}{2} q_1 + \frac{1}{3} q_2 \qquad (6\text{-}106)$$

$$x_C = q_3 + \frac{1}{2} q_1 + \frac{1}{3} q_2$$

The Cartesian momenta $\{P_A, P_B, P_C\}$ conjugate to the Cartesian coordinates $\{x_A, x_B, x_C\}$ can be expressed in terms of the momenta conjugate to the generalized coordinates $\{q_1, q_2, q_3\}$ using the chain rule for differentiation and Eq. (6-106). This procedure furnishes

$$P_A = P_1 \frac{\partial q_1}{\partial x_A} + P_2 \frac{\partial q_2}{\partial x_A} + P_3 \frac{\partial q_3}{\partial x_A}$$

$$= -P_2 + \frac{1}{3} P_3$$

Similarly,

$$P_B = -P_1 + \frac{1}{2} P_2 + \frac{1}{3} P_3 \qquad (6\text{-}107)$$

$$P_C = P_1 + \frac{1}{2} P_2 + \frac{1}{3} P_3$$

where $\{P_1, P_2, P_3\}$ are conjugate to $\{q_1, q_2, q_3\}$. This method is valid because what is known as a point-contact transformation is being performed. In terms of $\{q, P\}$ variables, H becomes, by substitution into Eq. (6-105),

$$H = \frac{1}{2m} [2P_1^2 + \frac{3}{2} P_2^2 + \frac{1}{3} P_3^2] + V(q_1, q_2, q_3) \qquad (6\text{-}108)$$

Hamilton's equations of motion are obtained directly from (6-108) by setting $\partial H/\partial P_j = \dot{q}_j$ and $\partial H/\partial q_j = -\dot{P}_j$, where the overdot indicates a time derivative.

From (6-108) there arise six first-order differential equations which describe the dynamics of this linear H_3 system:

$$\frac{dq_1}{dt} = \frac{\partial H}{\partial P_1} = \frac{2P_1}{m}$$

$$\frac{dq_2}{dt} = \frac{\partial H}{\partial P_2} = \frac{3P_2}{2m}$$

$$
\begin{aligned}
\frac{dq_3}{dt} &= \frac{\partial H}{\partial P_3} = \frac{P_3}{2m} \\
\frac{dP_1}{dt} &= -\frac{\partial H}{\partial q_1} = -\frac{\partial V}{\partial q_1} = -\left[\frac{\partial V}{\partial r_{AB}}\frac{\partial r_{AB}}{\partial q_1} + \frac{\partial V}{\partial r_{BC}}\frac{\partial r_{BC}}{\partial q_1} + \frac{\partial V}{\partial r_{AC}}\frac{\partial r_{AC}}{\partial q_1}\right] \\
\frac{dP_2}{dt} &= -\left[\frac{\partial V}{\partial r_{AB}}\frac{\partial r_{AB}}{\partial q_2} + \frac{\partial V}{\partial r_{BC}}\frac{\partial r_{BC}}{\partial q_2} + \frac{\partial V}{\partial r_{AC}}\frac{\partial r_{AC}}{\partial q_2}\right] = -\frac{\partial V}{\partial q_2} \\
\frac{dP_3}{dt} &= -\left[\frac{\partial V}{\partial r_{AB}}\frac{\partial r_{AB}}{\partial q_3} + \frac{\partial V}{\partial r_{BC}}\frac{\partial r_{BC}}{\partial q_3} + \frac{\partial V}{\partial r_{AC}}\frac{\partial r_{AC}}{\partial q_3}\right] = -\frac{\partial V}{\partial q_3}
\end{aligned}
\tag{6-109}
$$

In the last three equations the partial derivatives of the potential energy surface [Eq. (6-104)] are given in terms of the coordinates $\{r_{AB}, r_{BC}, r_{AC}\}$ of (6-104) and the derivatives of these coordinates with respect to the alternative set of coordinates $\{q_1, q_2, q_3\}$. The equation $dP_3/dt = -(\partial V/\partial q_3)$ is identically zero because the potential-energy surface is independent of the center of mass, q_3, and varies only with the interparticle coordinates. The motion of the center of mass of the linear H_3 system is of no particular interest in trajectory calculations because the interparticle potential energies do not involve this coordinate. Hence, the time derivative of q_3 can be chosen to have an arbitrary value without losing any physically significant information. The choice $dq_3/dt = 0$ is usually made and is equivalent to choosing a coordinate system which moves with the center of mass. The elimination of the center of mass motion leaves four first-order differential equations to solve. This set could be constrained further using the equations of conservation of energy and linear momentum to eliminate two additional differential equations. This is seldom done; rather, the numerical procedure used for integrating the equations of motion is tested for its accuracy by determining how well the total energy and linear momentum are maintained constant over many integration steps, each of which can introduce a small amount of error because of roundoff and approximations made in the numerical procedure. Refer to Section 1.6.

Further manipulation of the equation for dP_1/dt and dP_2/dt is necessary because the potential-energy surface is most naturally given in terms of interparticle distances [Eq. (6-104)], not q_1 and q_2. To effect this change of variables the chain rule for differentiation is used together with the algebraic relations between $\{r_{AB}, r_{BC}, r_{AC}\}$ and $\{q_1, q_2, q_3\}$. These are given below:

$$
\begin{aligned}
r_{AB} &= q_2 - \frac{1}{2} q_1 \\
r_{BC} &= q_1 \\
r_{AC} &= q_2 + \frac{1}{2} q_1
\end{aligned}
\tag{6-110}
$$

$$\frac{\partial V}{\partial q_1} = -\frac{1}{2}\left(\frac{\partial V}{\partial r_{AB}}\right) + \frac{\partial V}{\partial r_{BC}} + \frac{1}{2}\left(\frac{\partial V}{\partial r_{AC}}\right)$$

$$\frac{\partial V}{\partial q_2} = \frac{\partial V}{\partial r_{AB}} + \frac{\partial V}{\partial r_{AC}} \tag{6-111}$$

With Eq. (6-111), the four simultaneous first-order differential equations which must be integrated are

$$\frac{dq_1}{dt} = 2 \cdot \frac{P_1}{m}$$

$$\frac{dq_2}{dt} = 1.5 \cdot \frac{P_2}{m}$$

$$\frac{dP_1}{dt} = \frac{1}{2}\left(\frac{\partial V}{\partial r_{AB}}\right) - \frac{\partial V}{\partial r_{BC}} - \frac{1}{2}\left(\frac{\partial V}{\partial r_{AC}}\right) \tag{6-112}$$

$$\frac{dP_2}{dt} = -\frac{\partial V}{\partial r_{AB}} - \frac{\partial V}{\partial r_{AC}}$$

The potential-energy surface which we shall use is given in Eq. (6-104), where E_- becomes V in (6-112). To evaluate the right-hand side of the last two equations in (6-112) at any configuration $\{r_{AB}, r_{BC}, r_{AC}\}$ the analytical form of the derivatives must be known and can be obtained from (6-104) if the diatomic potentials from (6-101) are used.

To obtain the trajectories of the particles {A, B, C} as a function of time the four differential equations (6-112) are integrated numerically in a stepwise fashion beginning at $t = t_0$ with some given initial values of the coordinates and momenta of the three particles. The differential equations are then solved for a short time later, $t_0 + h$. This procedure is continued to $t_0 + 2h$, etc. Two numerical methods are used in the programming of the problem considered here: (1) the Runge-Kutta method is used to start the solution and (2) a predictor-corrector method is used to continue the solution. The predictor-corrector is, in general, a faster numerical procedure than the Runge-Kutta but it requires the coordinates and momenta at two or more (depending on the order of the method) previous time steps. The Runge-Kutta method is a self-starting method requiring only the coordinates and momenta at time $t - h$ in order to generate the solution at time t. Therefore, the Runge-Kutta method is used to find the numerical solution for the first few time steps, and then control is transferred to the predictor-corrector method for the completion of the solution. These methods are discussed in more detail in Section 1.6.

To start a trajectory calculation, the initial values of q_1, q_2, P_1, and P_2 must be chosen. There are several ways of selecting these parameters. One way is outlined as follows:

1. q_1 (t = 0): Choose the total initial energy of the BC molecule as E_{BC} and require that all of it be potential energy. The value of q_1 (t = 0) may then be found by solving the Morse equation

$$E_{BC} = D[1 - \exp\{-a(r_{BC} - r_{0BC})\}]^2$$

   for r_{BC} and recalling that $q_1 = r_{BC}$. This equation is similar to Eq. (6-101) except the origin of the potential energy scale is taken so that $E_{BC} = 0$ when $r_{BC} = r_{0BC}$.
2. q_2 (t = 0): Select this value so that the potential interaction of A with both B and C is negligible but not so large that numerical integration times are excessive.
3. P_1 (t = 0): Since the above choice of q_1 (t = 0) requires the BC molecule to be at one of its classical turning points,

$$P_1\ (t = 0) = 0$$

4. P_2 (t = 0): By selecting the initial relative velocity, V_R (t = 0), the initial value of P_2 is given by

$$P_2\ (t = 0) = \mu_{A,BC} V_R\ (t = 0)$$

   where $\mu_{A,BC}$ is the reduced mass of the pseudo-two-particle system, m_A and m_{BC}.

Once these initial values are selected, the trajectory is determined. Different sets of initial conditions may lead to different outcomes; herein lies the problem for investigation.

The above discussion may be extended to more complicated models having more than one dimension and more than three particles. In any case, initial values must be specified for all the coordinates and momenta appearing in Hamilton's equations of motion.

Using Hamilton's equations of motion and the initial conditions, the complete trajectory can be obtained by numerical integration. While the details of these trajectories are of considerable interest, their most significant part is the final coordinates and momenta. Once the interaction (collision) is complete we are interested in learning if a reaction has taken place and/or if energy transfer of some sort has occurred. From a procedural point of view criteria must be set up to determine when a collision is completed. For the simple linear ABC system being used here, the distance between A and B or B and C may be used to determine the stopping point of the numerical integration. That is, when either the A – B distance or the B – C distance exceeds a certain value (large enough to make the potential between the final atom and the final molecule small) the calculation is terminated. This test will be valid only if the total energy in the system is small compared to that required for it to dissociate into three H atoms.

With the above criteria, a reaction has occurred at the termination if A and B are close together and C is far away; otherwise there has been no reaction. A typical reactive encounter is shown in Fig. 6-21, where the variation of the coordinates with time is plotted and the BC distance is very large at the termination while the AB distance is short and oscillatory.

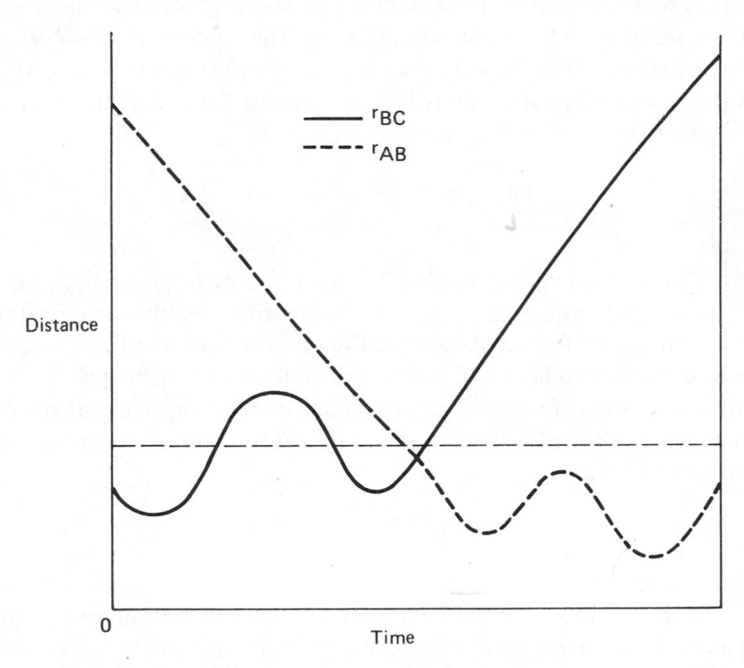

Figure 6-21. Trajectory illustrating a reactive collision between A and BC.

PROBABILITIES AND CROSS SECTIONS. Any dynamical feature of the products of a collision may be examined in a series of trajectory calculations. Here we shall restrict our attention to chemical reaction and ask for the probability that a hydrogen atom with a certain initial relative velocity and a hydrogen molecule with a certain initial vibrational energy will react upon colliding colinearly irrespective of the initial distance from the hydrogen atom to the center of mass of the molecule when the latter is started at its classical turning point. In the previous statement we have essentially outlined those initial conditions that we wish to keep fixed and those that we wish to vary. In the case described, E_{BC} and thus q_1 $(t = 0)$ and P_1 $(t = 0)$ will be fixed as will V_R $(t = 0)$ and P_2 $(t = 0)$, while q_2 $(t = 0)$ will be allowed to range at random over a range of distances. Loosely speaking, this range must allow the configuration of the H_2 molecule to take any one of its full range of values (between the classical turning points) at the time the H atom collides. For example, in some collisions the H_2 molecule will be expanding when the H atom attacks, while in others it will be contracting. In more precise terms, the selection of the initial value of q_2 from a range permits, in a set of trajectories, randomization of the phasing of the vibrational motion of H_2 with respect to the incoming H atom.

For the phasing to be completely random, the initial values of q_2 must be selected at random from their prescribed range. This may be accomplished through the correlation of random numbers (generally generated in a computer program subroutine) with various values of q_2. The random selection of all parameters that vary from trajectory to trajectory is essential in trajectory calculations if one wishes to obtain statistical estimates of the probability of reaction. The probability of reaction P_{rxn} may be defined as

$$P_{rxn} = \lim_{n \to \infty} \frac{n_{rxn}}{n} \tag{6-113}$$

where n is the total number of randomly selected trajectories and n_{rxn} the number of trajectories leading to reaction. Since an infinite number of trajectories cannot be run, we are forced to estimate P_{rxn} on the basis of a finite value of n. For such an estimate to be useful and amenable to statistical analysis for its uncertainty, the process of variable selection must be random. The statistical analysis is done by the Monte Carlo method. Suffice it to say that we can estimate P_{rxn} from the relation

$$P_{rxn} \cong \frac{n_{rxn}}{n} \tag{6-114}$$

provided q_2 is randomly selected.

For a linear system as discussed here, there is no cross section since there is no area in a one-dimensional world. However, the relationship between cross sections and probabilities is useful in the three-dimensional world. In what follows we relax, momentarily, the one-dimensional restriction and proceed to develop this relationship using the following hypothetical experimental considerations. Suppose that a beam of H atoms all moving with a single velocity v pass through a thin target of thickness Δx containing stationary hydrogen molecules whose number density is n_{H_2}. The target thickness is small enough to exclude multiple collisions; hence, each H atom makes either zero or one collision with target molecules. A schematic description is given in Fig. 6-22. The probability that

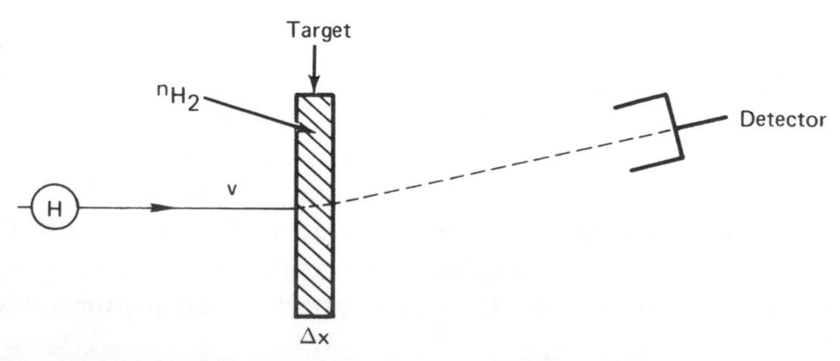

Figure 6-22. Scattering by a thin target.

a reaction will occur is, for each beam atom, proportional to the probability that it will make a collision, the latter probability being proportional to n_{H_2} and Δx. Thus, we write

$$P_{rxn} \propto n_{H_2} \Delta x \tag{6-115}$$

The proportionality constant must have units of square centimeters for the right-hand side of (6-115) to be unitless, as probabilities must be. The equation relating the probability of reaction to the cross section for reaction is thus

$$P_{rxn} = \sigma_{rxn}(v_R) n_{H_2} \Delta x \tag{6-116}$$

In differential form, the reaction cross section is defined as

$$\sigma_{rxn}(v_R) \equiv \frac{1}{n_{H_2}} \left| \frac{dP_{rxn}}{dx} \right| \tag{6-117}$$

This connection between the cross section and probability, while quite intuitive, is of little help in trajectory calculations because there are no parameters corresponding to n_{H_2} and Δx. Another relation may be developed between cross sections and probabilities using the knowledge that σ_{rxn} is expressed as an area. To establish this relationship we shall imagine a series of trajectories in which the impact parameter, b, is varied and for each impact parameter the probability of reaction at a certain velocity is determined. Figure 6-23 illustrates the important variables. For a given b and random orientation of BC we could obtain $P_{rxn}(b, v)$, the probability for reaction with speed v and impact parameter b. Note that the impact parameter is calculated assuming that BC does not interact with A rather than from the actual trajectory of A. Assuming cylindrical symmetry of the probability about an axis parallel to **v** and through the center of mass of BC the cross section for reaction at impact parameter b is just the probability of reaction when the trajectory has an impact parameter between

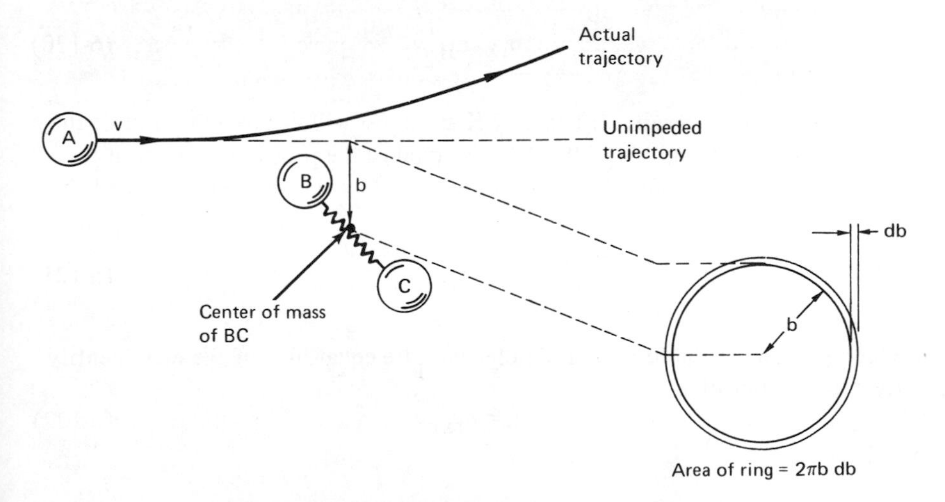

Figure 6-23. Scattering in a three-dimensional system.

b and b + db multiplied by the area of the circular ring shown in Fig. 6-23. The result is

$$\sigma_{rxn}(v, b) = 2\pi b P_{rxn}(v, b) \tag{6-118}$$

The total reaction cross section at velocity v is

$$\sigma_{rxn}(v) = 2\pi \int_0^{b_{max}} b P_{rxn}(v, b)\, db \tag{6-119}$$

where b_{max} is the impact parameter at which $P_{rxn}(v, b)$ has dropped to a negligible value. The actual value of b_{max} can be determined by plotting $P_{rxn}(v, b)$ obtained from the three-dimensional trajectory calculations versus b. Equation (6-119) makes clear the relationship between reaction probabilities and cross sections. The development we have made neglects several important effects such as the vibrational and rotational state of BC, but these can, in principle, be included in the calculation of the cross sections. It should also be borne in mind that our development has been classical. The quantum mechanical cross sections are beyond our scope.

BIMOLECULAR RATE COEFFICIENTS. The cross section is intimately related to the bimolecular rate coefficient as the following discussion suggests. To proceed we return to Eq. (6-115) and note that P_{rxn} is the probability that a single particle passing through dx with velocity v will undergo a reaction. Suppose that we keep the beam particle velocity equal to v and the target particle velocity fixed but imagine an experiment in which the target thickness is increased and a beam, whose cross-sectional area is A and number density is n_H, is passed through it. The probability that one beam particle will react in a unit time is calculated from (6-116)

$$\frac{dP_{rxn}}{dt} = \sigma_{rxn} n_{H_2} v \tag{6-120}$$

since x = vt and dx = v dt. The rate of H atom loss will be given in terms of the number density change, dn_H/dt, which is equal to the probability per unit time that one particle will undergo reaction multiplied by the number density of the beam or

$$-\frac{dn_H}{dt} = \sigma_{rxn} v n_H n_{H_2} \tag{6-121}$$

This equation has the form of a bimolecular rate equation, and we may identify the rate coefficient as

$$k = \sigma_{rxn} v \tag{6-122}$$

If we remove the uncomfortable restrictions of fixed velocities and assume that both H and H_2 move according to thermal velocity distributions, we can show (see Eliason and Hirschfelder in the References) that

$$k(T) = \frac{1}{\sqrt{\pi\mu}} \left(\frac{2}{kT}\right)^{3/2} \int_0^\infty E\sigma_{rxn}(E) \exp\left\{-\frac{E}{kT}\right\} dE \qquad (6\text{-}123)$$

The reaction cross section $\sigma_{rxn}(E)$ is an average over the internal states of the H_2 target. In Eq. (6-123), E is the relative translational energy of H with respect to H_2 and μ is the reduced mass of the colliding species considering only two particles H and H_2 or m_H and m_{H_2}. Then $\mu = (m_H m_{H_2})/(m_H + m_{H_2})$. Equation (6-123) may be regarded as the key relationship between theoretically calculated cross sections and experimentally determined thermal rate coefficients. It should be observed that (6-123) will not be valid if the reactants possess nonthermal energy distributions. The discussion of reaction cross sections and rate coefficients has been quite brief because in this experiment we are interested only in trajectory calculations of probabilities.

EXPERIMENT. The experiment, in this case, is carried out using a computer to furnish the trajectories of systems on the potential-energy surface developed in Eq. (6-122). A FORTRAN IV program suitable for studying individual trajectories is outlined on pages 321–325. The output as it stands is primarily a list of coordinates versus time. These are stored in the arrays T and Q1. Since most computer systems have plotting routines in their library, it should be straightforward to extend this code and furnish graphical output directly. This can be accomplished by calling the plotting subroutine after completing statement 10 of the code shown below.

The units are everywhere cgs and the program incorporates Hamilton's equations of motion for a linear three-particle system in which the masses are identical. In examining the program, give attention to those statements beginning with the letter C. As explained in Section 1.7 these are comment cards which help explain the parameters and methods employed. As the program stands it will compute only a single trajectory and then print the results. Additional trajectories may be easily computed in a single run by placing the trajectory initialization and computation portions in a loop. The instructor will furnish an indication of what parameters to utilize. A typical set is (the role of each parameter is given in the listing of the code)

$$K = 4$$

$$NITN = 5$$

$$H = 0.100 \times 10^{-15} \text{ sec}$$

$$\text{CONV} = 1.0 \times 10^{-4}$$

$$\text{ACONV} = 0.0$$

$$\text{C3(2)} = 3.0 \times 10^{-8} \text{ cm}$$

$$\text{EBC} = 0.5 \times 10^{-12} \text{ ergs}$$

$$\text{VREL} = -1.0 \times 10^{+6} \text{ cm/sec}$$

$$\text{D} = 7.6015 \times 10^{-12} \text{ ergs}$$

$$\text{RO} = 0.7416 \times 10^{-8} \text{ cm}$$

$$\text{A} = 1.9320 \times 10^{8} \text{ cm}^{-1}$$

$$\text{AVA} = 6.023 \times 10^{23} \text{ molecules/mole}$$

$$\text{NSTP2} = 5$$

$$\text{ALAM} = 0.20$$

$$\text{AM} = 0.1660 \times 10^{-23}$$

$$\text{RMAX} = 3.5 \times 10^{-8} \text{ cm}$$

Notice that VREL is given with a minus sign. This is necessary, because of the way in which the equations of motion were developed, if A is to move toward BC initially.

Several trajectory studies are possible through the variation of one parameter at a time; the most revealing studies will be those in which EBC, VREL, and C3(2) are varied. To obtain statistically meaningful results for the reaction probability as a function of VREL and EBC, the initial values of C3(2) must be randomly varied using more complicated programming than is desirable here.

After completing a set of trajectories, examine the printed output column ETOT, which is the total energy, and comment on how well the program preserves it. Examine the potential energy, POT, noting how it appears to fluctuate. Account for the observed fluctuations in terms of variations in the interparticle distances, RAB and RBC. From your trajectories discuss the following:

1. The influence of EBC on the reactive or nonreactive nature of the collision.
2. The extent of vibrational energy change in nonreactive collisions.
3. The adiabaticity or lack of it in the vibrational energy of a reactive encounter. A process is said to be vibrationally adiabatic if the vibrational quantum numbers do not change during the course of a reaction.
4. The influence of VREL and C3(2) on the reactivity of various collisions.

```
TRAJRY,1,100,52000,100.CMDJ0028,WHITE.
RUN(S)
MAP.
LGO.
'
      PROGRAM TRAJRY(INPUT, OUTPUT)
      DIMENSION CC(6), CNP(6), Y(6)
      DIMENSION RK(6), C(6), CDT(6)
      DIMENSION YDT(6), C1(6), C2(6), C3(6), C1DT(6), C2DT(6), C3DT(6),
     1CNTRL(6), CP(6)
      DIMENSION POT1(1000), ETOT1(1000)
      DIMENSION T(999,2), Q1(999,2), NDATA(2), XLINE(20), YLINE(20), ARE
     1A(13,301), ISYMBL(2), YSCALE(301)
      COMMON /DEQ/ CC, CNP, Y, RK, C, CDT, YDT, C1, C2, C3, C1DT, C2DT,
     1C3DT,3CNTRL,SCP
      COMMON /PARM/ RBC, RAB, RAC, RO, A, ALAM, AM, D
      COMMON /TST/ CONV, NITN, K, H, TIM
      COMMON /EXT/ XAB, XBC, XAC, X2AB, X2BC, X2AC, POT, ETOT
      READ 14, K
C  K IS THE NUMBER OF DIFFERENTIAL EQUATIONS.
      READ 14, NITN
C  NITN IS THE MAXIMUM NUMBER OF ITERATIONS IN SUBROUTINE PRECOR.
      READ 12, H
C  H IS THE STEP SIZE FOR THE INTEGRATION.
      READ 12, CONV, ACONV
C  CONV  IS THE CONVERGENCE CRITERION FOR PRECOR.
C  ACONV IS A FREE PARAMETER.
      READ 12, C3(2)
C  C3(2) IS THE INITIAL DISTANCE FROM A TO THE CENTER OF MASS OF BC.
      READ 20, EBC, VREL
C  EBC IS THE INITIAL TOTAL ENERGY IN BC.
C  VREL IS THE INITIAL RELATIVE VELOCITY OF A WITH RESPECT TO THE
C   CENTER OF MASS OF BC.
      READ 12, D, RO, A, AVA
C  D IS THE DISSOCIATION ENERGY OF THE MORSE FUNCTION FOR H2.
C  RO IS THE EQUILIBRIUM INTERNUCLEAR DISTANCE OF H2.
C  A IS THE CURVATURE PARAMETER FOR THE MORSE EQUATION FOR H2.
C  AVA IS AVOGADROS NUMBER.
      READ 14, NSTP2
C  NSTP2 IS THE CRITERION FOR DETERMINING THE NUMBER OF INTERVALS = H
C   BETWEEN POINTS PLOTTED OR PRINTED.
      READ 12, ALAM, AM
C  ALAM IS THE LAMBDA PARAMETER FOR THE POTENTIAL ENERGY FUNCTION.
C  AM IS THE MASS OF AN HYDROGEN ATOM IN GRAMS.
      READ 12, RMAX
C  RMAX IS THE END TEST CRITERION FOR CEASING INTEGRATION.
      KRNG = 1
      JOUT = 1
      RBC = (A*RO-ALOG(1.-SQRT(EBC/D)))/A
      TIM = T(1, 1) = T(1, 2) = 0.0
      C3(1) = RBC
      Q1(1, 1) = RBC
      Q1(1, 2) = C3(2)-0.5*RBC
      C3(3) = 0.0
      C3(4) = 0.6667*VREL/AVA
      PRINT 15
```

```
      PRINT 19
      PRINT 13, EBC, VREL, Q1(1, 2)
 1    CONTINUE
      DO 2 I = 1, K
      C(I) = C3(I)
 2    CONTINUE
      CALL RNGKUT
C  RNGKUT IS A FOURTH ORDER RUNGE-KUTTA INTEGRATION SUBROUTINE.
      DO 3 I = 1, K
      C2(I) = C(I)
      C3DT(I) = CDT(I)
 3    CONTINUE
      CALL RNGKUT
      DO 4 I = 1, K
      C1(I) = C(I)
      C2DT(I) = CDT(I)
 4    CONTINUE
      CALL RNGKUT
      DO 5 I = 1, K
      C1DT(I) = CDT(I)
      CP(I) = C(I)
 5    CONTINUE
 6    KRNG = KRNG+1
      CALL PRECOR
C  TRANSFER TO A FOURTH ORDER PREDICTOR-CORRECTOR INTEGRATION.
      DO 7 J = 1, K
C  TEST FOR CONVERGENCE AND IF NOT MAKE STEP SIZE SMALLER.
      IF ((CNTRL(J)) .EQ. 1.) GO TO 7
      TIM = TIM-4.*H
      H = H-H/5.
      PRINT 16
      GO TO 1
 7    CONTINUE
C  TEST FOR PRINTOUT WITH NSTP2. EVERY NSTP2ND. POINT IS PRINTED AND
C  PLOTTED.
      IF (((KRNG/NSTP2)*NSTP2) .EQ. KRNG) GO TO 8
      IF (RBC .GT. RMAX) GO TO 8
      IF (RAB .GT. RMAX) GO TO 8
      GO TO 6
 8    JOUT = JOUT+1
C  RAB AND RBC ARE PLOTTED VS. TIME
      Q1(JOUT, 1) = C(1)
      Q1(JOUT, 2) = C(2)-0.5*C(1)
      T(JOUT, 2) = TIM
      T(JOUT, 1) = TIM
      ETOT1(JOUT) = ETOT
      POT1(JOUT) = POT
      IF (RBC .GT. RMAX) GO TO 9
      IF (RAB .GT. RMAX) GO TO 9
      GO TO 6
 9    CONTINUE
C  PRINT OUT TRAJACTORIES.
      PRINT 17
      DO 10 J = 2, JOUT
      PRINT 18, T(J, 1), Q1(J, 1), Q1(J, 2), POT1(J), ETOT1(J)
 10   CONTINUE
```

```
      SUBROUTINE PRECOR
C  FOURTH ORDER PREDICTOR-CORRECTOR ROUTINE FOR K DIFFERENTIAL EQUNS.
      DIMENSION CC(6), CNP(6), Y(6)
      DIMENSION RK(6), C(6), CDT(6)
      DIMENSION YDT(6), C1(6), C2(6), C3(6), C1DT(6), C2DT(6), C3DT(6),
     1CNTRL(6), CP(6)
      COMMON /PARM/ RBC, RAB, RAC, RO, A, ALAM, AM, D
      COMMON /DEQ/ CC, CNP, Y, RK, C, CDT, YDT, C1, C2, C3, C1DT, C2DT,
     1C3DT,(CNTRL,(CP
      COMMON /TST/ CONV, NITN, K, H, TIM
      COMMON /EXT/ XAB, XBC, XAC, X2AB, X2BC, X2AC, POT, ETOT
      CALL DFEQS
      DO 1 J = 1, K
      CNP(J) = C3(J)+(4.*H/3.)*(2.*CDT(J)-C1DT(J)+2.*C2DT(J))
      Y(J) = C(J)
      C(J) = CNP(J)+(112./121.)*(Y(J)-CP(J))
C  CP(J) IS THE PREDICTED VALUE AT THE PREVIOUS POINT.
C  Y(J) IS TYHE VALUE OF C(J) AT THE PREVIOUS POINT. IT IS CONVERTED Y
C  IN ORDER THAT THE DFEQS SUBROUTINE MAY BE USED IN THE ITERATIVE LOOP
C  OF PRECOR( THE SUBROUTINE REQUIRES C(J) FROM THE PREVIOUS CYCLE OF
C  THE LOOP.
      YDT(J) = CDT(J)
      CP(J) = CNP(J)
      CNTRL(J) = 0.0
 1    CONTINUE
      TIM = TIM+H
      DO 6 I = 1, NITN
      CALL DFEQS
      DO 4 J = 1, K
      CNP(J) = C(J)
      C(J) = (1./8.)*(9.*Y(J)-C2(J))+(3.*H/8.)*(CDT(J)+2.*YDT(J)-C1DT(J)
     1)
      IF (C(J) .EQ. 0.0) CNTRL(J) = 1.
      IF (C(J) .EQ. 0.0) GO TO 4
      IF (ABS((C(J)-CNP(J))/C(J)) .LE. CONV) GO TO 2
      CNTRL(J) = 0.0
      GO TO 3
 2    CNTRL(J) = 1.0
 3    CONTINUE
 4    CONTINUE
      DO 5 J = 1, K
      IF ((CNTRL(J)) .EQ. 1.) GO TO 5
      GO TO 6
 5    CONTINUE
      GO TO 7
 6    CONTINUE
      RETURN
C NEED A TEST IN MAIN PROGRAM FOR TERMINATION.
 7    DO 8 J = 1, K
      C3(J) = C2(J)
      C2(J) = C1(J)
      C1(J) = Y(J)
      C3DT(J) = C2DT(J)
      C2DT(J) = C1DT(J)
      C1DT(J) = YDT(J)
 8    CONTINUE
C SETS VALUES FOR NEXT LOOP.
      RETURN
      END
```

```
      SUBROUTINE RNGKUT
C   FOURTH ORDER RUNGE-KUTTA ROUTINE FOR K DIFFERENTIAL EQUNS.
      DIMENSION CC(6), CNP(6), Y(6)
      DIMENSION RK(6), C(6), CDT(6)
      DIMENSION YDT(6), C1(6), C2(6), C3(6), C1DT(6), C2DT(6), C3DT(6),
     1CNTRL(6), CP(6)
      DIMENSION A1(6), A2(6), A3(6), A4(6)
      COMMON /DEQ/ CC, CNP, Y, RK, C, CDT, YDT, C1, C2, C3, C1DT, C2DT,
     1C3DT,(CNTRL,(CP
      COMMON /PARM/ RBC, RAB, RAC, R0, A, ALAM, AM, D
      COMMON /TST/ CONV, NITN, K, H, TIM
      COMMON /EXT/ XAB, XBC, XAC, X2AB, X2BC, X2AC, POT, ETOT
      CALL DFEQS
      TIM1 = TIM
      DO 1 I = 1, K
      CC(I) = C(I)
      A1(I) = H*CDT(I)
      C(I) = CC(I)+0.4*H*CDT(I)
 1    CONTINUE
      TIM = TIM1+0.4*H
      CALL DFEQS
      DO 2 I = 1, K
      A2(I) = H*CDT(I)
      C(I) = CC(I)+0.29697760*A1(I)+0.15875966*A2(I)
 2    CONTINUE
      TIM = TIM1+0.45573726*H
      CALL DFEQS
      DO 3 I = 1, K
      A3(I) = H*CDT(I)
      C(I) = CC(I)+0.21810038*A1(I)-3.05096470*A2(I)+3.83286432*A3(I)
 3    CONTINUE
      TIM = TIM1+H
      CALL DFEQS
      DO 4 I = 1, K
      A4(I) = H*CDT(I)
      C(I) = CC(I)+0.17476028*A1(I)-0.55148053*A2(I)+1.20553547*A3(I)+0.
     117118478*A4(I)
 4    CONTINUE
      RETURN
      END


C
C
 12   FORMAT  (8E10.4)
 13   FORMAT (10X,*INTERNAL BC ENERGY =*,E10.3,*  RELATIVE VELOCITY =*,F
     110.3,*  AB DISTANCE =*,E10.3)
 14   FORMAT  (8I10)
 15   FORMAT  (1H1)
 16   FORMAT  (5X,* NO CONVERGENCE *)
 17   FORMAT  (5/,10X,*   TIME     *,10X,*   RBC    *,10X,*   RAB     *,1
     110X,*  POT    *,10X,*  ETOT    *)
 18   FORMAT  (10X,E10.3,10X,E10.3,10X,E10.3,2(10X,E10.3))
 19   FORMAT (50X,*STARTING CONDITIONS*,2/)
 20   FORMAT  (4E15.5)
      END
```

```
      SUBROUTINE DFEQS
      DIMENSION CC(6), CNP(6), Y(6)
      DIMENSION RK(6), C(6), CDT(6)
      DIMENSION YDT(6), C1(6), C2(6), C3(6), C1DT(6), C2DT(6), C3DT(6),
     1CNTRL(6), CP(6)
      COMMON /DEQ/ CC, CNP, Y, RK, C, CDT, YDT, C1, C2, C3, C1DT, C2DT,
     1C3DT,(CNTRL,(CP
      COMMON /PARM/ RBC, RAB, RAC, RO, A, ALAM, AM, D
      COMMON /TST/ CONV, NITN, K, H, TIM
      COMMON /EXT/ XAB, XBC, XAC, X2AB, X2BC, X2AC, POT, ETOT
      RAB = C(2)-0.5*C(1)
      RAC = C(2)+0.5*C(1)
      RBC = C(1)
C    TO BE USED IN SUBROUTINE DFEQS
      XAB = EXP(-A*(RAB-RO))
      XBC = EXP(-A*(RBC-RO))
      XAC = EXP(-A*(RAC-RO))
      X2AB = EXP(-2.*A*(RAB-RO))
      X2BC = EXP(-2.*A*(RBC-RO))
      X2AC = EXP(-2.*A*(RAC-RO))
      DA2 = D*A*2.
C THEN GET PARTS OF EQUNS. IN DFEQS
      GAB = DA2*(XAB-X2AB)
      GBC = DA2*(XBC-X2BC)
      GAC = DA2*(XAC-X2AC)
      VAB = D*(X2AB-2.*XAB)
      VBC = D*(X2BC-2.*XBC)
      VAC = D*(X2AC-2.*XAC)
      F = SQRT(VAB*(VAB-VBC-VAC)+VBC*(VBC-VAC)+VAC*VAC)
      POT = ALAM*(VAB+VBC+VAC)-(1.-ALAM)*F
      ETOT = 0.5*(2.*C(3)*C(3)+1.5*C(4)*C(4))/AM+POT
C  HAMILTONS EQUATIONS OF MOTION.
      CDT(1) = 2.0*C(3)/AM
      CDT(2) = 1.5*C(4)/AM
      CDT(3) = -ALAM*GBC+0.5*(1.-ALAM)*(2.*VBC-VAB-VAC)*GBC/F+0.5*ALAM*(
     1GAB-GAC)-0.25*(1.-ALAM)*((2.*VAB-VBC-VAC)*GAB-(2.*VAC-VAB-VBC)*GAC
     2)/F
      CDT(4) = -ALAM*(GAB+GAC)+0.5*(1.-ALAM)*((2.*VAB-VBC-VAC)*GAB+(2.*V
     1AC-VAB-VBC)*GAC)/F
C
      RETURN
C
 1    FORMAT  (8(3X,E12.4))
      END
```

DATA DECK

```
            4
            5
0.1000E-15
1.0000E-041.0000E-04
4.0000E-08
    2.00000E-12    -1.00000E+06
7.6015E-120.7416E-081.9320E+086.0230E+23
            15
0.2000E+000.1660E-23
4.0000E-08
```

PROBLEMS

Several of these problems assume familiarity with the Appendix which follows.

1. Consider two hydrogen nuclei A and B each having one electron in a 1s-orbital. Prove that it is possible to treat, for this system, valence bond wave functions of different total spin separately if the Hamiltonian is spin-free.

2. For the system outlined in Problem 1, write the Slater determinant corresponding to

$$A(1)\,\alpha(1)\,B(2)\,\beta(2)$$

Show for this case that the Slater determinant is antisymmetric under the operation of exchanging electrons and that indistinguishability is guaranteed.

3. Derive Eq. (A6-9) by applying the variation principle to the equation

$$W = \frac{\langle \psi_{VB} | \mathcal{H} | \psi_{VB} \rangle}{\langle \psi_{VB} | \psi_{VB} \rangle}$$

4. Show that Eq. (A6-13) arises from the expansion of Eq. (A6-3).

5. Show that $\langle \psi_2 | \psi_2 \rangle = 1 - \langle B|C \rangle^2$. Refer to Eq. (A6-14).

6. Develop Eq. (A6-17).

7. Look up in G. Herzberg, *Spectra of Diatomic Molecules* (Van Nostrand, New York, 1950) or some other suitable source, the necessary parameters for the Morse equation for H_2.

8. What is the difference in the forms of the Morse Eq. (6-101) and that used in establishing the initial conditions for the trajectory?

9. Using the initial conditions outlined in the text and a potential energy diagram, describe the initial configuration of the BC molecule. What does the term *classical turning point* imply?

Appendix to Trajectory Calculations in Chemical Kinetics

For its full development the foregoing experiment requires the construction of a potential-energy surface for three hydrogen atoms. This is a quantum mechanical problem and a block diagram of the procedures is given in Fig. A6-1. Assuming that only one electron per nucleus participates in bonding, the valence bond description of the system ABC at any internuclear configuration makes use

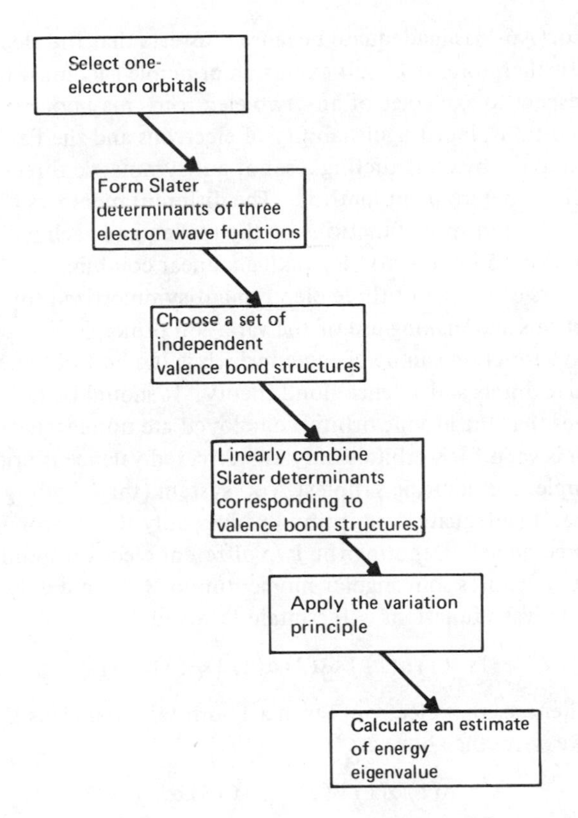

Figure A6-1. Valence bond approach to the electronic Schrödinger equation for three electrons and three nuclei.

of one-electron atomic orbitals, which are solutions (or approximate solutions) to a one-electron Schrödinger equation for each of the separated atoms. If the system consists of three noninteracting atoms, the total wave function will be the product of three one-electron atomic orbitals. If there are interactions between the atoms, this product will be only an approximate solution. For example, assume that the bonding properties of the ABC system are determined by three electrons, one from each nucleus. Then the valence bond description of the molecular system will involve approximate three-electron wave functions of the form

$$\Psi'_e = \eta_A(1)\eta_B(2)\eta_C(3) \tag{A6-1}$$

where the η_is are one-electron atomic orbitals and include both space and spin parts. The numbers in Eq. (A6-1) denote electrons and the subscripts denote nuclei.

As it stands Eq. (A6-1) is inadequate because it asserts that the electrons are distinguishable; furthermore, the Pauli exclusion principle (Ψ_e must be antisymmetric with respect to exchange of any two electrons) may not be satisfied. These two conditions, indistinguishability of electrons and the Pauli principle, can be accounted for by constructing a set of antisymmetric three-electron functions by the Slater determinant method. The different members of this set may have different spin and space functions on the nuclei, and each will be an approximate solution to Eq. (6-96). By taking a linear combination of the independent members of the set of three-electron antisymmetrized (by Slater determinants) functions and making use of the variation principle, an approximation to the true wave function can be obtained which is the best obtainable from the starting atomic orbitals and valence bond theory. It should be noted that what follows assumes that the atomic orbitals employed are nondegenerate; that is, this formalism is valid for s-orbitals only and directed valence is not considered.

As an example, consider the simplest ABC system (three hydrogen atoms) and assume that an adequate description involves only the 1s atomic orbitals on each of the three nuclei. Denoting the two different electron spin functions by α and β, where α denotes spin angular momentum of $+\frac{1}{2}\hbar$ and β denotes $-\frac{1}{2}\hbar$, the approximate wave functions will contain terms such as

$$1s_A(1)\,\alpha(1)\,1s_B(2)\,\alpha(2)\,1s_C(3)\,\alpha(3)$$

where $1s_B(2)$ denotes that electron 2 is in a 1s-orbital on nucleus B. To shorten the notation we write the above as

$$A(1)\,\alpha(1)\,B(2)\,\alpha(2)\,C(3)\,\alpha(3)$$

Note that the total spin angular momentum of this configuration is $\frac{3}{2}\hbar$ since the spin of each electron is $+\frac{1}{2}$. For the example we are considering, the space parts all have the same functional form (e.g., 1s) so that the only difference from term to term in the Slater determinants will be in the total spin and in the electron ordering. For example, another term will be

$$A(2)\,\beta(2)\,B(1)\,\alpha(1)\,C(3)\,\alpha(3)$$

In this case the total spin angular momentum is $+\frac{1}{2}\hbar$, and electrons are ordered differently than above. Now consider all the possible total spin arrangements as outlined in Table A6-1 for a three-electron system.

At this point in valence bond theory use is made of what is known about the reactants and products; in the three H atom case both the reactants and the products are H and H_2. This suggests that Ψ'_7 and Ψ'_8 are unimportant because in these functions all the electrons are unpaired (e.g., no bonding). Furthermore, because of the form taken by the total secular equation for a spin-free Hamiltonian, it is possible to treat different total spin functions separately prior to taking the linear combination which furnishes the total solution. Table A6-1 then suggests combining Ψ'_1, Ψ'_2, and Ψ'_5 or Ψ'_3, Ψ'_4, and Ψ'_6 (after they are

Table A6-1.

Function (arbitrarily assigned)	Spin of electron on nucleus A	Spin of electron on nucleus B	Spin of electron on nucleus C	Total spin angular momentum
Ψ'_1	α	β	α	$\frac{1}{2}\hbar$
Ψ'_2	β	α	α	$\frac{1}{2}\hbar$
Ψ'_3	β	α	β	$-\frac{1}{2}\hbar$
Ψ'_4	α	β	β	$-\frac{1}{2}\hbar$
Ψ'_5	α	α	β	$+\frac{1}{2}\hbar$
Ψ'_6	β	β	α	$-\frac{1}{2}\hbar$
Ψ'_7	α	α	α	$+\frac{3}{2}\hbar$
Ψ'_8	β	β	β	$-\frac{3}{2}\hbar$

antisymmetrized). Taking Ψ'_1, Ψ'_2, and Ψ'_5, note that the reactants $H_A + H_B - H_C$ could be described by Ψ'_1 and Ψ'_5, whereas the products $H_A - H_B + H_C$ could be described by Ψ'_1 and Ψ'_2. A third possible structure, $H_A - H_C + H_B$, would be described by Ψ'_2 and Ψ'_5; however, it can be represented mathematically as a linear combination of the other two and is therefore not independent of them. In any valence bond calculation it is necessary to consider only a linear combination of independent structures; for the present case we choose $H_A + H_B - H_C$ and $H_A - H_B + H_C$. The approximate wave functions corresponding to these two structures are themselves linear combinations of antisymmetrized three-electron functions. Denoting by Θ'_1 the wave function corresponding to $H_A + H_B - H_C$ and by Θ'_2 the structure $H_A - H_B + H_C$ we have

$$\begin{aligned} \Theta'_1 &= C_1\Psi_1 + C_2\Psi_5 \\ \Theta'_2 &= C_3\Psi_1 + C_4\Psi_2 \end{aligned} \tag{A6-2}$$

where Ψ_1 is the antisymmetrized form of Ψ'_1 and is given by

$$\Psi_1 = \frac{1}{\sqrt{3!}} \begin{vmatrix} A(1)\,\alpha(1) & A(2)\,\alpha(2) & A(3)\,\alpha(3) \\ B(1)\,\beta(1) & B(2)\,\beta(2) & B(3)\,\beta(3) \\ C(1)\,\alpha(1) & C(2)\,\alpha(2) & C(3)\,\alpha(3) \end{vmatrix} \tag{A6-3}$$

The determinant representing Ψ_1 is called the Slater determinant and guarantees that Ψ_1 is antisymmetric and that the electrons are indistinguishable. Similar expressions can be written for Ψ_2 and Ψ_5. The substitution of these into (A6-2) furnishes Θ'_1 and Θ'_2, which are themselves then linearly combined to furnish an approximate overall description of the H_3 system for any internuclear configuration in which all three atoms are not separated from one another by infinite distance. When the expectation value of the Hamiltonian operator is minimized, the resulting energy eigenvalues furnish the "best" estimate of E_{el} which can be obtained using this set of functions. If we chose the set $\{\Psi_3, \Psi_4, \Psi_6\}$ the same result would obtain because the spatial parts are identical and the spin parts degenerate with $\{\Psi_1, \Psi_2, \Psi_5\}$. If we write an electron exchange operator as P_{12}, then

$$P_{12}\Theta'_2 = -\Theta'_2 \tag{A6-4}$$

according to the Pauli principle. We have already defined Θ'_2 in Eq. (A6-2), and studying how P_{12} effects the right-hand side of the equation furnishes

$$P_{12}(C_3\Psi_1 + C_4\Psi_2) = C_3P_{12}\Psi_1 + C_4P_{12}\Psi_2 = C_3\Psi_2 + C_4\Psi_1 \tag{A6-5}$$

where the last equality arises by noting that, according to Table A6-1, electron exchange in Ψ_1 just changes spins on A and B; the spin of C does not change (that is, electron 3) and the orbital (space) parts are all identical 1s functions. Thus, we have

$$-\Theta'_2 = C_3\Psi_2 + C_4\Psi_1 \quad \text{or} \quad \Theta'_2 = -C_3\Psi_2 - C_4\Psi_1 \tag{A6-6}$$

Hence, $C_3 = -C_4$. Similarly from Eq. (A6-3), $C_1 = -C_2$, and the following arises:

$$\begin{aligned} \Theta'_1 &= C_1(\Psi_1 - \Psi_5) \\ \Theta'_2 &= C_3(\Psi_1 - \Psi_2) \end{aligned} \tag{A6-7}$$

Linearly combining Θ'_1 and Θ'_2 furnishes the total valence bond wave function:

$$\Psi_{VB} = K_1\Theta_1 + K_2\Theta_2 \tag{A6-8}$$

where $\Theta_1 = \Theta'_1/C_3$ and $\Theta_2 = \Theta'_2/C_1$. The valence bond wave function (A6-8) need not be an eigenfunction of any particular Hamiltonian; it must only be well behaved (finite, single-valued, and continuous). Employing the variation principle, K_1 and K_2 can be chosen so that $\langle\Psi_{VB}|\mathcal{H}|\Psi_{VB}\rangle/\langle\Psi_{VB}|\Psi_{VB}\rangle \equiv W$ is minimized.* Minimization is accomplished by differentiating W with respect to K_1 and K_2 and setting the result equal to zero to provide two linear equations in which the unknowns are assumed to be K_1 and K_2. The solution of this pair of equations gives rise to the secular determinant which must vanish for a nontrivial result according to the theory of linear equations. As shown below the potential-energy surface arises from solution of the secular determinant, and since it involves neither K_1 nor K_2, we never find it necessary to evaluate them directly in determining the potential-energy surface.

The secular determinant arising from the application of the variation principle and (A6-8) has the form

$$\begin{vmatrix} \langle\Theta_1|\mathcal{H}|\Theta_1\rangle - E\langle\Theta_1|\Theta_1\rangle & \langle\Theta_1|\mathcal{H}|\Theta_2\rangle - E\langle\Theta_1|\Theta_2\rangle \\ \langle\Theta_1|\mathcal{H}|\Theta_2\rangle - E\langle\Theta_1|\Theta_2\rangle & \langle\Theta_2|\mathcal{H}|\Theta_2\rangle - E\langle\Theta_2|\Theta_2\rangle \end{vmatrix} = 0 \tag{A6-9}$$

where the bracket notation signifies integration over both space and spin parts.

*The notation $\langle\Psi_{VB}|\mathcal{H}|\Psi_{VB}\rangle$ is typical bracket notation and in the Schrödinger formulation of quantum mechanics symbolizes the operation $\int\Psi^*_{VB}\mathcal{H}\Psi_{VB}\,d\tau$ where integration is over all space and spin coordinates.

Expanding gives a quadratic equation in E whose roots are

$$E_{\pm} = \frac{1}{\gamma_1}[-\gamma_2 \pm (\gamma_2^2 - \gamma_1\gamma_3)^{1/2}] \tag{A6-10}$$

where

$$\gamma_1 \equiv \langle\Theta_1|\Theta_1\rangle\langle\Theta_2|\Theta_2\rangle - \langle\Theta_1|\Theta_2\rangle^2$$

$$\gamma_2 \equiv \langle\Theta_1|\mathcal{H}|\Theta_2\rangle\langle\Theta_1|\Theta_2\rangle - \frac{1}{2}\langle\Theta_1|\mathcal{H}|\Theta_1\rangle\langle\Theta_2|\Theta_2\rangle - \frac{1}{2}\langle\Theta_2|\mathcal{H}|\Theta_2\rangle\langle\Theta_1|\Theta_1\rangle$$

$$\gamma_3 \equiv \langle\Theta_1|\mathcal{H}|\Theta_1\rangle\langle\Theta_2|\mathcal{H}|\Theta_2\rangle - \langle\Theta_1|\mathcal{H}|\Theta_2\rangle^2$$

Obviously, the roots of Eq. (6-105) can be found only if the integrals involved in the definitions of γ_1, γ_2, and γ_3 are evaluated. While these are quite complicated it is worthwhile to consider the various types of terms. To do so we introduce a shorthand notation for the space part of the three-electron function functions:

$$ABC \equiv A(1)B(2)C(3)$$

$$BAC \equiv B(1)A(2)C(3)$$

$$\vdots$$

Also note that the two spin functions α and β are orthonormal.

To evaluate γ_1, the normalization integrals $\langle\Theta_1|\Theta_1\rangle$ and $\langle\Theta_2|\Theta_2\rangle$ and the overlap integral $\langle\Theta_1|\Theta_2\rangle$ are needed:

$$\begin{aligned} \langle\Theta_1|\Theta_1\rangle &= \langle(\Psi_1 - \Psi_5)|(\Psi_1 - \Psi_5)\rangle \\ &= \langle\Psi_1|\Psi_1\rangle + \langle\Psi_5|\Psi_5\rangle - 2\langle\Psi_1|\Psi_5\rangle \\ \langle\Theta_2|\Theta_2\rangle &= \langle\Psi_1|\Psi_1\rangle + \langle\Psi_2|\Psi_2\rangle - 2\langle\Psi_1|\Psi_2\rangle \end{aligned} \tag{A6-11}$$

$$\langle\Theta_1|\Theta_2\rangle = \langle\Psi_1|\Psi_1\rangle + \langle\Psi_5|\Psi_2\rangle - \langle\Psi_5|\Psi_1\rangle - \langle\Psi_1|\Psi_2\rangle \tag{A6-12}$$

The antisymmetrized wave functions required for the right-hand side of these equations are of the form of Eq. (A6-3). Expanding the determinant which represents Ψ_1 furnishes, in the shorthand notations,

$$\Psi_1 = \frac{1}{\sqrt{3!}}[(\alpha\beta\alpha)(ABC - CBA) + (\alpha\alpha\beta)(CAB - ACB) + (\beta\alpha\alpha)(BCA - BAC)] \tag{A6-13}$$

Since $\langle\alpha\beta\alpha|\alpha\alpha\beta\rangle = \langle\alpha\beta\alpha|\beta\alpha\alpha\rangle = 0$ and $\langle\alpha\beta\alpha|\alpha\beta\alpha\rangle = \langle\alpha\alpha\beta|\alpha\alpha\beta\rangle = \langle\beta\alpha\alpha|\beta\alpha\alpha\rangle = 1$, the term $\langle\Psi_1|\Psi_1\rangle$ reduces to

$$\begin{aligned} \langle\Psi_1|\Psi_1\rangle = \frac{1}{6}[&\langle(ABC - CBA)|(ABC - CBA)\rangle + \langle(CAB - ACB)|(CAB - ACB)\rangle \\ &+ \langle(BCA - BAC)|(BCA - BAC)\rangle] \end{aligned}$$

Assuming that $\langle A|A\rangle = \langle B|B\rangle = \langle C|C\rangle = 1$ and noting that $\langle CBA|ABC\rangle = \langle C|A\rangle\langle B|B\rangle\langle A|C\rangle$, etc. (see below), this equation becomes

$$\langle\Psi_1|\Psi_1\rangle = 1 - \langle A|C\rangle^2$$

Similarly,

$$\begin{aligned}
\langle\Psi_2|\Psi_2\rangle &= 1 - \langle B|C\rangle^2 \\
\langle\Psi_5|\Psi_5\rangle &= 1 - \langle A|B\rangle^2 \\
\langle\Psi_1|\Psi_2\rangle &= \langle A|C\rangle\langle A|B\rangle\langle B|C\rangle - \langle A|B\rangle^2 \\
\langle\Psi_2|\Psi_5\rangle &= \langle A|C\rangle\langle A|B\rangle\langle B|C\rangle - \langle A|C\rangle^2 \\
\langle\Psi_1|\Psi_5\rangle &= \langle A|C\rangle\langle A|B\rangle\langle B|C\rangle - \langle B|C\rangle^2
\end{aligned} \tag{A6-14}$$

Note that terms like $\langle CBA|ABC\rangle$ are integrals over the space of three different electrons, whereas those like $\langle A|C\rangle$ are over the space of a single electron, e.g.,

$$\begin{aligned}
\langle CBA|ABC\rangle &= \iiint C^*(1)B^*(2)A^*(3)A(1)B(2)C(3)\, d\tau_1\, d\tau_2\, d\tau_3 \\
&= \left(\int C^*(1)A(1)\, d\tau_1\right)\left(\int B^*(2)B(2)\, d\tau_2\right)\left(\int A^*(3)C(3)\, d\tau_3\right) \\
&= \langle C|A\rangle\langle B|B\rangle\langle A|C\rangle \\
&= \langle A|C\rangle^2
\end{aligned} \tag{A6-15}$$

With (A6-14), Eqs. (A6-11) and (A6-12) can be evaluated for use in the secular equation.

The terms of the secular equation (A6-9) which involve $\mathcal{H}$ are quite complex because the Hamiltonian includes several types of kinetic- and potential-energy terms. The appropriate $\mathcal{H}$ to use in (A6-9) is the fixed-nuclei Hamiltonian.

Considering one of the terms $\langle\Theta_1|\mathcal{H}|\Theta_1\rangle$ we have

$$\langle\Theta_1|\mathcal{H}|\Theta_1\rangle = \langle\Psi_1|\mathcal{H}|\Psi_1\rangle + \langle\Psi_5|\mathcal{H}|\Psi_5\rangle - 2\langle\Psi_1|\mathcal{H}|\Psi_5\rangle \tag{A6-16}$$

Expanding the determinant representing Ψ_1, the first term on the right-hand side of (A6-16) becomes

$$\begin{aligned}
\langle\Psi_1|\mathcal{H}|\Psi_1\rangle = \frac{1}{6}[&\langle(ABC - CBA)|\mathcal{H}|(ABC - CBA)\rangle \\
&+ \langle(CAB - ACB)|\mathcal{H}|(CAB - ACB)\rangle \\
&+ \langle(BCA - BAC)|\mathcal{H}|(BCA - BAC)\rangle]
\end{aligned}$$

Note that in the $\mathcal{H}$ we are using each nucleus is identical and each electron is identical so that $\langle ABC|\mathcal{H}|ABC\rangle = \langle CBA|\mathcal{H}|CBA\rangle$, etc. For the same reason

$\langle ABC|\mathcal{H}|CBA\rangle = \langle CAB|\mathcal{H}|ACB\rangle = \langle BCA|\mathcal{H}|BAC\rangle$. With this the above element reduces to

$$\langle\Psi_1|\mathcal{H}|\Psi_1\rangle = \langle ABC|\mathcal{H}|ABC\rangle - \langle ABC|\mathcal{H}|CBA\rangle \qquad \text{(A6-17)}$$

The first term is called a Coulomb integral and the second an exchange integral (electrons on nuclei A and C are exchanged). Similarly,

$$\begin{aligned}
\langle\Psi_2|\mathcal{H}|\Psi_2\rangle &= \langle ABC|\mathcal{H}|ABC\rangle - \langle ABC|\mathcal{H}|ACB\rangle \\
\langle\Psi_5|\mathcal{H}|\Psi_5\rangle &= \langle ABC|\mathcal{H}|ABC\rangle - \langle ABC|\mathcal{H}|BAC\rangle \\
\langle\Psi_1|\mathcal{H}|\Psi_2\rangle &= \langle ABC|\mathcal{H}|CAB\rangle - \langle ABC|\mathcal{H}|BAC\rangle \qquad \text{(A6-18)} \\
\langle\Psi_1|\mathcal{H}|\Psi_5\rangle &= \langle ABC|\mathcal{H}|CAB\rangle - \langle ABC|\mathcal{H}|ACB\rangle \\
\langle\Psi_2|\mathcal{H}|\Psi_5\rangle &= \langle ABC|\mathcal{H}|CAB\rangle - \langle ABC|\mathcal{H}|CBA\rangle
\end{aligned}$$

With the information at hand we can now evaluate the γs of Eq. (A6-10) in terms of integrals involving the basis orbitals. The components of γ_1, γ_2, and γ_3 are

$$\begin{aligned}
\langle\Theta_1|\Theta_2\rangle &= 4 + 4\langle A|B\rangle^2 - 2\langle B|C\rangle^2 - 2\langle A|C\rangle^2 - 4\langle A|B\rangle\langle B|C\rangle\langle A|C\rangle \\
\langle\Theta_2|\Theta_2\rangle &= 4 - 2\langle A|B\rangle^2 + 4\langle B|C\rangle^2 - 2\langle A|C\rangle^2 - 4\langle A|B\rangle\langle B|C\rangle\langle A|C\rangle \\
\langle\Theta_1|\Theta_2\rangle &= -2 - 2\langle A|B\rangle^2 - 2\langle B|C\rangle^2 + 4\langle A|C\rangle^2 + 4\langle A|B\rangle\langle B|C\rangle\langle A|C\rangle \\
\langle\Theta_1|\mathcal{H}|\Theta_1\rangle &= 4\langle ABC|\mathcal{H}|ABC\rangle + 4\langle ABC|\mathcal{H}|BAC\rangle - 2\langle ABC|\mathcal{H}|ACB\rangle \\
&\quad -2\langle ABC|\mathcal{H}|CBA\rangle - 4\langle ABC|\mathcal{H}|CAB\rangle \qquad \text{(A6-19)} \\
\langle\Theta_2|\mathcal{H}|\Theta_2\rangle &= 4\langle ABC|\mathcal{H}|ABC\rangle - 2\langle ABC|\mathcal{H}|BAC\rangle + 4\langle ABC|\mathcal{H}|ACB\rangle \\
&\quad -2\langle ABC|\mathcal{H}|CBA\rangle - 4\langle ABC|\mathcal{H}|CAB\rangle \\
\langle\Theta_1|\mathcal{H}|\Theta_2\rangle &= -2\langle ABC|\mathcal{H}|ABC\rangle - 2\langle ABC|\mathcal{H}|BAC\rangle - 2\langle ABC|\mathcal{H}|ACB\rangle \\
&\quad +4\langle ABC|\mathcal{H}|CBA\rangle + 2\langle ABC|\mathcal{H}|CAB\rangle
\end{aligned}$$

The above formal development of the expressions needed to evaluate the secular equation (A6-9) is tedious but straightforward. With the γs at hand Eq. (6-10) can be solved to furnish the potential-energy surface. The coefficients K_1 and K_2 of Eq. (A6-8) can be determined by solving the linear system of two equations in K_1 and K_2 which arises in the linear variational treatment leading to (A6-9). In this solution, two sets of coefficients are found, one corresponding to E_+ and the other to E_-. The lowest energy state for H_3 is E_-. Remember that $E_\pm$ both vary with internuclear configuration.

Many of the integrals indicated in (A6-19) are complicated multicenter integrals solvable only numerically. With a high-speed digital computer, it is possible to calculate accurate values for these integrals in a reasonable time. However, this may not be necessary and indeed may even be a waste of time because the formal development is itself only an approximate solution to the problem. Earlier in this section, the London equation was mentioned, and we shall now proceed to obtain it by simplifying γ_1, γ_2, and γ_3. London assumed that the orbitals in the basis set were orthogonal; mathematically this means that

$$\langle A|B\rangle = \langle A|C\rangle = \langle B|C\rangle = 0$$

The normalization and exchange integrals of Eq. (A6-18) then reduce to

$$\langle\Theta_1|\Theta_1\rangle = \langle\Theta_2|\Theta_2\rangle = 4$$

$$\langle\Theta_1|\Theta_2\rangle = -2$$

The γs reduce to

$$\gamma_1 = 12$$

$$\gamma_2 = -2[\langle\Theta_1|\mathcal{H}|\Theta_2\rangle + \langle\Theta_1|\mathcal{H}|\Theta_1\rangle + \langle\Theta_2|\mathcal{H}|\Theta_2\rangle]$$

$$= -12[\langle ABC|\mathcal{H}|ABC\rangle - \langle ABC|\mathcal{H}|CAB\rangle]$$

$$\gamma_3 = \langle\Theta_1|\mathcal{H}|\Theta_1\rangle\langle\Theta_2|\mathcal{H}|\Theta_2\rangle - \langle\Theta_1|\mathcal{H}|\Theta_2\rangle^2$$

$$\frac{1}{12}\gamma_3 = \langle ABC|\mathcal{H}|ABC\rangle^2 - \langle ABC|\mathcal{H}|BAC\rangle^2 - \langle ABC|\mathcal{H}|ACB\rangle^2$$

$$-\langle ABC|\mathcal{H}|CBA\rangle^2 + \langle ABC|\mathcal{H}|CAB\rangle^2$$

$$-2\langle ABC|\mathcal{H}|ABC\rangle\langle ABC|\mathcal{H}|CAB\rangle + 2\langle ABC|\mathcal{H}|CAB\rangle\langle ABC|\mathcal{H}|ACB\rangle$$

$$-\langle ABC|\mathcal{H}|CAB\rangle\langle ABC|\mathcal{H}|BAC\rangle + \langle ABC|\mathcal{H}|BAC\rangle\langle ABC|\mathcal{H}|ACB\rangle$$

$$+\langle ABC|\mathcal{H}|ACB\rangle\langle ABC|\mathcal{H}|CBA\rangle + \langle ABC|\mathcal{H}|BAC\rangle\langle ABC|\mathcal{H}|CBA\rangle$$

If the multiple exchange integral $\langle ABC|\mathcal{H}|CAB\rangle$ is neglected in the above expressions for the γs, Eq. (A6-10) reduces to

$$E_\pm = \langle ABC|\mathcal{H}|ABC\rangle \pm [\langle ABC|\mathcal{H}|BAC\rangle^2 + \langle ABC|\mathcal{H}|ACB\rangle^2 + \langle ABC|\mathcal{H}|CBA\rangle^2$$

$$-\langle ABC|\mathcal{H}|BAC\rangle\langle ABC|\mathcal{H}|ACB\rangle - \langle ABC|\mathcal{H}|ACB\rangle\langle ABC|\mathcal{H}|CBA\rangle$$

$$-\langle ABC|\mathcal{H}|BAC\rangle\langle ABC|\mathcal{H}|CBA\rangle]^{1/2}$$

which on rearranging furnishes the London equation in the form

$$E_\pm = \langle ABC|\mathcal{H}|ABC\rangle \pm [\tfrac{1}{2}(\langle ABC|\mathcal{H}|BAC\rangle - \langle ABC|\mathcal{H}|ACB\rangle)^2$$

$$+\frac{1}{2}(\langle ABC|\mathcal{H}|ACB\rangle - \langle ABC|\mathcal{H}|CBA\rangle)^2$$

$$+\frac{1}{2}(\langle ABC|\mathcal{H}|CBA\rangle - \langle ABC|\mathcal{H}|BAC\rangle)^2]^{1/2}$$

In the case of H_3, E_- gives the lowest energy.

REFERENCES

Section 6.1

D. A. House, *Chem. Rev.* **62**, 185 (1962). Reviews reactions of the persulfate ion.

P. C. Moews, Jr., and R. H. Petrucci, *J. Chem Educ.* **41**, 549 (1964); H. F. Shurnell, *J. Chem. Educ.* **43**, 555 (1966). Literature references dealing with the specific reaction under study here.

K. J. Morgan, M. G. Beard, and C. F. Cullis, *J. Chem. Soc.* 1865 (1951); W. C. Bray, *J. Am. Chem. Soc.* **52**, 3580 (1930). Contains detailed discussion of the mechanism involved in this reaction.

S. Petrucci (ed.), *Ionic Interactions* (Academic Press, New York, 1971). A recent two-volume set dealing with a wide range of ionic phenomena in systems ranging from dilute electrolyte solutions to fused salts.

C. Capellos and B. H. J. Bielski, *Kinetic Systems* (Wiley-Interscience, New York, 1972). Presents the mathematical apparatus for describing reactions in solution.

E. A. Moelwyn-Hughes, *The Chemical Statics and Kinetics of Solutions* (Academic Press, New York, 1971); K. J. Laidler, *Reaction Kinetics,* Vol. II (Pergamon, Elmsford, N.Y., 1963). Solution reaction texts.

Section 6.2

F. A. Cotton and G. Wilkinson, *Advanced Inorganic Chemistry,* 3rd ed. (Wiley-Interscience, New York, 1972); A. E. Martell (ed.), *Coordination Chemistry* (Van Nostrand Reinhold, New York, 1971). Among the inorganic chemistry texts that discuss coordination complexes in some detail.

F. Basolo and R. G. Pearson, *Mechanisms of Inorganic Reactions,* 2nd ed. (Wiley, New York, 1967). Thoroughly presents reactions of inorganic complexes.

G. Brumfitt, *J. Chem. Educ.* **46**, 250 (1969); R. H. Holyer, C. D. Hubbard, S. F. A. Kettle, and R. G. Wilkins, *Inorg. Chem.* **4**, 929 (1965); P. Ellis and R. G. Wilkins, *J. Chem. Soc.* 299 (1959); P. Ellis, R. Hogg, and R. G. Wilkins, *J. Chem. Soc.* 3308 (1959); R. H. Linnell and A. Kaczmarczyk, *J. Phys. Chem.* **65**, 1198 (1961); K. Sone, P. Krumholz, and H. Stammreich, *J. Am. Chem. Soc.* **77**, 777 (1955). Literature references on the experiments performed here.

P. B. Dorain, *Symmetry in Inorganic Chemistry* (Addison-Wesley, Reading, Mass., 1965); J. P. Fackler, *Symmetry in Coordination Chemistry* (Academic Press, New York, 1971). Texts dealing particularly with symmetry in inorganic chemistry.

Section 6.3

Literature references to the molecules recommended for study include

1. Cyclopentene: D. W. Vanas and W. D. Walters, *J. Am. Chem. Soc.* **70**, 4035 (1948).
2. Di-*t*-butyl peroxide: A. F. Trotman-Dickenson, *J. Chem. Educ.* **46**, 396 (1969).
3. 1,1-Dichloroethane: G. R. DeMare, P. Goldfinger, G. Huybrechts, J. Olbregts, and M. Toth, *J. Chem. Educ.* **46**, 684 (1969).
4. Norbornylene: B. C. Roquitte, *J. Phys. Chem.* **69**, 1351 (1965).
5. 2,5-Dihydrofuran: C. A. Wellington and W. D. Walters, *J. Am. Chem. Soc.* **83**, 4888 (1960), and J. A. Rubin and S. V. Filseth, *J. Chem. Educ.* **46**, 57 (1969).

J. G. Calvert and J. N. Pitts, Jr., *Photochemistry* (Wiley, New York, 1966); A. Maccol in S. L. Friess, E. S. Lewis, and A. Weissberger (eds.), *Technique of Organic Chemistry,* Vol. VIII (Wiley-Interscience, New York, 1961), Part I, p. 428. Deal with the construction of suitable furnaces.

A. O. Allen, *J. Am. Chem. Soc.* **56**, 2053 (1934). Deals with pressure corrections.

L. S. Kassel, *J. Phys. Chem.* **32**, 225 (1928); R. A. Marcus, *J. Chem. Phys.* **20**, 359 (1952). Original literature references to the theory of unimolecular decomposition. They contain references to related work.

D. L. Bunker, *Theory of Elementary Gas Reaction Rates* (Pergamon, Elmsford, N.Y., 1966); W. C. Gardiner, Jr., *Rates and Mechanisms of Chemical Reactions* (Benjamin, Reading, Mass., 1969); H. S. Johnston, *Gas Phase Reaction Rate Theory* (Ronald, New York, 1966); L. S. Kassel, *The Kinetics of Homogeneous Gas Reactions* (Chemical Catalog Co., New York, 1932); K. J. Laidler, *Theories of Chemical Reaction Rates* (McGraw-Hill, New York, 1969); G. L. Pratt, *Gas Kinetics* (Wiley, New York, 1969); R. P. Wayne in C. H. Bamford and C. F. H. Tipper (eds.), *Comprehensive Chemical Kinetics,* Vol. 2 (Elsevier, Amsterdam, 1969), Ch. 3; R. E. Weston, Jr., and H. A. Schwarz, *Chemical Kinetics* (Prentice-Hall, Englewood Cliffs, N.J., 1972). Treat unimolecular reaction rate theory.

S. W. Benson and H. E. O'Neal *Kinetic Data on Gas Phase Unimolecular Reactions* (U.S. National Bureau of Standards, Superintendent of Documents, U.S. Government Printing Office, Washington, D.C., 1970). A recent tabulation of unimolecular reaction rate data.

Section 6.4

J. G. Calvert and J. N. Pitts, Jr., *Photochemistry* (Wiley, New York, 1966); R. P. Wayne, *Photochemistry* (Butterworth's, London, 1970); R. B. Cundall and

A. Gilbert, *Photochemistry* (Nelson, London, 1970). Excellent general texts on photochemistry.

R. A. Fass, *J. Phys. Chem.* **74**, 984 (1970); R. A. Fass, J. W. Hoover, and L. M. Simpson, *J. Phys. Chem.* **76**, 2801 (1972); J. M. White and H. Y. Su, *J. Chem. Phys.* **57**, 2344 (1972); G. O. Wood and J. M. White, *J. Chem. Phys.* **52**, 2613 (1970); J. L. Holmes and P. Rodgers, *Trans Faraday Soc.* **64**, 2348 (1968); R. D. Penzhorn and B. de B. Darwent, *J. Phys. Chem.* **72**, 1639 (1968). Literature references on hot atom chemistry of H and D systems.

A. A. Paschier, J. D. Christian, and N. W. Gregory, *J. Phys. Chem.* **71**, 937 (1967). Includes a detailed absorption spectrum of Br_2.

B. J. Heubert and R. M. Martin, *J. Phys. Chem.* **72**, 3046 (1968); C. F. Goodeve and A. W. C. Taylor, *Proc. Roy. Soc.*, **A152**, 221 (1935); C. F. Goodeve and A. W. C. Taylor, *Proc. Roy. Soc.*, **A154**, 181 (1936). Include the absorption spectra of HBr and HI.

Section 6.5

S. Glasstone, K. J. Laidler, and H. Eyring, *The Theory of Rate Processes* (McGraw-Hill, New York, 1941). Includes a good discussion of potential energy surfaces.

A. Ralston, *A First Course in Numerical Analysis* (McGraw-Hill, New York, 1962); S. D. Conte, *Elementary Numerical Analysis* (McGraw-Hill, New York, 1965). Numerical analysis texts.

F. London, *Z. Elektrochem.* **35**, 552 (1929). Proposes the London equation.

H. Eyring and M. Polanyi, *Z. Physik. Chem.* **B12**, 279 (1931). Proposes the method of relating Coulomb and exchange integrals in Eq. (6-102).

S. Sato, *J. Chem. Phys.* **23**, 592 (1955); L. M. Raff, L. Stivers, R. N. Porter, D. L. Thompson, and L. B. Sims, *J. Chem. Phys.* **52**, 3449 (1970). Propose improved methods of relating Coulomb and exchange integrals.

F. T. Wall, L. A. Hiller, and J. Mazur, *J. Chem. Phys.* **29**, 255 (1958); M. Karplus, R. N. Porter, and R. D. Sharma, *J. Chem. Phys.* **43**, 3259 (1965). Classical trajectory studies on H_3.

E. A. McCullough and R. E. Wyatt, *J. Chem. Phys.* **54**, 3578, 3592 (1971); D. G. Truhlar and A. Kuppermann, *J. Chem. Phys.* **56**, 2232 (1972); M. Karplus and K. T. Tang, *Discussions Faraday Soc.* **44**, 56 (1968). Quantum mechanical reactive scattering calculations on the H_3 system.

M. A. Eliason and J. O. Hirschfelder, *J. Chem. Phys.* **30**, 1426 (1959). Establishes the connection between cross sections and rate coefficients.

Chapter 7

Spectroscopy

7.1 Atomic Emission Spectra

The purpose of this experiment is to examine both visually and electronically the emission spectra of atomic hydrogen and helium and to compare the observed spectra with those predicted from quantum mechanical models.

THEORY. The theory of quantum mechanics has been applied with great success to atomic and molecular problems. One formulation of this theory was developed by Schrödinger. He proposed that atomic and molecular systems can be described by a function Ψ which is a solution to the second-order differential equation

$$\mathcal{H}\Psi = -\frac{\hbar}{i}\frac{\partial\Psi}{\partial t} \tag{7-1}$$

where $\mathcal{H}$ is the Hamiltonian operator for the system and contains, among other things, second-order differentials with respect to the coordinates of the system. The i appearing in Eq. (7-1) is the imaginary root of -1, i.e., $i \equiv \sqrt{-1}$. Equation (7-1) is similar to the wave equation of classical electromagnetic theory—hence, the name Schrödinger wave equation.

The function Ψ is called the state function and it furnishes a complete description of the system (i.e., all one can learn about the system is obtained from operations on Ψ). According to the postulates of quantum mechanics the product $\Psi^*\Psi\, d\tau$ is the probability that the coordinates describing the system lie in $d\tau$ at time t. The notation $d\tau$ symbolizes a differential volume element in the complete coordinate space of the system (i.e., $d\tau = dq_1 dq_2 \ldots dq_N$, where q_1 denotes one of the coordinates and N is the total number of coordinates required to describe the system). Ψ^* denotes the complex conjugate of Ψ. Since the coordinates of the system must lie somewhere in the space available, we can write

$$\int \Psi^*\Psi\, d\tau = 1 \tag{7-2}$$

From these considerations it is clear that in the Schrödinger formulation of quantum mechanics, the state function describing the system furnishes a probability density function (i.e., $\Psi^*\Psi\, d\tau$).

The Hamiltonian operator, $\mathcal{H}$, which appears in Eq. (7-1) must be specified before a solution Ψ can be found. Neglecting electronic and nuclear spin, this operator can be constructed formally from the classical Hamiltonian function, which is given by

$$H = T + V \tag{7-3}$$

where T is the kinetic energy and V the potential energy. One of the postulates of quantum mechanics states that for every classical mechanical dynamical variable there is a corresponding quantum mechanical operator. The dynamical variables appearing in H are q_j, p_j and t, where p_j is the momentum associated with (conjugate to) the coordinate q_j. To construct the Hamiltonian operator in Cartesian coordinates, one makes the following formal replacements:

$$H \rightarrow \mathcal{H}$$

$$q_j \rightarrow q_j$$

$$t \rightarrow t$$

$$p_j \rightarrow \frac{\hbar}{i}\frac{\partial}{\partial q_j}$$

Consider, for example, the hydrogen atom, which is a two-particle system consisting of an electron and proton. Figure 7-1 shows a diagram of this system using a general set of coordinate axes. Referring to this figure, the Hamiltonian function may be developed as follows:

$$H = T + V$$

$$T = \frac{1}{2} m_e \left(\frac{dr_2}{dt}\right)^2 + \frac{1}{2} m_p \left(\frac{dr_1}{dt}\right)^2 \tag{7-4}$$

where m_e and m_p are the mass of the electron and proton, respectively.

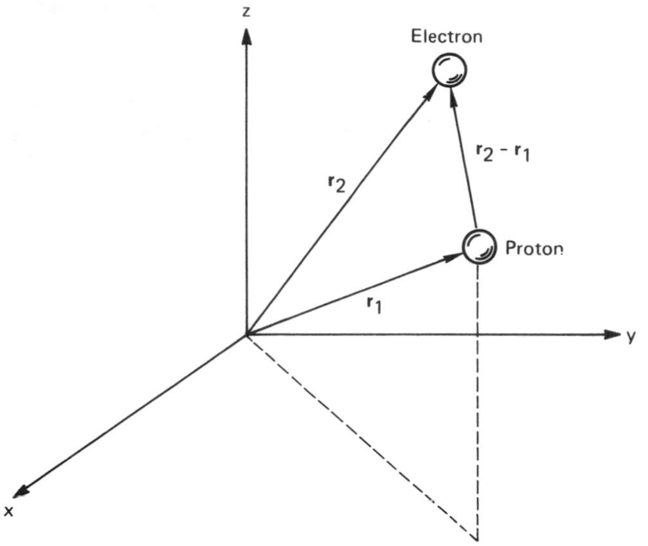

Figure 7-1. Cartesian coordinates for hydrogen atom.

The potential is Coulombic so that we may write

$$V = -\frac{Ze^2}{|\mathbf{r}_{12}|} = -\frac{Ze^2}{|\mathbf{r}_2 - \mathbf{r}_1|}$$

where Z is the number of nuclear charge units.

$$H = \frac{1}{2} m_e \left(\frac{d\mathbf{r}_2}{dt}\right)^2 + \frac{1}{2} m_p \left(\frac{d\mathbf{r}_1}{dt}\right)^2 - \frac{Ze^2}{|\mathbf{r}_{12}|}$$

Replacing $m_e(d\mathbf{r}_2/dt)$ by $\mathbf{p}_2$ and $m_p(d\mathbf{r}_1/dt)$ by $\mathbf{p}_1$ gives rise to

$$H = \frac{\mathbf{p}_1 \cdot \mathbf{p}_1}{2m_p} + \frac{\mathbf{p}_2 \cdot \mathbf{p}_2}{2m_e} - \frac{Ze^2}{|\mathbf{r}_{12}|} \tag{7-5}$$

Making the formal operator substitutions furnishes

$$\mathcal{H} = -\frac{\hbar^2}{2m_p}\frac{\partial^2}{\partial \mathbf{r}_1{}^2} - \frac{\hbar^2}{2m_e}\frac{\partial^2}{\partial \mathbf{r}_2{}^2} - \frac{Ze^2}{|\mathbf{r}_{12}|} \tag{7-6}$$

where

$$\frac{\partial^2}{\partial \mathbf{r}_i{}^2} \equiv \frac{\partial^2}{\partial x_i{}^2} + \frac{\partial^2}{\partial y_i{}^2} + \frac{\partial^2}{\partial z_i{}^2} \tag{7-7}$$

Using the operator given by relation (7-6) in Eq. (7-1), one arrives at the second-order differential equation to be solved.

The coordinate system used in the previous derivation is not the only one which may be used. In fact, from a theoretical point of view the relative coordinate system is much more convenient because it permits a simplification of the mathematics. It is illustrated in Fig. 7-2. The vector $\mathbf{r}_{12}$ is the relative position vector specifying the location of the electron with respect to the nucleus. The vector **R** specifies the position of the center of mass with respect to the origin of the system of axes.

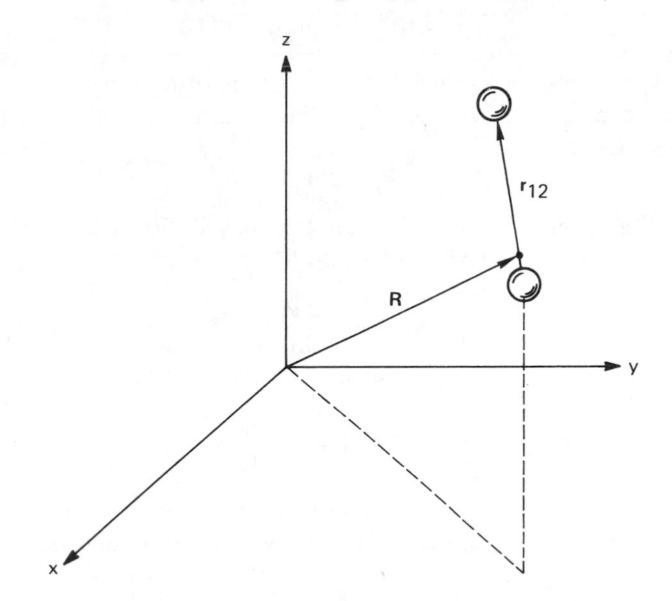

Figure 7-2. Relative coordinates for hydrogen atom.

By definition, the position of the center of mass is

$$\mathbf{R} \equiv \frac{m_1\mathbf{r}_1 + m_2\mathbf{r}_2}{m_1 + m_2} \tag{7-8}$$

and comparing Figs. 7-1 and 7-2 we find that

$$\mathbf{r}_{12} = \mathbf{r}_2 - \mathbf{r}_1 \tag{7-9}$$

Equations (7-8) and (7-9) may be solved simultaneously for $\mathbf{r}_1$ and $\mathbf{r}_2$ to give

$$\mathbf{r}_1 = \mathbf{R} - \frac{m_2}{m_1 + m_2}\mathbf{r}_{12} \tag{7-10}$$

$$\mathbf{r}_2 = \mathbf{R} + \frac{m_1}{m_1 + m_2}\mathbf{r}_{12} \tag{7-11}$$

Taking time derivatives and substituting into Eq. (7-4), we obtain

$$H = \frac{1}{2}M\left(\frac{d\mathbf{R}}{dt}\right)^2 + \frac{1}{2}\mu\left(\frac{d\mathbf{r}_{12}}{dt}\right)^2 - \frac{Ze^2}{|\mathbf{r}_{12}|} \tag{7-12}$$

The Hamiltonian operator in these coordinates is then

$$\mathcal{H} = -\frac{\hbar^2}{2M}\frac{\partial^2}{\partial \mathbf{R}^2} - \frac{\hbar^2}{2\mu}\frac{\partial^2}{\partial \mathbf{r}_{12}{}^2} - \frac{Ze^2}{|\mathbf{r}_{12}|} \tag{7-13}$$

where $M = m_1 + m_2$ and $\mu \equiv$ reduced mass $= m_1 m_2/(m_1 + m_2)$.

If the system is in an eigenstate Ψ_j of the Hamiltonian operator, the following relation is obeyed:

$$\mathcal{H}\Psi_j = E_j\Psi_j \tag{7-14}$$

where Ψ_j is an eigenfunction of the Hamiltonian and the total energy of the system is E_j. Comparing Eqs. (7-14) and (7-1) it is observed that

$$E_j\Psi_j = -\frac{\hbar}{i}\frac{\partial \Psi_j}{\partial t} \tag{7-15}$$

Solving Eq. (7-15) yields

$$\Psi_j = \Psi_j(q_i)\exp\left\{-\frac{iE_jt}{\hbar}\right\} \tag{7-16}$$

Note that $\Psi_j(q_i)$ does not depend on time. From the above equation and Eq. (7-1) we see in searching for the energy eigenstates that we only have to solve the equation

$$\mathcal{H}\Psi_j(q_i) = E_j\Psi_j(q_i) \tag{7-17}$$

which is independent of the time.

Using the Hamiltonian operator expressed in terms of the relative coordinates [Eq. (7-13)], Eq. (7-17) becomes

$$-\frac{\hbar^2}{2M}\frac{\partial^2\Psi(\mathbf{R}, \mathbf{r}_{12})}{\partial \mathbf{R}^2} - \frac{\hbar^2}{2\mu}\frac{\partial^2\Psi(\mathbf{R}, \mathbf{r}_{12})}{\partial \mathbf{r}_{12}{}^2} - \left(E_j + \frac{Ze^2}{|\mathbf{r}_{12}|}\right)\Psi(\mathbf{R}, \mathbf{r}_{12}) = 0 \tag{7-18}$$

This equation is dissected by the separation of variables method in which the total wave function is written as the product of two parts, namely,

$$\Psi(\mathbf{R}, \mathbf{r}_{12}) = \theta(\mathbf{R})\phi(\mathbf{r}_{12}) \tag{7-19}$$

After substitution into (7-18) and separation we find two equations, one in $\mathbf{R}$ and the other in $\mathbf{r}_{12}$:

$$\frac{\hbar^2}{2M}\frac{\partial^2\theta}{\partial \mathbf{R}^2} + (E'' - E)\theta = 0 \tag{7-20}$$

$$\frac{\hbar^2}{2\mu}\frac{\partial^2\phi}{\partial \mathbf{r}_{12}{}^2} + \left(E'' + \frac{Ze^2}{|\mathbf{r}_{12}|}\right)\phi = 0 \tag{7-21}$$

where $E'' - E$ is the total energy associated with the center of mass vector **R** and E'' the total energy associated with the relative position vector $\mathbf{r}_{12}$. Note that Eq. (7-21) is the physically interesting one because it contains the interaction term $Ze^2/|\mathbf{r}_{12}|$. An equation of this form with μ replaced by m_e is often used in introductory physical chemistry texts as the wave equation describing the hydrogen atom. Since $m_e \ll m_p$, $\mu \cong m_e$ and the approximation is very good. The purpose of the above development is to show how the reduced mass comes into the equations describing the system. In passing it is worthwhile to note why Eq. (7-18) is solvable by the separation of variables method: The separate terms in Eq. (7-13) depend on one or the other of the variables $\mathbf{r}_{12}$ and **R** but not on both.

Solutions to Eq. (7-21) subject to the boundary conditions of the problem can be found. These solutions obtained by separation of variables in a spherical polar coordinate system depend on three quantum numbers: n, ℓ, and m_l. To a first approximation the energy eigenstates of the hydrogen atom depend only on the quantum number n. If relativistic effects are taken into account, the energy eigenvalues depend slightly on ℓ. Neglecting the latter, the allowed energies are

$$E_n = \frac{2\pi^2\mu Z^2 e^4}{h^2 n^2} \tag{7-22}$$

where n is an integer.

Observed electronic energy changes (in the present experiment these changes are observed as spectral lines) are given by differences in two energy levels characterized by quantum numbers n and n′:

$$\Delta E = E_n - E_{n'} = -\frac{2\pi^2\mu Z^2 e^4}{h^2}\left[\frac{1}{n^2} - \frac{1}{(n')^2}\right] \tag{7-23}$$

If this electronic energy is released as electromagnetic radiation, its frequency is governed by the Einstein relation:

$$\Delta E = h\nu \tag{7-24}$$

The Rydberg constant is obtained experimentally by an empirical analysis of the hydrogen atom emission spectrum. The observed spectral lines are fit to a formula having the form

$$\bar{\nu} = R\left[\frac{1}{n_1^2} - \frac{1}{n_2^2}\right] \tag{7-25}$$

where $\bar{\nu} = 1/\lambda$ and R is the Rydberg constant. The spectral lines are analyzed in series having n_1 fixed and $n_2 > n_1$; both n_1 and n_2 are integers. Comparing Eqs. (7-25) and (7-24) and noting that $\Delta E = hc/\lambda$, we find the theoretical expression for R to be

$$R = \frac{2\pi^2\mu Z^2 e^4}{ch^3} \tag{7-26}$$

While the energy of an electron bound to the nucleus of an atom is always of interest, other dynamical variables also play an important role in both experimentally observed phenomena and theoretical models. One of the most important of these is the angular momentum. Since it is a vector quantity, its components along particular directions are also of fundamental importance. Classical mechanics predicts that angular momentum is a continuous variable; quantum mechanics predicts that it is discrete (quantized). The significance of this distinction is illuminated by the simple Bohr model of the hydrogen atom. It is also borne out in all more sophisticated treatments of atomic and molecular problems. The basic unit of angular momentum is, of course, 6.62×10^{-28} erg-sec, which is given the name Planck's constant. In any mathematical expression, the appearance of Planck's constant is symptomatic of the introduction of quantum theory in some way.

The angular momentum of an electron in the hydrogen atom is specified by the quantum number ℓ, which arises in Schrödinger theory from the solution of the angular part of the hydrogen atom problem. The result of such a derivation demonstrates that the total angular momentum is quantized and has a magnitude given by

$$L = \sqrt{\ell(\ell + 1)}\,\hbar \tag{7-27}$$

For a given n in Eq. (7-22), the ℓ values are limited to 0, 1, ..., n − 1. If an external magnetic field is applied, the component of the total angular momentum along the field direction (generally defined as the z axis) is quantized. The quantitative expression is

$$L_z = m_l \hbar \tag{7-28}$$

where m_l takes on integer and zero values according to the bounds

$$-\ell \leqslant m_l \leqslant +\ell \tag{7-29}$$

According to Eq. (7-29), if $\ell = 1$, then m_l may be ±1, 0. In this discussion we are neglecting the influence of the intrinsic magnetic moments (spins) of both the electron and proton.

The eigenfunctions of different electronic states of the hydrogen atom are functions of three quantum numbers n, ℓ, and m_l. Formally we write $\Psi(n, \ell, m_l)$. These are eigenfunctions of three operators: $\mathcal{H}$, $\mathcal{L}^2$, and $\mathcal{L}_z$. $\mathcal{L}^2$ is the operator for the total orbital angular momentum and $\mathcal{L}_z$ is the operator for its z component. The following relations are satisfied:

$$\begin{aligned} \mathcal{H}\Psi(n, \ell, m_l) &= E_n\Psi(n, \ell, m_l) \\ \mathcal{L}^2\Psi(n, \ell, m_l) &= \ell(\ell + 1)\hbar^2\Psi(n, \ell, m_l) \\ \mathcal{L}_z\Psi(n, \ell, m_l) &= m_l\hbar\Psi(n, \ell, m_l) \end{aligned} \tag{7-30}$$

In atomic spectroscopy different electronic states are denoted with symbols of the form

$$^{x}L_J \tag{7-31}$$

L denotes the value of ℓ for one electron atoms or the vector sum over the ℓ values for multielectron atoms. L takes on the values S, P, D, F, and G according to $\ell = 0, 1, 2, 3, 4$. x denotes the electron spin multiplicity, which is calculated with the relation $2S + 1$, where S is the total spin of the electrons calculated by applying the operator for the square of the total spin angular momentum to an appropriate spin function. This is discussed further below. J denotes the vector sum of L and S, which can take on only integer or half-integer values bounded by the relation

$$|L - S| \leq J \leq |L + S| \tag{7-32}$$

These labels may be used to denote the possible states of an electron in hydrogen. Since only one electron is involved, we need not concern ourselves with the Pauli exclusion principle or indistinguishability. The lowest energy level (term value) of hydrogen has $n = 1$, $\ell = 0$, $m_l = 0$, and $m_s = \frac{1}{2}$ or what is called a 1s-electron configuration. Since $m_s = \frac{1}{2}$, $S = \frac{1}{2}$. In spectroscopy this state is denoted $^2S_{1/2}$. Excited states of hydrogen must have $n \geqslant 2$.

While the one-electron hydrogen atom problem is solvable analytically, any multielectron problem including helium is not. Helium appears simpler than the heavier elements; however, this is a bit deceptive in that the essential complexities of the heavier elements are also present in helium. In Problem 4 at the end of this section the quantum mechanical Hamiltonian operator for the helium atom is developed and compared with the operator for hydrogen. Because the two electrons are not independent, the Hamiltonian takes a nonseparable form. To calculate, from a theoretical point of view, the eigenstate energies for the electrons in helium we are thus forced to use numerical methods or some approximate analytical method.

The simplest approach is to assume that the electrons are approximately independent and ascribe to each of them a set of four quantum numbers obtained by building up the total state function out of hydrogen-like functions, one for each electron. This leads quite naturally to the ground-state electron configuration for He of $1s^2$, where the superscript denotes the number of electrons in the 1s orbital. Since we are dealing with two electrons, the Pauli principle and electron indistinguishability principle must be satisfied. The former requires that the total wave function be antisymmetric when any two electrons are interchanged. Since the space parts 1s are the same, we may write approximately for the ground state of helium

$$\Psi = \phi_{1s}(1)\alpha(1)\phi_{1s}(2)\beta(2) \tag{7-33}$$

where $\alpha(1)$ implies $m_s = \frac{1}{2}$ for electron 1 and $\beta(2)$ that $m_s = -\frac{1}{2}$ for electron 2. This function is not satisfactory because (1) it is not antisymmetric as required by the Pauli principle and (2) it asserts that electrons 1 and 2 are distinguishable on the basis of their spins. A satisfactory approximate wave function is

$$\Psi = \phi_{1s}(1)\phi_{1s}(2)\left[\frac{1}{\sqrt{2}}\{\alpha(1)\beta(2) - \alpha(2)\beta(1)\}\right] \tag{7-34}$$

Equation (7-34) is antisymmetric since when the electrons are interchanged $\Psi \to -\Psi$. The spectroscopic notation for this state is $^1S_0$.

The first excited state of helium would have the configuration 1s2s and acceptable hydrogen-like wave functions include

$$\Psi_1 = \frac{1}{\sqrt{2}}[\phi_{1s}(1)\phi_{2s}(2) - \phi_{1s}(2)\phi_{2s}(1)]\,\alpha(1)\alpha(2)$$

$$\Psi_2 = \frac{1}{\sqrt{2}}[\phi_{1s}(1)\phi_{2s}(2) - \phi_{1s}(2)\phi_{2s}(1)]\,\beta(1)\beta(2)$$

$$\Psi_3 = \frac{1}{\sqrt{2}}[\phi_{1s}(1)\phi_{2s}(2) - \phi_{1s}(2)\phi_{2s}(1)]\frac{1}{\sqrt{2}}[\alpha(1)\beta(2) + \alpha(2)\beta(1)] \qquad (7\text{-}35)$$

$$\Psi_4 = \frac{1}{\sqrt{2}}[\phi_{1s}(1)\phi_{2s}(2) + \phi_{1s}(2)\phi_{2s}(1)]\frac{1}{\sqrt{2}}[\alpha(1)\beta(2) - \alpha(2)\beta(1)]$$

Each of these satisfies the Pauli principle (antisymmetric on exchange of any two electrons) and asserts indistinguishability. The first three functions have space parts which are identical and to a good approximation are energetically degenerate. The spin parts of these three functions furnish a total spin whose magnitude is 1. While this is obvious for the states $\alpha(1)\alpha(2)$ and $\beta(1)\beta(2)$, it is instructive to consider the third spin function. The magnitude of the total spin is calculable by applying the spin operators $\mathcal{S}_1{}^2$ and $\mathcal{S}_2{}^2$ to the function. These operators act on spin eigenfunctions in a fashion analogous to the operation of $\mathcal{L}^2$, the total orbital angular momentum operator, on its eigenfunctions. The results may be summarized for two-electron spin functions as

$$S = S_1 + S_2$$

$$\mathcal{S}_1{}^2\alpha(1) = \frac{1}{2}\left(\frac{1}{2} + 1\right)\hbar^2\alpha(1) = S_1(S_1 + 1)\hbar^2\alpha(1)$$

$$\mathcal{S}_1{}^2\alpha(2) = 0$$

$$\mathcal{S}_1{}^2\beta(1) = \frac{1}{2}\left(\frac{1}{2} + 1\right)\hbar^2\beta(1) \qquad (7\text{-}36)$$

$$\mathcal{S}_1{}^2[\alpha(1)\beta(2) + \alpha(2)\beta(1)] = \beta(2)\hbar^2 \cdot \frac{1}{2}\left(\frac{1}{2} + 1\right)\alpha(1) + \alpha(2)\hbar^2 \cdot \frac{1}{2}\left(\frac{1}{2} + 1\right)\beta(1)$$

$$= \frac{1}{2}\left(\frac{1}{2} + 1\right)\hbar^2[\alpha(1)\beta(2) + \alpha(2)\beta(1)]$$

$$\therefore S_1 = \frac{1}{2}$$

where the subscript 1 on $S_1{}^2$ denotes the spin coordinates of electron 1. Similarly $S_2 = \frac{1}{2}$, and the total spin S is $S_1 + S_2 = 1$. Hence, the three functions Ψ_1,

Ψ_2, and Ψ_3 each have a total spin quantum number of 1, while Ψ_4 has S = 0. The three spin-1 functions are energetically degenerate and form what is known as a triplet state, while the spin-0 function forms a singlet. The spectroscopic notations are $^3S_1$ and $^1S_0$, respectively.

For helium the electron configuration 1s2p is not energetically degenerate with 1s2s because the interaction which occurs between the two electrons varies with the ℓ value of the orbitals as well as the n values. For the 1s2p configuration two states (terms) are possible: ^3P and $^1P_1$. The ^3P term can be split into three sublevels if an external magnetic field is applied. The three sublevels correspond to different couplings of L, the orbital angular momentum, and S, the spin angular momentum. The possible couplings J = L + S are such that the component of J along the field direction is equal to 2, 1, or 0 times $\hbar$, as specified by Eq. (7-32).

Considering other possible configurations of helium in which only one of its electrons is excited it becomes clear that a set of singlet states and a set of triplet states is built up. In the emission spectrum the electric dipole selection rules prohibit transitions between states of different spin. The rule is $\Delta S = 0$ for an allowed transition. Similarly, for electric dipole transitions,

$$\Delta L = \pm 1$$

For example, consider the following transitions:

$^1F \rightarrow {}^3P$, forbidden, $\Delta S \neq 0$, $\Delta L \neq \pm 1$

$^1F \rightarrow {}^1S$, forbidden, $\Delta L \neq \pm 1$

$^1F \rightarrow {}^1D$, allowed

$^3S \rightarrow {}^3P$, allowed

Figure 7-3 gives an energy level diagram for the helium atom. The separation between singlets and triplets is indicated. These states are used in analysis of the observed spectrum.

It is possible, at least approximately, to fit the experimental data for each series of terms in Fig. 7-3 to a Rydberg-type formula of the form

$$\Delta E = RZ_a{}^2\left[\frac{1}{\{n - \delta(n, L)\}^2} - \frac{1}{\{n' - \delta(n', L')\}^2}\right] \tag{7-37}$$

where R is Rydberg's constant for the atom and $\delta(n, L)$ an empirical parameter called the quantum defect. Z_a is an effective nuclear charge calculated as $Z_a = A + 1 - n_e$, where n_e is the number of electrons. The quantity $[n - \delta(n, L)]$ is called the effective quantum number for the state. If in Eq. (7-37) n' goes to infinity, the resulting transition corresponds to the ionization potential of the state labeled (n, L). If the ionization potential for a given state is known, its quantum defect and effective quantum number can be calculated.

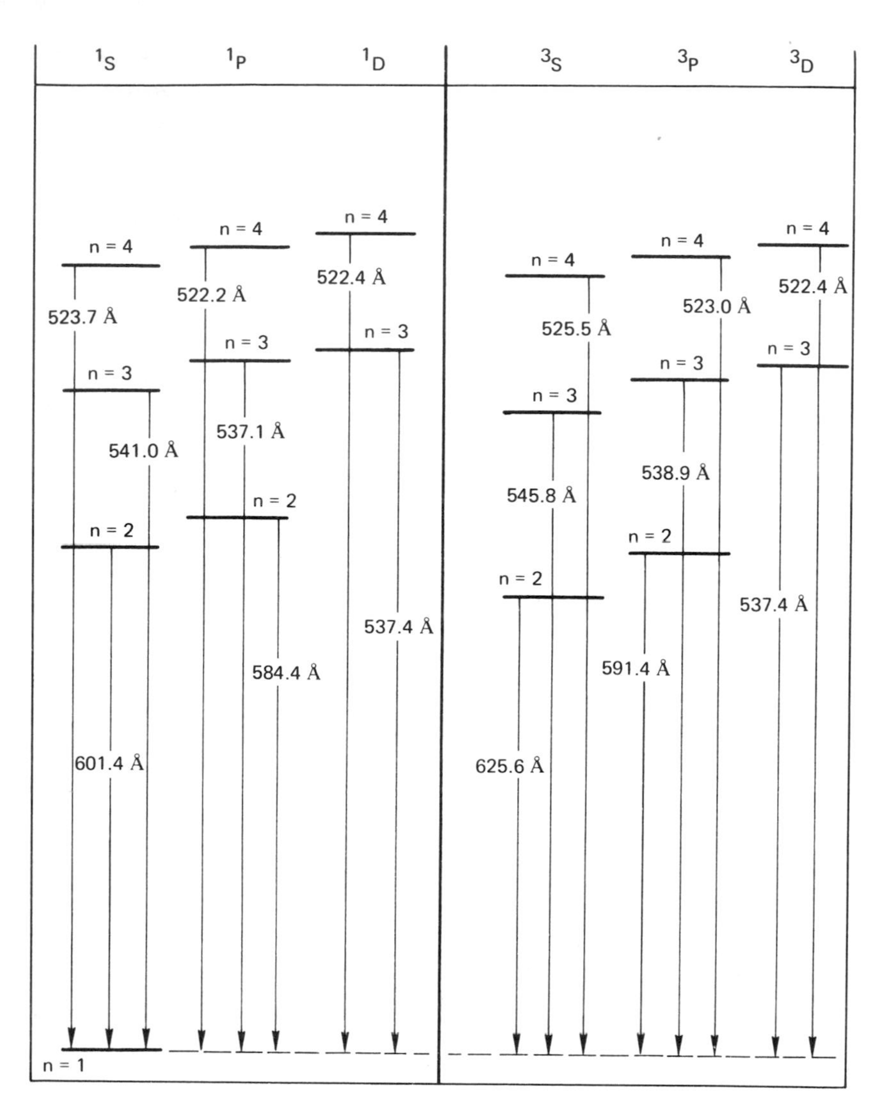

Figure 7-3. Abbreviated energy level diagram for helium.

EXPERIMENTAL.

1. Examine the hydrogen spectrum using the spectroscope provided. Record the wavelengths of the lines you can observe. These lines in the visible region of the spectrum correspond to $n_1 = 2$ in Eq. (7-25). The line in the red portion of the spectrum corresponds to $n_2 = 3$. This series is known as the Balmer series for hydrogen.

2. Record the spectra for hydrogen and helium in the wavelength region 300 to 700 nm using a spectrograph or spectrometer. Data gathering in this experiment is quite straightforward, and a spectrometer or spectrograph capable of resolving lines 50 Å apart is satisfactory if one does not pursue the H atom spectra into the 3000-Å region where the lines become closely spaced. Accurate assignment of wavelengths to the observed spectra requires calibration of the photographic or photoelectronic system. One way is to use the spectrum of some arc source, say iron, and use literature values for its emission lines. A less accurate but simpler way is to assume literature values for two of the lines of the hydrogen spectrum–one at each end of the range. These same lines can be used as calibration lines for the helium spectrum.

The sources of the emission spectra can be any high-voltage, low-pressure discharge tube. These tubes and the appropriate power supplies are commercially available. *Note:* Most inexpensive hydrogen discharge tubes contain impurities which also give rise to spectra.

ANALYSIS.

1. Analyze the hydrogen atom spectrum in terms of Eq. (7-25) by plotting $\bar{\nu}$ versus $[(1/n_1{}^2) - (1/n_2{}^2)]$.
2. From a least-squares analysis of the plot, determine the value of R and compare it to the theoretical value given by Eq. (7-26).
3. Calculate a Rydberg constant for He in cm^{-1}.
4. Using the wavelengths given in Fig. 7-3, compute wavelengths of the first few transitions from ^1P to ^1S which end in n = 2. Are these transitions allowed? Why?
5. Consider the set of triplet terms. The ionization potentials for a few of the states are

$$2^3S: \quad 38{,}460.5 \text{ cm}^{-1}$$

$$2^3P: \quad 29{,}228.6 \text{ cm}^{-1}$$

$$3^3D: \quad 12{,}214.9 \text{ cm}^{-1}$$

   Calculate quantum defects and effective quantum numbers for each of these states. Comment on the apparent variation of $\delta(n, L)$ with n and L.
6. Using the results of step 5, estimate the energies and wavelengths of the following transitions:

$$3^3S \rightarrow 2^3S$$

$$3^3S \rightarrow 2^3P$$

$$4^3S \rightarrow 2^3P$$

$$3^3D \rightarrow 2^3P$$

$$4^3D \rightarrow 2^3P$$

$$3^3P \rightarrow 2^3P$$

Which of these are allowed? Why? From your estimates, make assignments of the lines in the observed spectrum.

7. How can a helium atom in the 2^3S state return to the ground state?

PROBLEMS

1. Show that Eqs. (7-20) and (7-21) are obtainable from Eq. (7-18) when ψ takes the form given in Eq. (7-19).

2. For the lowest energy of the H atom, compute, using Eq. (7-22), the theoretical energy in cm^{-1} using two different mass values. First use the reduced mass of the system, and then use the mass of just the electron. Discuss the difference in these two values and the accuracy required to distinguish between the two systems.

3. Hydrogen has three isotopes: normal hydrogen with an atomic weight of 1 amu, deuterium with an atomic weight of 2 amu, and tritium with an atomic weight of 3 amu. The isotopes each have, of course, one electron and one proton and differ only in the number of neutrons. Why, in terms of variations in specific equations in the text, will the spectra of hydrogen, deuterium, and tritium differ? Calculate the theoretically predicted position of the red line of the Balmer series for both hydrogen and deuterium.

4. Write the classical Hamiltonian of the helium atom. From it develop the quantum mechanical Hamiltonian operator. Remember that helium has two protons in its nucleus. What type of term appears for helium that does not appear for hydrogen? The complexity of the helium spectrum is related, in part, to these differences.

5. Consider the following sets of quantum numbers for an electron in an hydrogen atom:

$$n = 2, \quad \ell = 1, \quad m_\ell = 1, \quad m_s = \frac{1}{2}$$

$$n = 2, \quad \ell = 0, \quad m_\ell = 0, \quad m_s = \frac{1}{2}$$

$$n = 3, \quad \ell = 2, \quad m_\ell = -2, \quad m_s = \frac{1}{2}$$

Label these states using both electron configuration notation and spectroscopic notation.

6. Show that the spin function, $(1/\sqrt{2})[\alpha(1)\beta(2) - \alpha(2)\beta(1)]$, has a total spin of zero.

7.2 Energy States of the Mercury Atom by Electron Impact

The purpose of this experiment is to measure the energy of the electronic transition in a mercury atom from its ground state to the first excited state using a beam of electrons to excite the mercury atoms.

THEORY. From a theoretical point of view, the electronic structure of atoms may be described in principle by finding solutions of the appropriate Schrödinger wave equation, as discussed in Section 7.1, or the equivalent Heisenberg matrix equation. It is well known that analytical solutions exist only for one-electron atoms and molecules, whereas all multielectron atoms and molecules can be treated only approximately, albeit in some cases very accurately.

From an experimental point of view, two widely used methods of probing the electronic structure of atoms are photon and electron spectroscopy, the latter being less familiar to chemists. The atomic emission spectra experiment illustrates the former method. The emission spectrum of complex atoms is often very complicated, and the empirical construction of a compatible set of energy levels is a difficult task. Electron spectra serve as complementary data which can be used advantageously. A second reason electron spectra are desirable arises because of selection rules. Optical spectra are subject, at least approximately, to the single-triplet selection rule; that is, the optically induced transitions from singlet to triplet electronic states are forbidden. Electron-induced transitions are not subject to this optical selection rule. Hence, low-lying triplet electronic energy levels, which are very important in many chemical reactions, can be observed by electron spectroscopy.

In this experiment an elementary electron scattering apparatus is used which Franck and Hertz (see the References) developed in 1919–hence, the name Franck-Hertz experiment. Using the technique they were able to demonstrate that the ideas of Bohr were at least qualitatively correct.

Figure 7-4 outlines the apparatus, which consists essentially of a vacuum tube containing, in this case, mercury vapor. The tube is heated to about 150°C so that the pressure of mercury vapor is between 5 and 20 Torr. Electrons are emitted thermally from the cathode by heating it with a heater. These electrons are accelerated toward grid 2 under the influence of an attractive potential. Grid 1 is located very near the cathode and at a slightly positive voltage with respect to the cathode. This grid controls the space charge (the electric field between the cathode and grid 2) and thus the emission current. As grid 1 is made more positive the emission current increases because the space charge near the cathode is reduced in magnitude. The space charge effect arises because electrons in the space between the cathode and plate repel the electrons being emitted by the cathode. The electrons which pass through grid 1 are accelerated by the voltage V applied between the cathode and grid 2. Electrons which pass through grid 2 with a voltage greater than about 1 V are collected at the plate and give rise to a current in an external circuit containing an electrometer current meter, A. This is the plate current.

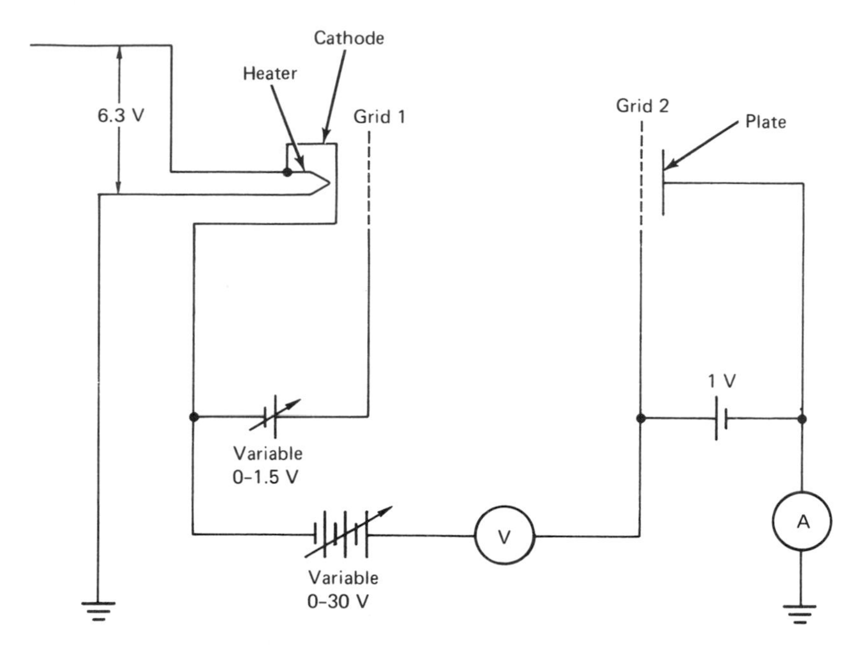

Figure 7–4. Franck-Hertz apparatus.

When the tube is cold, very little mercury is in the vapor phase, and the emitted electrons traverse the region between cathode and plate without collision with any mercury atoms. However, when the tube is heated and the mercury is vaporized to an equilibrium pressure between 5 and 20 Torr, the emitted electrons undergo many collisions with the mercury atoms in the course of passing from cathode to the plate.

Two kinds of collisions may occur: (1) inelastic collisions in which the electron loses energy to the electrons of the mercury atoms and (2) elastic collisions in which translational energy is exchanged between the electron and the mercury atom as a whole. The latter collisions are very ineffective in changing the energy of the accelerated electrons because the electron is very light compared to the mercury atom (like hitting a house with a ping-pong ball—little energy is transferred to the house). The first kind of collision is, however, very effective in lowering the energy of the electron because a relatively large amount of energy is involved. Because of the discrete nature of the electronic energy states of an atom, inelastic collisions are not possible unless the accelerated electron possesses sufficient energy to excite the atom from one state to another (usually from the ground state to a low-lying electronic excited state). Referring to Fig. 7-4 we expect the current reaching the plate to increase slowly as we increase the accelerating voltage beginning near zero because the increased voltage will overcome space charge effects. When the accelerating voltage gives the electrons sufficient energy to excite the atoms, the current will drop sharply because the postcollision electron energy is too low to overcome the small repulsive potential between grid 2 and the plate. Hence, we expect the plate current to vary with

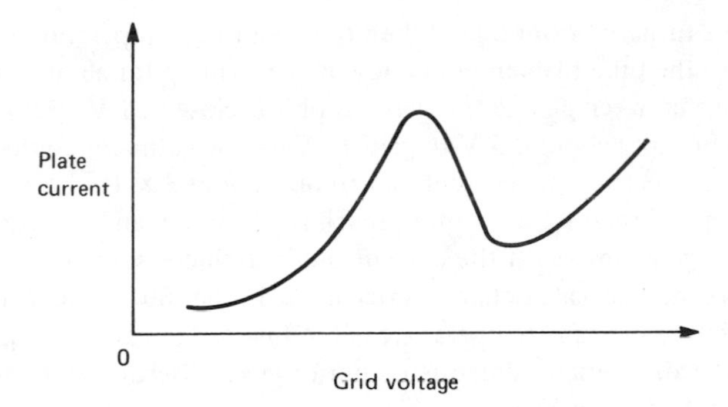

Figure 7-5. Variation of plate current with grid voltage.

grid voltage as shown in Fig. 7-5. The current increases again at voltages to the right of the peak because the electrons are reaccelerated as they approach grid 2. Another peak in the current-voltage curve will occur when the electrons can excite another state or can undergo two inelastic collisions.

Much more sophisticated electron spectroscopy equipment has been developed which is capable of gathering detailed data on the interaction of electrons with atoms and molecules (see Trajmar et al. in the References).

PROCEDURE. The apparatus should be connected electrically as shown in Fig. 7-6. With the apparatus connected but the tube power supplies off, bring the furnace and tube up to a constant temperature near 160°C by adjusting the

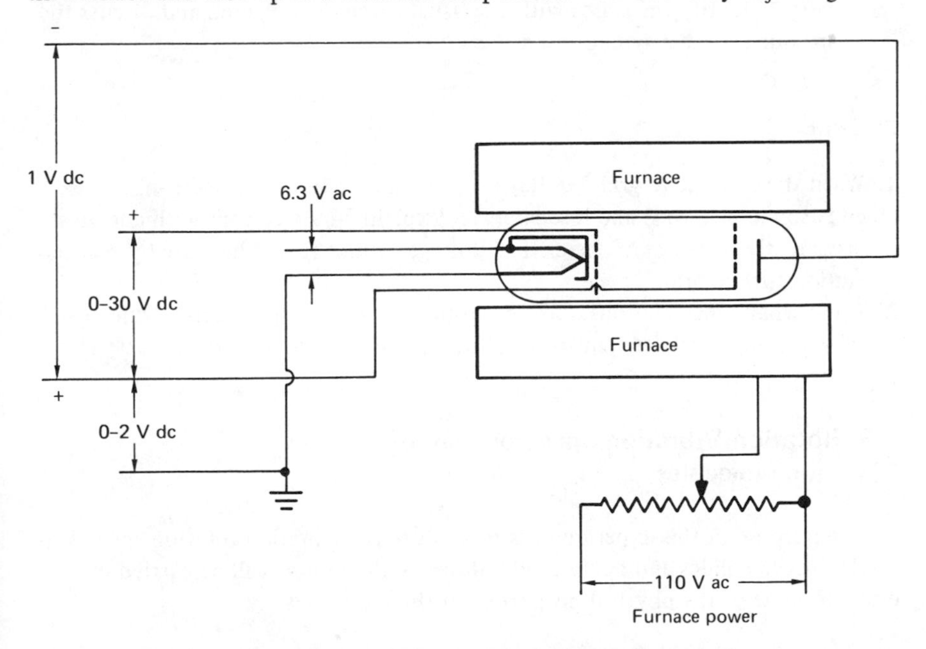

Figure 7-6. Franck-Hertz apparatus.

rheostat in the furnace power line. When the operating temperature is reached, apply 6.3 V to the tube filament and allow it to warm up for about 1 min. Adjust the voltage between g_2 and the plate until it is close to 1 V with the plate negative. Then apply about 0.5 V to grid 1. Vary the cathode-to-grid-2 voltage and observe the electrometer current (set to read about 3×10^{-9} A full scale). The current should increase uniformly, reach a maximum, and then decrease as the grid 2 voltage increases. If the current suddenly increases without going through a maximum, a gas discharge is occurring in the tube. Under these circumstances, either the tube temperature is too low or the voltage on grid 1 is too high. Make the appropriate adjustments until the gas discharge does not occur over the range 0–25 V on grid 2.

With the apparatus properly adjusted, measure currents and voltages using a very narrowly spaced set of voltages near the peaks in the spectrum. Repeat these measurements to ascertain the reproducibility that can be achieved.

ANALYSIS

1. From a graphical analysis of the spectrum, calculate the energy associated to the inelastic collision process being observed.
2. If this transition were to occur in an optical absorption or emission spectrum, what wavelength would be involved? Is this a prominent line in the emission spectrum of mercury? Reference your source of information.
3. Using references on atomic structure (see the last two entries in the References), construct an energy level diagram for the low-lying states of mercury. Identify the states with the usual symbols S, P, etc., and discuss the meaning of each symbol used.

PROBLEMS

1. When the cathode-to-grid-2 voltage reaches a certain value, electrons can excite atomic electrons and lose energy. Explain why electrons with energies just slightly in excess of the critical voltage cannot reach the plate by reacceleration toward grid 2.
2. From what considerations does the optical singlet-triplet selection rule arise? Why are singlet-triplet transitions allowed in electron-induced excitation?

7.3 Rotation-Vibration Spectroscopy of Diatomic Molecules

The purpose of this experiment is to analyze the vibration-rotation spectra of the hydrogen halides and deuterated halides. Calculations will be carried out to evaluate some of the physical properties of the molecules.

THEORY. In the previous two experiments nuclear motion played a minor role and to a first approximation could be neglected. We were asking questions about the behavior of electrons bound to nuclei, and since the center of mass for any atom lies essentially at the position of the nucleus, we could neglect nuclear motion in the center of mass coordinate system. Passing now to the dynamics of molecules it is immediately clear that the center of mass of the system no longer coincides even approximately with the position of all the nuclei and in most cases coincides with none of them. Thus, nuclear motion becomes of vital concern in the center of mass coordinate system. Remember that the center of mass system is useful because the resulting equations of motion do not depend on how far the system is from some arbitrary reference point outside it but depend only on the positions of the particles of the system with respect to a well-defined reference point within it.

Since nuclear motion is important in molecular dynamics, we shall proceed to treat the simplest class of molecules, the diatomics. In particular the discussion will detail those features of nuclear motion which give rise to observed vibration-rotation spectra. While electron motion is crucial in molecules just as in atoms, we shall relegate it to the background in this discussion by assuming that it provides a potential-energy field in which the nuclei move. The quantum mechanical calculation of such a potential-energy relation is a formidable task. An indication of how to go about it is given in Section 6.5. In this section we shall circumvent this problem with various assumptions about the form of the potential-energy function and thereby eliminate the need for explicitly considering the electrons.

Focusing attention on the nuclei of a diatomic molecule, recall that classical mechanics requires three coordinates for each particle in order to specify its position. Thus, for a diatomic molecule, six coordinates or degrees of freedom must be specified. If we are interested only in the internal motions (vibration and rotation) of the molecule and not in the translational motion of the molecule as a whole, then center of mass coordinates are useful. In this coordinate system, the origin is attached to the center of mass of the molecule and the axes are parallel to the laboratory axes (see Fig. 7-7). The motion of the position of the center of mass characterizes the motion of the molecule as a whole through the laboratory. Three coordinates are needed to specify the center of mass position, leaving three other coordinates to describe rotations and vibrations. One of these, the internuclear distance, is used in describing the vibrational motions of the molecule; the other two (angles θ and ϕ in Fig. 7-7) are used to describe the rotation of the molecule AB about its center of mass. Each group of coordinates has energy associated with it; thus, we speak about translational, vibrational, and rotational energy. For the remainder of this discussion, we shall assume the use of center of mass coordinates. Note that an observer fixed in the laboratory sees the center of mass as moving but that in the center of mass system he sees the center of mass as stationary.

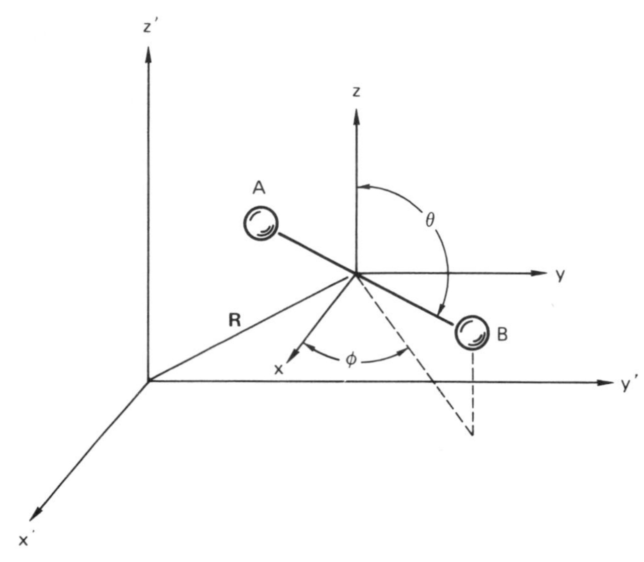

Figure 7-7. Diatomic molecule AB in laboratory and center of mass coordinates.

The center of mass **R** of a diatomic molecule is given by

$$\mathbf{R} = \frac{m_A \mathbf{r}_1 + m_B \mathbf{r}_2}{m_A + m_B} \tag{7-38}$$

where $\mathbf{r}_1$ and $\mathbf{r}_2$ are the position vectors of A and B, respectively, in the laboratory system of coordinates. In the center of mass system, the distance of particle A from the origin (center of mass) is denoted by $\mathbf{r}_A$ and is given mathematically by

$$\mathbf{r}_A = \frac{m_B}{m_A + m_B} \mathbf{r} \tag{7-39}$$

where **r** is the internuclear distance and m_i is the mass of nucleus i. Another important property of a two-particle system is the reduced mass, μ, which is defined as

$$\mu \equiv \frac{m_1 m_2}{m_1 + m_2} \tag{7-40}$$

Using the reduced mass and the center of mass coordinate system, all the features of the internal energies can be discussed in terms of a single pseudoparticle of mass = μ located a distance r away from the origin of the center of mass coordinate system as shown in Fig. 7-8 (Problem 1). Here, as in the hydrogen atom problem, the utility of the center of mass coordinate system is apparent.

There are several models which may be employed to describe the motions of a diatomic molecule. All of them are approximate, and they all implicitly account for the effect of the electrons. The simplest model considers the diatomic

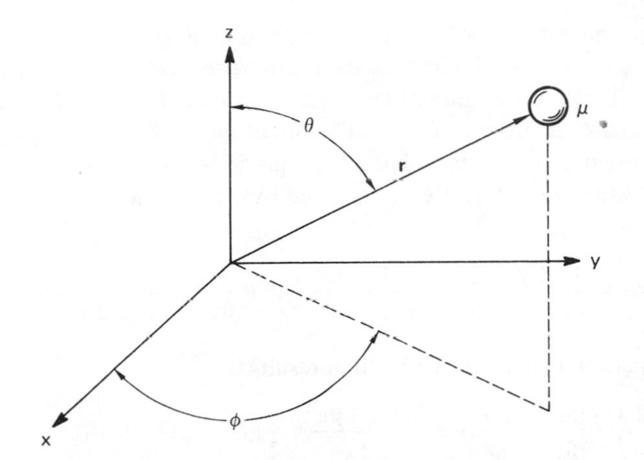

Figure 7-8. Center of mass coordinate system after reduction of a two-particle system to a pseudo-one-particle system.

molecule to be a rigid rotor and harmonic oscillator. Rigid rotor means that the internuclear distance, or r, does not change as the molecule rotates. This is an obvious approximation because the chemical bond will stretch as the rotational speed increases. The harmonic oscillator model requires the vibrational motion to be like two balls connected by a Hooke's law spring. The potential for such a model is parabolic in shape, as shown by Eq. (7-41) below. This is also an obvious approximation since the molecule could never be dissociated if this model were rigorously applicable. However, it does give good results for molecules that possess vibrational energies which are only a small fraction of the dissociation energy. The classical Hamiltonian for this model is

$$H = T_{rot} + T_{vib} + V_{vib}$$

$$H = T + V_{vib}$$

$$T = \frac{1}{2}\mu V^2 \tag{7-41}$$

$$V_{vib} = +\frac{1}{2}k(r - r_e)^2$$

where T denotes kinetic energy of the particle of mass μ and V_{vib} denotes potential energy. Note that there is no rotational potential energy other than the rigid rotor assumption. Making standard formal substitutions for T and V_{vib}, the time-independent Schrödinger equation may be written as

$$-\frac{\hbar^2}{2\mu}\nabla^2\Psi + \frac{1}{2}k(r - r_e)^2\Psi = (E_{rot} + E_{vib})\Psi \tag{7-42}$$

where k is the harmonic force constant and r_e the equilibrium internuclear distance.

Notice that the potential-energy term contains no angular dependence. This means that the above Schrödinger equation can be separated into r, θ, and ϕ components following procedures identical to those used to find the electronic states in the hydrogen atom problem. The quantum mechanically allowed energy levels are determined by solving the equations which result from the separation of variables. The latter is achieved by writing the Laplacian in spherical polar coordinates; namely

$$\nabla^2 = \frac{1}{r^2}\frac{d}{dr}\left(r^2\frac{d}{dr}\right) + \frac{1}{r^2 \sin\theta}\frac{\partial}{\partial\theta}\left(\sin\theta\frac{\partial}{\partial\theta}\right) + \frac{1}{r^2\sin^2\theta}\frac{\partial^2}{\partial\phi^2}$$

The radial equation (r dependence) which results is

$$\frac{1}{r^2}\frac{d}{dr}\left\{r^2\frac{d\Psi}{dr}\right\} + \frac{2\mu}{\hbar^2}\left\{E - \frac{J(J+1)\hbar^2}{2I_e} - \frac{1}{2}k(r - r_e)^2\right\}\Psi = 0 \tag{7-43}$$

where I_e is the equilibrium moment of inertia given by μr_e^2, $E = E_{rot} + E_{vib}$, and J is the rotational quantum number and is a positive integer or zero.

Acceptable solutions, $\Psi(v, J)$, of this equation can be found when the energy is given by

$$E(v, J) = \hbar\sqrt{\frac{k}{\mu}}\left(v + \frac{1}{2}\right) + \frac{\hbar^2 J(J+1)}{2I_e} \tag{7-44}$$

where v = 0, 1, 2, ... This equation is often rewritten as

$$\bar{E}(v, J) = h\nu_e\left(v + \frac{1}{2}\right) + B_e J(J+1) \tag{7-45}$$

where

$$\nu_e = \frac{1}{2\pi}\sqrt{\frac{k}{\mu}}$$

$$B_e = \frac{\hbar^2}{2I_e} \tag{7-46}$$

B_e is called the rotational constant and ν_e the fundamental frequency. The energies given by Eq. (7-45) are in ergs. A more common unit is cm^{-1} (wavenumber), which is obtained from ergs by dividing by hc (Planck's constant multiplied by the cgs speed of light). Equation (7-45) then becomes

$$\bar{E}(v, J) = \bar{\nu}_e(v + \frac{1}{2}) + \bar{B}_e J(J+1), \qquad v = 0, 1, 2, \ldots \text{ and } J = 0, 1, 2, \ldots \tag{7-47}$$

where the overbar indicates units of cm^{-1}. Keep in mind that Eq. (7-47) is an approximation to the real situation.

A better approximation can be obtained if we use a model where (1) the internuclear distance is allowed to change during rotation, (2) the vibrational potential energy is allowed to depart from harmonicity, and (3) the vibrational and rotational motions are allowed to influence (couple with) each other.

Considering point (1) above, note that $I = \mu r^2$ will change with r; hence, $\overline{B} \equiv \hbar^2/2hcI$ will not be precisely equal to $\overline{B}_e$ unless $r = r_e$. Thus, we conclude that the energy levels of a nonrigid rotor must contain a variable $\overline{B}$. To a very good approximation, detailed calculations show that $\overline{B}_eJ(J + 1)$ in Eq. (7-47) should be replaced by $\overline{B}_eJ(J + 1) + \overline{D}J^2(J + 1)^2$ to account for this effect. The parameter $\overline{D}$ is assumed to be constant and is called the centrifugal stretching constant.

Considering point (2) above, we note that, especially when $r > r_e$, departures from the harmonic oscillator potential are expected. The actual potential will be lower than the harmonic potential at a given internuclear distance. This anharmonicity is introduced by expanding the potential in a Taylor series about r_e to obtain

$$V(r) = V(r_e) - \left(\frac{\partial V}{\partial r}\right)_{r=r_e} (r - r_e) + \left(\frac{\partial^2 V}{\partial r^2}\right)_{r=r_e} \frac{(r - r_e)^2}{2} + \left(\frac{\partial^3 V}{\partial r^3}\right)_{r=r_e} \frac{(r - r_e)^3}{6} + \ldots$$

We may arbitrarily set $V(r_e) = 0$; further, $(\partial V/\partial r)|_{r=r_e} = 0$ since the potential has zero slope at $r = r_e$. The above expression, retaining only second- and third-order terms, thus becomes

$$V(r) \cong f(r - r_e)^2 + g(r - r_e)^3 \tag{7-48}$$

Analysis of this potential-energy function shows that $\overline{\nu}_e(v + \frac{1}{2})$ in Eq. (7-47) should be replaced by $\overline{\nu}_e(v + \frac{1}{2}) + x_e\overline{\nu}_e(v + \frac{1}{2})^2$, where $x_e\overline{\nu}_e$ is called the anharmonicity constant.

Part (3) of the nonrigid rotor-anharmonic oscillator model permits coupling of the rotational and vibrational motions (i.e., the vibrational motion influences the rotational motion and vice versa). The correction for this interaction is written $-\alpha_e(v + \frac{1}{2})J(J + 1)$ and α_e is called the coupling constant. Notice how the magnitude of this term depends on the positive integers v and J. To get a picture of how this interaction occurs, think of a rotating and vibrating system and note that, during a vibration, r is changing and consequently $\overline{B}$ is changing.

Taking into account all three of these terms furnishes a more realistic picture of a diatomic molecule. Equation (7-47) becomes

$$\overline{E}(v, J) = \overline{\nu}_v\left(v + \frac{1}{2}\right) + \overline{B}_vJ(J + 1) - \overline{D}J^2(J + 1)^2 \tag{7-49}$$

$$\overline{\nu}_v \equiv \overline{\nu}_e - x_e\overline{\nu}_e\left(v + \frac{1}{2}\right) \tag{7-50}$$

$$B_v \equiv \overline{B}_e - \alpha_e\left(v + \frac{1}{2}\right) \tag{7-51}$$

OBSERVED SPECTRUM. In the infrared region of the spectrum the photon energy ($E = h\nu$) is sufficient to cause a simultaneous change of both the vibrational and rotational states (quantum numbers) of most molecules. This change occurs between E(v′, J′) and E(v″, J″), where single primes denote the initial state and double primes denote the final state. Assuming that the model furnishing Eq. (7-49) is valid we write for the infrared absorption process

$$\begin{aligned}\Delta\bar{E} &= \bar{E}(v'', J'') - \bar{E}(v', J') \\ &= \bar{\nu}_{v''}\left(v'' + \frac{1}{2}\right) - \bar{\nu}_{v'}\left(v' + \frac{1}{2}\right) \\ &\quad + \bar{B}_{v''}J''(J'' + 1) - \bar{B}_{v'}J'(J' + 1) \\ &\quad - \bar{D}[J''^2(J'' + 1)^2 - J'^2(J' + 1)^2]\end{aligned} \tag{7-52}$$

If the vibrating and rotating molecule has a time-dependent dipole moment or a permanent dipole moment, then electromagnetic radiation can interact, in principle, with the dipoles to give rise to electric dipole transitions. Other types of interaction between radiation and matter can occur (electric quadrupole and magnetic dipole), but transitions induced because of these interactions are generally weaker than electric dipole transitions. If the latter are absent because the molecule has no electric dipole, the quadrupole or magnetic dipole transitions may be observed. The hydrogen and deuterium halides all possess permanent dipole moments and thus they are infrared-active.

The values of the quantum numbers in Eq. (7-52) are subject to electric dipole selection rules; that is, some transitions between states are allowed (observable), while others are forbidden (not observable). Dipole selection rules are based on integrals of the type $\langle\Psi''|\mathbf{er}|\Psi'\rangle$, where **er** is the dipole moment and Ψ'' and Ψ' are the total wave functions for the final and initial states, respectively. In most cases it is valid to approximate both Ψ'' and Ψ' as products of the form $\Psi_e\Psi_{vr}$, where e, v, and r denote electronic, vibrational, and rotational, respectively. Using this approximation we say that a transition is allowed if $\langle\Psi_e''\Psi_{vr}''|\mathbf{er}|\Psi_e'\Psi_{vr}'\rangle$ is nonzero and forbidden if it is equal to zero.

In the present experiment, $\Psi_e'' = \Psi_e'$ since there is no change in the electronic state. Furthermore, only the electronic ground state participates in the observed spectra. This state is characterized by (1) the component of the total electronic angular momentum along the internuclear axis, (2) the total spin angular momentum of the electrons, (3) the vector sum of characteristics (1) and (2), and (4) the symmetry of the electronic wave function upon reflection through any plane containing the nuclei. The notation is discussed in Section 7.6. For present purposes it is only important to note that the electronic ground states of the hydrogen halides have no net electronic angular momentum along the internuclear axis and that under these circumstances the electric dipole selection rules for the anharmonic nonrigid rotor model may be summarized as

$$\begin{aligned}\Delta v &= v'' - v' = 1, 2, 3, \ldots \\ \Delta J &= J'' - J' = \pm 1\end{aligned} \tag{7-53}$$

Most diatomic molecules follow these rules. A notable exception is NO, which has, in its electronic ground state, a nonzero component of electronic angular momentum along the internuclear axis. In this case the ΔJ selection rule changes to read $\Delta J = 0, \pm 1$.

In the present experiment we shall deal only with $\Delta v = +1$ and $\Delta J = \pm 1$. Note that the selection rule on J prohibits any transition for which $\Delta J \neq \pm 1$. Making use of these selection rules in Eq. (7-52) furnishes two equations: one for $J'' = J' + 1$, which is called the R branch, and the other for $J'' = J' - 1$, which is called the P branch. If $\Delta J = 0$ is allowed, a Q branch appears. For which branch is $\Delta \bar{E}$ largest?

For the R branch ($J'' = J' + 1$) we have from (7-52) that

$$\Delta \bar{E} = \bar{\nu}_{v''}\left(v'' + \frac{1}{2}\right) - \bar{\nu}_{v'}\left(v' + \frac{1}{2}\right) + 2\bar{B}_{v''} + (3\bar{B}_{v''} - \bar{B}_{v'})J' + (\bar{B}_{v''} - \bar{B}_{v'})J'^2 - 4\bar{D}(J' + 1)^3, \quad J' = 0, 1, 2, \ldots \tag{7-54}$$

For the P branch ($J'' = J' - 1$),

$$\Delta \bar{E} = \bar{\nu}_{v''}\left(v'' + \frac{1}{2}\right) - \bar{\nu}_{v'}\left(v' + \frac{1}{2}\right) - (\bar{B}_{v''} + \bar{B}_{v'})J' + (\bar{B}_{v''} - \bar{B}_{v'})J'^2 + 4\bar{D}J'^3, \quad J' = 1, 2, 3, \ldots \tag{7-55}$$

Note that in Eq. (7-55), $J' = 0$ is not permitted because J'' would then be -1. These two equations serve as starting points for the analysis of the vibration-rotation spectrum of a diatomic molecule. Making the substitutions $v' = 0$ and $v'' = 1$ in Eqs. (7-54) and (7-55) furnishes two equations applicable to the observed spectrum insofar as the model is valid. For the R branch,

$$R(J') \equiv \Delta \bar{E} = \bar{\nu}_{01} + 2\bar{B}_1 + (3\bar{B}_1 - \bar{B}_0)J' + (\bar{B}_1 - \bar{B}_0)J'^2 - 4\bar{D}(J' + 1)^3, \quad J' = 0, 1, 2, \ldots \tag{7-56}$$

For P branch,

$$P(J') \equiv \Delta \bar{E} = \bar{\nu}_{01} - (\bar{B}_1 + \bar{B}_0)J' + (\bar{B}_1 - \bar{B}_0)J'^2 + 4\bar{D}J'^3, \quad J' = 1, 2, 3, \ldots \tag{7-57}$$

where $\bar{\nu}_{01} \equiv \bar{\nu}_1(\frac{3}{2}) - \bar{\nu}_0(\frac{1}{2})$ and is called the vibration-rotation band origin or just the band origin. Keep in mind that Eqs. (7-56) and (7-57) represent observed absorptions in cm^{-1}.

Using Eqs. (7-56) and (7-57) and the absorption spectrum, we can find values for $\bar{\nu}_{01}$, $\bar{B}_1$, $\bar{B}_0$, and $\bar{D}$ which will fit the observed data. Two very useful combinations of Eqs. (7-56) and (7-57) are

$$R(J') - P(J') = 4\bar{B}_1(J' + \frac{1}{2}) - 4\bar{D}[(J' + 1)^3 + J'^3] \tag{7-58}$$

$$R(J' - 1) - P(J' + 1) = 4\bar{B}_0(J' + \frac{1}{2}) - 4\bar{D}[(J' + 1)^3 + J'^3] \tag{7-59}$$

where $R(J' - 1)$ is obtained mathematically by substituting $J' - 1$ for J' into Eq. (7-56). Spectroscopically, $R(J' - 1)$ is the energy of the transition in the R branch, which has a J value equal to $J' - 1$. Hence, the left-hand side of Eq. (7-58) or (7-59) can be obtained from the observed spectrum. If two adjacent J' values are used in Eq. (7-58), there will result a set of two simultaneous equations in two unknowns $\bar{B}_1$ and $\bar{D}$. A similar procedure can be carried out using Eq. (7-59) to furnish $\bar{B}_0$ and $\bar{D}$. Keep in mind that the right-hand sides of these equations result from a particular approximate model. Therefore, we would not expect to obtain precisely the same values for $\bar{B}_0$, $\bar{B}_1$, and $\bar{D}$ if we used another set of J' values. Any discrepancies will indicate the accuracy of the model and/or the data.

EXPERIMENT. Obtain the infrared vibration-rotation spectrum of the molecules provided in the laboratory. Pairs such as HBr–DBr and HCl–DCl are suggested. Both molecules of a pair can be placed in the absorption cell since their spectra do not overlap. The bands are located in the following wave-number regions: HBr, 2560 cm^{-1}; DBr, 1820 cm^{-1}; HCl, 2890 cm^{-1}; and DCl, 2090 cm^{-1}. A standard infrared gas cell about 10 cm long and 2 in. in diameter with NaCl windows cemented on the ends is quite adequate. Pressures of each gas should be on the order of 100 Torr. If the windows are sealed on with picein cement (black wax), a mixture of DCl and HCl can easily be made by introducing 200 Torr of DCl into the cell. The wax undergoes an exchange reaction with the DCl to form some HCl and deuterated wax. The rotational lines are spaced on the order of every 20 cm^{-1} in the hydrogen halides, while in the deuterated halides they occur about every 10 cm^{-1}. The spectrometer should have the capability of resolving lines 3 cm^{-1} apart for good results in this experiment. Instructions for operation of the spectrometer will be given by the instructor. If suitable infrared spectrometers or samples are not available, the analysis may be carried out using spectra provided by the instructor.

ANALYSIS

1. Examine the spectrum; observe that as the wave number (cm^{-1}) increases the transition energy [Eq. (7-52)] increases. Further note from the selection rules that the transition $J' = 0 \rightarrow J'' = 0$ is forbidden. There will, therefore, be a gap in the spectrum at this energy. From these observations, label the R branch and the P branch, stating why you believe your assignments are correct.
2. Assign J' and J'' values to each line in these branches making use of the allowed J values given with Eqs. (7-54) and (7-55).
3. Assign wave numbers to these lines [$\bar{\nu} = 1/\lambda$].
4. Using the pairs of adjacent J' values (1, 2), (5, 6), and (7, 8), compute $\bar{B}_0$, $\bar{B}_1$, and $\bar{D}$ for each pair and compare them. Comment on the accuracy of the model used. Convert the resulting averages to ergs and kilocalories per mole.

5. Compute α_e and B_e from the averages found in step 4 making use of the relation

$$\bar{B}_v = \bar{B}_e - \alpha_e\left(v + \frac{1}{2}\right)$$

6. Compute $\bar{\nu}_{01}$.
7. The anharmonicity constants for hydrogen halides are

$$\text{HCl and DCl:}\quad 52.05\ \text{cm}^{-1}$$

$$\text{HBr and DBr:}\quad 45.21\ \text{cm}^{-1}$$

$$\text{HI and DI:}\quad 39.73\ \text{cm}^{-1}$$

   Using these values, compute the force constant for the molecule you are studying. Predict $\bar{\nu}_{01}$ for the isotopic partner.
8. Using tables in Herzberg (see the References), compare the computed values with the literature values.
9. Compute and plot to scale the energy levels for your molecule for v = 0, J = 0 to 5 and v = 1, J = 0 to 5 under the following assumptions:
   a. Rigid rotor-harmonic oscillator model
   b. Nonrigid rotor-anharmonic oscillator with rotation-vibration coupling model.

   Comment on differences and how they depend on the molecular spectroscopic properties and J.
10. Making use of the plot constructed in step 9, compare rotational energy level spacings. If you wanted to induce the transition from J = 0 to J = 1 in the v = 0 state, what wavelength in Angstroms would you use?
11. Referring to one of the texts in the References, explain why the intensities of the peaks go through a maximum as one moves away from the band origin in either direction.
12. Explain why there appear to be two absorption peaks in each major region of absorption if you have a hydrogen chloride spectrum.
13. Calculate the isotope effect on $\bar{B}_e$.

PROBLEMS

1. Derive the classical Hamiltonian function for the two-particle system $\{m_A, m_B\}$ in the center of mass system and demonstrate that the internal motion can indeed be described in terms of a pseudo-one-particle system of mass μ.

2. Beginning with Eq. (7-52), derive both Eq. (7-54) and Eq. (7-55).

3. Derive, from Eqs. (7-56) and (7-57), Eqs. (7-58) and (7-59).

7.4 Infrared Vibration Spectra of Polyatomic Molecules

The purpose of this experiment is to obtain infrared spectra of several polyatomic molecules and by comparison with various theoretical models deduce some of their structural properties. The rudiments of group theory, normal mode analysis, and selection rules are introduced as theoretical background.

THEORY. Previous experiments in this text have emphasized the physical properties of diatomic molecules and how they influence the spectra observed in the infrared, visible, and ultraviolet. Basically, the analysis of the nuclear motions responsible for the observed spectra was made straightforward through the use of center of mass coordinates and the reduction of the two-particle diatomic problem to a pseudo-one-particle problem of reduced mass μ. Although many of the notions which were useful in describing the nuclear rotation and vibration of diatomic molecules may be used qualitatively in describing polyatomic molecules, the quantitative details become very complex because of the number of particles involved. Lest we despair it should be pointed out that considerable progress has been made in understanding the vibration-rotation spectra of polyatomic molecules, especially those possessing some elements of symmetry. Through the use of the mathematical discipline of group theory, as applied to the symmetry elements of a given model for a molecule, a great deal of qualitative information about vibrational motion and molecular geometry can be obtained with relatively little effort. The keys which link the properties of mathematical groups to molecular vibrations are the following:

1. The complete set of symmetry operations on the symmetry elements of a molecular model generate a mathematical point group.
2. The normal modes of molecular vibration transform under the symmetry operations exactly as one of the irreducible representations of the point group to which the molecule belongs.

These points suggest that the abstract properties of mathematical groups may be brought to bear on problems concerning the structure of molecules. Both these points will be developed more fully later. First, however, we shall turn our attention to the classical mechanics of nuclear motion in polyatomic molecules.

To begin we shall recall repetitiously that by assuming the validity of the Born-Oppenheimer approximation we may separate electronic and nuclear motion and find, in principle, a potential-energy surface on which the nuclei move. This is, on an ab initio basis, a quantum mechanical problem of great complexity, and therefore, generally speaking, semiempirical methods are employed. With the potential surface at hand we inquire about the nuclear dynamics and generally approach the quantum mechanical problem through the method of the classical analog. The position of a single particle is describable by three coordinates implying that for N particles 3N coordinates are needed. Assuming the N particles form a nonlinear molecule, three of the coordinates may be used to specify the translation of the molecular center of mass and an additional three coordinates may be used to describe the rotations of the molecule. There remain 3N - 6 coordinates to specify the vibrations, and it is these in which we are particularly interested here.

The route to the quantum mechanical Hamiltonian operator is through the classical Hamiltonian function H = T + V, where T is the kinetic and V the potential energy. Choosing Cartesian coordinates to describe the nuclear positions, the total kinetic energy of N atoms is given by

$$T = \sum_{i=1}^{N} \frac{1}{2m_i} (\dot{x}_i^2 + \dot{y}_i^2 + \dot{z}_i^2) \qquad (7\text{-}60)$$

where $\dot{x}_i \equiv dx_i/dt$ and m_i is the mass of particle i. Frequently an alternative set of coordinates, mass-weighted displacement coordinates, is used. These are defined to measure the displacement of an atom i from its equilibrium position $\{a_i, b_i, c_i\}$. The definition is $q_j = \sqrt{m_j}(x_j - a_j)$, etc., and the kinetic energy becomes

$$2T = \sum_{j=1}^{3N} \dot{q}_j^2 \qquad (7\text{-}61)$$

Since these q_js form a complete set of coordinates which span the mathematical space of the molecular system, we may formally write the potential energy as $V = V(q_1, q_2, \ldots, q_{3N})$ assuming that the potential-energy function is time-independent. To make further progress toward the actual form of the classical Hamiltonian we employ an assumption: the dynamics are such that only small displacements from equilibrium occur. This permits an expansion of the potential in a Taylor series about the equilibrium configuration and neglect of high-order terms in the expansion. Proceeding we write

$$V = V(q_1, q_2, \ldots, q_{3N})$$

$$2V = 2V_0 + 2\sum_{j=1}^{3N}\left(\frac{\partial V}{\partial q_j}\right) q_j + \sum_{j,k=1}^{3N}\left(\frac{\partial^2 V}{\partial q_j\, \partial q_k}\right) q_j q_k + \cdots \qquad (7\text{-}62)$$

The partial derivatives are evaluated at the equilibrium configuration. The small displacement assumption allows neglect of terms in the qs of higher order than those of Eq. (7-62). Equation (7-62) may be further simplified by noting that the function $(\partial V/\partial q_j)$ evaluated at the equilibrium configuration will be identically zero for all q_j (by the definition of equilibrium). Furthermore, we may choose the origin of the potential arbitrarily; $V_0 = 0$ is often chosen. It should be noted that this choice renders nonzero the potential between the atoms when they are infinitely separated from one another. With the origin chosen at the equilibrium configuration Eq. (7-62) becomes

$$2V \cong \sum_{j,k=1}^{3N} \left(\frac{\partial^2 V}{\partial q_j \, \partial q_k}\right) q_j q_k = \sum_{j,k=1}^{3N} f_{jk} q_j q_k \tag{7-63}$$

where f_{jk} is called a force constant.

In terms of the qs, H becomes

$$H = \frac{1}{2} \sum_{j=1}^{3N} \dot{q}_j^2 + \frac{1}{2} \sum_{j,k=1}^{3N} f_{jk} q_j q_k \tag{7-64}$$

and the Lagrangian L

$$L = \frac{1}{2} \sum_{j=1}^{3N} \dot{q}_j^2 - \frac{1}{2} \sum_{j,k=1}^{3N} f_{jk} q_j q_k \tag{7-65}$$

Lagrange's classical equations of motion furnish 3N simultaneous second-order differential equations of the form

$$\frac{d}{dt}\left(\frac{\partial L}{\partial \dot{q}_j}\right) + \frac{\partial L}{\partial q_j} = 0 \quad \text{or} \quad \ddot{q}_j + \sum_{k=1}^{3N} f_{kj} q_k = 0, \qquad j = 1, 2, \ldots, 3N \tag{7-66}$$

One possible solution for these equations has the form

$$q_j = A_j \cos(\sqrt{\lambda}\, t + \phi) \tag{7-67}$$

where A_j is an amplitude. Note that $\sqrt{\lambda}$, the frequency in radians per second, and ϕ, the phase shift, are the same for all values of j (e.g., the same for all nuclei). Assuming a solution having the form of Eq. (7-67), substitution of it into Eq. (7-66) furnishes a set of simultaneous linear algebraic equations each of which has the form

$$\lambda A_j - \sum_{i=1}^{3N} f_{ij} A_i = 0 \tag{7-68}$$

or

$$\sum_{i=1}^{3N} A_i (f_{ij} - \lambda \delta_{ij}) = 0 \tag{7-69}$$

where δ_{ij} is the Kronecker delta function which takes the value zero for all pairs {i, j} except when i = j in which case it takes the value unity. A nontrivial solution (the A_is not all equal to zero) is obtained for Eq. (7-69) only if the determinant of the coefficients of the A_is is equal to zero. The determinant of the coefficients represents a polynomial of order 3N in λ and in general has 3N roots, six of which may be shown to be identically zero; these correspond to rotation and translation, as previously discussed. There remain 3N - 6 roots corresponding to the vibrations. Some of these roots may be degenerate, and additional mathematical techniques must be used to handle them. The determinant of the coefficients, called the secular determinant, is shown in the following equation:

$$\begin{vmatrix} f_{11}-\lambda & f_{12} & f_{13} & \cdots f_{1,3N} \\ f_{21} & f_{22}-\lambda & f_{23} & \cdots f_{2,3N} \\ f_{31} & \cdot & f_{33}-\lambda & \cdots \quad \cdot \\ \cdot & \cdot & \cdot & \cdot \\ \cdot & \cdot & \cdot & \cdot \\ \cdot & \cdot & \cdot & \cdot \\ f_{3N,1} & f_{3N,2} & f_{3N,3} & \cdots f_{3N,3N}-\lambda \end{vmatrix} = 0 \tag{7-70}$$

Suppose that we have determined one of the roots of Eq. (7-70); call it λ_m. With it we shall now return to the set of equations, Eq. (7-69), and attempt to find the A_is. In principle, this set is solvable by the method of elimination of variables, and the net result is an elimination of all the A_is except one, say A_1, which remains arbitrary. In passing we note that there is in general a different set of A_i values for each λ_m; therefore, we attach another subscript, m, to the amplitudes, making them A_{im}. If we choose $A_{1m} = 1$ for all m and then define a_{im} as

$$a_{im} = \frac{A_{im}}{\sqrt{\sum_i A_{im}^2}} \tag{7-71}$$

then the set $\{a_{im}\}$ is normalized. That is,

$$\sum_{i=1}^{3N} a_{im}^2 = 1 \tag{7-72}$$

As usual the boundary conditions (initial coordinates and velocities) must be brought to bear in finding the actual solution. They are used to find the coefficients B_m of the equations

$$A_{im} = B_m a_{im}, \qquad i = 1, 2, 3, \ldots, 3N \tag{7-73}$$

This formally completes the solution to Lagrange's equation of motion. However, in a direct sense the results appear to be of little use in solving the quantum mechanical problem because we have not progressed toward finding a useful and explicit form of the classical Hamiltonian.

Further consideration, however, illustrates how the above formulation points the way toward a quantum mechanical solution of the polyatomic small-vibration problem. We shall begin by noting again that a given root, λ_k, of the secular determinant (7-70) furnishes the time dependence of the oscillatory motion of each nucleus. Equation (7-67) demonstrates that, for λ_k fixed, all the nuclei move together in the sense that they pass through their equilibrium configuration at the same instant in time and also move to positions of minimum or maximum extension at the same instant. While the frequency and phase of the motion is identical for each nucleus, the amplitudes in general are not. What we have uncovered here are the normal modes of vibrational motion. That is, each distinct root of the secular equation furnishes the frequency of a normal mode.

As an extension of the foregoing development it is possible to show that in the small-vibration approximation a set of coordinates $\{Q_i\}$ exists which render the Hamiltonian separable. That is, we may write

$$T = \frac{1}{2} \sum_{i=1}^{3N} \dot{Q}_i^2 \tag{7-74}$$

and

$$V = \frac{1}{2} \sum_{i=1}^{3N} \beta_k Q_k^2 \tag{7-75}$$

The form of the wave equation may be shown, after taking into account translation and rotation, to be given to a high degree of approximation by

$$-\frac{\hbar^2}{2} \sum_{i=1}^{3N-6} \frac{\partial^2 \Psi}{\partial Q_k^2} + \frac{1}{2} \sum_{i=1}^{3N-6} \beta_k Q_k^2 \Psi = E\Psi \tag{7-76}$$

This equation shows the distinct advantage of using normal coordinates; namely, Eq. (7-76) may be decomposed into 3N - 6 separate equations, one for each normal mode. The form of the resulting equations is identical to the harmonic oscillator model. The boundary conditions are also the same; hence, solutions take the form

$$\psi(Q_k; v_k) = N_{v,k} \exp\{-\frac{1}{2} \gamma_k Q_k^2\} H(\sqrt{\gamma_k} Q_k, v_k) \tag{7-77}$$

where $\gamma_k \equiv \omega_k/\hbar$, v_k is the vibrational quantum number, N_{vk} is a normalizing factor, H is a Hermite polynomial, and ω_k is the fundamental frequency in radians per second.

Formally, this completes the development of the quantum mechanical small-vibration problem. For a complex molecule containing several nuclei, normal mode analysis is not an easy task. Fortunately, many qualitative features of the small vibrations of molecules may be ascertained using group theory. To that end we shall proceed with a brief description of mathematical groups and several of their properties which are particularly useful in normal mode analysis.

The fundamental ingredients of a mathematical group are (1) a list of elements and (2) an operation connecting the elements. The operation can be anything well defined, for example, ordinary multiplication or addition. In the case of molecular vibrations, the operations will be specified in terms of the symmetry of the molecular model. Whatever the list of elements and the operation connecting them, four basic postulates must be satisfied:

1. Identity: The existence of an identity element E such that the expressions A op E = A and E op A = A are valid for every element A in the list. The label op is taken to represent the mathematical operation.
2. Closure: For every pair of elements {A, B} in the list the relation A op B = C furnishes another element, C, in the list.
3. Inverse: For every element A in the list there exists another element A^{-1} in the list such that the following relation is satisfied: A^{-1} op A = E. A^{-1} is called the inverse of A.
4. Associative: For every triad of elements {A, B, C} the following relation is valid: A op B op C = (A op B) op C = A op (B op C); that is, either pair {A, B} or {B, C} may be combined first. Note that the sequence (left to right) is crucial and is preserved throughout.

When any list and any operation connecting them satisfies these four postulates we have a mathematical group at hand. We shall now summarize without proof several of the useful theorems and definitions from group theory. The application of these notions will be illustrated later using ammonia as an example.

Order. The order, h, of a group is defined as the number of elements it contains.

Classes. Group elements may be divided into classes which are defined in terms of the following relations. If we perform the operation X^{-1} op Y op X = Z using the single element Y and all elements X from the group operations, then the list {Y, Z, ...} is said to form a class.

For example, suppose that {E, A, B, C, D} forms a group; then we may form the class containing A by evaluating the relations (the symbol op is implied and the results are hypothetical)

$$E^{-1}AE = \boxed{A}$$
$$A^{-1}AA = \boxed{B}$$
$$B^{-1}AB = \boxed{A}$$
$$C^{-1}AC = \boxed{A}$$
$$D^{-1}AD = \boxed{B}$$

The quantities A and B appearing in the box form a class. Therefore E, C, and D belong to other classes.

Symmetry Groups. The complete set of symmetry operations which is possible for a given geometric molecular model forms a group.

Symmetry Elements and Symmetry Operations. The elements of molecular symmetry may be divided into five categories and to each element there corresponds one or more symmetry operations:

1. Identity, E: This symmetry element, possessed by every molecular model, is related to the symmetry operation of making no change in any coordinates.
2. Center of symmetry, i: This element is possessed by a molecular model whenever inversion of all the coordinates through the center of gravity produces a system which is indistinguishable from the original. The planar form of benzene is a good example.
3. Axis of symmetry, C_n: A model possesses an n-fold axis of symmetry if rotation through $2\pi/n$ radians or any integer multiple thereof forms an indistinguishable model. A molecule may possess several rotational axes. The highest-order axis (biggest n) is called the principal axis. As an example, consider benzene again. It possesses a C_6 axis perpendicular to the molecular plane and six C_2 axes lying in the molecular plane. The C_6 axis is principal.
4. Plane of symmetry, σ: A given model may possess several symmetry planes. The symmetry operation is a reflection of all the coordinates through the plane. If the plane contains the principal axis of rotation, it is denoted σ_v (vertical symmetry plane); if it is perpendicular to the principal axis, it is denoted σ_h (horizontal symmetry plane). Sometimes there are two types of vertical planes, one passing through nuclei and the other bisecting the angle between two nuclei. The former are denoted σ_v and the latter σ_d. In benzene, for example, ther are $3\sigma_v$ planes and $3\sigma_d$ planes.
5. Rotation-reflection axis, S_n: A molecule possessing this axis will give rise to an indistinguishable reproduction of itself if it is rotated by $2\pi/n$ radians and then reflected through a plane perpendicular to the axis of rotation. Benzene, for example, possesses both an S_3 and an S_6 axis perpendicular to the molecular plane.

This completes the list of symmetry elements and operations. From among them the group elements arise.

Matrix Representations. A representation of a molecular symmetry group is defined as a set of mathematical operators or functions which combine in precisely the same way as the symmetry operations. The representations we search for are square matrices of dimension n which under matrix multiplication behave in the same way as the symmetry operations. For example, if C_3 op $\sigma_v = S_3$, then the matrix representatives of C_3 and σ_v must multiply to give the matrix representative of S_3. Matrix representations may be of two types: reducible or irreducible. The latter are of fundamental significance. A matrix representation is said to be irreducible if no transformation of the type $X^{-1}AX = B$ can be found which renders A in the same blocked out diagonal form for every matrix A in the representation.

Characters of Representations. The character of a matrix is defined as the sum of its diagonal elements. The characters of the representations of symmetry groups are of great utility in working out the symmetry aspects of many molecular properties, including vibrations.

A few rules:

1. The number of irreducible representations is equal to the number of classes.
2. In a given irreducible representation the characters of the matrices corresponding to all elements in one class are equal.
3. All the matrices in a given irreducible representation have the same dimension.
4. The sum over all the different irreducible representations of the squares of the dimensions of the matrices equals h, the order of the group (one term in the sum for each irreducible representation).
5. In a single irreducible representation, the sum over all the matrix representatives of the squares of their characters is equal to h, the order of the group.
6. Given two different irreducible representations, m and n, the products of corresponding characters from m and n when summed over all the symmetry operations are equal to zero. The characters of the matrices of two irreducible representations are thus said to be orthogonal.
7. Given an irreducible representation k and a reducible one j, the number of times the irreducible representation is contained in the reducible one may be calculated by taking 1/h of the sum of products of corresponding characters from the two representations.

An example of the application of group theory concepts is given in the appendix at the end of this section.

EXPERIMENTAL. Obtain the infrared absorption spectra of methane and one or more of NH_3, CO_2, and SO_2 in accordance with directions provided by the instructor. The samples may be introduced into the infrared cell using a standard vacuum line such as that described in Section 6.4. The actual pressure to be used will depend on the spectrometer characteristics and the cell length. A nominal starting point is to make the product of pressure times cell length equal to 200 Torr-cm. If at all possible, the spectrometer should be operated so that the rotational structure associated with each vibrational band can be resolved. Refer to Section 2.3 for additional details of infrared optics and instrumentation.

ANALYSIS

1. Working with a tetrahedral model for methane, carry out a group theoretical analysis of the molecule including
   a. Symmetry elements
   b. Symmetry operations
   c. Group multiplication table
   d. Irreducible matrix representations
   e. Cartesian displacement coordinate representation
   f. Number of normal modes
   g. Determination of infrared activity of fundamental normal modes
   h. Determination of the contribution of various internal coordinates to the normal modes
2. Compare the observed infrared absorption spectrum with that predicted from the above analysis with regard to the number of observed infrared-active modes.
3. In a similar way, compare the vibrational absorption frequencies for ammonia with those predicted in the text. Predict on the basis of a bond length-bond angle representation whether the normal modes involve stretching or bending or both. *Procedure for starting:* We need $3N - 6 = 6$ basis vectors to describe the normal modes. In the bond length-bond angle formulation one generally selects linear stretches of the bonds as members of independent basis vectors. Consider the bond length vectors and how they transform under the symmetry operations of the group. Then consider the bond angle increments and how they transform. The results asked for above follow.
4. What energy separates the major peaks in the ammonia spectrum at 950 cm^{-1}?
5. SO_2 is nonlinear, while CO_2 is linear. What differences does this introduce into the symmetry operations?

O—C—O

Appendix: An Example of the Application of Group Theory

As an example of the use of group theory in the elucidation of the infrared activity of molecular vibrations, consider the molecule ammonia. By hypothesis we shall select the triangular pyramid structure shown in Fig. A7-1 in which the N—H bond lengths are all equal, as are the H—N—H bond angles. This model obviously possesses some symmetry, and we shall now search for its symmetry elements from among the types E, σ_v, σ_h, σ_d, i, C_n, and S_n. First we note that any molecular model has E, the identity, as one of its symmetry elements. The model we have chosen has no center of symmetry, i, because the operation of inversion through the center of mass would turn the ammonia molecule upside

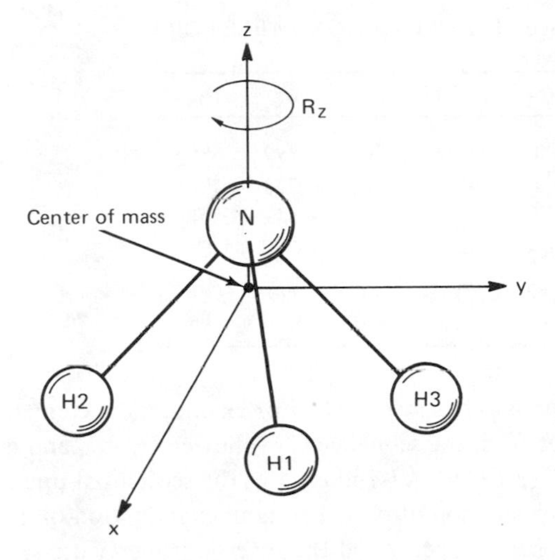

Figure A7-1. Geometric model for ammonia.

down–a condition distinguishable from the starting orientation. The model does, however, possess an axis of symmetry passing through the nitrogen atom and perpendicular to the plane containing the three hydrogen atoms. This is a threefold axis since rotation about it by $2\pi/3$ or $4\pi/3$ furnishes a system which is indistinguishable from the starting configuration. There is no symmetry plane perpendicular to this axis; therefore, there are no σ_h elements. Neither are there any S_n elements. This model for ammonia does have three vertical symmetry planes; each passes through one of the hydrogen atoms and contains the C_3 rotation axis. A list of the symmetry elements for the proposed model is thus $\{E, C_3, \sigma_{v1}, \sigma_{v2}, \sigma_{v3}\}$ and a list of all the symmetry operations is $\{E, C_3, C_3{}^2, \sigma_{v1}, \sigma_{v2}, \sigma_{v3}\}$, where $C_3{}^2$ is the symbol used to denote rotation through $4\pi/3$ radians and the subscripts 1, 2, and 3 on the σ_vs denote vertical planes containing H1, H2, and H3 of Fig. A7-1, respectively. Note the distinction between symmetry elements and operations: it is the complete set of symmetry operations which forms a group. If the above set of elements is complete, it should satisfy all the group postulates. This can be confirmed by working out the group multiplication table to ascertain closure and the existence of inverses. Multiplication is here used in the sense of applying sequentially two operations to the proposed model. The very nature of point symmetry operations ensures the existence of an identity and the associative property. It should be noted that the multiplication table may prove to provide closure and inverses and yet the set of symmetry operations may be incomplete. This situation will arise if the incomplete set of operations is a subgroup of the symmetry group for the molecule. We may, however, be certain that a multiplication table which does not show closure is based on an incomplete set of operations. The multiplication table for the symmetry operations of the ammonia model is given in Table A7-1. By convention

Table A7-1. GROUP MULTIPLICATION TABLE FOR C_{3v}

C_{3v}	E	C_3	$C_3{}^2$	σ_{v1}	σ_{v2}	σ_{v3}
E	E	C_3	$C_3{}^2$	σ_{v1}	σ_{v2}	σ_{v3}
C_3	C_3	$C_3{}^2$	E	σ_{v2}	σ_{v3}	σ_{v1} (circled)
$C_3{}^2$	$C_3{}^2$	E	C_3	σ_{v3}	σ_{v1}	σ_{v2}
σ_{v1}	σ_{v1}	σ_{v2}	σ_{v3}	E	$C_3{}^2$	C_3
σ_{v2}	σ_{v2}	σ_{v3}	σ_{v1}	C_3	E	$C_3{}^2$
σ_{v3}	σ_{v3}	σ_{v1}	σ_{v2}	$C_3{}^2$	C_3	E

the row operation is performed first. For example, the element σ_{v1} in the second row arises from the sequence C_3 followed by σ_{v3} and is denoted symbolically as $\sigma_{v3}C_3$. Figure A7-2 illustrates the sequential operations together with the direct application of σ_{v1}. From an examination of Table A7-1 it is clear that the closure property and the inverse property are satisfied. This completes the list of four requirements for the formation of a group. By examination it becomes clear that the list of six operations on the ammonia model is indeed complete. This set of operations is known as the C_{3v} group and ammonia is said to possess C_{3v} symmetry.

With the group multiplication table at hand we shall begin to search for matrices which form irreducible representations of the group of symmetry operations. To progress along these lines we first take cognizance of the order of the group and then determine its classes. The order h of the group C_{3v} is equal to 6 since there are six distinct symmetry operations. The classes may be developed

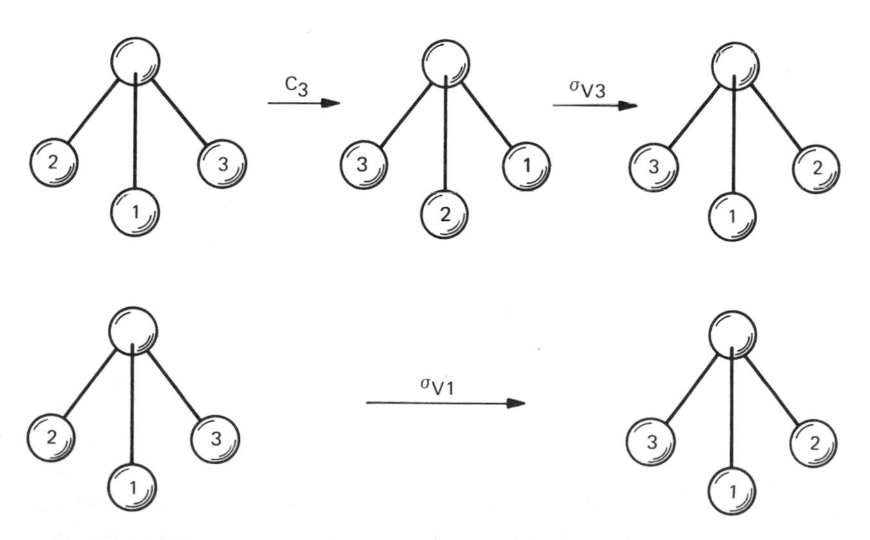

Figure A7-2. Sequential application (product) of C_3 and σ_{v3} and the direct application of σ_{v1} to the ammonia model.

as follows. First we note that E is in a class by itself since for every symmetry operation Y the following relation holds:

$$Y^{-1}EY = Y^{-1}Y = E$$

The remaining classes are developed as follows. Beginning with C_3 we write

$$E^{-1}C_3E = EC_3 = C_3$$
$$C_3^{-1}C_3C_3 = C_3^2C_3^2 = C_3$$
$$(C_3^2)^{-1}C_3C_3^2 = C_3E = C_3$$
$$\sigma_{v1}^{-1}C_3\sigma_{v1} = \sigma_{v1}\sigma_{v2} = C_3^2$$
$$\sigma_{v2}^{-1}C_3\sigma_{v2} = \sigma_{v2}\sigma_{v3} = C_3^2$$
$$\sigma_{v3}^{-1}C_3\sigma_{v3} = \sigma_{v3}\sigma_{v1} = C_3^2$$

To get the inverse elements we have made use of the multiplication table. For example, $C_3^{-1} = C_3^2$ because $C_3^2C_1 = E$. From the above results we conclude that C_3 and C_3^2 constitute a second class. Proceeding again with σ_{v1} we find

$$E^{-1}\sigma_{v1}E = E\sigma_{v1} = \sigma_{v1}$$
$$C_3^{-1}\sigma_{v1}C_3 = C_3^2\sigma_{v2} = \sigma_{v1}$$
$$C_3^{2\text{-}1}\sigma_{v1}C_3^2 = C_3\sigma_{v3} = \sigma_{v1}$$
$$\sigma_{v1}^{-1}\sigma_{v1}\sigma_{v1} = \sigma_{v1}E = \sigma_{v1}$$
$$\sigma_{v2}^{-1}\sigma_{v1}\sigma_{v2} = \sigma_{v2}C_3 = \sigma_{v3}$$
$$\sigma_{v3}^{-1}\sigma_{v1}\sigma_{v3} = \sigma_{v3}C_3^2 = \sigma_{v2}$$

From this result we have a third class consisting of the operations σ_{v1}, σ_{v2}, and σ_{v3}. This completes the placement of all the operations into classes, and in summary we have

class 1: E

class 2: C_3, C_3^2

class 3: $\sigma_{v1}, \sigma_{v2}, \sigma_{v3}$

Since there are three classes, there are also three irreducible representations. Using this result we may proceed by recalling that the sum over all the irreducible matrix representations of the squares of their dimensions ℓ_i^2 must furnish the order of the group. For the ammonia model this sum is written as

$$\ell_1^2 + \ell_2^2 + \ell_3^2 = h$$

where ℓ_i is a positive integer. Obviously the triad $\{1, 1, 2\}$ for $\{\ell_1, \ell_2, \ell_3\}$ is the only set which is satisfactory. Hence, we search for matrices with dimensions 1, 1, and 2 which multiply together as do the group operations. One representation is always valid; namely, the totally symmetric one, as shown in the first row of

Table A7-2. CHARACTER TABLE FOR GROUP C_{3v}

	E	C_3	C_3^2	σ_{v1}	σ_{v2}	σ_{v2}
A_1	1	1	1	1	1	1
A_2	1	1	1	−1	−1	−1
E	2	−1	−1	0	0	0

REDUCED CHARACTER TABLE FOR C_{3v}

	E	$2C_3$	$3\sigma_v$	Coordinate or Rotation
A_1	1	1	1	z
A_2	1	1	−1	Rz
E	2	−1	0	(x, y)(Rx, Ry)
Γ_1	12	0	2	

Note: Γ_1 is a reducible representation developed later in the text.

Table A7-2 in which all the representatives are unit matrices of dimension 1. All other representations must satisfy three requirements in addition to the proper dimension: (1) the sum of the squares of their characters must equal the order of the group, (2) the summation over all the operations in the group of the products of corresponding characters from any two irreducible representations must equal zero, and (3) the characters in any given representation corresponding to operations in the same class must be equal. Armed with these restrictions we search for the second first-order irreducible representation. By direct examination it must be the set $\{1, 1, 1, -1, -1, -1\}$ as shown in the second row of Table A7-2. The third representation must be constructed of two-dimensional matrices. The character of E (always the unit matrix) is 2, and the characters of the other representatives may be worked out as in Table A7-2 by trial and error using the above three restrictions. This completes the construction of the character table, and it should be noted that we have not yet been forced to develop the actual elements of the two-dimensional matrix representations. The character table itself is normally given in a reduced form with all the elements in one class lumped together. The lower part of Table A7-2 illustrates this.

With the character table in hand we may now ask how the application of the group operations to various coordinates and functions describing the NH_3 model transforms these coordinates and functions. First consider any point on the z axis of Fig. A7-1. By straightforward application of the symmetry operations it is clear that we may write

$$Ez = [1]z$$

$$C_3z = [1]z$$

$$C_3{}^2 z = [1]\, z$$

$$\sigma_{v1} z = [1]\, z$$

$$\sigma_{v2} z = [1]\, z$$

$$\sigma_{v3} z = [1]\, z$$

Clearly there is an isomorphism (one-to-one correspondence) between the operators on the left and the matrices on the right. The set of matrices generated are identical to the totally symmetric representation A_1 of the group C_{3v}. We say then that z transforms according to (or belongs to) the A_1 representation.

Examining both x and y under these operations we immediately find a more complex situation because all the rotation and reflection operations mix x and y coordinates. However, z never changes, as we have shown above. We conclude therefore that the pair of coordinates {x, y} transforms according to the E representation. These results normally appear in conjunction with group character tables as shown in the lower part of Table A7-2. The properties of rotations about the x, y, and z axes are also normally listed. These may be obtained by examining the result of applying the group operators to a circle with a sense (direction) about the axes, for example, R_z of Fig. A7-1. Utilizing R_z we write

$$ER_z = [1]\, R_z$$

$$C_3 R_z = [1]\, R_z$$

$$C_3{}^2 R_z = [1]\, R_z$$

$$\sigma_{v1} R_z = [-1]\, R_z$$

$$\sigma_{v2} R_z = [-1]\, R_z$$

$$\sigma_{v3} R_z = [-1]\, R_z$$

We note that none of these operations change the orientation of the plane of the rotational motion; however, the reflections change the direction of the rotation. Therefore, rotation about the z axis belongs to the A_2 representation. In a similar way it is clear that the pair of rotations about the x and y axes $\{R_x, R_y\}$ transforms as the E representation.

We may use these results in detailing which normal modes of vibration of a molecule will be infrared-active, that is, which normal modes will be excited by absorption of infrared radiation. The process of absorption from the ground vibrational state involves the interaction of the instantaneous dipole moment operator of the molecule with the electromagnetic field, and the key absorption probability integral arising from a perturbation theory treatment of this problem is

$$I = \int \Psi_j{}^*(0)\mathbf{er}\Psi_j(1)d\tau \qquad \text{(A7-1)}$$

where **er** is the instantaneous dipole moment and $\Psi_j^*(0)$ and $\Psi_j(1)$ are the vibrational eigenfunctions of the ground and first excited states of the jth normal coordinate, respectively. Clearly, if **er** is identically zero (i.e., no dipole moment at any time), there will be no infrared absorption. While in some cases it may be quite apparent that either the dipole moment is or is not identically zero, in many other instances it is not. Detailed analysis of normal modes of vibration shows that each of them belongs to an irreducible representation of the molecule. Therefore, the symmetry of the normal modes can be determined in a quite straightforward way using the group character table. Once these symmetries are at hand, their infrared activity may be determined, assuming that the transitions are fundamentals of a normal mode. The integral in Eq. (A7-1) contains two wave functions; $\Psi_j^*(0)$ in the harmonic oscillator approximation belongs to the totally symmetric representation of the molecule, while $\Psi_j(1)$ has the symmetry of the jth normal coordinate ($\Psi_j(1) \propto Q_j \exp\{-bQ_j^2\}$). Therefore, in order that the integral I be nonzero, the normal coordinate Q_j must belong to the same representation as the dipole moment; in particular, Q_j must belong to the same representation as any one of the Cartesian coordinates because the dipole moment belongs to these representations (i.e., A_1 and E for ammonia).

We shall now proceed to examine the ammonia model by attaching Cartesian displacement vectors to each atom and applying the group operations—the result is called a Cartesian displacement coordinate basis. Figure A7-3 illustrates the model and Table A7-3 summarizes the results for the C_3^2 operation. The entries in Table A7-3 may be regarded as forming a matrix which when applied to the column vector $\{x_N, y_N, ..., z_3\}$ furnishes a transformed set of coordinates $\{x_N', y_N', ..., z_3'\}$. As an illustrative example, consider the C_3^2 operation on the set $\{x_3, y_3, z_3\}$. After transformation these become related as shown in Fig. A7-4 to the set $\{x_1', y_1', z_1'\}$. Directly from Table A7-3 we find that

$$\begin{bmatrix} -\frac{1}{2} & +\frac{\sqrt{3}}{2} & 0 \\ -\frac{\sqrt{3}}{2} & -\frac{1}{2} & 0 \\ 0 & 0 & 1 \end{bmatrix} \begin{bmatrix} x_3 \\ y_3 \\ z_3 \end{bmatrix} = \begin{bmatrix} x_1' \\ y_1' \\ z_1' \end{bmatrix}$$

or

$$x_1' = -\frac{1}{2}x_3 + \frac{\sqrt{3}}{2}y_3$$

$$y_1' = -\frac{\sqrt{3}}{2}x_3 - \frac{1}{2}y_3$$

$$z_1' = z_3$$

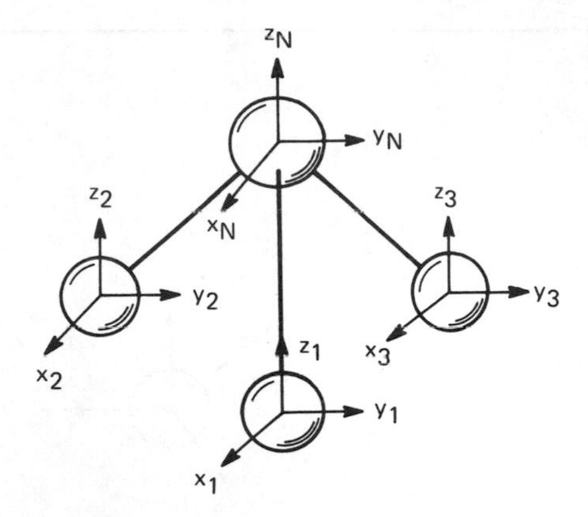

Figure A7-3. Cartesian displacement coordinates for the ammonia molecule.

Table A7-3.

	x_N	y_N	z_N	x_1	y_1	z_1	x_2	y_2	z_2	x_3	y_3	z_3
x_N'	$-\frac{1}{2}$	$+\frac{\sqrt{3}}{2}$	0	0	0	0	0	0	0	0	0	0
y_N'	$-\frac{\sqrt{3}}{2}$	$-\frac{1}{2}$	0	0	0	0	0	0	0	0	0	0
z_N'	0	0	1	0	0	0	0	0	0	0	0	0
x_1'	0	0	0	0	0	0	0	0	0	$-\frac{1}{2}$	$+\frac{\sqrt{3}}{2}$	0
y_1'	0	0	0	0	0	0	0	0	0	$-\frac{\sqrt{3}}{2}$	$-\frac{1}{2}$	0
z_1'	0	0	0	0	0	0	0	0	0	0	0	1
x_2'	0	0	0	$-\frac{1}{2}$	$+\frac{\sqrt{3}}{2}$	0	0	0	0	0	0	0
y_2'	0	0	0	$-\frac{\sqrt{3}}{2}$	$-\frac{1}{2}$	0	0	0	0	0	0	0
z_2'	0	0	0	0	0	1	0	0	0	0	0	0
x_3'	0	0	0	0	0	0	$-\frac{1}{2}$	$+\frac{\sqrt{3}}{2}$	0	0	0	0
y_3'	0	0	0	0	0	0	$-\frac{\sqrt{3}}{2}$	$-\frac{1}{2}$	0	0	0	0
z_3'	0	0	0	0	0	0	0	0	1	0	0	0

The character of this matrix representation of C_3^2 is equal to zero. Proceeding in a similar way with the other operations of the group C_{3V} the set of characters of the representations denoted Γ_1 is developed and is given in Table A7-2. The characters may be obtained using a shortcut method if we note that any atoms moved by a given symmetry operation can contribute nothing to the character

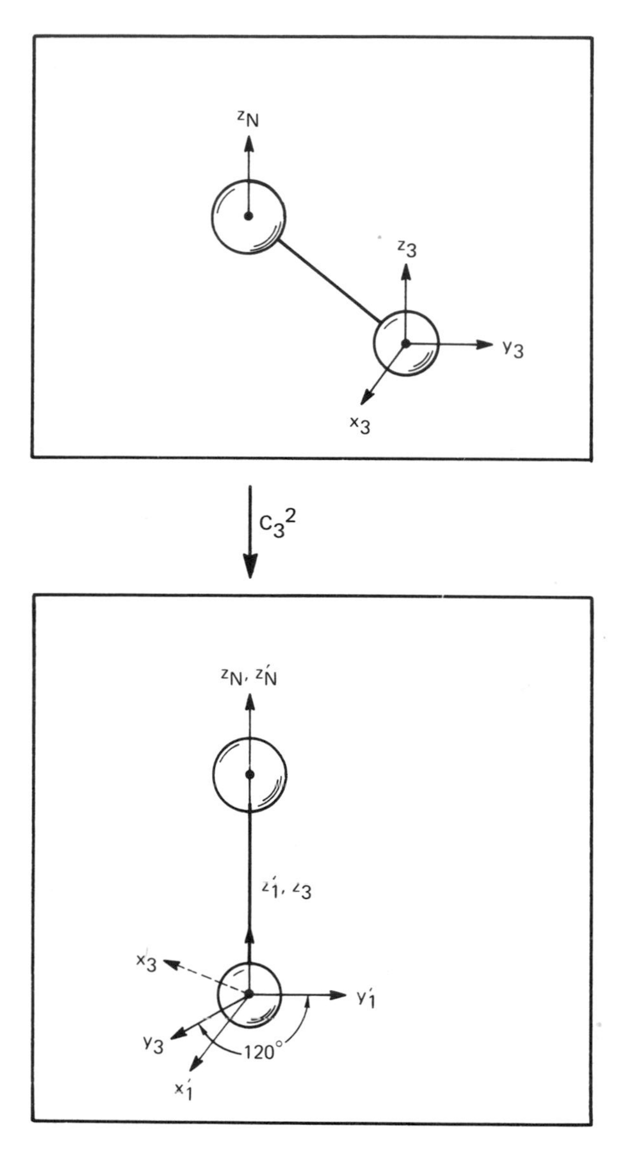

Figure A7-4. $C_3{}^2$ rotation of $\{x_3, y_3, z_3\}$.

of the matrix representing that operation. Only those atoms remaining unchanged need be considered. In the example we have just completed only the nitrogen atom is unmoved; therefore, the only coordinates contributing to the character are $\{x_N, y_N, z_N\}$. The reducible representation Γ_1 may be decomposed into its component irreducible parts using the group theory theorem which states that the number of times a given irreducible representation appears in a reducible one is given by the sum of the products of the elements of the reducible representation with their counterparts in the irreducible representation, the sum being divided by the order of the group. Applying this rule to Γ_1 and A_1 for C_{3V} we write, making use of the lower portion of Table A7-2,

$$(12)(1)(1) + (0)(1)(2) + (2)(1)(3) = 18$$

$$\frac{18}{6} = 3$$

In the summation, the first number in each product is from Γ_1, the second from A_1, and the third from the top line of the reduced character table, which gives the number of elements in each class. Hence, the Cartesian displacement coordinate basis, Γ_1, contains 3 degrees of freedom which correspond to or transform according to the totally symmetric representation, A_1. Similarly we find one A_2 representation and 4E representations in Γ_1. Summarizing we find that

$$\Gamma_1 = 3A_1 + A_2 + 4E$$

From this list we must remove those representations which correspond to nonvibrational degrees of freedom. Translation corresponds to 1 A_1 representation and 1 E representation because translation involves shifts of x, y, and z. Similarly rotation corresponds to 1 A_2 and 1 E representation. This is made clear by examining the right-hand column of the lower portion of Table A7-2. For example, translation along x and y must correspond to an E representation. The remainder of terms in Γ_1 correspond to vibrational degrees of freedom and have symmetries 2 A_1 + 2 E. These four representations correspond to 6 degrees of vibrational freedom because each E corresponds to 2 degrees while each A_1 corresponds to 1.

Having established the symmetries of the normal modes of vibration of the ammonia model, we may now inquire as to their infrared activity. This question is readily answered from Table A7-1. Since z transforms as A_1, any normal mode which transforms as A_1 will be infrared-active. Similarly since {x, y} transforms as E, any mode whose symmetry is the same as E will be infrared-active. For the ammonia model we conclude therefore that all four normal modes will be infrared-active. It should be remembered that these symmetry considerations have nothing to say about the relative absorption intensities.

Other bases besides Cartesian displacement coordinates can be used. For example, the participation of bond length changes and bond angle changes in various normal modes can be ascertained by determining to which irreducible representations the stretching and bending of the bonds belong.

7.5 Electronic Absorption Spectra of I_2 and Br_2

The purpose of this experiment is to obtain the electronic absorption spectra of I_2 and Br_2 and from the results derive features of the ground electronic state.

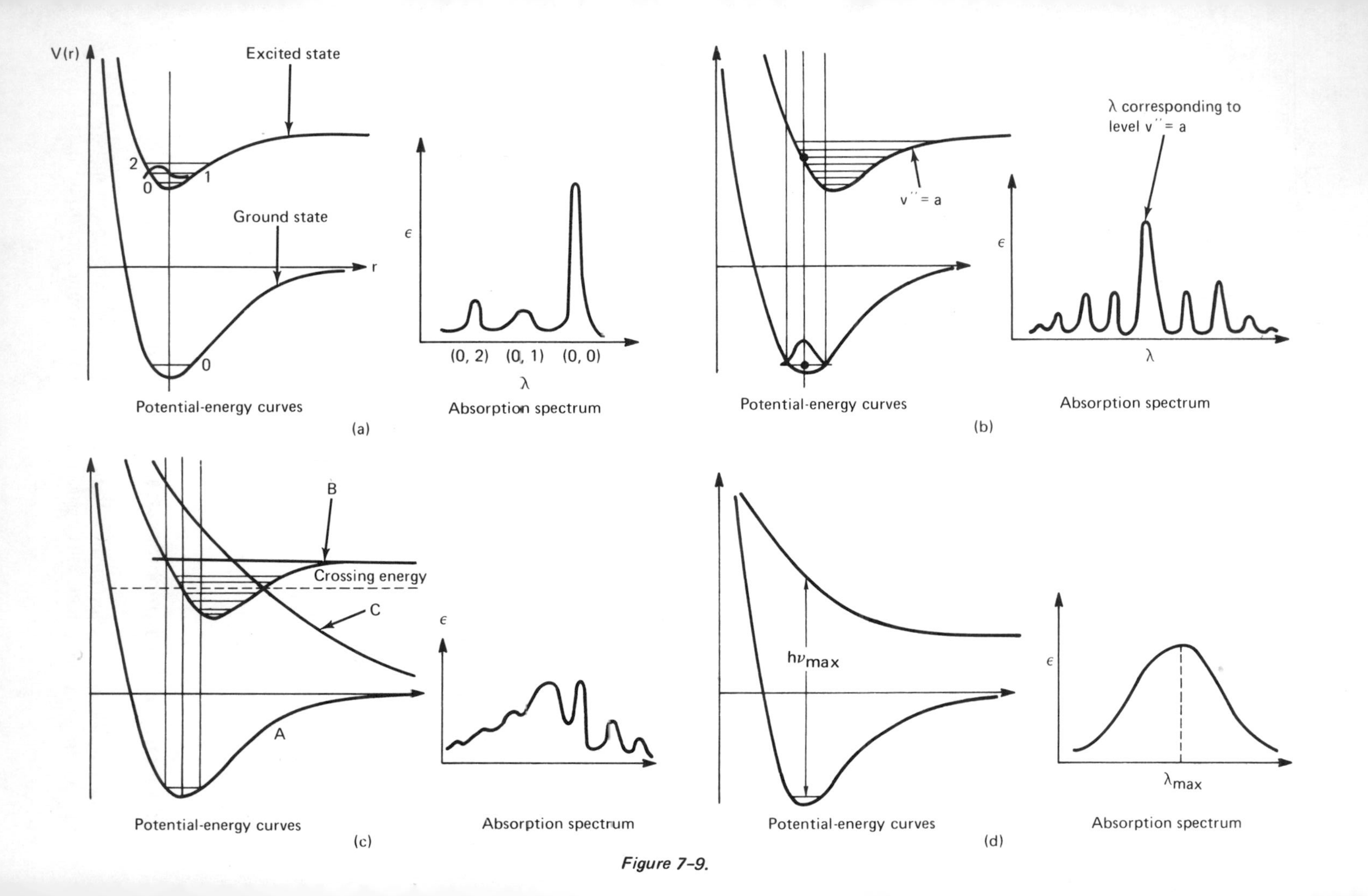
V(r)
Excited state
2
0
1
Ground state
r
0
Potential-energy curves
ϵ
(0, 2)
(0, 1)
(0, 0)
λ
Absorption spectrum
(a)
v'' = a
Potential-energy curves
λ corresponding to level v'' = a
ϵ
λ
Absorption spectrum
(b)
B
Crossing energy
C
A
Potential-energy curves
ϵ
Absorption spectrum
(c)
hν max
Potential-energy curves
ϵ
λ max
Absorption spectrum
(d)

Figure 7–9.

THEORY. The spectroscopy of atomic and molecular systems covers a broad range of wavelengths and physical phenomena. One widely used spectroscopic tool is the visible-ultraviolet absorption spectrum of a molecule, which covers, roughly speaking, the wavelength range 2000–8000 Å, the ultraviolet portion extending from 2000 to 4000 Å and the visible region from 4000 Å (violet) to 8000 Å (red). In this region, the photon energies are sufficient to excite the electrons of an atom or molecule into one of its low-lying electronically excited states. Frequently vibrational and rotational excitation accompanies this change in total energy which occurs upon photon absorption (refer to Section 7.6). This, however, is not necessarily the case; often the electronically excited state has no vibrational or rotational levels associated with it at all. This is another way of saying that it is either unbound or only very weakly bound. These considerations are illuminated below.

In this experiment the absorption spectra of both Br_2 and I_2 are observed and their features are discussed in terms of the potential-energy curves for the two molecules. By means of simple numerical analysis and a simple molecular model the dissociation energy of the ground state of each molecule is calculated.

In its principles, this experiment has a great deal in common with Section 7.6; the experiment depends on absorption of light, one of the molecules involved is I_2, and the potential-energy curves are the same. Suffice it to say that review of the above experiment is desirable. There are differences, however, two of which are essential: First, the present experiment detects absorption rather than fluorescence, and, second, the exciting light is nearly monochromatic but of variable wavelength rather than of a fixed wavelength.

The features of the absorption spectrum of I_2 and Br_2 can be understood by studying Fig. 7-9, which shows schematically several sets of potential-energy curves and the associated electronic absorption spectra. Figure 7-9(a), for example, shows a hypothetical pair of electronic potential-energy curves providing the potential for nuclear motion. In this case both the ground and the excited molecular electronic states are bound. Furthermore, the minima of both states occur at very nearly the same internuclear distance. Recalling that the intensity of absorption is proportional to the square of the vibrational overlap integral it is relatively easy to see that the transition $v' = 0 \rightarrow v'' = 0$ will be the strongest, assuming all transitions start in $v' = 0$ of the ground state. On the other hand, $v' = 0 \rightarrow v'' = 1$ will be very weak, while $v' = 0 \rightarrow v'' = 2$ will be a bit stronger. Problem 1 at the end of this section requires a qualitative study of these features. A spectrum which plots absorption coefficient or extinction coefficient versus wavelength would appear as shown in Fig. 7-9(a).

Figure 7-9(b) is similar to Fig. 7-9(a) except the potential-energy minima are at different internuclear distances; the excited state has the larger equilibrium internuclear distance. In this case the absorption spectrum will be predominately in the region near $v'' = a$, because this is near the most probable internuclear distance in the ground state and according to the Franck-Condon principle the internuclear distance does not change much during absorption. Other wavelengths may be absorbed but not so strongly. Figure 7-9(b) shows by vertical lines the range of r values which are significant in $v' = 0$. By intersection with the electronic excited state these same vertical lines show the range of vibrational states which will dominate the observed absorption spectrum schematically shown in Fig. 7-9(b).

Another arrangement of electronic states is shown in Fig. 7-9(c) in which a second excited state crossing the first and, in this case, completely repulsive plays a role. Note that any molecule on curve C will dissociate; that is, the interatomic distance will tend toward infinity. For clarity, the range of r values important in absorption from $v' = 0$ is again shown. These lines cross both excited states and so we may expect absorption into both; however, absorption into B occurs at much lower energies and we assume here that it is responsible for the observed absorption spectrum. The role of state C is then not to absorb light but to perturb the energy levels of state B and most strongly so when the total system energy is near that level where the two states B and C intersect. Insofar as absorption is concerned the result is a broadening of the energy levels of state B by the continuum of states associated with C. This broadening makes the absorption spectrum diffuse (loss of sharpness of the vibrational structure) for transitions at energies near the crossing energy.

The absorption spectrum associated with the pair of potential-energy curves of Fig. 7-9(d) is continuous because the excited state has no discrete energy levels associated with it–rather a continuum of translational energy levels whose range of r values is governed by the total energy and the steeply rising potential at small internuclear distances.

Turning now to I_2 and Br_2, the important potential energy curves are positioned like those of Fig. 7-9(c) and so we expect the spectrum to show some diffuseness, at least in some wavelength regions, but yet show vibrational structure in absorption throughout the spectrum until the photon energy exceeds the asymptotic value of potential energy curve B. When this occurs the spectrum will become continuous. Thus, we expect the Br_2 and I_2 spectra to appear qualitatively as shown in Fig. 7-10, which shows both the discrete and continuous parts of the spectrum. By appropriate analysis of the data the convergence limit of the measured spectrum can be determined. This is simply the energy corresponding to a transition from $v' = 0$ to the asymptotic value of potential curve B of Fig. 7-9(c). From the energy of the convergence limit the dissociation energy

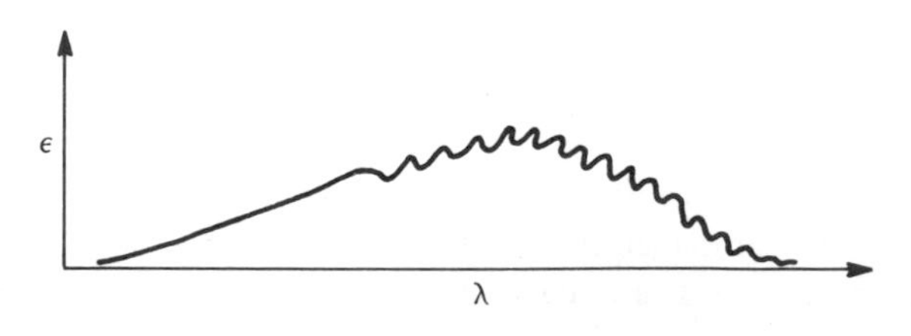

Figure 7-10.

of the ground state, D_0, can be calculated, provided the energy difference between the asymptote of state B and the asymptote of the ground state is known. For Br_2 this difference is 3860 cm^{-1}, while it is 7599 cm^{-1} for I_2. These differences correspond to the energy required to excite either one Br or one I atom from the $^2P_{3/2}$ to the $^2P_{1/2}$ state.

It should be borne in mind that we have taken here a rather simplistic view of one case; namely, the case where the excited-state potential-energy curve is shifted to sufficiently large values of r that the measurable absorption spectrum (which extends only over a limited range of r values) includes significant parts of both the discrete and continuous spectra of the molecules. That the convergence limit may not always appear directly in the spectrum becomes obvious from consideration of the potential-energy curves of Fig. 7-9(a). In this case the convergence limit is obtainable only by mathematical treatment of the data based on some molecular model.

We shall now proceed to develop a method of extrapolation to define the convergence limit using an anharmonic oscillator model (compare Section 7.3) for the molecular electronic excited state. The energies involved are delineated in Fig. 7-11. For any given wavelength of absorbed light, the following equations

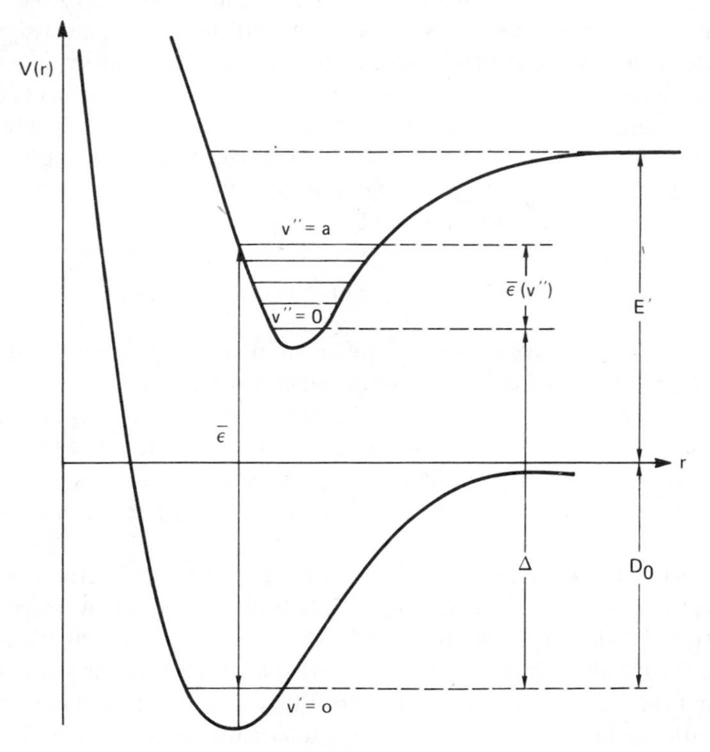

Figure 7-11.

follow from conservation of energy and the anharmonic oscillator model. Expressing energies units of cm^{-1} we have

$$\begin{aligned}\bar{\epsilon} &= \bar{\nu} \\ &= \Delta + \bar{\epsilon}(v'') \\ &= \Delta + v''\bar{\nu}_e - x_e\bar{\nu}_e(v''^2 + v'')\end{aligned} \tag{7-78}$$

where v'' is the vibrational quantum number of the excited state. The vibrational energy of the excited oscillator with respect to $v'' = 0$ is

$$\bar{\epsilon}(v'') = v''\bar{\nu}_e - x_e\bar{\nu}_e(v''^2 + v'') \tag{7-79}$$

As v'' increases, the convergence limit is approached, and because of anharmonicity, the energy difference between two adjacent vibrational levels decreases. As the convergence limit is approached this difference becomes zero. Expressed mathematically this notion provides an extrapolation procedure for locating the convergence limit. The energy difference of two adjacent lines is

$$\Delta\bar{\epsilon} = \bar{\nu}_e - 2x_e\bar{\nu}_e(v + 1) \tag{7-80}$$

where v is the smaller of the two vibrational quantum numbers involved. By analyzing the spectrum, $\Delta\bar{\epsilon}$ values are available, but the appropriate assignment of v values is not clear and so we begin with an arbitrary assignment of $v = 0$ to one of the lines and proceed from there to higher energies in increments of $\Delta v = 1$. Plotting $\Delta\bar{\epsilon}$ versus v should then furnish a linear relationship which falls to zero at some value of v, namely, at the convergence limit for our arbitrarily assigned scale of vs. The energy, measured from $v' = 0$ of the ground state, required to reach this asymptote is given by

$$\bar{\epsilon} = \bar{\epsilon}(v = 0) + \frac{1}{2}v_{c.l.}\Delta\bar{\epsilon}(v = 0) \tag{7-81}$$

where $\bar{\epsilon}(v = 0)$ is the wave number of the line to which $v = 0$ is arbitrarily assigned, $\Delta\bar{\epsilon}(v = 0)$ the difference in energy between $v = 0$ and $v = 1$, and $v_{c.l.}$ the graphically determined vibrational quantum number at the convergence limit. It should be noted that because of experimental uncertainties the experimental point assigned to $v = 0$ may not lie precisely on the plot of $\Delta\bar{\epsilon}$ versus v. For best results the value predicted by the curve should be used in calculations.

EXPERIMENTAL PROCEDURE. Obtain the absorption spectra of I_2 and Br_2 over the wavelength intervals 4500-6500 Å and 3500-6000 Å, respectively. Instructions for the spectrophotometer will be provided. It is advisable first to scan rapidly the above intervals, noting where the structure of the spectra appears, and then to scan the region of interest very slowly. The cell containing I_2 may need to be heated in order to obtain a sufficiently high pressure. For good results the products of the bromine pressure in Torr and the cell length in centimeters should be on the order of 100 to furnish optical densities near unity at the maximum in the Br_2 spectrum. For I_2 good results can be obtained by heating a 5-cm-long cell containing I_2 crystals to about 70°C. At this temperature the vapor of I_2 is a little less than 10 Torr and in a 5-cm cell the optical density will be on the order of unity at the maximum in the I_2 spectrum. A cell constructed of Pyrex is quite satisfactory. The spectral lines will be spaced about 10-Å apart, and for reasonable data to be obtained the spectrometer should be set to transmit bands no more than 3 Å wide.

ANALYSIS

1. Compare the spectra obtained for Br_2 and I_2 and offer an explanation for their differences with regard to wavelength range and complexity. Note that the I_2 spectrum appears to have more than one series of absorption peaks. For example, near 5450 Å two sets are evident.
2. From each spectrum, try to determine the convergence limit by direct examination. Compare this observation with that obtained using the extrapolation procedure outlined above. Use a linear least-squares fitting procedure for the extrapolation. Using the latter as the most reliable means of estimating the convergence limit, calculate the dissociation energy of the ground state (i.e., D_0 of Fig. 7-11) and compare with values generally accepted in the scientific literature.
3. For I_2, analyze each series of lines prevalent in the spectrum and from the results determine the energy difference between the vibrational states $v' = 0 \rightarrow v' = 1$ and $v' = 1 \rightarrow v' = 2$. We leave the detailed development of the appropriate equations to the student. Compare the results of your experiment with the accepted literature values.
4. Assuming that the cell containing the Br_2 is 100% transmitting when the Br_2 is not present and that the extinction coefficient for Br_2 at 4200 Å is 165.5 (see Seery and Britton in the References) liters mole^{-1} cm^{-1}, calculate the concentration of Br_2 in the cell.

PROBLEMS

1. Making use of rough sketches of vibrational wave functions, explain the absorption spectrum of Fig. 7-9(a).
2. Considering the potential-energy curves of Fig. 7-9(a), present an argument supporting the notion that the convergence limit may not be detectable directly from the observed absorption spectrum.
3. Explain why the absorption spectrum of Fig. 7-9(d) is bell-shaped.
4. Derive Eqs. (7-78) and (7-80).

5. Why is it possible to assign arbitrary quantum numbers to the observed lines for the purposes of arriving at Eq. (7-81)?
6. By graphical means, demonstrate the validity of Eq. (7-81).

7.6 Fluorescence of Molecular Iodine

The purpose of this experiment is to examine the fluorescence from an excited electronic state of the iodine molecule. From the observed spectra some of the spectroscopic constants of the I_2 molecule will be calculated.

THEORY. When a ground-state molecule absorbs an ultraviolet or visible photon, the photon's energy is transferred initially to the electronic energy of the molecule and, after shuffling, appears as electronic, vibrational, and rotational energy of molecular excitation. This energy is dissipated in various ways, and the molecule either undergoes a chemical reaction or eventually returns to the ground state. One of the ways in which it can return to the ground state is by fluorescence in which the molecule emits an ultraviolet or visible photon. Fluorescence of I_2 molecules from the $B^3\Pi_{0u}{}^+$ state is the process under study in this experiment.

In discussing the properties of diatomic molecules the notion of a potential-energy curve is a very useful concept. Such a curve furnishes the potential field in which the nuclei move as a function of the internuclear distance. This field is, to a very good approximation, determined by the kinetic and potential energy of the electrons and the internuclear potential energy. The Born-Oppenheimer approximation, as this approximation is known, is discussed more fully in Section 6.5. Figure 7-12 shows several of the low-energy potential curves for the I_2 molecule. The different curves correspond to different configurations of the electrons in the I_2 molecule, with the higher curves having more electronic energy than the lower ones. Of particular interest in this experiment are the two electronic states $A^1\Sigma_g{}^+$ and $B^3\Pi_{0u}{}^+$. The symbols associated with each of these states summarize the characteristics of the wave functions for that state. Σ and Π denote the component of the total electronic angular momentum which lies along the internuclear axis, with Σ denoting zero units of $\hbar$ and Π one unit. This type of symbol is analogous to the S, P, D, etc., symbols used in atoms. The superscript to the left of the Greek letter denotes the spin multiplicity, just as for atoms. The 1 denotes a singlet state in which the spins of all the electrons are paired, while 3 denotes a triplet state. A 2 in this position would denote a doublet state in which one electron is unpaired. The superscript to the right of the Greek letter denotes the symmetry of the electronic wave function on reflection of it through any plane containing the two nuclei. A positive sign denotes no change in the wave function and a minus sign a change in the sign of the wave function when this operation is carried out. The subscripts g and u are

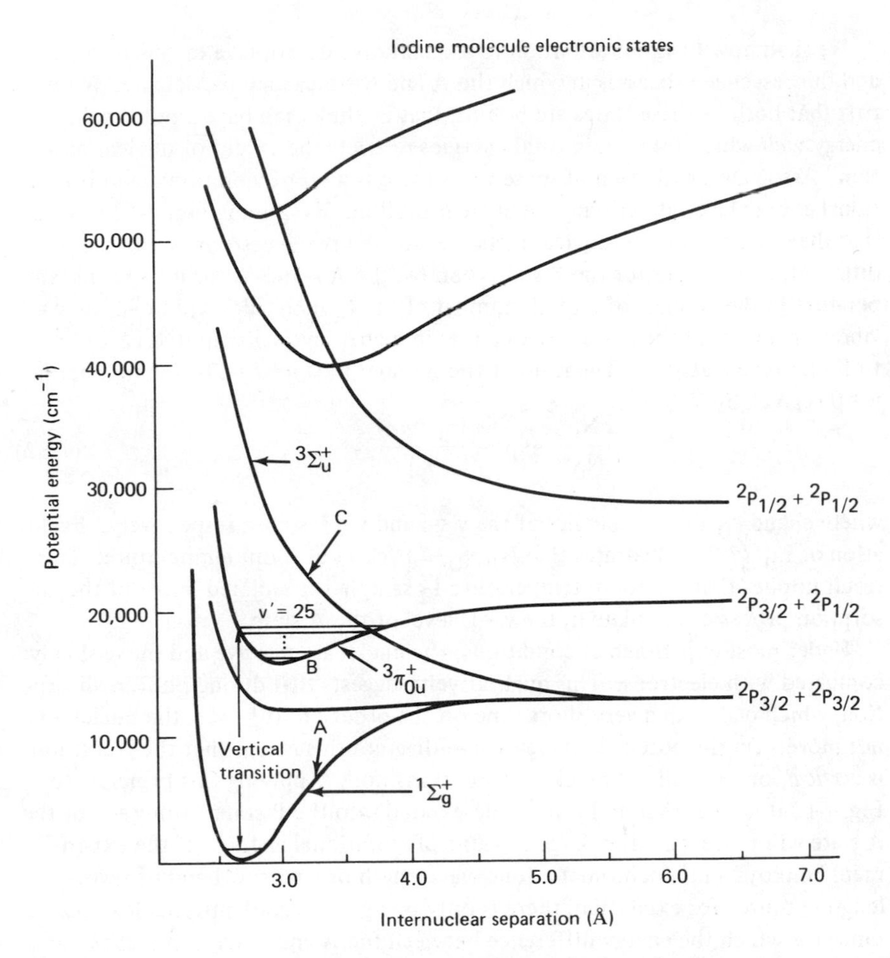

Figure 7-12. Potential-energy diagram for I_2 molecule.

used in describing states of homonuclear diatomic molecules and denote the change of sign (u) or the lack of it (g) when the wave function is reflected through the center of symmetry of the molecule (i.e., through a point on the internuclear axis halfway between the nuclei). The subscript 0 to the right of the Greek Π denotes the coupling between the orbital angular momentum of the electrons and the orbital angular momentum of the nuclei. The symbols A and B arise from group theoretical considerations, with A denoting a totally symmetric state and B denoting, in this case, the lack of symmetry on reflection through the molecular center. These symbols are given in the character tables corresponding to different symmetry groups into which molecules fit (see Cotton and Eyring et al. in the References). Group theoretical considerations are discussed in greater detail in Section 7.4.

We shall now turn our attention to a qualitative description of the absorption and fluorescence processes in which the A and B states may participate. Note first that both of these states are bound; that is, they each have a potential-energy *well* which for certain total energies restricts the extent of nuclear motion. Associated with each of these wells there is a set of vibrational and rotational energy levels describing the nuclear motion. Examining Fig. 7-12 reveals that the equilibrium internuclear distances for the two electronic states are quite different, being larger for the B state than for the A state. At or near room temperature in the absence of radiation, most of the I_2 molecules will be in the v = 0 vibrational level of the A state with a few in higher levels distributed according to Boltzmann statistics. The ratio of the number in state v = i to the number in v = 0 is given by

$$\frac{N_i}{N_0} = \exp\{-\frac{(\epsilon_i - \epsilon_0)}{kT}\} \tag{7-82}$$

where ϵ_i and ϵ_0 are the energies of the v = i and v = 0 states, respectively. Evaluation of Eq. (7-82) illustrates that $N_i/N_0 \ll 1$ for I_2 at room temperature. This result implies that if a room temperature I_2 sample is irradiated, most of the absorption processes originate in the v = 0 level of the A state.

Under most experimental conditions, the nuclei are massive and move slowly compared with electrons. This qualitatively suggests that during photon absorption, which occurs in a very short time on the order of 10^{-17} sec, the nuclei do not move. On the potential-energy curve diagram this means that the transition is *vertical,* or a so-called Franck-Condon transition. Applying this principle to Fig. 7-12 it is clear that an I_2 molecule excited into the B state from v = 0 of the A state will possess a rather large amount of vibrational energy. If the experiment employs a monochromatic (one wavelength or a narrow band of wavelengths) source for excitation, there is only one ground-state internuclear distance for which the energy difference between the A and B states matches the photon energy. Absorption can occur at this wavelength. In the experiment described here the 546.1 nm (5461-Å) mercury green line provides the excitation. The energy of this line produces B-state molecules with 25 quanta of vibrational energy.

Once the I_2 is electronically excited it can lose its energy in a variety of ways, including (1) fluorescence, (2) vibrational relaxation, and (3) collisionally induced dissociation or quenching. In the last process the electronically excited I_2 molecule suffers a collision with either I_2 or some added gas and in the process "crosses over" from the B state to the repulsive $C^3\Sigma_u{}^+$ state shown in Fig. 7-12. Within one vibrational period the result is dissociation. To the extent to which this occurs the fraction of molecules which fluoresce is reduced. Although fluorescence can occur from many vibrational levels of the B state after partial vibrational relaxation, the most intense bands arise from the v′ = 25 state. From this state a wide variety of v″ levels of the A state can be populated. Figure 7-13 illustrates schematically some possible transitions. Energy-wise these begin at the energy of excitation and extend to lower energies (longer wave-

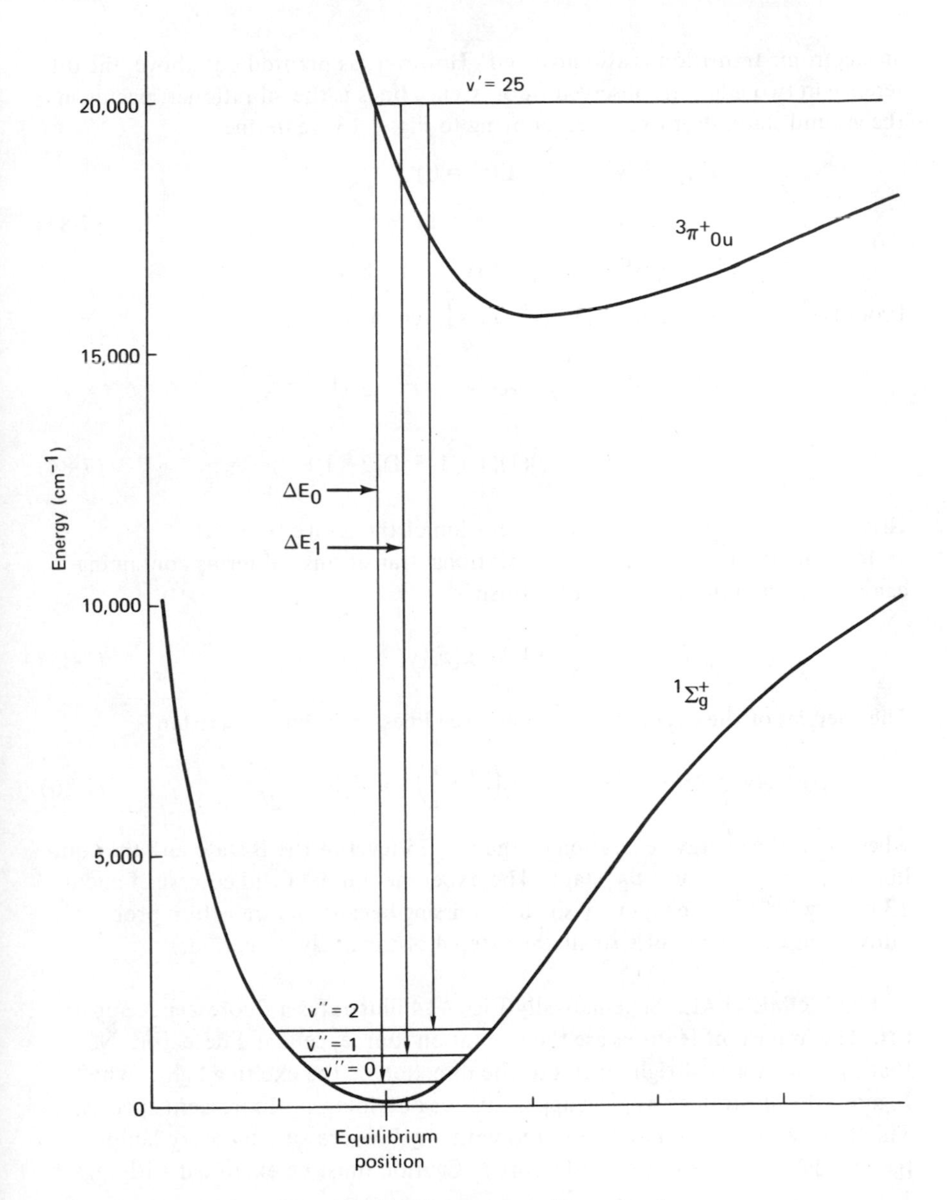

Figure 7-13. Fluorescence of I_2 molecule.

lengths). Notice that the ordinate of Fig. 7-13 is given in cm^{-1}, which is the reciprocal of λ. This may be used as a unit of energy since it is related to ergs by the factor hc (i.e., $E = h\nu = hc/\lambda$). Also note that the observable lines differ by the vibrational energy level spacing of the ground state. Hence, an analysis of the lines should furnish information about the ground state of I_2.

In Section 7.3, a discussion of the vibration-rotation spectrum of a diatomic molecule is given. In the present case, the experimental situation differs in that an electronic transition is also involved. However, as pointed out above, the difference in two adjacent observed fluorescence lines is the vibrational spacing in the ground state. For example, referring to Fig. 7-13 we define

$$\begin{aligned} \Delta E_0 &= E(v' = 25) - E(v'' = 0) \\ \Delta E_1 &= E(v' = 25) - E(v'' = 1) \\ \Delta E_{01} &= \Delta E_0 - \Delta E_1 = E(v'' = 1) - E(v'' = 0) \end{aligned} \tag{7-83}$$

From Eq. (7-49) of Section 7.3 $E(v'', J)$ is given in cm^{-1} as

$$\bar{\nu}(v'', J) = \bar{\nu}_e\left(v'' + \frac{1}{2}\right) + \bar{B}_e J(J+1) - x_e\bar{\nu}_e\left(v'' + \frac{1}{2}\right)^2 - \alpha_e\left(v'' + \frac{1}{2}\right)(J)(J+1) - D_e J^2(J+1)^2 \tag{7-84}$$

Give attention to the physical interpretation of the above terms.

If we are interested only in the vibrational transitions, all terms containing J can be neglected in Eq. (7-84) to furnish

$$\bar{\nu}(v'') = \bar{\nu}_e\left(v'' + \frac{1}{2}\right) - x_e\bar{\nu}_e\left(v'' + \frac{1}{2}\right)^2 \tag{7-85}$$

The energies of the experimentally observed lines may thus be written

$$\bar{\nu}(v' = 25 \longrightarrow v'') = \bar{\nu}_0 - \bar{\nu}_e\left(v'' + \frac{1}{2}\right) + x_e\bar{\nu}_e\left(v'' + \frac{1}{2}\right)^2 \tag{7-86}$$

where $\bar{\nu}_0$ is the energy separation of the $v' = 25$ level of the B state and the equilibrium position of the $^1\Sigma_g{}^+$ state. The experimental data will consist of about 13 lines, with $v'' = 9$ being very small or missing because its transition probability turns out to be quite small. See step 4 of the analysis section.

EXPERIMENTAL. Schematically Fig. 7-14 illustrates a fluorescence apparatus. The important features are the excitation source, cell, and detector. Note that the detector is at right angles to the direction of the exciting light. Why? Figure 7-15 illustrates a typical apparatus for obtaining I_2 fluorescent spectra. The fluorescence lines may be excited with any low-pressure mercury lamp (germicidal lamps are quite satisfactory). Caution must be exercised with regard

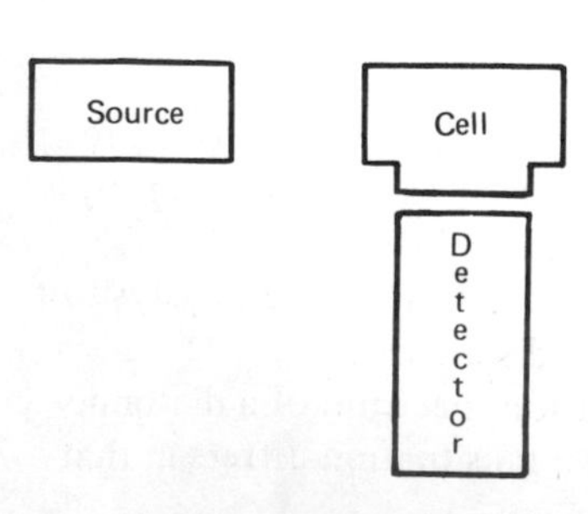

Figure 7-14. Schematic arrangement of fluorescence apparatus.

the intensity of the observed fluorescence line is proportional to the integral:

$$\left[\int_{r=0}^{r=\infty} \Psi^*(v')\Psi(v'')dr\right]^2 \tag{7-91}$$

This is called an overlap integral or Franck-Condon factor and predicts that the observed fluorescent intensity is proportional to the square of an integral whose integrand depends on the product of the vibrational wave function of the ground state with that of the excited state. The integration is over all values of the internuclear distance when applied to a diatomic molecule. Two graphs like Fig. 7-16 will be provided on which the potential curves are plotted and the vibra-

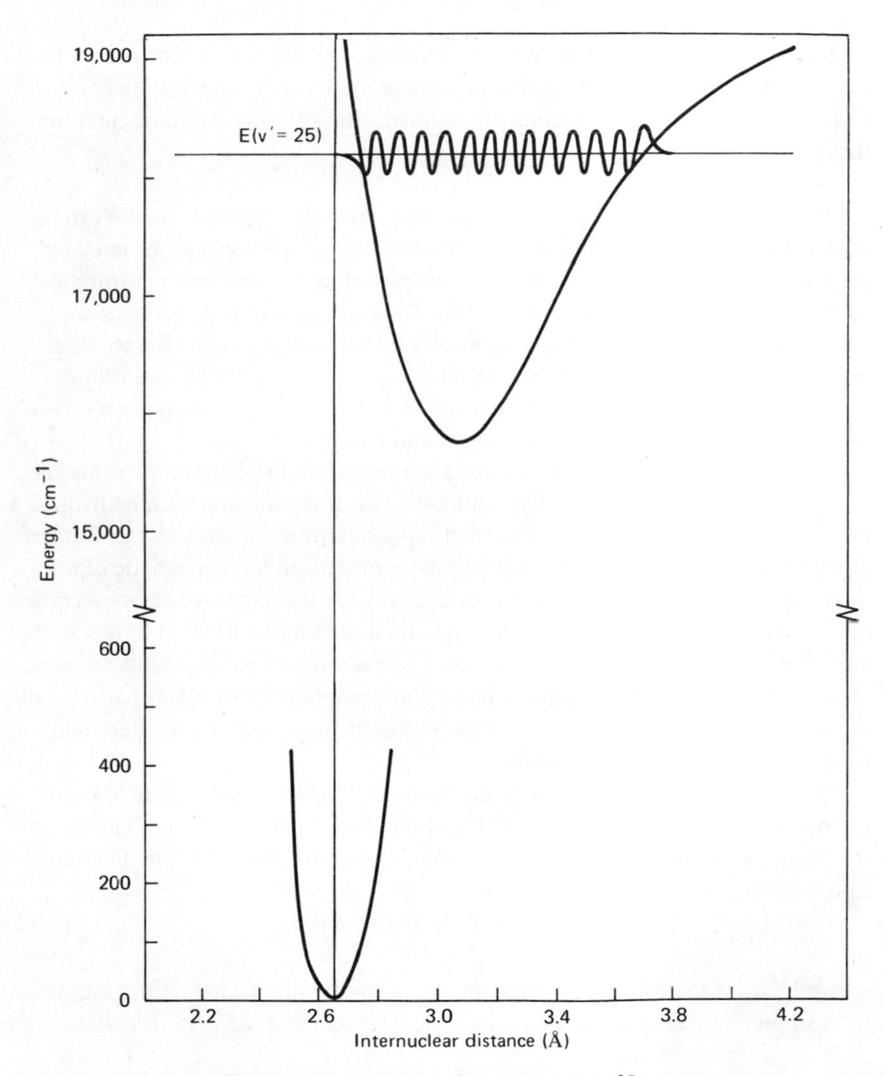

Figure 7-16. Ground and excited states of I_2.

tional wave function of $v' = 25$ is also sketched. On one plot the $v'' = 1$ vibrational wave function and on the other the $v'' = 2$ function. Examine the overlap (i.e., the product) of $\Psi(v' = 25)$ with $\Psi(v'' = 1)$ and also with $\Psi(v'' = 2)$. Predict qualitatively from these considerations which of the two emitted fluorescence lines will be most intense. Does this agree with your experimental observations?

Hint: In making the above evaluation it may be very helpful to transfer the vibrational wave functions onto a single line by plotting amplitude versus r in such a way as to preserve the relative positions of r_e in the two electronic states.

7.7 Nuclear Magnetic Resonance

The purpose of this experiment is to examine the dependence of energy level spacing on magnetic field for different nuclear spin states. In addition some NMR spectra of various molecules are studied. Background based on quantum theory is presented.

THEORY. Just as the fine details of the electronic spectra of atoms can be explained if we postulate an intrinsic electronic spin, the behavior of nuclei in magnetic fields can also be explained if we postulate protons and neutrons as having intrinsic spin angular momentum. By intrinsic we mean that this spin angular momentum is inherent and belongs to the very nature of the particle. An intrinsic property can then be considered as belonging to the definition of the particle. For example, the definition of a proton, a neutron, or an electron should include its spin as well as its mass and charge.

These intrinsic spin angular momenta are postulated to behave quantum mechanically just like orbital angular momenta. Each proton and each neutron in a nucleus has associated with it an intrinsic spin quantum number of $\pm\frac{1}{2}$. Further, any particle having an intrinsic spin angular momentum has a magnetic dipole moment (we postulate intrinsic spin to account for the experimental observation of magnetic dipole moments). This magnetic dipole moment will interact with any other magnetic fields in the system. For example, if we superimpose an external magnetic field H_0 on a particle with a magnetic moment, this particle will experience a force arising from the interaction of the externally applied field with the nuclear magnetic moment.

The potential energy which a magnetic dipole has in a magnetic field is defined as being zero when the dipole and the field are perpendicular. The potential energy as a function of position of the dipole with respect to the field direction is then

$$V_{mag} = -\mathbf{m} \bullet \mathbf{H} = -|\mathbf{m}||\mathbf{H}| \cos \theta \tag{7-92}$$

where V_{mag} is the potential energy, $\mathbf{m}$ the magnetic dipole, and $\mathbf{H}$ the magnetic field strength at the nucleus being studied. The sign of Eq. (7-92) is chosen negative by convention. Examination of Eq. (7-92) shows that the angle between **m** and **H** is important in determining the potential energy. When **m** and **H** are parallel, $\theta = 0$ and the potential energy takes on its smallest possible value.

In a complex nucleus the magnetic moments of the neutrons and protons must be taken together and summed vectorially to give the net intrinsic spin which interacts with the magnetic field at the nucleus, and in many cases that net spin is nonzero. As a result, these nuclei possess an intrinsic magnetic moment which interacts with a magnetic field.

Atomic particles, of which the nucleus is one, are described by quantum mechanics. Such particles behave in nonclassical ways. In the present case when an external magnetic field is applied to the nucleus, the intrinsic angular momentum vector can align itself only in certain directions with respect to the field. These directions are such that the component of the intrinsic angular momentum along the direction of the field is an integer multiple of $\hbar/2$. For example, consider the hydrogen nucleus, which, consisting of a single proton, has a nuclear spin quantum number I of $\frac{1}{2}$. The total intrinsic angular momentum is then $\sqrt{I(I+1)}\hbar = 0.866\hbar$. In the presence of a magnetic field there are only two directions this vector can take with respect to the field, and the resulting angular momentum components along the field are $\pm\hbar/2$. Notice carefully that the component of the angular momentum along the field direction is never as large as the total angular momentum. If it were, the uncertainty principle would be violated.

Now let us consider the problem of a hydrogen atom in a magnetic field. According to quantum mechanics, the description of this system is completely contained in a wave function which depends on the coordinates of the nucleus and electron, the nuclear and electronic spin, and the time. Formally we can write that

$$\mathcal{H}\Psi = -\frac{\hbar}{i}\frac{\partial\Psi}{\partial t} \qquad (7\text{-}93)$$

In Eq. (7-93), $\mathcal{H}$ is the Hamiltonian operator, which we can construct in the usual way from the classical Hamiltonian analog provided we add the interactions of the intrinsic magnetic moments which have no classical analog. Taking the coordinate system as located on the hydrogen nucleus gives

$$\mathcal{H} = (T + V)_{op} \qquad (7\text{-}94)$$

$$T_{op} = -\frac{\hbar^2}{2m}\nabla_e^2 \qquad (7\text{-}95)$$

$$V_{op} = -\frac{e^2}{r} - \mathbf{m}_s \bullet \mathbf{H}_e - \mathbf{m} \bullet \mathbf{H} \qquad (7\text{-}96)$$

where m is the mass of the electron, ∇_e^2 the square of the Laplacian involving the electronic coordinates, e the electronic charge, r the distance of the electron from the nucleus, **m** the nuclear magnetic moment, **H** the magnetic field at the nucleus, $\mathbf{m}_s$ the magnetic moment of the electron, and $\mathbf{H}_e$ the magnetic field strength at the electron. Other interactions could be included in Eq. (7-96); for example, there is an interaction between **m** and $\mathbf{m_s}$. For the purpose of this discussion these will be neglected completely.

Substituting (7-95) and (7-96) into (7-93) gives

$$\frac{\hbar^2}{2m}\nabla_e^2\Psi - \frac{e^2}{r}\Psi - \mathbf{m} \bullet \mathbf{H}\Psi - \mathbf{m_s} \bullet \mathbf{H_e}\Psi = -\frac{\hbar}{i}\frac{\partial \Psi}{\partial t} \tag{7-97}$$

Equation (7-97) is the time-dependent Schrödinger equation for a hydrogen atom in a magnetic field.

Our purpose is to find out what effect the magnetic field at the nucleus has on the energy of the hydrogen atom. That is, we would like to determine the magnetic interaction with the nuclear magnetic dipole. To approach this goal, let us first examine a few particular cases. If we were to remove the magnetic field at the nucleus, Eq. (7-97) becomes

$$\frac{\hbar^2}{2m}\nabla_e^2\Psi' - \frac{e^2}{r}\Psi' = -\frac{\hbar}{i}\frac{\partial \Psi'}{\partial t} \tag{7-98}$$

This equation shows that in the absence of a magnetic field, the hydrogen atom will behave just as though it had no nuclear magnetic moment. As far as we would be able to detect, the energy of a nucleus with $I = \frac{1}{2}$ would be precisely the same as the energy of a nucleus with $I = -\frac{1}{2}$. Since this is the case, we have what is called a twofold degeneracy. Notice that if we now turn on the magnetic field so that our system is again described by Eq. (7-97) this degeneracy no longer exists because the presence of the magnetic field gives rise to an interaction energy (potential energy).

As a second particular case, suppose that we turned on the magnetic field to a fixed strength and looked for the energy levels of the hydrogen atom. Our problem is to find the stationary states of the hydrogen atom in this field. Later we shall discuss some of the factors besides the external fields which determine the field at the nucleus. An hydrogen atom in one of these states is completely characterized by a wave function which does not depend on time. Under these conditions we can write that

$$\Psi(x, y, z, \text{spins}, t) = \Psi(x, y, z, \text{spins}) \exp\left\{-\frac{iEt}{\hbar}\right\} \tag{7-99}$$

Note that $\Psi(x, y, z, \text{spins})$ does not depend on time.

Substituting this equation into Eq. (7-97) furnishes

$$-\frac{\hbar^2}{2m}\nabla_e^2\Psi - \frac{e^2}{r}\Psi - \mathbf{m} \bullet \mathbf{H}\Psi - \mathbf{m_s} \bullet \mathbf{H_e}\Psi = E\Psi \tag{7-100}$$

There are a few important observations which one should make in Eq. (7-100). First, notice that the term $\mathbf{m} \bullet \mathbf{H}$ does not depend on any of the variables associated with the motion of the electrons. Second, the terms $-(\hbar^2/2m)\nabla_e^2$, e^2/r, and $-\mathbf{m}_s \bullet \mathbf{H}_e$ do not depend on any of the variables associated with the nucleus. This kind of situation occurs frequently in quantum mechanical problems, and it permits one to write the total wave function as the product of two separate wave functions; one of these depends on the nuclear coordinates and the other depends on the electronic terms. Stated in another way we sometimes say that there is no interaction between the nuclear and electronic coordinates and that the variables associated with these two are then said to be separable. Let us then write Ψ as the product of $\Psi_{nuc}\Psi_{el}$, where Ψ_{nuc} is the nuclear spin part and Ψ_{el} depends on the electronic coordinates and electronic spin. In addition, since the two kinds of motion are separable, the energies associated to them are separable so that we may write

$$E = E_{nuc} + E_{el}$$

where E_{nuc} depends on nuclear coordinates (space and spin). Substituting these expressions into Eq. (7-100) furnishes

$$\left[-\frac{\hbar^2}{2m}\nabla_e^2 - \frac{e^2}{r} - \mathbf{m} \bullet \mathbf{H} - \mathbf{m}_s \bullet \mathbf{H}_e\right]\Psi_{nuc}\Psi_{el} = [E_{nuc} + E_{el}]\,\Psi_{nuc}\Psi_{el} \tag{7-101}$$

$$\Psi_{nuc}\left[-\frac{\hbar^2}{2m}\nabla_e^2\Psi_{el} - \frac{e^2}{r}\Psi_{el} - \mathbf{m}_s \bullet \mathbf{H}_e\Psi_{el} - E_{el}\Psi_{el}\right] = \Psi_{el}[\mathbf{m} \bullet \mathbf{H}\Psi_{nuc} + E_{nuc}\Psi_{nuc}] \tag{7-102}$$

Equation (7-102) is obtained from (7-101) making use of the facts we just mentioned concerning the variables associated with the operators which are used. For example, ∇_e^2 does not depend on the coordinates of Ψ_{nuc} so that Ψ_{nuc} appears as a constant to this operator. This means that Ψ_{nuc} can be carried to the left of the ∇_e^2 operator.

Examining Eq. (7-102) we observe that since Ψ_{nuc} and Ψ_{el} are independent, their coefficients (quantities in brackets) must both be zero for a nontrivial solution of Eq. (7-102). This leads to two equations, both separately equal to zero. Remember that we were able to form these two equations because of the way in which the coordinates of the operators in the Schrödinger equation depended on the variables of the problem.

We have now succeeded in obtaining what we sought, namely, the energy of the hydrogen atom as a result of the presence of the magnetic field. Since the coefficient of Ψ_{el} in Eq. (7-102) must be zero, we have

$$-\mathbf{m} \bullet \mathbf{H}\Psi_{nuc} = E_{nuc}\Psi_{nuc} \tag{7-103}$$

Note that E_{nuc} arises only because the magnetic field is present.

In the second paragraph of this discussion, we pointed out that the intrinsic nuclear spin angular momenta were postulated to behave exactly like the mechanical orbital angular momenta, for example, the orbital angular momentum of an electron about a nucleus. This means that **J**, the nuclear spin angular momentum vector, and **m**, the nuclear magnetic moment vector, are parallel or antiparallel. For protons, **m** and **J** are parallel and thus are related to each other by a scalar γ, which is called the gyromagnetic ratio. This is written as

$$\mathbf{m} = \gamma \mathbf{J} \tag{7-104}$$

for electrons **m** and **J** are antiparallel. Postulating that the spin angular momentum behaves exactly as its orbital counterpart also means that the vector **J** can be written as an operator multiplied by $\hbar$:

$$\mathbf{J} = \mathscr{I}\hbar \tag{7-105}$$

where $\mathscr{I}$ is the nuclear spin angular momentum operator. Furthermore, if we define the direction of the magnetic field to be along the z axis, we can write that

$$\mathbf{m} \bullet \mathbf{H} = \gamma\hbar\mathscr{I} \bullet \hat{k}H = \gamma\hbar\mathscr{I}_z H \tag{7-106}$$

where $\mathscr{I}_z$ is the z component of the spin angular momentum operator and $\hat{k}$ a unit vector in the z direction.

Using Eq. (7-106) in Eq. (7-103) gives

$$-\gamma\hbar H \mathscr{I}_z \Psi_{nuc} = E_{nuc}\Psi_{nuc} \tag{7-107}$$

Multiplying on the left by Ψ^*_{nuc} and integrating, we obtain

$$-\gamma\hbar H \int \Psi^*_{nuc}\mathscr{I}_z\Psi_{nuc}\, d\tau = E_{nuc} \int \Psi^*_{nuc}\Psi_{nuc}\, d\tau \tag{7-108}$$

Now $\mathscr{I}_z$ operating on the nuclear wave function gives the wave function back multiplied by an integer or half-integer multiple of $\hbar$. This is really what we mean when we say that the spin of a proton is $\pm\frac{1}{2}$. For our example of the hydrogen nucleus this means that $\mathscr{I}_z$ operating on Ψ_{nuc} will give either $\pm\frac{1}{2}\Psi_{nuc}$ depending on which state the wave function is describing. Suppose that the total spin were $\frac{3}{2}$. In this case, the presence of the magnetic field would split the degenerate levels into four distinct levels having $m = +\frac{3}{2}, +\frac{1}{2}, -\frac{1}{2}$, and $-\frac{3}{2}$. Here m denotes the eigenvalue arising from $\mathscr{I}_z$ operating on the nuclear spin function. Making use of the above-described property of the z component of the spin angular momentum operator we can rewrite Eq. (7-108) as

$$-\gamma\hbar Hm \int \Psi^*_{nuc}\Psi_{nuc}\, d\tau = E_{nuc} \int \Psi^*_{nuc}\Psi_{nuc}\, d\tau \quad \text{or} \quad -\gamma\hbar Hm = E_{nuc} \tag{7-109}$$

where

$$m = \pm\frac{1}{2}$$

Notice how Eq. (7-109) predicts that E_{nuc} will depend on the magnetic field at the nucleus of the hydrogen atom (see Fig. 7-17).

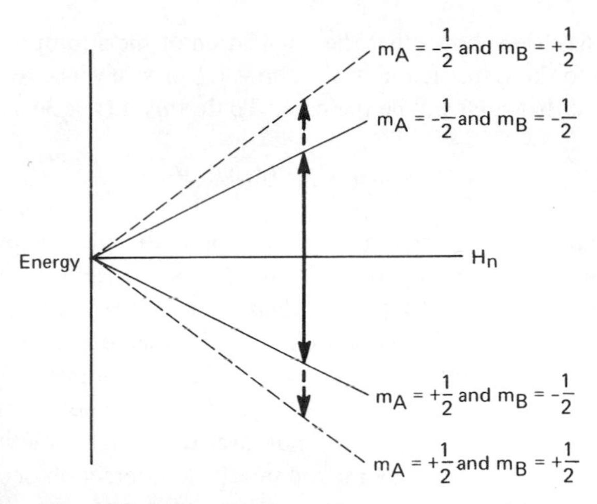

Figure 7-17. Change in energy level spacing of protons in magnetic environment A as a result of spin of neighboring proton. External magnetic field is assumed constant.

Now consider the following question: How much energy is required in going from the state of lower energy to the state of higher energy? This is simply given by the difference in E_{nuc} for $m = +\frac{1}{2}$ and $m = -\frac{1}{2}$. Hence, we can write

$$\begin{aligned} \Delta E &= E_{nuc}(m = -\tfrac{1}{2}) - E_{nuc}(m = +\tfrac{1}{2}) \\ \Delta E &= \gamma \hbar H \end{aligned} \tag{7-110}$$

Now suppose that we have our hydrogen nucleus in a fixed magnetic field and that we would like to induce transitions from $m = +\frac{1}{2}$ or vice versa. To do this we can superimpose on the sample and the magnetic field an electromagnetic wave of just the proper frequency to match the ΔE required for the transition to take place. The electromagnetic wave must be perpendicular to H for transitions to take place. This frequency is called the resonant frequency:

$$\Delta E = \hbar\omega = h\nu = \gamma\hbar H \quad \text{or } \omega = \gamma H \tag{7-111}$$

Note that here, as in all spectroscopic experiments, the experimental observable is the energy difference between two states. Equation (7-111) predicts that ω, the frequency required, is directly proportional to the value of H, the field at the nucleus. Further notice that, in principle, one can choose any H and find a frequency such that Eq. (7-111) is satisfied. In practice only a certain range of values of H and ω is experimentally available, and compatible values of the two must be chosen in order to observe the transition $m = -\frac{1}{2}$ to $m = +\frac{1}{2}$. Typically H is on the order of 10,000 G and ω is on the order of 40 MHz (megacycles per second) for a hydrogen atom. This ω is in the rf (radio-frequency) region of the electromagnetic spectrum.

Thus far it has been shown that the application of the appropriate frequency could give rise to the transition $m = -\frac{1}{2}$ to $m = +\frac{1}{2}$ or vice versa. However, it has not been proved that such will be the case. To do this it is necessary to show that

$$\int \Psi^*_{nuc}(m_i)\mathcal{H}_{pert}\Psi_{nuc}(m_j)\, d\tau \neq 0 \tag{7-112}$$

In Eq. (7-112) the Ψs are wave functions containing the nuclear spin coordinates. $\mathcal{H}_{pert}$ is the perturbation we apply to the system using the electromagnetic radiation. It is simply an additional term in the total Hamiltonian function and would be added to Eq. (7-96) to obtain the Hamiltonian operator for the system when both the magnetic and electromagnetic fields were applied.

If integration of Eq. (7-112) over the range of nuclear coordinates gives zero as the result; this means that $\mathcal{H}_{pert}$ will not give rise to an interaction between the nuclear spin states denoted by m_i and m_j. If no interaction occurs, then no transitions will result from applying the perturbation (in this case the electromagnetic radiation) even if the proper frequency to satisfy Eq. (7-111) was used. The solution to this problem is beyond the scope of this discussion, and only results are given here. At the outset, it was pointed out that the intrinsic spin angular momentum of the nucleus is postulated to behave exactly as orbital angular momenta. The selection rule for magnetic dipole radiation (the kind of interest here) is $\Delta m = m_i - m_j = \pm 1$. Application of this rule indicates that both the $-\frac{1}{2} \rightarrow +\frac{1}{2}$ and $+\frac{1}{2} \rightarrow -\frac{1}{2}$ transitions are "allowed" for the hydrogen atom. From an experimental point of view this selection rule has been shown empirically to be true in NMR. On the other hand, if $I = \frac{3}{2}$, then the transitions between the states

$$m(\tfrac{1}{2}) \rightarrow m(\tfrac{3}{2}), \qquad \Delta m = +1$$

$$m(\tfrac{3}{2}) \rightarrow m(\tfrac{1}{2}), \qquad \Delta m = -1$$

$$m(\tfrac{1}{2}) \rightarrow m(-\tfrac{1}{2}), \qquad \Delta m = -1$$

$$m(-\tfrac{1}{2}) \rightarrow m(-\tfrac{3}{2}), \qquad \Delta m = -1$$

are allowed [i.e., they furnish nonzero values for Eq. (7-112)] but the transitions between

$$m(\tfrac{3}{2}) \rightarrow m(-\tfrac{1}{2}), \qquad \Delta m = -2$$

$$m(\tfrac{3}{2}) \rightarrow m(-\tfrac{3}{2}), \qquad \Delta m = -3$$

$$m(\tfrac{1}{2}) \rightarrow m(-\tfrac{3}{2}), \qquad \Delta m = -2$$

are "forbidden" [i.e., they furnish vanishing values for Eq. (7-112)] since $\Delta m \neq \pm 1$. In summary, Eq. (7-112) gives the selection rules for transitions. That is,

$$\int \Psi^*_{nuc}(m_i)\mathcal{H}_{pert}\Psi_{nuc}(m_j)\, d\tau \neq 0$$

only if $m_i = m_j \pm 1$. The selection rules show that the application of an electromagnetic wave to a sample which is in a static magnetic field will cause transitions between $m = -\frac{1}{2}$ and $m = +\frac{1}{2}$ when the frequency of the electromagnetic wave is chosen to satisfy Eq. (7-111).

At this point, consider what determines H in Eq. (7-111). H has already been defined as being the magnetic field at the nucleus under study and so the problem is to find what factors determine the field at the nucleus.

As a starting point consider an isolated atom which has nuclear spin (for example, a hydrogen atom which has nuclear spin $I = \frac{1}{2}$). In the presence of a magnetic field at the nucleus, the spin degeneracy will be split energetically, and two energy levels will be present. The separation between them will be given by Eq. (7-111). There are three contributions to the magnetic field at the isolated nucleus. The first is due to the presence of a uniform magnetic field applied externally (for example, by placing the sample between the pole faces of an electromagnet). The second arises because of the interaction of the external magnetic field with the electrons. Since the electrons are charged particles, their motion will be influenced by the magnetic field according to the relation

$$\mathbf{F} = -e\mathbf{v} \times \mathbf{H}_0 \tag{7-113}$$

Here **F** is the force, e the electronic charge, **v** the velocity, and $\mathbf{H}_0$ the externally applied field strength. Equation (7-113) predicts that the electrons will rotate as a group in a direction perpendicular to $\mathbf{H}_0$ and will therefore circulate around the direction of the applied magnetic field. Now a circulating electric charge is a closed current loop and gives rise to a magnetic field perpendicular to the plane of rotation. In the present case, since the charge is negative, the direction of this induced magnetic field is opposite the direction of the externally applied magnetic field. The magnitude of this effect is directly proportional to the applied magnetic field. This is called diamagnetic shielding because its effect is to reduce the applied field. The field at the nucleus is written as

$$\mathbf{H} = \mathbf{H}_0(1 - \sigma) \tag{7-114}$$

where σ is called the diamagnetic shielding constant.

The third contribution to the field at the nucleus arises only when an atom or molecule possesses electrons having unpaired spins. When unpaired electrons are present, the atom possesses a net electronic spin angular momentum. Associated with the net spin angular momentum is a magnetic moment which is present even in the absence of an applied magnetic field. This moment is polarized (becomes aligned parallel or antiparallel to the magnetic field direction) when an external magnetic field is applied and serves to change the magnetic field at the nucleus. The interaction of this moment with the field is given in Eq. (7-96). At this point it is important to understand that magnetic moments and magnetic fields are inseparably connected (i.e., the existence of one implies the presence of the other). The effect of unpaired electron spins is called paramagnetic shielding.

In summary, three factors determine the magnetic field at the nucleus of an isolated atom: the externally applied field, diamagnetic shielding, and the effects of unpaired electrons.

To continue, consider an isolated molecule which contains atoms with nuclear spins (by isolated one means that this molecule is subjected to no forces other than those of the applied magnetic field). The addition of other atoms into the neighborhood of a particular nucleus influences the field at that nucleus because they participate in both the diamagnetic and paramagnetic shielding. Note that in most molecules there is no net electronic spin (i.e., all the electrons are *paired*) so that paramagnetic effects will usually not be present. However, motion of the electrons as a group in a molecule occurs under the influence of a magnetic field. Such motion gives rise to diamagnetic shielding. The magnitude of the diamagnetic shielding at a particular nucleus depends on the electronic charge density in the region of that nucleus. The electronic charge density depends on the type of chemical bonds present in the molecule, the geometry of the molecule, and the kinds of atoms present. Recognition of these effects leads one to conclude that the magnetic field at a nucleus should depend on the chemical environment in which it is located. Since Eq. (7-111) predicts that ω, the frequency required to cause the nuclear spin change, is proportional to H, the field at the nucleus, nuclei in different chemical environments should show different transition frequencies or energies.

For example, consider an isolated ethanol molecule:

```
    H       H
     \     /
  H — C ——— C — OH
     /     \
    H       H
```

In this molecule, assume that only the hydrogen nuclei have magnetic moments. There are three distinct positions (or environments). The three hydrogens bonded to the terminal carbon are in magnetically equivalent environments, the two hydrogens bonded to the central carbon are equivalent, and the single hydrogen bonded to the oxygen is in a third distinct magnetic environment. This simple analysis leads to the conclusion that there should be three different resonant frequencies at which transitions should occur.

In addition to the diamagnetic effects discussed above the magnetic field at a particular nucleus is affected by the nuclear spin state of neighboring nuclei which have net magnetic moments. The strongest effects are observed from nuclei no more than three bonds away, and only these will be considered here.

The nuclear magnetic moment of the neighboring nucleus (call it B) interacts with the local electron density, and the effect is transmitted to the nucleus (call it A) being considered. To a first approximation the effect is independent of the externally applied magnetic field and depends only on the nuclear spin state of the neighboring nucleus, because the applied magnetic field does not affect the magnitude of the magnetic moment associated with the nuclear spin. The effect of this interaction is to split the A resonance into a number of lines of slightly different frequency. For example, if the nucleus B has a nuclear spin quantum number of $\frac{1}{2}$, then $m = +\frac{1}{2}$ and $m = -\frac{1}{2}$ characterize the possible nuclear spin states. The field at nucleus A is different for $m = +\frac{1}{2}$ and $m = -\frac{1}{2}$. Therefore, two different resonant frequencies are expected for each allowed transition in nucleus A. This effect is called spin-spin splitting.

For example, consider $CHBr_2CHCl_2$ (Table 7-1).

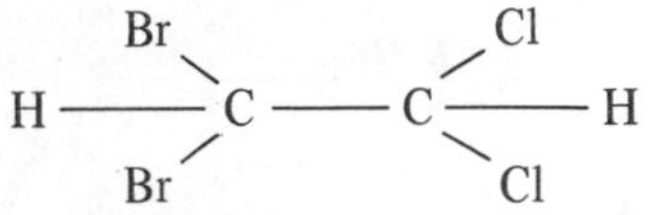

```
        Br              Cl
          \            /
H -------- C -------- C -------- H
          /            \
        Br              Cl
```

Assume that only the hydrogen nuclei have net nuclear spin. Hence, they are the only ones which can interact with an externally applied magnetic field. There are two distinct chemical environments; hence, two resonances are expected if only chemical environments are considered. If the spin-spin splitting effect is taken into account, it is expected that four resonant frequencies will result, which one can characterize as in Table 7-1. Label the hydrogen atom connected with the $-CHBr_2$ group as A and that connected with the $-CHCl_2$ group as B.

Table 7-1. EXPECTED RESONANCES IN $CHBr_2CHCl_2$

Chemical environment	Spin of H_A	Spin of H_B
A	$+\frac{1}{2} \rightarrow -\frac{1}{2}$ $-\frac{1}{2} \rightarrow +\frac{1}{2}$	$+\frac{1}{2}$
A	$+\frac{1}{2} \rightarrow -\frac{1}{2}$ $-\frac{1}{2} \rightarrow +\frac{1}{2}$	$-\frac{1}{2}$
B	$+\frac{1}{2}$	$+\frac{1}{2} \rightarrow -\frac{1}{2}$ $-\frac{1}{2} \rightarrow +\frac{1}{2}$
B	$-\frac{1}{2}$	$+\frac{1}{2} \rightarrow -\frac{1}{2}$ $-\frac{1}{2} \rightarrow +\frac{1}{2}$

Consider what effect the intrinsic magnetic moment (spin) of the proton in environment B has on the resonant frequencies in environment A. In environment A, the frequency required to go from $m = -\frac{1}{2} \rightarrow m = +\frac{1}{2}$ is always equal to the frequency required to go from $m = +\frac{1}{2}$ to $-\frac{1}{2}$. However, the difference in energy between the $m = +\frac{1}{2}$ state and the $m = -\frac{1}{2}$ state will be influenced by the spin state of the proton in environment B. Figures 7-17, 7-18, and 7-19 and Table 7-1 illustrate these phenomena.

As a more complicated case, consider ethanol. Denote the magnetic environment at the terminal carbon as A, that at the central carbon as B, and that at the oxygen end as C. (See Table 7-2 for allowed transitions and indication of spin-spin splitting.)

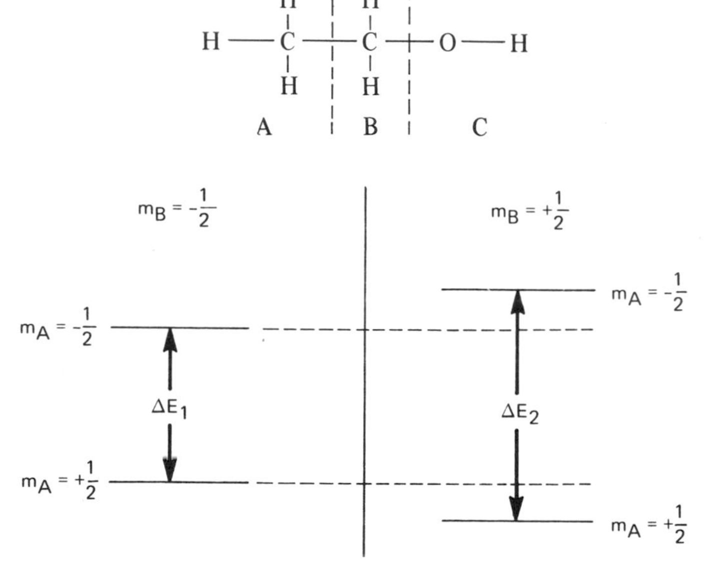

Figure 7-18. Variation of energy of nuclear spin states of A with spin of neighboring nucleus (B).

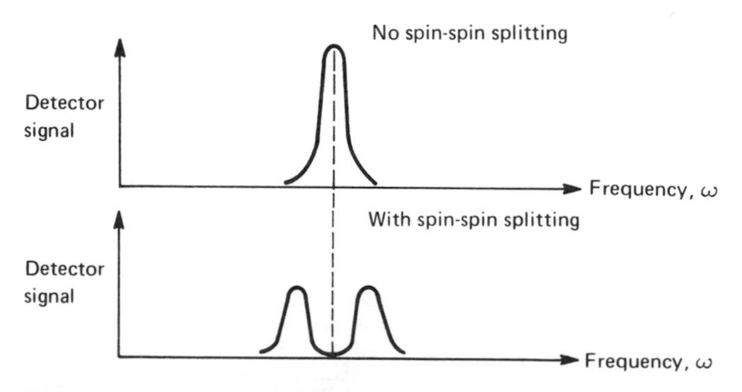

Figure 7-19. Spin-spin splitting of a proton resonance by a single neighboring proton. The external field is assumed fixed.

Table 7-2. EXPECTED RESONANCES IN ETHANOL

Chemical environment	Spin of H_A	Spin of H_B	Spin of H_C
A	$+\frac{1}{2} \longrightarrow -\frac{1}{2}$ $-\frac{1}{2} \longrightarrow +\frac{1}{2}$	$+\frac{1}{2}+\frac{1}{2}=1$	No effect
A	$+\frac{1}{2} \longrightarrow -\frac{1}{2}$ $-\frac{1}{2} \longrightarrow +\frac{1}{2}$	$+\frac{1}{2}-\frac{1}{2}=0$	No effect
A	$+\frac{1}{2} \longrightarrow -\frac{1}{2}$ $-\frac{1}{2} \longrightarrow +\frac{1}{2}$	$-\frac{1}{2}-\frac{1}{2}=-1$	No effect
B	$+\frac{1}{2}+\frac{1}{2}+\frac{1}{2}=\frac{3}{2}$	$+\frac{1}{2} \longrightarrow -\frac{1}{2}$ $-\frac{1}{2} \longrightarrow +\frac{1}{2}$	$-\frac{1}{2}$
B	$+\frac{1}{2}+\frac{1}{2}+\frac{1}{2}=\frac{3}{2}$	$+\frac{1}{2} \longrightarrow -\frac{1}{2}$ $-\frac{1}{2} \longrightarrow +\frac{1}{2}$	$+\frac{1}{2}$
B	$+\frac{1}{2}+\frac{1}{2}-\frac{1}{2}=\frac{1}{2}$	$+\frac{1}{2} \longrightarrow -\frac{1}{2}$ $-\frac{1}{2} \longrightarrow +\frac{1}{2}$	$-\frac{1}{2}$
B	$+\frac{1}{2}+\frac{1}{2}-\frac{1}{2}=\frac{1}{2}$	$+\frac{1}{2} \longrightarrow -\frac{1}{2}$ $-\frac{1}{2} \longrightarrow +\frac{1}{2}$	$+\frac{1}{2}$
B	$+\frac{1}{2}-\frac{1}{2}-\frac{1}{2}=-\frac{1}{2}$	$+\frac{1}{2} \longrightarrow -\frac{1}{2}$ $-\frac{1}{2} \longrightarrow +\frac{1}{2}$	$-\frac{1}{2}$
B	$+\frac{1}{2}-\frac{1}{2}-\frac{1}{2}=-\frac{1}{2}$	$+\frac{1}{2} \longrightarrow -\frac{1}{2}$ $-\frac{1}{2} \longrightarrow +\frac{1}{2}$	$+\frac{1}{2}$
B	$-\frac{1}{2}-\frac{1}{2}-\frac{1}{2}=-\frac{3}{2}$	$+\frac{1}{2} \longrightarrow -\frac{1}{2}$ $-\frac{1}{2} \longrightarrow +\frac{1}{2}$	$-\frac{1}{2}$
B	$-\frac{1}{2}-\frac{1}{2}-\frac{1}{2}=-\frac{3}{2}$	$+\frac{1}{2} \longrightarrow -\frac{1}{2}$ $-\frac{1}{2} \longrightarrow +\frac{1}{2}$	$+\frac{1}{2}$
C	No effect	$+\frac{1}{2}+\frac{1}{2}=1$	$+\frac{1}{2} \longrightarrow -\frac{1}{2}$ $-\frac{1}{2} \longrightarrow +\frac{1}{2}$

Table 7-2. EXPECTED RESONANCES IN ETHANOL (Cont.)

Chemical environment	Spin of H_A	Spin of H_B	Spin of H_C
C	No effect	$+\frac{1}{2}-\frac{1}{2}=0$	$+\frac{1}{2}\longrightarrow-\frac{1}{2}$ $-\frac{1}{2}\longrightarrow+\frac{1}{2}$
C	No effect	$-\frac{1}{2}-\frac{1}{2}=-1$	$+\frac{1}{2}\longrightarrow-\frac{1}{2}$ $-\frac{1}{2}\longrightarrow+\frac{1}{2}$

All the splittings suggested in Table 7-2 are not normally observed because the H atom on the OH group is spatially far removed, on the average, from protons in environment B. As a result the interaction is very weak. As a result we may regard the environments A and B as coupled to give spin-spin splitting, while environment C is isolated. Figure 7-20 illustrates the splitting of a CH_3 resonance by a CH_2 group.

In summary, for an isolated molecule, we expect to find a dependence of the resonant frequencies on chemical environment and interaction with neighboring nuclear spins.

Most of the molecules studied by nuclear magnetic resonance are in the liquid phase and are dissolved in a solvent for study. The effect of the solvent and the total magnetization of the sample must be accounted for. The difficulties encountered in determining the effect of these parameters is overcome by making relative measurements of resonances with respect to some standard. These are called chemical shift measurements. To compare the separations of resonant frequencies for different molecules, each observed resonance is measured with respect to the resonant frequency of some standard molecule. In this very way accurate determinations of the separations between resonances can be made. If the rf electromagnetic field is fixed and the magnetic field varied, then the chemical shift, δ, is defined by the following relation:

$$\delta = \frac{H_s - H_r}{H_r} \times 10^6 \qquad (7\text{-}115)$$

where H_s is the magnitude of the external field corresponding to resonance in the sample and H_r the field for resonance in the reference. A suitable reference is one having a single sharp resonance absorption. A very commonly used reference is tetramethyl silane. Water can also be used.

Since the solvent has approximately the same effect on the value of H_r and H_s, the chemical shift is nearly independent of the solvent. Observe that the parameters of the chemical shift equation [Eq. (7-115)] may also be expressed in frequency units through use of Eq. (7-111). Hence, the separation between the reference and sample signals may be measured in gauss or in hertz. From the determination of all the chemical shifts and the separations between resonances as a result of spin-spin splitting a magnetic characterization of the molecule becomes possible.

To a first approximation all similar nuclei have the same probability of undergoing a transition. This is the equal a priori probability rule. Furthermore, all nuclear spin states are equally probable. Making use of these two approximations we can reexamine Table 7-2 and attempt to determine the relative intensities of the observed resonance absorptions in ethanol. First, since there are three protons on the terminal carbon, two on the center carbon, and one on the oxygen, the overall relative intensities associated with environments A, B, and C neglecting spin-spin splitting is 3:2:1. Taking into account spin-spin splitting of the A environment, Table 7-2 predicts three resonance absorptions resulting from the effect of the nuclear spins of the protons of the CH group. The relative intensities of this group of three absorptions is determined by the total number of ways each total spin can be obtained. For the protons of the CH_2 group, spin = −1 and +1 can be obtained only in one way, while spin = 0 can be obtained in two ways [i.e., $(+\frac{1}{2}, -\frac{1}{2})$ and $(-\frac{1}{2}, +\frac{1}{2})$] assuming that the protons are distinguishable. Hence, the intensity ratio in the resulting group of three resonances associated with the CH_3 group is 1:2:1, as shown in Figure 7-20.

Using the kind of analysis described in the preceding sections one can determine the number of protons in equivalent chemical environments; the magnitude of the chemical shift, δ; and the spin-spin splitting, which is given the symbol J_{AB} and measures the coupling between nuclei in chemical environments A and B. This information furnishes data which are very useful in determining molecular structures. For example, given the empirical formula $C_2H_2Cl_2Br_2$ and the

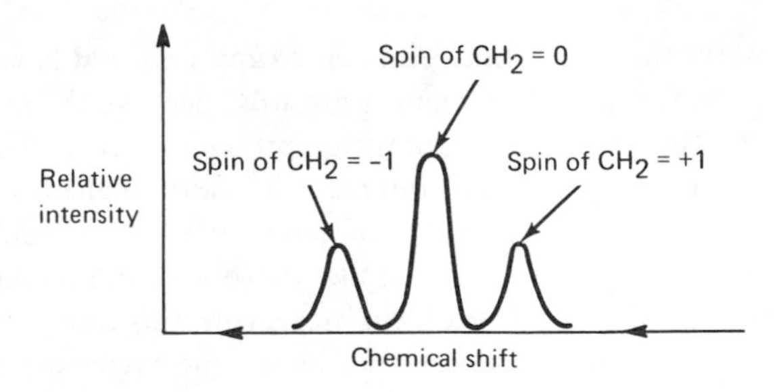

Figure 7-20. Spin-spin splitting of a CH_3 resonance by a CH_2 group.

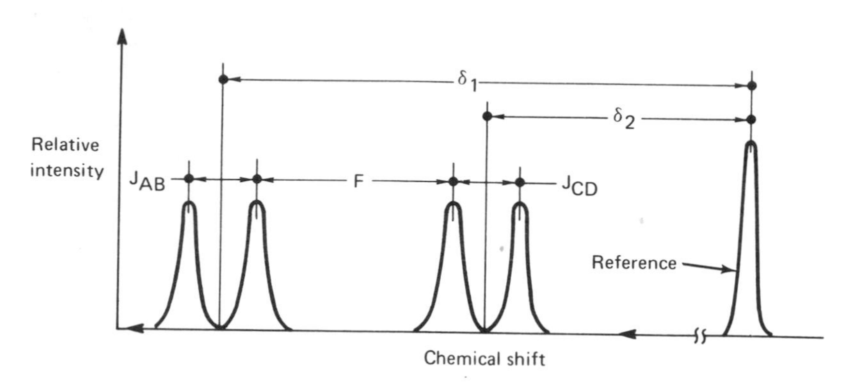

Figure 7-21. Proton NMR spectrum of $C_2H_2Br_2Cl_2$.

proton NMR spectrum shown in Fig. 7-21, the molecular formula can be determined. The two groups of peaks tell one there are two distinct proton environments, A and B. The splitting of each group into two equally intense peaks tells one that the two distinct proton environments are separated by three or less bonds and that each environment contains equal number of protons. Hence, the structure must be

$$\begin{array}{ccccc} & H & & Cl & \\ & \diagdown & & \diagup & \\ Br - & C & - & C & - H \\ & \diagup & & \diagdown & \\ & Br & & Cl & \end{array}$$

Furthermore, note that $J_{AB} \neq J_{CD}$ in general and that the difference in chemical shifts of the two groups is

$$\delta^1 = \delta_1 - \delta_2 = F + \frac{1}{2} J_{AB} + \frac{1}{2} J_{CD}$$

Carefully observe the assumptions and approximations that have been made. When $\delta^1 \cong J_{AB}$ the foregoing treatment is not valid because the intensities are no longer given by the equal a priori transition probability rule. The breakdown occurs because the nuclear spin states interact (in quantum chemistry this is called *mixing*) when $J_{AB} = \delta^1$. This interaction may be thought of as linking the nuclear spin states together, which implies that they can no longer be treated as being independent. One of the problems deals with this case.

Finally, keep in mind that NMR spectra can become extremely complex. Complexity can arise from instrumental limitations, molecular complexities, and molecular interactions. As an example of a moderately complex spectrum, refer to Fig. 7-22.

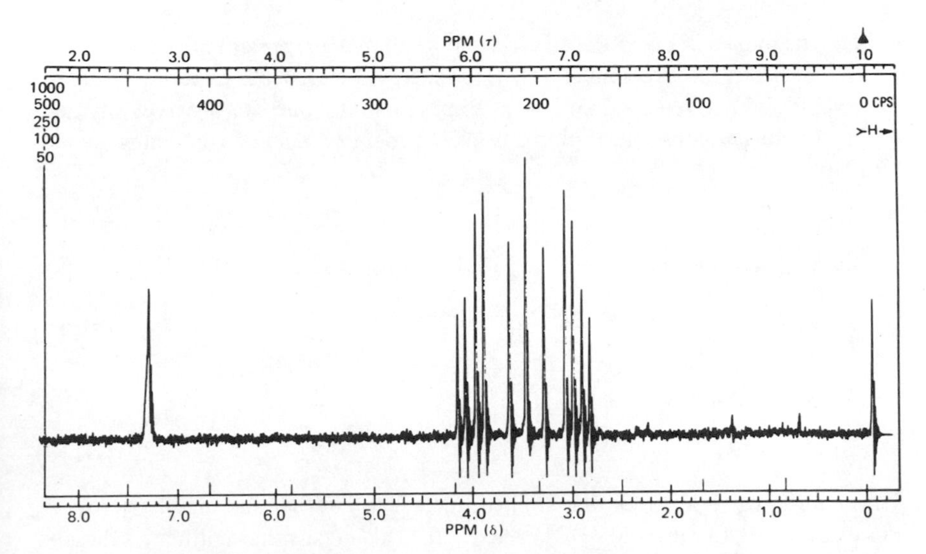

Figure 7-22. The NMR spectrum of dibromopropionic acid and an impurity. The single peak at $\delta = 7.2$ is due to an impurity. The eleven clearly resolved peaks arise from the acid.

EXPERIMENTAL APPARATUS.* In this experiment a relatively simple NMR instrument is used to study some of the magnetic properties of nuclei. In the previous section, the theoretical framework has been laid for understanding this experiment. Recall that nuclear spin states are separated energetically when they interact with an external magnetic field. Further, the magnitude of this splitting is a linear function of the field at the nucleus. Figure 7-23 shows a plot of the energies of the two spin states of a proton as a function of the magnetic

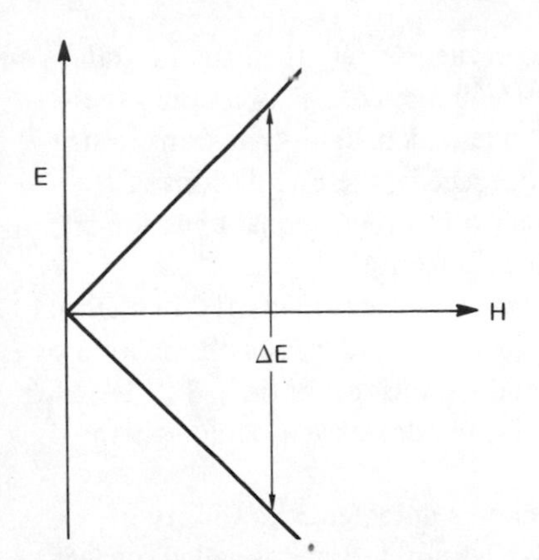

Figure 7-23. Variation of energy states with magnetic field.

*Part of this material follows closely the work of M. H. Profitt and W. C. Gardiner, Jr., of the University of Texas at Austin. See the References.

field at the nucleus. From the figure we note that ΔE represents the energy difference between the two spin states. This energy difference is experimentally observable and is given by Eq. (7-96). Experimentally one can observe only net absorption or net emission of photons having energy ΔE and a frequency given by

$$\nu = \frac{\Delta E}{h} \tag{7-116}$$

Eliminating ΔE from Eqs. (7-96) and (7-116) furnishes

$$\nu = \frac{kH}{h} \quad \text{or} \quad \omega = \frac{kH}{\hbar} \tag{7-117}$$

where $k = \gamma\hbar$. Defining $\alpha_s \equiv k/h$ gives

$$\nu = \alpha_s H \tag{7-118}$$

The quantity α_s varies from one kind of nucleus to another and may be zero. For example, both ^{16}O and ^{12}C have $\alpha_s = 0$. This means that neither of these nuclei have net magnetic moments.

An important consideration in nuclear magnetic resonance experiments is the relative populations of the various states involved. At thermal equilibrium the Boltzmann distribution is applicable and predicts that nondegenerate higher-energy states will have a lower population than the lowest energy state. If the magnetic field and applied frequency are such that Eq. (7-117) is satisfied, two transitions can occur in a proton system:

1. $m = +\frac{1}{2} \rightarrow m = -\frac{1}{2}$
2. $m = -\frac{1}{2} \rightarrow m = +\frac{1}{2}$

Since at thermal equilibrium more species are in the $+\frac{1}{2}$ state than the $-\frac{1}{2}$ state, there will be a net absorption of radiation of frequency, ω. This assumes the applicability of the equal a priori probability rule which states that transitions 1 and 2 are equally likely. As the absorption proceeds, the populations of the two states will become equal, at which point absorption will equal emission and no net change will be detectable. This is called saturation.

If the system is saturated and the applied frequency is turned off, the thermal populations will, in time, be restored. The time required depends on the kind of sample and its physical condition. The longitudinal relaxation time, T_1, describes this phenomena. For a given sample, T_1 will decrease if paramagnetic ions are added to it.

Nuclear magnetic energy level transitions can be detected by a variety of methods. The method we shall use is possibly the simplest and is called nuclear resonance. A detailed description of the electronics of the instrument has been given by Profitt and Gardiner (see the References). Because of the limited resolution of the instrument, it is not suitable for most chemical research problems.

For example, it is of no use in measuring chemical shifts and spin-spin splittings. The instrument is of the *marginal oscillator* type and detects energy absorption. It has four main components, shown schematically in Fig. 7-24:

1. The magnet
2. The marginal oscillator and probe assembly
3. A sawtooth generator
4. An oscilloscope

Some additional details are given in Section 2.4.

The magnet is a variable field permanent magnet which has a range of about 300-3000 G. The field strength can be determined by reading the number of shunt rings from the shunt ring indicator on the magnet, and then translating this information into gauss by using a calibration curve supplied by the instructor. The hysteresis effect in the magnet is taken into account in the calibration by showing two different fields for a given shunt ring setting; one if the field is

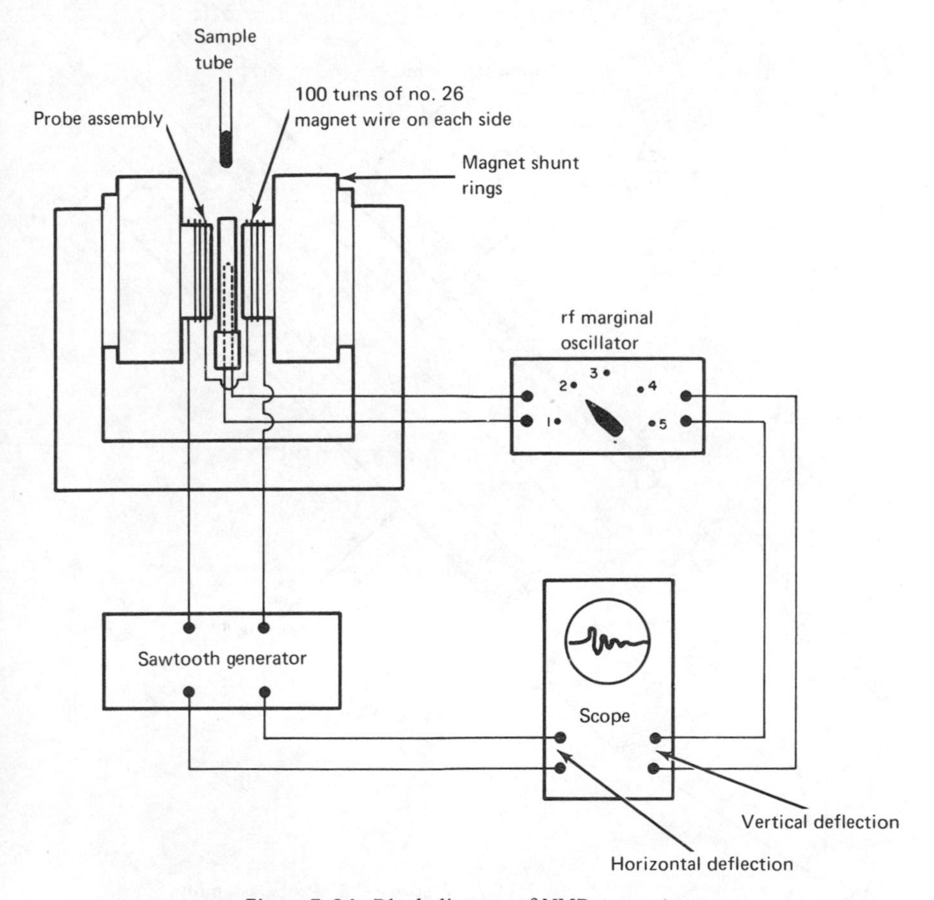

Figure 7-24. Block diagram of NMR apparatus.

increased from 0 to the required field and the other if the field is decreased. Since the magnet pole faces are precision ground for maximum field homogeneity, care must be taken to avoid damaging them.

Samples are contained in thin-walled glass NMR tubes for insertion into the probe assembly. The latter consists of a metal tube containing a Teflon guide for the sample tubes and is shown in Fig. 7-25. The length of the brass tube is not critical and should be designed to allow easy insertion of the sample tube to a position near the center of the magnetic field and so the 34 turns of magnet wire is also near the center of the magnetic field. This coil couples rf energy from the marginal oscillator into the sample. It emits energy in the form of an alternating magnetic field perpendicular to the larger field of the permanent magnet. When the frequency of the oscillator and the magnetic field satisfy Eq.

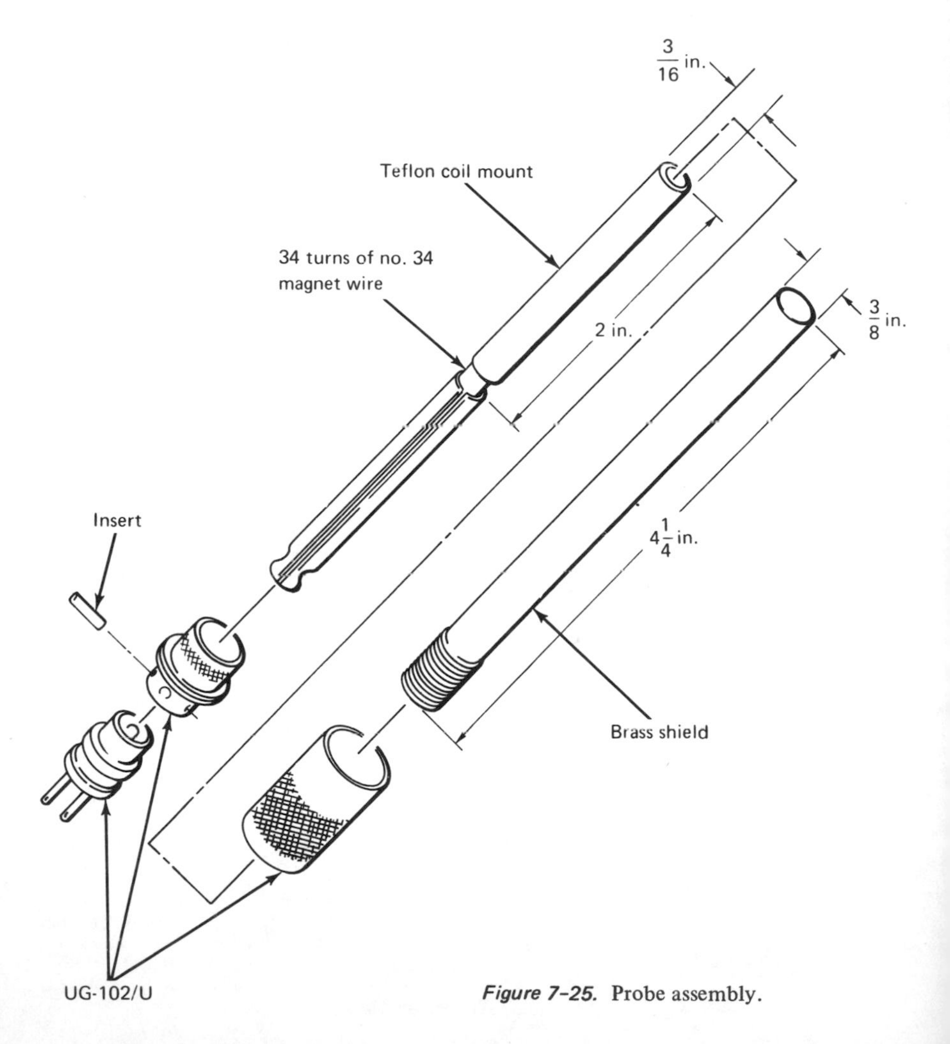

Figure 7-25. Probe assembly.

(7-117), some of the energy will be absorbed, causing a reduction in the voltage across the coil and change in current. The current is carried back to the marginal oscillator where it is balanced with a second current of equal magnitude but 180° out of phase with the initial current sent to the coil. The difference between the two currents is read on the oscilloscope. There are several frequency settings for the oscillator covering the range 11.5–12.5 MHz, and their values will be provided by the laboratory instructor. A circuit diagram of the marginal oscillator is given in Section 2.4 (Fig. 2-66). The output of the marginal oscillator described there will be on the order of 0.2 V when resonance is reached in a solution of aqueous $CuNO_3$. Thus, a very inexpensive oscilloscope is quite adequate. A little thought makes clear what capabilities this apparatus possesses. We can use it at several fixed frequencies and vary the magnetic field strength continuously. The data collected should then consist of a few points off a plot such as in Fig. 7-23. The probe assembly is centered in the magnetic field in order to obtain maximum field homogeneity about the sample. The student should avoid moving it from this position. If the coil in the probe assembly is not centered, a decrease in signal size occurs.

Since for a given marginal oscillator frequency there is only one magnetic field where nuclear resonance will occur, the magnetic field must be passed through the *resonance* condition repeatedly in order that the NMR peak can be displayed on an oscilloscope. A sawtooth generator is used for sweeping (or changing) the magnetic field by small amounts (1–10 G) for this purpose. Coils (100 turns each) are wrapped around the pole faces to carry the current necessary to provide the *sweeping field*. These coils are connected across the pole faces and are wound continuously in the same direction on both sides so that they reinforce. The effect is expressed graphically in Fig. 7-26. Essentially the sawtooth generator provides a short "ramp" which repeats itself many times per second. As a result the magnetic field can be correlated with time as shown in Fig. 7-26. This permits a repetitious timed sweep of the x axis of the oscilloscope to reproduce the absorption "peak" on each cycle.

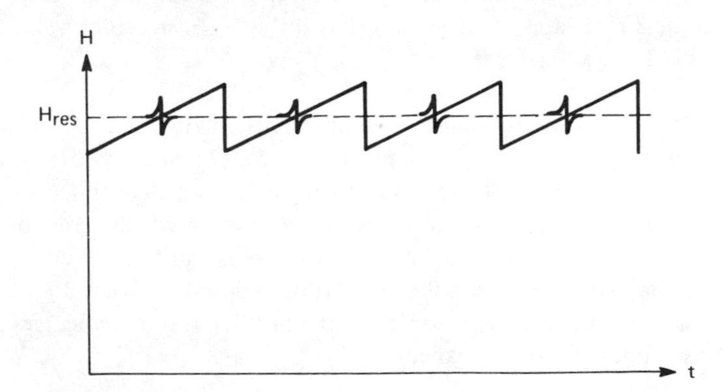

Figure 7-26. Sawtooth wave superimposed on fixed magnetic field.

EXPERIMENT. Using a dilute aqueous solution of copper nitrate and the lowest fixed frequency of the marginal oscillator, vary the magnetic field from lower to higher values. At some particular field the oscilloscope display should show an absorption something like the one in Fig. 7-27 because of the protons of water. Center the peak on the scope face and record the magnitude of the magnetic field. Moving momentarily to higher fields and then back, locate the resonance as the field is scanned downward and record the magnetic field. Repeat these measurements at the other fixed oscillator frequencies.

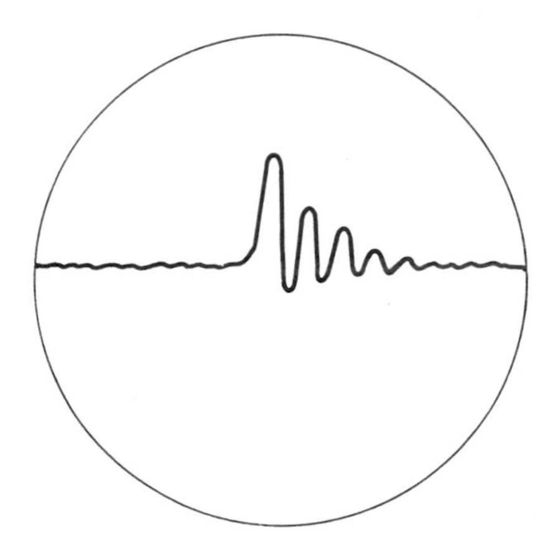

Figure 7-27. Oscilloscope signal at resonance.

Continue by performing the same experiments with a sample of distilled water. These peaks will not be as strong and will appear transiently. It will be interesting to note the change in the peaks as the sample tube is moved in and out of the probe assembly.

With the oscillator at its highest frequency, search for resonances caused by fluorine nuclei in a sample of trifluoroacetic acid. Compute and compare α_S for H and F in these two cases. Why should they differ? The accepted value of α_S for protons is 2.6753×10^{-4} sec^{-1} G^{-1}.

ANALYSIS. Using appropriate Boltzmann statistical relations, calculate the ratio at thermal equilibrium of the number of protons in water whose spins are $-\frac{1}{2}$ to those whose spins are $+\frac{1}{2}$. Is this ratio dependent on the applied field? If so calculate the population ratio for the water resonances which were observed in the aqueous copper nitrate solution at various fields and frequencies.

At saturation, what value does the population ratio take? What does this have to do with the differences observed in the distilled water resonances and the aqueous copper nitrate resonances?

PROBLEMS

The following problems utilize the spectra of Figs. 7-28, 7-29, 7-30, and 7-31.

1. Match the spectra labeled (a), (b), (c), and (d) in Fig. 7-28 with one of the molecules and explain the observed spectra in terms of simple chemical shifts and spin-spin splittings. In all these spectra $\delta' \gg J$.

   molecules: methyl ethyl ketone
   ethylene oxide
   ethanol
   formaldehyde
   acetone
   methyl formate
   N-butane
   methyl acetate

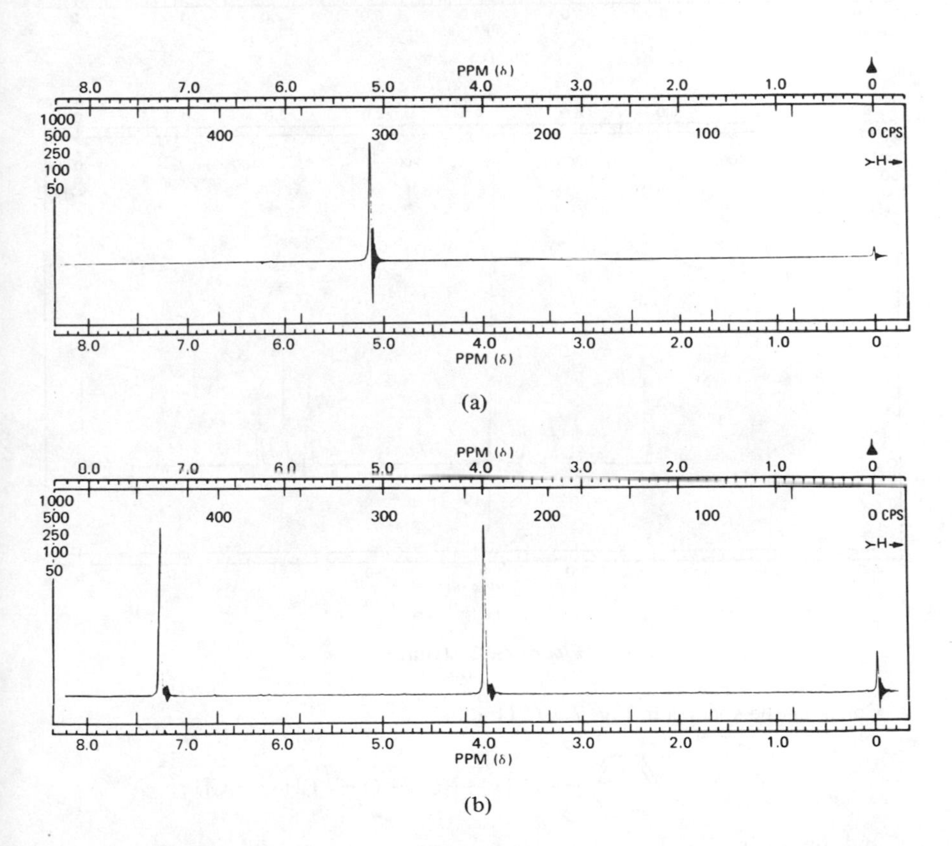

Figure 7–28.

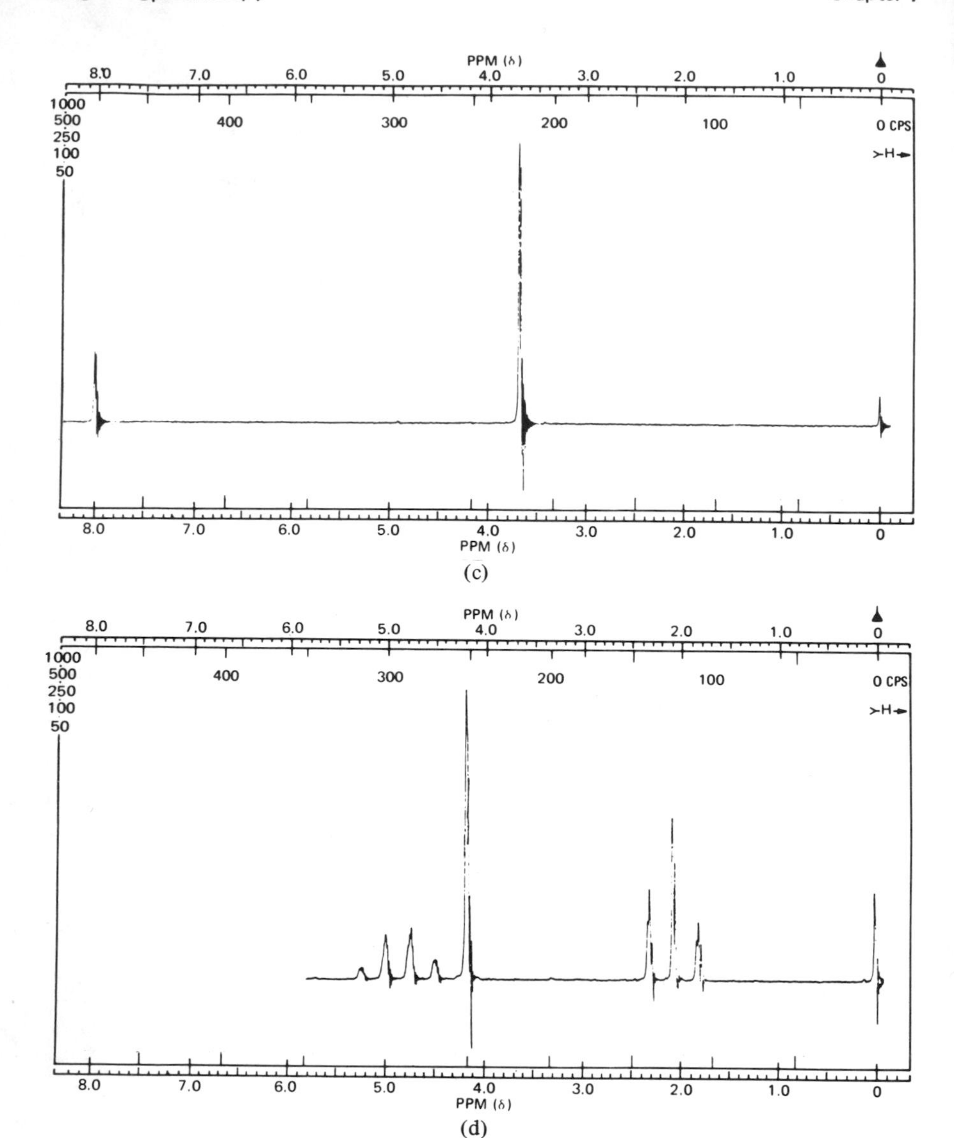

Figure 7–28. (cont.)

2. Consider the spectra in Fig. 7-29. One is

$$C_6H_5-CH_2-\overset{O}{\overset{\|}{C}}-O-CH_2-CH_3$$

and the other is

$$C_6H_5-CH_2CH_2-O-\overset{O}{\overset{\|}{C}}-CH_3$$

Which is which, and why do you think so?

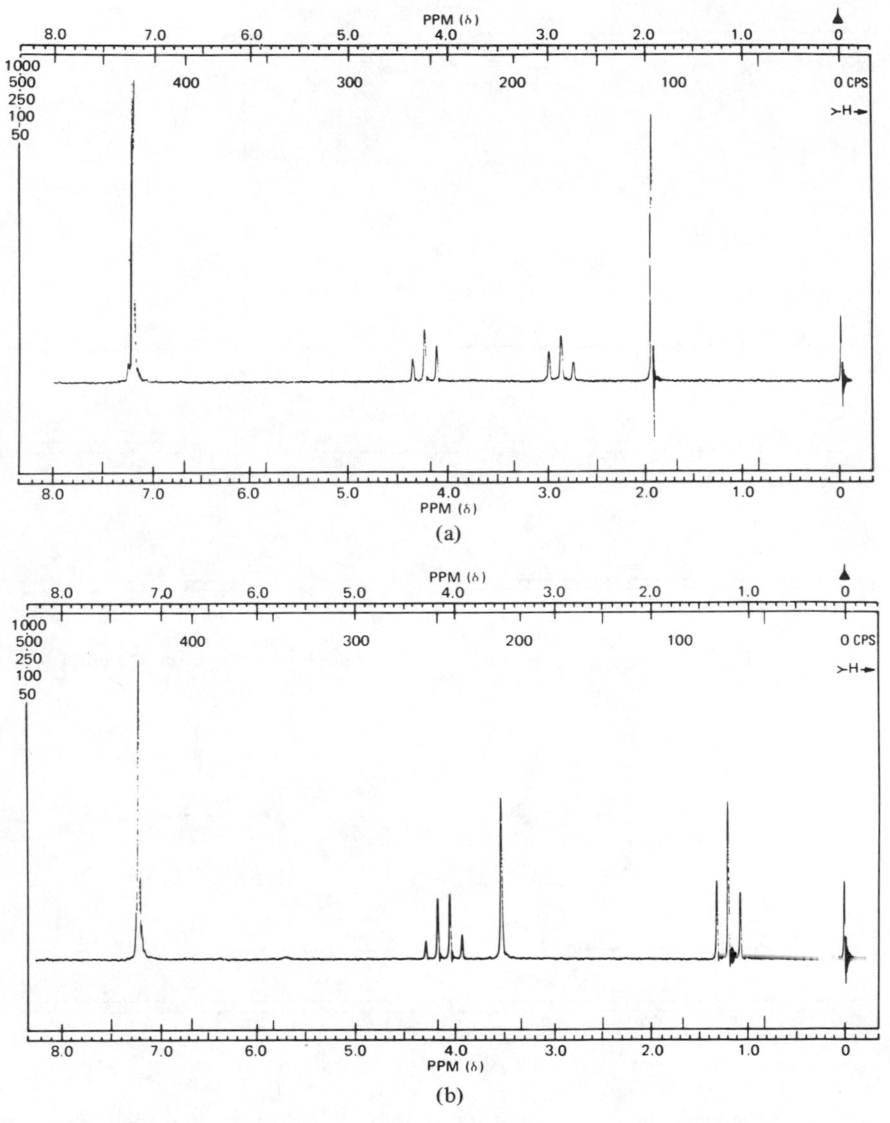

Figure 7-29.

3. Suggest an appropriate structure for the molecule producing the spectrum in Fig. 7-30. It is a six-carbon-ring compound which has the empirical formula $C_{10}H_{12}O_2$.

4. The spectrum in Fig. 7-31 is 2,3-dibromothiophene, a two-proton system for which $J \simeq \delta'$ (i.e., coupling of spins is about as large as the chemical shift). From the spectrum it is clear that four different nuclear energy levels are present. The energy levels are calculable from a quantum mechanical model which permits mixing of different nuclear spin states. The four eigenfunc-

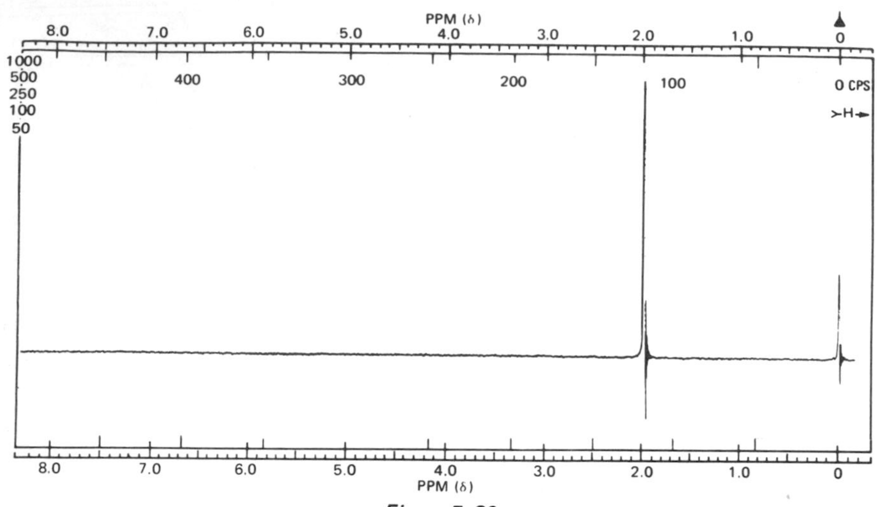

Figure 7-30.

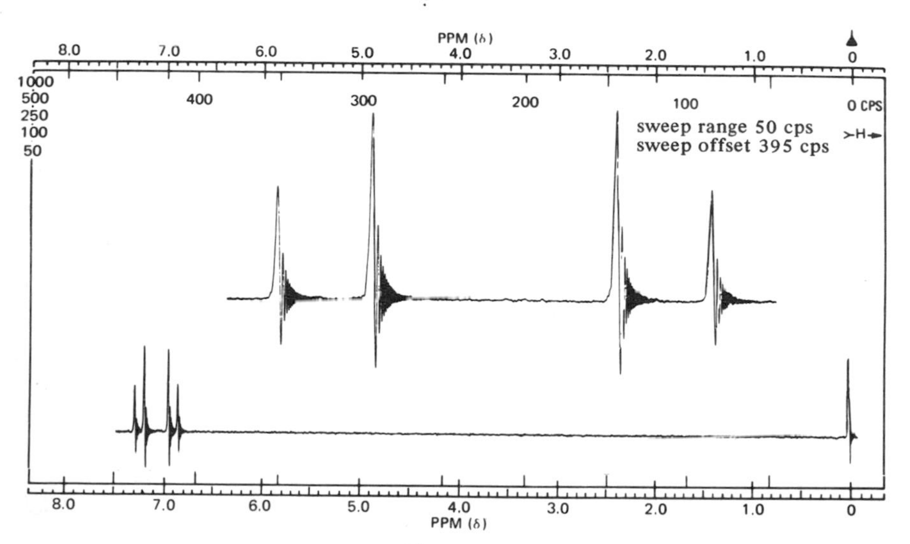

Figure 7-31.

tions of the magnetic part of the total Hamiltonian operator and their corresponding energies are

n	Ψ	Energy
1	$\alpha\alpha$	$\gamma\hbar H(1 - \frac{1}{2}\sigma_A - \frac{1}{2}\sigma_B) + \frac{1}{4}J)$
2	$\alpha\beta \cos\theta + \beta\alpha \sin\theta$	$-\frac{1}{4}J + C$
3	$-\alpha\beta \sin\theta + \beta\alpha \cos\theta$	$\gamma \quad -\frac{1}{4}J - C$
4	$\beta\beta$	$\gamma\hbar H(-1 + \frac{1}{2}\sigma_A + \frac{1}{2}\sigma_B) + \frac{1}{4}J)$

where $\alpha \equiv$ wave function for $+\frac{1}{2}$ spin, $\beta \equiv$ wave function for $-\frac{1}{2}$ spin, and C is defined by the relation

$$C^2 = \frac{1}{2}\gamma\hbar H(\sigma_A - \sigma_B)^2 + \frac{1}{2}J^2$$

$$C \cos 2\theta = \frac{1}{2}\gamma\hbar H(\sigma_A - \sigma_B)$$

$$C \sin 2\theta = \frac{1}{2}J$$

The selection rules forbid transitions between the pairs of states 2-3 and 1-4. From the information given, derive expressions for the energies of the allowed transitions. There are four of these transitions, one corresponding to each observed resonant frequency. Assuming that $\sigma_A > \sigma_B$ and $J > 0$, assign the observed frequencies to the transitions and calculate J, C, and $\sigma_A - \sigma_B$. Are J and $\sigma_A - \sigma_B$ molecular parameters or do they depend on the external environment (i.e., solvent, applied field, etc.)? *Note:* On the chart of the spectrum there are two spectra. The upper one is an expanded version of the lower one. The sweep range of 50 Hz means that the hertz scale located at the top and inside the graph is 50 Hz in length for the upper curve instead of 500 Hz as it is for the lower curve. The sweep offset of 395 Hz means that the origin of the chart is 395 Hz for the upper curve rather than 0 Hz as it is for the lower curve. Find the values of the energy difference in hertz. The following references may be helpful: H. J. Bernstein, J. A. Pople, and W. G. Schneider, *Can. J. Chem.* **35**, 65 (1957); H. M. McConnell, A. D. McLean, and C. A. Reilly, *J. Chem. Phys.* **23**, 1152 (1955); and K. F. Kuhlmann and C. L. Braun, *J. Chem. Educ.* **46**, 750 (1969).

5. The uncertainty principle requires that $\Delta x \cdot \Delta p_x \geqslant \hbar$, where x is a coordinate and p_x is the momentum conjugate to that coordinate. Using this principle, prove that the component of the total angular momentum along the field direction can never be as large as the total angular momentum.

6. Prove that a wave function having the form of Eq. (7-99) leads to the time-independent Schrödinger equation [Eq. (7-100)].

7. Describe in your own words how selection rules can be obtained theoretically.

8. Suppose that the following neutral molecule is stable. Draw a graph of what you would predict for its NMR spectrum assuming that only the protons have magnetic moments.

```
        H     H
        |     |
   H —— C ——— C —— S —— N
       / \    |
      H   H   H
```

REFERENCES

Section 7.1

J. Davis, *Advanced Physical Chemistry* (Ronald, New York, 1965); R. M. Eisberg, *Fundamentals of Modern Physics* (Wiley, New York, 1961); M. W. Hanna, *Quantum Mechanics in Chemistry,* 2nd ed. (Benjamin, Reading, Mass., 1969). Most introductory physical chemistry texts discussing quantum chemistry treat both H and He. In addition to the References of Appendix XI these may be useful.

G. Herzberg, *Atomic Spectra and Atomic Structure* (Dover, New York, 1944). A full text devoted to atomic spectra.

Handbook of Chemistry and Physics (Chemical Rubber Publishing Company, Cleveland), any edition; C. E. Moore, "Atomic Energy Levels" (in three volumes), *Document NSRDS-NBS 35,* available from Superintendent of Documents, U.S. Government Printing Office, Washington, D.C. 20402. Tabulate emission lines and energy levels for atoms.

Section 7.2

H. S. W. Massey and E. H. S. Burhop, *Electronic and Ionic Impact Phenomena* (Oxford University Press, Inc., New York, 1952). A classic text in the area of electron impact phenomena.

J. Franck and G. Hertz, *Physik Z.* **20**, 132 (1919); J. Franck and G. Hertz, *Verk. Deut. Phys. Ges.* **16**, 457 (1914). References to the original work.

S. Trajmar, J. K. Rice, and A. Kuppermann, *Advan. Chem. Phys.* **18**, 15 (1970). A review of electron impact spectroscopy.

Handbook of Chemistry and Physics (Chemical Rubber Publishing Company, Cleveland), any edition; C. E. Moore, "Atomic Energy Levels" (in three volumes), *Document NSRDS-NBS 35,* available from Superintendent of Documents, U.S. Government Printing Office, Washington, D.C. 20402. Tabulate emission lines and energy levels.

G. Herzberg, *Atomic Spectra and Atomic Structure* (Dover, New York, 1944); D. C. Peaslee, *Elements of Atomic Physics* (Prentice-Hall, Englewood Cliffs, N.J., 1955). Discuss the mercury transitions.

Section 7.3*

G. M. Barrow, *Introduction to Molecular Spectroscopy* (McGraw-Hill, New York, 1962); G. Herzberg, *Spectra of Diatomic Molecules* (Van Nostrand Reinhold, New York, 1950); L. Pauling, *The Nature of the Chemical Bond,* 3rd. ed. (Cornell University Press, Ithaca, N.Y., 1960); L. Pauling and E. B. Wilson, *Intro-*

*See any of the references in Appendix XI.

duction to Quantum Mechanics (McGraw-Hill, New York, 1935). Discuss vibration-rotation spectra of diatomic molecules.

D. N. Kendall, *Applied Infrared Spectroscopy* (Van Nostrand Reinhold, New York, 1966); G. K. T. Conn, *Infrared Methods: Principles and Applications* (Academic Press, New York, 1960); A. E. Martin, *Infrared Instrumentation and Techniques* (Elsevier, Amsterdam, 1966). Reference texts on infrared techniques.

"Tables of Wavenumbers for the Calibration of Infra-red Spectrometers," *Pure Appl. Chem.* **1**, 537 (1961). An excellent source of tables of infrared spectra of small molecules.

Section 7.4

F. A. Cotton, *Chemical Applications of Group Theory* (Wiley-Interscience, New York, 1963); G. Herzberg, *Infrared and Raman Spectra of Polyatomic Molecules* (Van Nostrand Reinhold, New York, 1945); G. W. King, *Spectroscopy and Molecular Structure* (Holt, Rinehart and Winston, Inc., New York, 1964); M. Avram and Gh. Mateescu, *Infrared Spectroscopy* (Wiley-Interscience, New York, 1972); E. B. Wilson, Jr., J. C. Decius, and P. C. Cross, *Molecular Vibrations: The Theory of Infrared and Raman Vibrational Spectra* (McGraw-Hill, New York, 1955). Discuss infrared spectroscopy of polyatomic molecules.

J. R. Miller, *Essays Chem.* **3**, 59 (1972). A very readable discussion of symmetry in chemistry.

Standard Infrared Spectra (Sadtler Research Laboratories, 3316 Spring Garden St., Philadelphia, Pa. 19104). A catalog of infrared spectra of an enormous number of compounds.

"Tables of Wavenumbers for the Calibration of Infra-red Spectrometers," *Pure Appl. Chem.* **1**, 537 (1961). A good source of detailed spectra of the small molecules suggested for this experiment.

Section 7.5

G. Herzberg, *Spectra of Diatomic Molecules* (Van Nostrand Reinhold, New York, 1950) (an excellent source of spectroscopic data on diatomic molecules); J. G. Calvert and J. N. Pitts, *Photochemistry* (Wiley, New York, 1966); G. M. Barrow, *Introduction to Molecular Spectroscopy* (McGraw-Hill, New York, 1962); G. W. King. *Spectroscopy and Molecular Structure* (Holt, Rinehard and Winston, Inc., New York, 1964). Describe the absorption spectra of diatomic molecules.

D. W. Seery and D. Britton, *J. Phys. Chem.* **68**, 2263 (1964); J. I. Steinfeld, R. N. Zare, L. Jones, M. Lesk, and W. Klemperer, *J. Chem. Phys.* **42**, 25 (1965); J. Tellinghuisen, *J. Chem. Phys.* **57**, 2397 (1972). Describe the optical absorption properties of iodine and bromine.

J. Franck, *Trans. Faraday Soc.* **21**, 536 (1925). A very early paper on optical absorption of iodine.

Section 7.6

G. Herzberg, *Spectra of Diatomic Molecules* (Van Nostrand Reinhold, New York, 1950); G. M. Barrow, *Introduction to Molecular Spectroscopy* (McGraw-Hill, New York, 1962). Discuss diatomic molecules in detail.

J. I. Steinfeld and W. Klemperer, *J. Chem. Phys.* **42**, 3475 (1965); J. Tellinghuisen, *J. Chem. Phys.* **57**, 2397 (1972); L. Brewer, R. A. Berg, and G. M. Rosenblatt, *J. Chem. Phys.* **38**, 1381 (1963). Discuss the properties of excited iodine molecules.

J. Franck, *Trans. Faraday Soc.* **21**, 536 (1925). One of the earliest studies of the optical absorption properties of iodine.

J. I. Steinfeld, *J. Chem. Educ.* **42**, 85 (1965). Describes the experimental approach used here.

R. N. Zare, *J. Chem. Phys.* **40**, 1934 (1964). Describes Franck-Condon calculations.

F. A. Cotton, *Chemical Applications of Group Theory* (Wiley-Interscience, New York, 1963); H. Eyring, J. Walter, and E. W. Kimball, *Quantum Chemistry* (Wiley, New York, 1944). Treat group theory with chemical applications.

Section 7.7

M. H. Profitt and W. C. Gardiner, *J. Chem. Educ.* **43**, 152 (1966); M. H. Profitt and W. C. Gardiner, *Am. J. Phys.* **34**, 163 (1966). The details of construction of the following: The experimental apparatus assumed in the text can be built for about $60 of electronic parts not including the magnet and oscilloscope. The magnet is a variable flux device and is relatively inexpensive (about $600). It could be replaced with a magnetron fixed flux magnet costing about $30 from surplus suppliers. This substitution results in loss of the variable flux capabilities. If the latter route is chosen, the rf oscillator should be made continuously variable by adding a tuning capacitor in the rf circuit.

A. Abragam, *The Principles of Nuclear Magnetism* (Oxford University Press, London, 1961); E. R. Andrew, *Nuclear Magnetic Resonance* (Cambridge University Press, New York, 1958); A. Carrington and A. D. McLachlin, *Introduction to Magnetic Resonance* (Harper & Row, New York, 1967); J. A. Pople, W. G. Schneider, and H. J. Bernstein, *High Resolution Nuclear Magnetic Resonance* (McGraw-Hill, New York, 1959); J. D. Roberts, *Nuclear Magnetic Resonance* (McGraw-Hill, New York, 1959); C. P. Slichter, *Principles of Magnetic Resonance* (Harper & Row, New York, 1963). On NMR.

D. G. Howery, *J. Chem. Educ.* **48**, A327, A389 (1971). A review of NMR instrumentation.

Chapter 8

Bulk Electric and Magnetic Properties

8.1 Paramagnetic Susceptibility of Transition Metal Ions in Aqueous Solution

The purpose of this experiment is to measure the perturbation of a proton resonance by a paramagnetic ion. This is accomplished by measuring the difference in proton resonance frequencies with and without the paramagnetic ion present. In this experiment *t*-butyl alcohol is used as a source of protons. The theory section discusses the quantitative description of susceptibility and how it can be calculated from the above measurements. In addition the relation between bulk magnetization and molecular level magnetic moments is discussed.

THEORY. Section 7.7 discusses the essential background required to discuss the proton resonance spectrum of *t*-butyl alcohol. With the structure

$$\begin{array}{c} CH_3 \\ | \\ CH_3{-}C{-}OH \\ | \\ CH_3 \end{array}$$

we expect the methyl proton resonance in this molecule to be a strong singlet since each methyl is in the same chemical environment and spin-spin splitting resulting from the —OH proton is negligible. The position of this singlet is controlled by the magnetic field at the methyl protons. In turn this local field can be dissected into two parts: (1) that caused by the external field and (2) that caused by the magnetic moments of the local chemical environment. It is on the latter that we wish to focus.

The magnetic properties of the proton's environment can be thought of as arising from two sources: induced and permanent magnetic moments. These moments, **m**, are the magnetic analogs of the electric dipole moments discussed in Section 8.2. The magnetic moments of interest in this experiment are the permanent moments associated with the nonzero net electron spin of certain transition metal ions in aqueous solution. The experiment involves a bulk sample, and the measurements permit a determination of the bulk magnetization **M**, which is defined as the net average magnetic moment per unit volume. This definition carries with it the notion that **M** and **m** are connected. For normally available fields **M** and **H** are proportional as

$$\mathbf{M} = \chi \mathbf{H} \tag{8-1}$$

The parameter χ, called the volume magnetic susceptibility, obviously measures the magnetic response of a substance to the applied field. For diamagnetic materials χ is negative, while for paramagnetic materials it is positive. The magnitude of χ varies with the kind of material. In this experiment we are especially interested in the change that occurs in the volume magnetic susceptibility of an aqueous dilute solution of *t*-butyl alcohol when paramagnetic ions are added. This change in susceptibility is equivalent to a change of the local magnetic field at the protons and thus a shift of their resonant frequency. In terms of a shift of $\Delta\omega$ Hz the observations are connected to the theory by

$$\frac{\Delta\omega}{\omega_0} = \frac{2\pi}{3}(\chi_p - \chi_0) \tag{8-2}$$

where ω_0 is the methyl group resonance frequency in hertz when no paramagnetic ions are present and χ_p and χ_0 are the volume susceptibilities of the bulk solutions with and without paramagnetic ions, respectively.

Since the detailed development of Eq. (8-2) is relatively obscure, we shall inquire briefly as to its origin. First recall that the resonant frequency ω of a proton in a certain environment can be written as

$$\omega = \gamma \mathrm{H} \tag{8-3}$$

where γ is the gyromagnetic ratio and H the time-average magnetic field at the proton in question. The gyromagnetic ratio is fixed for protons, but H varies depending on the physical and chemical characteristics of the surroundings. The problem is to dissect H into separately treatable parts. In general this is done by writing H as a sum of terms:

$$\mathrm{H} = \mathrm{H}_0 + \mathrm{H}_1 + \mathrm{H}_2 \tag{8-4}$$

where H_0 is the externally applied field, H_1 the induced diamagnetic field, and H_2 the field resulting from paramagnetic species. In Section 7.7 H_0 and H_1 are treated with $\mathrm{H}_1 = -\sigma \mathrm{H}_0$, where σ is the diamagnetic shielding constant. In the present experiment two samples are compared in identical external field strength; in one the paramagnetic species are present, while in the second they are absent. The difference ΔH in the field for the two samples is

$$\Delta H = H_2 \quad \text{or} \quad \Delta\omega = \gamma H_2 \tag{8-5}$$

The interpretation of the experimental result thus becomes an interpretation of the magnitude of H_2. This field at the proton of a bulk sample, because of a small number of paramagnetic species, may be analyzed in terms of three components: (1) the Lorentz or cavity field caused by the oriented magnetic moments on the surface of a hypothetical small sphere centered on the proton of interest, (2) a depolarization field which depends on the sample geometry and the diamagnetic properties of the paramagnetic species, and (3) a contribution resulting from the small number of magnetic moments within the hypothetical sphere. The geometry-dependent factor is, for cylindrical samples, approximately equal to $-(2\pi/3)M$, while the cavity field is $(4\pi/3)M$. The last factor (3) is very small and is neglected here. As a result H_2 becomes

$$H_2 \simeq \frac{2\pi}{3}M \tag{8-6}$$

The magnetization appearing in Eq. (8-6) is just that resulting from the paramagnetic species. Experimentally it is approximated closely by the difference in magnetization of the sample with the paramagnetic species, M_p, and that without, M_0

$$M = M_p - M_0 \tag{8-7}$$

Hence, from Eq. (8-1)

$$M \approx H_0(\chi_p - \chi_0) \tag{8-8}$$

where H has been approximated by H_0. Combining Eqs. (8-8), (8-6), (8-5), and (8-3) furnishes the desired result: Eq. (8-2). The approximations employed include the insertion of H_0 for H in Eq. (8-8). This is not serious since H_0 is much larger than H_1 and H_2. Of greater seriousness is the neglect of magnetic moments within the Lorentz cavity and the cylindrical geometry approximation for the demagnetization field. For the present experiment these are not serious.

The magnetic susceptibility is generally reported in terms of pure materials and as mass susceptibility, which has dimensions of $cm^3\text{-}g^{-1}$ and is defined as

$$\bar{\chi}_i = \frac{\chi_i}{m_i} \tag{8-9}$$

where m_i is the *density* in g cm^{-3} of material i. The term density must be used cautiously in that $\bar{\chi}_i$ can be defined for a solution or any one of the components making up that solution. For the latter the density is not that of the pure material i but the weight of material i in 1 cm^3 of the actual solution. In terms of the mass susceptibilities Eq. (8-2) becomes

$$\frac{\Delta\omega}{\omega_0} = \frac{2\pi}{3}(\bar{\chi}_p m_p - \bar{\chi}_0 m_0) \tag{8-10}$$

For the pure paramagnetic species alone and for the aqueous *t*-butyl alcohol solution we define two mass susceptibilities as $\bar{\chi}$ and $\bar{\chi}_s$, respectively. In these terms

$$m_p\bar{\chi}_p = m\bar{\chi} + (m_1 - m)\bar{\chi}_s \tag{8-11}$$

where m and m_1 are the mass of paramagnetic species per cubic centimeter and the mass per cubic centimeter of the paramagnetic solution, respectively. Equation (8-10) thus becomes, after rearrangement to set out the variable of interest $\bar{\chi}$,

$$\bar{\chi} = \frac{3}{2\pi}\frac{\Delta\omega}{\omega_0}\frac{1}{m} + \bar{\chi}_s\left[1 + \left(\frac{m_0 - m_1}{m}\right)\right] \tag{8-12}$$

In most cases $(m_0 - m_1)/m << 1$ and can be neglected. To use Eq. (8-12), $\Delta\omega/\omega_0$ is measured using a nuclear magnetic resonance apparatus; m, m_0, and m_1 are calculated from known properties of the solutions; and $\bar{\chi}_s$ is assumed known.

Another form of the susceptibility is often used, namely, the molar susceptibility, defined as

$$\chi_{Mi} = M_i\bar{\chi}_i \tag{8-13}$$

For the paramagnetic species of interest here we thus have $\chi_M = M\chi$. The molar susceptibility of a substance is a bulk property. Using statistical mechanical arguments this bulk property can be connected to the magnetic moment of an individual molecule, atom, or ion. If the molar susceptibility is considered to arise only because the paramagnetic ions have a net electron spin (this neglects their diamagnetic properties), the connection is

$$\chi_M = \frac{N_0}{3kT}\langle m^2\rangle \tag{8-14}$$

where $\langle m^2\rangle$ is the expectation value of the square of the magnetic moment and N_0 is Avogadro's number. Considering only electron spin magnetic moments,

$$\langle m^2\rangle = 4\beta^2 S(S + 1) \tag{8-15}$$

where β is the Bohr magneton and S the total spin quantum number of the paramagnetic species. The source of the temperature dependence in Eq. (8-14) is the competition between the orienting influence of the applied magnetic field on the permanent magnetic dipoles and the disorienting influence associated with the random thermal motion of which the temperature is a measure. The form of Eq. (8-14) is established by statistical mechanical methods (see Problem 1 at end of Section 8.2).

If the diamagnetic properties of the sample are incorporated into Eq. (8-14), then χ_M becomes

$$\chi_M = \frac{N_0}{3kT}\langle m^2\rangle - N_0\alpha \tag{8-16}$$

where α denotes the temperature-independent diamagnetic properties. Equation (8-14) or (8-16) is known as Curie's law. The plotting of χ_M versus 1/T is used to determine $\langle m^2\rangle$ and α. The quantity $N_0\langle m^2\rangle/3k$ is called the Curie constant.

EXPERIMENTAL TECHNIQUE. The essential requirements for this experiment are a moderately high-resolution NMR instrument (30 MHz or better) which can be temperature-programmed. A large part of the experiment can be done at a constant temperature, but determination of the Curie constant requires temperature variation. It is a general feature of these instruments that the sample is rotated about an axis coincident with the cylinder axis of the sample tube. In this way samll inhomogeneities in the applied magnetic field are averaged over the sample in an attempt to make the value of H in Eq. (8-3) identical for every proton.

The basic requirements for detecting proton resonance have been outlined in Section 7.7. The apparatus called for here is much more sophisticated than is outlined there, but the essential ingredients are the same. We assume that a commercial instrument is available for this experiment.

Two approaches may be used. In the first and preferred approach the resonant frequencies of the methyl protons in *t*-butyl alcohol are measured in two samples simultaneously. A coaxial pair of cylindrical sample tubes is used, the outer tube containing the aqueous *t*-butyl alcohol solution and the inner tube containing the same solution with some added paramagnetic species. These tubes are available commercially from a variety of suppliers.

If coaxial tubes are not available, then reasonable data can be obtained, if the NMR apparatus is stable, by scanning alternately samples with and without the paramagnetic ion. The shift of the methyl resonance depends on the paramagnetic ion concentration and the type of paramagnetic ion. A typical shift for the conditions of this experiment is about 20 Hz or 0.3 ppm.

EXPERIMENT. Prepare about 200 ml of approximately 2% *t*-butyl alcohol in water. Prepare paramagnetic-ion-containing solutions (20 ml of each) using this stock solution with the following concentrations:

$NiCl_2$: 0.16 M

$NiCl_2$. 0.08 M

$CuSO_4$: 0.08 M

$FeSO_4$: 0.08 M

$K_3Fe(CN)_6$: 0.08 M

It is not important that these concentrations be exactly met, but they should be in this range and known to three significant figures if possible.

Using a syringe with a long needle, prepare the NMR sample tubes and scan the spectra of each sample. If coaxial cells are not used, scan the reference sample and the paramagnetic-ion-containing sample at least twice. If the temperature of the NMR instrument sample holder differs from room temperature remember to allow several minutes for thermal equilibrium to be reached.

Using the 0.08-M $NiCl_2$ solution, determine the temperature dependence by making observations at 20° intervals from 5° to 65°C. Several minutes will be required for the sample holder temperature to equilibrate.

ANALYSIS. For each sample, calculate the molar susceptibility of the paramagnetic ion using Eqs. (8-12) and (8-13). Assume that $\bar{\chi}_s = -0.72 \times 10^{-6}$ cm^3 g^{-1}. Neglect the last term in Eq. (8-12) unless there is sufficient time to measure the required solution densities as outlined in Section 5.6. Compare the $\bar{\chi}_M$ values among themselves. From $\bar{\chi}_M$, compute $\langle m^2 \rangle$ and compare it with the value predicted by Eq. (8-15). Use $\beta = 0.927 \times 10^{-20}$ erg g^{-1}, and to determine S, write out the electronic structure of the paramagnetic ion, determine the number of unpaired electrons, and add $\frac{1}{2}$ for each of them to get a sum for S. For the $Fe(CN)_6^{-3}$ complex ion the iron is in a low spin state (see Section 6.2). Compare the calculated susceptibilities with literature values. Compare and comment on any influence of concentration observed in the $NiCl_2$ experiments.

From the temperature dependence of $\bar{\chi}_M$ for the $NiCl_2$ solutions, compute the Curie constant and the diamagnetism parameter α by a least-squares analysis of the results. Estimate statistical uncertainties in these two quantities.

At 20°C literature values for the mass susceptibilities $\bar{\chi}$ are

Material	$\bar{\chi}$ cm^3 g^{-1}
$NiCl_2$	34.2×10^{-6}
$CuSO4$	10.1×10^{-6}
$FeSO_4$	70×10^{-6}
$K_3Fe(CN)_6^{-3}$	8.24×10^{-6}

PROBLEMS

1. How could paramagnetic susceptibility measurements be used to determine the number of unpaired spins in a transition metal complex ion?
2. If one wants to assign parts of the total susceptibility to paramagnetic and diamagnetic aspects of a given species, why is it important to make temperature a variable?
3. Why does the diamagnetic contribution depend on sample geometry?

8.2 Dielectric Properties of Nonionic Molecules in Dilute Solution

The purpose of this experiment is to measure the change in capacitance when a dilute solution of a polar molecule in a nonpolar solvent is inserted as a dielectric in a capacitor formerly evacuated or filled with ambient air. From these measurements at a variety of concentrations, temperatures, and electrical conditions, the dielectric properties of the polar molecule will be calculated, and in some cases an evaluation of the standard Gibbs free energy of a process will be made. In the theory section the relations between capacitance, dielectric constant, polarization, and dipole moment are developed. It is important to note that in this development and in the experiment, macroscopic properties (dielectric constant) and microscopic properties (dipole moment) are interrelated.

THEORY. In this experiment the dielectric constant of a solution is used with a theory to determine the dipole moment of a molecule or the average dipole moment of a group of molecules. The dipole moment is of interest because it can be related to molecular structure of a single type of molecule and to conformational equilibria of isomers. Attention on the latter includes the former, and the experiment outlined below involves conformational equilibria.

A molecule contains both positive and negative charge, and when these are distributed so that their average first moments do not coincide in magnitude and/or direction, the molecule is said to possess a permanent dipole moment. The first moment of a negative charge distribution can be calculated by picking an arbitrary origin, drawing vectors $\mathbf{r}_i$ to each negative charge, and then computing the vector sum, $\mathbf{R}_n = \Sigma f_i \mathbf{r}_i$, where f_i is the fraction of the total negative charge appearing on particle i. The same procedure from the same origin is followed for the positive charges to give a vector sum $\mathbf{R}_p$. If $\mathbf{R}_n \neq \mathbf{R}_p$, then the system has an instantaneous dipole moment. In the actual treatment of molecules, the prediction of whether an observable dipole moment exists is much more involved than would appear from the above discussion. This situation arises because the charged particles are in constant rapid motion (especially the electrons) and one must average the first moments over these motions before arriving at any conclusions. If these averages, $\langle \mathbf{R}_n \rangle$ and $\langle \mathbf{R}_p \rangle$, respectively, are not equal, the molecule is said to possess a dipole moment $\boldsymbol{\mu}$, which is a vector quantity having an orientation given by the vector $\boldsymbol{\rho}$ connecting $\langle \mathbf{R}_n \rangle$ to $\langle \mathbf{R}_p \rangle$ and a magnitude given by $|\langle \mathbf{R}_p \rangle - \langle \mathbf{R}_p \rangle| \times q$, where q is the total positive charge. Figure 8-1 illustrates this. The cgs units of the dipole moment are esu cm, or, expressing esu in force

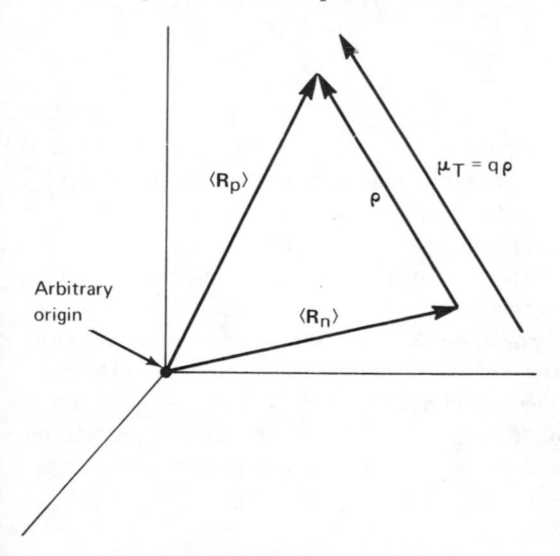

Figure 8-1. Diagram relating moments of charge distributions to dipole moment.

and length, the units are $dyn^{1/2}cm^2$. Commonly, dipole moments are listed in Debye units: $1 \text{ D} = 10^{-18} \text{ dyn}^{1/2}\text{cm}^2$.

The above route to dipole moments is a theoretical one; often we use intuition and chemical experience to decide whether or not a molecule has a dipole. For example, we expect HCl to have a dipole because in considering the valence electrons participating in the bond we recognize that the greater electronegativity of Cl will tend to increase the electron density at this end of the molecule and reduce it at the H end. Thus, we expect HCl to have a permanent dipole, which in fact it does. A distinction is made between permanent and induced dipole moments. A permanent dipole is one which exists in a molecule in the absence of an external electric field, whereas an induced dipole arises because of the field-molecule interaction. For complex molecules and excited states of molecules the situation is not always so easy and our intuition may prove faulty. Symmetry, and thus group theory, is often useful. For example, methane has a center of symmetry (see Section 7.4) and will not possess a permanent dipole moment.

Experimentally, we determine the existence and magnitude of dipole moments by determining the dielectric constant of a material. A theory is used to connect the two. Dielectric constants are generally measured by determining the capacitance of a condenser that contains the dielectric material between its plates. It is important to note that the theory which follows is not a theory of the dipole moment but rather a theory which assumes the existence of a permanent dipole moment, whatever the reason, and attempts to relate it to the dielectric constant. Since the dipole moment is a molecular property while the dielectric constant is a bulk property, we can anticipate that the necessary theory must postulate, explicitly or implicitly, the connections to be made between molecular and bulk properties. Generally such postulates incorporate statistics of some kind.

To get started in connecting dipole moments and dielectric constants, we shall begin with a relation between experimentally measured quantities and the dielectric constant, ϵ, namely

$$\epsilon = \frac{C}{C_0} \tag{8-17}$$

where C is the capacitance measured for a condenser with the dielectric (insulator) between the plates and C_0 the capacitance when the space between the plates is evacuated. As defined here ϵ is the dimensionless ratio of two capacitances.

Why should the capacitance change when some material is inserted, and in which direction should it change? Restricting our attention to nonionic species, the experimental situations corresponding to C and C_0 are examined. When the capacitor is evacuated and a certain voltage V is applied to the capacitor, a positive charge Q_0 accumulates on the positive plate and a negative charge $-Q_0$ accumulates on the negative plate. The capacitance C_0 is the ratio $C_0 = Q_0/V$. Figure 8-2 illustrates this. If the nonconducting medium is now introduced between the plates, the surface layer of the material next to the positive plate becomes nega-

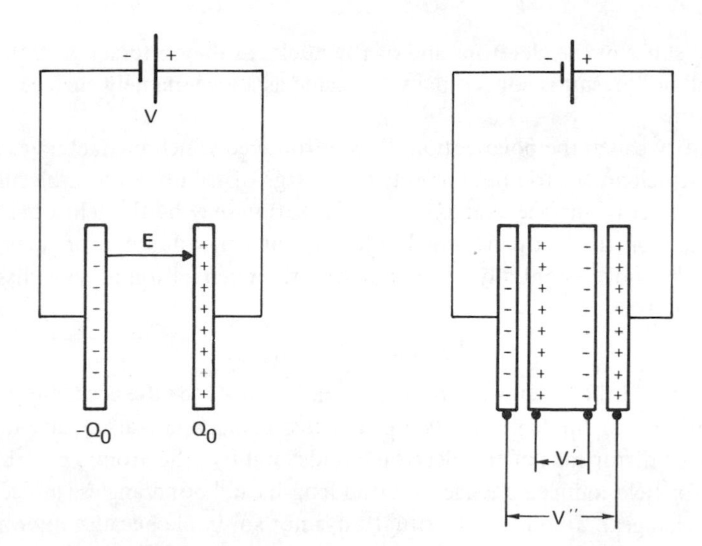

Figure 8-2. Capacitor with and without dielectric.

tively charged, while the opposite occurs at the negative plate. Looking just across the dielectric itself there now appears to be a voltage drop V' which opposes the voltage between the plates and acts momentarily to reduce the potential between the capacitor plates to V''. If the source of the potential is maintained at a constant voltage, additional charge must flow onto the plates to restore the voltage to its original value and in effect nullify the surface layer charges resulting from the dielectric medium. Thus, $C = Q/V$ is larger than $C_0 = Q_0/V$, and according to Eq. (8-17), $\epsilon \geqslant 1$. In effect the dielectric acts to reduce the electric field strength **E** between the capacitor plates. In answer to the question posed at the beginning of this paragraph we conclude that capacitance changes and increases on insertion of a dielectric and does so because of the electrical influence of surface charge layers in the dielectric (polarization charge). The magnitude of the surface charge is governed by the voltage applied to the capacitor and the properties of the molecules making up the dielectric.

What molecular properties of nonconducting media are responsible for this surface charge? Since nonconducting media are insulators and are therefore not useful for transporting electrical current, we must focus our attention on the charge distributions in individual molecules to find the source of these surface charges. In this context the permanent molecular dipole enters because the electric field will tend to orient (polarize) these dipoles. The interior of the dielectric has its dipoles arranged in a head-to-tail fashion which leaves any interior layer neutrally charged, but the surface layers take on a net charge, one surface negative and the other positive. Even if the molecules of a dielectric possess no permanent dipole moment, the external electric field of the capacitor will induce a dipole because of shifts of the electrons and of the nuclei as they interact with the field. The overall bulk result is conceptually the same as above, namely surface polarization.

A quantity called the polarization, P, is introduced which characterizes the extent to which an electric field orients the charge distributions of molecules and thereby produces a surface charge. The polarization may be thought of as the total dipole moment (permanent and induced) per unit volume. For gases, where the molecules move essentially independently, the polarization may be dissected into three parts as

$$P = P_M + P_E + P_A \tag{8-18}$$

where P_M, called the orientation polarization, accounts for the contribution of permanent electric dipole moments; P_E, the electronic polarization, accounts for field-induced distortions of the electron clouds; and P_A, the atomic polarization, accounts for field-induced changes in bond lengths and bond angles (molecular geometry changes). In liquids the situation is not so simple because intermolecular interactions take place, and they introduce their own influences on the electrical characteristics of the bulk sample.

It is important to note that P_M arises because of a permanent electrical characteristic of the molecule, while P_E and P_A are induced by the external electric field and are therefore present only when the field is present. The primary object of our attention in this experiment is to calculate P_M. This is accomplished by relating the dielectric constant ϵ to the polarization P and then by using further theory dissecting P into components, one component being P_M.

Having now laid the intuitive physical basis for the origin of dielectric effects, we proceed to a brief resumé of the theoretical considerations involved in connecting dielectric constants with polarization. Limiting our attention to electrically isotropic media (eliminating many crystalline materials), the polarization, which is really a vector quantity **P**, is oriented in the same direction as the field and is, for the relatively small fields employed here, proportional to the field **E**. Since both are oriented in the same direction, the vector character may be dropped with no loss of generality, and P and E can be related by a proportionality constant χ, the susceptibility, as

$$P = \chi E \tag{8-19}$$

From the above definition of polarization, P is also given by

$$P = N\bar{\mu}_T \tag{8-20}$$

where N is the number of dipoles per unit volume and $\bar{\mu}_T$ the average contribution along the field direction of a dipole. The dielectric constant and susceptibility are related by

$$\epsilon = 1 + 4\pi\chi \tag{8-21}$$

so that

$$P = \frac{\epsilon - 1}{4\pi} E \tag{8-22}$$

where E is the field strength with the dielectric in place. The relation between dielectric constant and susceptibility is established by considering Maxwell's electric displacement D, which accounts for the decrease of the electric field strength when the dielectric is introduced:

$$D = E + 4\pi P \tag{8-23}$$

But D is also related to the field strength through the dielectric constant ϵ as

$$D = \epsilon E \tag{8-24}$$

Thus, we have two alternative ways of viewing the influence of the dielectric. One way looks microscopically through the polarization P, and the other macroscopically through the dielectric constant ϵ. The combination of Eqs. (8-19), (8-23), and (8-24) furnishes Eqs. (8-21) and (8-22). With Eqs. (8-20) and (8-22) we have two relations for P, but they are not in particularly useful form. Equation (8-20) contains the average value of a component of $\bar{\mu}_T$, while Eq. (8-22) contains the measurable quantity ϵ, but the presence of E in Eq. (8-22) frustrates further progress. The way out is to relate $\bar{\mu}_T$ linearly to the field as follows:

$$\bar{\mu}_T = \left(\frac{\mu^2}{3kT} + a_E + a_A\right)E_{int} \tag{8-25}$$

where E_{int} is the local field at a molecule within the dielectric. The three parts of this equation correspond to the three types of polarization in Eq. (8-18): orientation, electronic, and atomic, respectively. The first term is temperature-dependent because the thermal energy of the molecules of the dielectric operates to randomize dipole orientations, while the electric field tries to orient all of them. These opposing tendencies compromise, and the result for a given permanent dipole, μ; temperature, T; and electric field, E, is given by the first term of Eq. (8-25). According to Eq. (8-25), as T decreases, the contribution to $\bar{\mu}_T$ from the permanent dipoles goes up as expected since there is now less thermal energy. On the other hand, the second two terms are temperature-independent. The quantities inside the parentheses are called polarizabilities. The problem with Eq. (8-25) is that it calls for the local internal field rather than the bulk or macroscopic field E of Eq. (8-22). The relation between E and E_{int} is

$$E = \frac{3}{\epsilon + 2}E_{int} \tag{8-26}$$

which requires the bulk field to be somewhat less than the local internal field since $\epsilon \geqslant 1$. The derivation of this equation is found in most texts on electricity and magnetism.

Combining Eqs. (8-20), (8-22), (8-25), and (8-26) allows the elimination of E and P to give

$$\frac{\epsilon - 1}{\epsilon + 2} = \frac{4\pi N}{3}\left(\frac{\mu^2}{3kT} + a\right) \tag{8-27}$$

where $a = a_E + a_A$. Recalling that N is the number of dipoles per unit volume, we can relate N to the molecular weight M and density d by $N = N_0 d/M$, N_0 being Avogadro's number. Equation (8-27) can thus be rearranged to

$$\frac{\epsilon - 1}{\epsilon + 2}\frac{M}{d} = \frac{4\pi N_0}{3}\left(\frac{\mu^2}{3kT} + a\right) \tag{8-28}$$

The left-hand side of this equation is known as the molar polarization $\overline{P}$, and the connection between $\overline{P}$ and ϵ given in Eq. (8-28) is called the Clausius-Mosotti equation. The detailed treatment of $\overline{P}$ in terms of polarizability and dipole moments is attributed to P. W. Debye.

Strictly speaking, Eq. (8-28) is valid only for dilute gases because no account has been taken of interdipole interactions, that is, the influence one molecular dipole has on a neighboring dipole. High-density gases, especially those with permanent dipole moments, cannot be adequately described with this equation. Very dilute solutions of polar molecules (permanent dipole moment) in nonpolar (no permanent dipole) solvents can be described reasonably well by Eq. (8-28) provided account is taken of the polarizability of the pure solvent. Realistically, however, we are almost always forced to consider interactions between the polar solute molecules. Since the experiment here deals with dilute solutions, we shall proceed to consider methods for treating them.

If we regard the total molar polarization of a solution as comprised of two parts depending only on the fractional number of dipoles of each substance, we can write, for a two-component system,

$$\overline{P} = \overline{P}_1 X_1 + \overline{P}_2 X_2 \tag{8-29}$$

where $\overline{P}_i$ is the effective molar polarization of species i and X_i the mole fraction of species i. This equation can be made to account for interdipole interactions provided we regard $\overline{P}_1$ and $\overline{P}_2$ as dependent on concentration. The polarization of an isolated solute molecule would then be available by extrapolation to infinite dilution of an equation for $\overline{P}_2$. With these considerations we write the Clausius-Mossoti equation as

$$\overline{P} = \overline{P}_1 X_1 + \overline{P}_2 X_2 = \frac{\epsilon - 1}{\epsilon + 2}\frac{X_1 M_1 + X_2 M_2}{d} \tag{8-30}$$

where ϵ is the dielectric constant of the solution and d its density. There is a variety of extrapolation methods for extracting $\overline{P}_2^\circ$, the molar polarization of pure solute. Once $\overline{P}_2^\circ$ is obtained the dipole moment of the polar molecule can be estimated. Two extrapolation procedures are discussed briefly. In the first procedure we rewrite Eq. (8-30) as

$$\overline{P}_1^\circ X_1 + K X_2 = \frac{\epsilon - 1}{\epsilon + 2}\frac{X_1 M_1 + X_2 M_2}{d} \tag{8-31}$$

where $\overline{P}_1^\circ$ is the molar polarization of pure solvent (nonpolar) which we can measure and K an effective polarization which is selected so that Eq. (8-31) holds (i.e., every other parameter is measurable). Solving for K furnishes

$$K = \frac{[(\epsilon - 1)/(\epsilon + 2)]\,[X_1M_1 + X_2M_2)/d] - \overline{P}_1^\circ X_1}{X_2} \tag{8-32}$$

To find $\overline{P}_2^\circ$ we extrapolate a graph of K versus X_2 to infinite dilution. Mathematically

$$\overline{P}_2^\circ = \lim_{X_2 \to 0} K \tag{8-33}$$

This method relies strongly on data for the pure solvent $\overline{P}_1^\circ$.

The second method proposed by Halverstadt and Kumler and modified by Estok (see the References) assumes that both the dielectric constant and the reciprocal of the density are linear functions of the solute weight fraction. That is,

$$\epsilon = \epsilon_1 + bw_2 \tag{8-34}$$

$$\frac{1}{d} = \frac{1}{d_1} + hw_2 \tag{8-35}$$

where w_2 is the weight fraction of solute and ϵ_1 and d_1 refer to solvent. Using weight fractions w_i instead of mole fractions X_i, Eq. (8-30) becomes

$$\overline{P} = \overline{P}_1 w_1 + \overline{P}_2 w_2 = \frac{\epsilon - 1}{\epsilon + 2}\,\frac{w_1M_1 + w_2M_2}{d} \tag{8-36}$$

Equations (8-30) and (8-36) are not identical but just different approximations to the same quantity, $\overline{P}$. If we substitute $w_1 = 1 - w_2$ into Eq. (8-36) and then differentiate the result with respect to w_2, assuming that P_1 and P_2 are constant, we arrive at

$$\begin{aligned}\overline{P}_2 - \overline{P}_1 &= \frac{3(d\epsilon/dw_2)}{(\epsilon + 2)^2}\,\frac{M_1(1 - w_2) + w_2M_2}{d} \\ &+ \frac{\epsilon - 1}{\epsilon + 2}\left\{\frac{d(1/d)}{dw_2}[(1 - w_2)M_1 + w_2M_2] + \frac{1}{d}(M_2 - M_1)\right\}\end{aligned} \tag{8-37}$$

The derivatives of ϵ and $1/d$ with respect to w_2 are available from Eqs. (8-34) and (8-35) as b and h, respectively. Hence Eq. (8-37) becomes

$$\begin{aligned}\overline{P}_2 - \overline{P}_1 &= \frac{3b}{(\epsilon + 2)^2}\,\frac{M_1(1 - w_2) + w_2M_2}{d} \\ &+ \frac{\epsilon - 1}{\epsilon + 2}\left\{h[(1 - w_2)M_1 + w_2M_2] + \frac{1}{d}[M_2 - M_1]\right\}\end{aligned} \tag{8-38}$$

Experimentally every quantity on the right-hand side of Eq. (8-38) is available if plots are made of the dielectric constant and the reciprocal of the density versus the solute weight fraction. From these plots b and h can be determined. As the solution is diluted $\overline{P}_1$, ϵ, and d all approach closely the values taken for the pure solvent, while $\overline{P}_2$ approaches $\overline{P}_2^\circ$ and $X_2 \rightarrow 0$. If we make these approximations, Eq. (8-38) becomes

$$\overline{P}_2^\circ - \frac{\epsilon_1 - 1}{\epsilon_1 + 2}\frac{M_1}{d_1} = \frac{3b}{(\epsilon_1 + 2)^2}\frac{M_1}{d_1} + \frac{\epsilon_1 - 1}{\epsilon_1 + 2}\left[hM_1 + \frac{M_2 - M_1}{d_1}\right] \tag{8-39}$$

where ϵ_1 and d_1 refer to pure solvent. Equation (8-39) is readily solved for $\overline{P}_2^\circ$ to give

$$\overline{P}_2^\circ = \frac{3b}{(\epsilon_1 + 2)^2}\frac{M_1}{d_1} + \frac{\epsilon_1 - 1}{\epsilon_1 + 2}\left[hM_1 + \frac{M_2}{d_1}\right] \tag{8-40}$$

which is the expression of interest. This equation contains the essence of Halverstadt and Kumler's method. To use it, values for b and h must be determined. One can do so by measuring the dielectric constant and the reciprocal solution density versus the weight fraction and finding b and h from the slopes. Estok has suggested a method for obtaining h which does not rely on solution density measurements but assumes additivity of volumes of solvent and solute for the dilute solutions. In this method only the densities of the pure materials are required, and h is given by

$$h = \frac{1}{d_2} - \frac{1}{d_1} \tag{8-41}$$

We are thus left experimentally with the problem of finding d_1, d_2, and b in order to evaluate $\overline{P}_2^\circ$ according to Eq. (8-40).

In either of these two methods [Eq. (8-33) or (8-40)], once $\overline{P}_2^\circ$ is determined it remains to find the dipole moment μ_2 from the equation

$$\overline{P}_2^\circ = \frac{4\pi N_0}{3}\left(\frac{\mu_2^2}{3kT} + a_2\right) \tag{8-42}$$

Electronic and atomic polarization both contribute to a_2. There are several ways to estimate a_2. One widely used way is to measure the index of refraction, n_2, of pure solute using the sodium D line and the equation

$$a_2 \approx \frac{3}{4\pi N_0}\frac{M_2}{d_2}\frac{n_2^2 - 1}{n_2^2 + 1} \tag{8-43}$$

This expression is derived along the same lines as Eq. (8-28), and it may be criticized for not distinguishing the electronic and atomic parts of a_2 properly. However, we shall use it here as a reasonable approximation.

Frequency Dependence of the Dielectric Constant. Throughout the foregoing discussion we have tacitly assumed that the voltage applied to the capacitor plates is a dc voltage. Experimental methods for measuring dielectric constants, however, rely on ac voltages in which the capacitor plates are charged alternately positively and negatively. It is important therefore to ask what variations are expected in the measured dielectric constant if the frequency of the applied voltage changes. For simplicity assume that the applied voltage is sinusoidal:

$$V(t) = V_0 \sin(\omega t + \phi) \tag{8-44}$$

where ω is the frequency in radians per second, t the time, ϕ a phase factor, and V_0 the amplitude. This ac voltage gives rise to an ac electric field in the capacitor and in the dielectric medium.

How does a polar molecule in the liquid or gas phase respond to such a field? Basically it tries to orient itself with the field. At very low frequencies there is sufficient time during one cycle of the field for the polar molecule to rotate and keep itself aligned with the field as the latter changes sign from positive to negative. At higher frequencies, especially in liquid solutions where the solvent viscosity hinders free rotation, the molecules lose their ability to follow the field. They rotate at some maximum frequency governed by their rotational energy and the viscosity of the medium, but this maximum can be exceeded by the frequency of the applied voltage. When this occurs the permanent dipoles of the molecules are no longer aligned by the field and the polar molecule behaves as a nonpolar species. This generally occurs at about 10^8 Hz. Thus, the dielectric constant falls off with increasing frequency. If the frequency is made enormous by going to infrared light (10^{13} Hz), the permanent part of the polarization contributes nothing to the measured dielectric constant, while the electronic and atomic polarizabilities remain. If the visible light is used (10^{15} Hz), the atomic polarizability drops out because of the inability of the nuclei to move quickly enough, while the electronic polarizability remains. In the latter cases where electromagnetic radiation is used, the index of refraction is measured—not the dielectric constant. Both, however, are intimately connected to the polarizability [see Eq. (8-43)], and in fact when there is no strong magnetic polarizability, $\epsilon = n^2$.

An important property of dielectrics, called dielectric loss, is associated with the ability of the polarization to follow the changes in an electric field. As the frequency passes into a region where the time required for polarization is of the same order of magnitude as the period (time for one oscillation) of the electric field then the polarization vector lags behind the electric field vector. In electromagnetic theory the result is some absorption of energy from the field and dissipation of that energy in the dielectric as heat. The form of the mathematical expression for this loss is similar to Ohm's law. Based on the qualitative discussion given above we expect the magnitude of the dielectric loss to vary significantly with the frequency of the applied electric field.

EXPERIMENTAL METHODS. There are several methods for determining the capacitance in the presence of a dielectric for use in Eq. (8-17). Three widely used approaches, the bridge method, the resonance method, and the heterodyne beat method, will be described here. All employ ac voltages.

Bridge Method. In the bridge method a circuit similar to the Wheatstone bridge (see Section 2.4) is used but one which incorporates capacitance. The so-called Schering bridge is shown in Fig. 8-3. In this circuit the alternating voltage (30–300 kHz) is applied across the bridge, and the impedance is balanced by adjusting an accurately calibrated variable capacitor C_B.

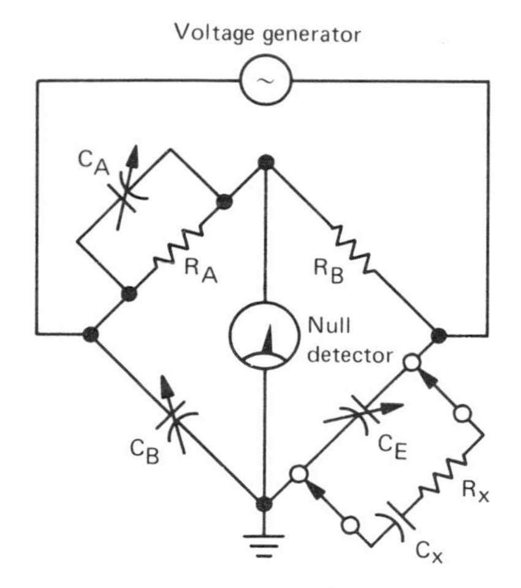

Figure 8-3. Schering bridge circuit for dipole moment measurements.

The balance condition is detected with a high-impedance voltage meter, the null detector in Fig. 8-3. The arms containing R_A and R_B are called the ratio arms. The balance condition is achieved when the impedance ratio of the two resistance arms is equal to the impedance ratio of the two capacitance arms. The ratio R_A/R_B is generally variable in steps of 1, 10, 100, and 1000, while the precision capacitor C_B is adjustable over the range 100–1000 $\mu\mu$F. The unknown with its capacitance, C_x, and effective resistance, R_x, is placed in parallel with a known capacitance C_E. Adjustments of both resistance and capacitance are then made until the minimum current is registered by the null detector. If the unknown effective resistance R_x is very large, there will be very little dielectric loss, and only the capacitor C_B need be adjusted. When dielectric loss is important the bridge balance condition will be strongly dependent on the impedance of the R_A - C_A combination. At the balance condition two equations are valid:

$$C_x + C_E = C_B \frac{R_A}{R_B} \tag{8-45}$$

and

$$D_x = 2\pi f R_A C_A = 2\pi f R_x (C_x + C_E) \tag{8-46}$$

where f is the frequency in hertz. The latter equation (providing the dissipation factor D_X) is related to dielectric loss and will not be considered in this experiment. The unknown capacitance C_X is determined by the substitution method, meaning that we balance the bridge, $C_B = C_{B1}$, once with only C_E in the unknown position and a second time, $C_B = C_{B2}$, with both C_E and C_X in the unknown position. The difference of these two measurements furnishes C_X as

$$C_X = (C_{B2} - C_{B1})\frac{R_A}{R_B} \tag{8-47}$$

The substitution method helps eliminate systematic errors caused by stray capacitance and inductance in the sense that both measurements include these stray effects and the difference removes them, at least partially. Remember that C_X will be determined for the empty dielectric constant cell and for the filled dielectric cell. These two values are the C_0 and C to be used in Eq. (8-17).

Resonance Method. In the resonance method two almost independent electrical circuits are involved—almost but not completely independent. One circuit is a high-frequency oscillator driven by an appropriate power supply and contains a coupling coil which provides the mechanism for loose coupling into a nearby coil which forms part of second circuit containing capacitance, inductance, and a very high-impedence voltmeter (vacuum tube voltmeter). The inductive coupling between the two coils gives rise to an alternating current in the secondary circuit, the magnitude of which is determined by the inductance and capacitance of the secondary circuit and the frequency of the driving oscillator. At a given frequency ω of the oscillator, there is a condition in the secondary circuit for which the current amplitude (and thus voltage) is maximized. This resonance condition is given mathematically as

$$\omega = \frac{1}{\sqrt{L_r C_r}} \tag{8-48}$$

where L_r and C_r are the equivalent inductance and equivalent capacitance, respectively, at the resonance condition. The frequency ω in Eq. (8-48) is in radians per second. To convert to f Hz, use $\omega = 2\pi f$.

In the resonance method a substitution procedure is normally used; that is, resonance is determined at a given frequency both with and without the cell present. A typical resonance apparatus is depicted in Fig. 8-4. The oscillator provides an alternating voltage in coil 1, which by mutual inductance couples into coil 2, inducing a current there. The *coarse* and *fine* capacitors are adjusted to maximize the voltage read at the meter. To avoid loading, the meter has an impedance much higher than any other part of the circuit. Working at a fixed frequency ω and a fixed value for L_2, the resonance conditions measured are (1) without the cell, (2) with the empty cell, and (3) with the full cell. To each of

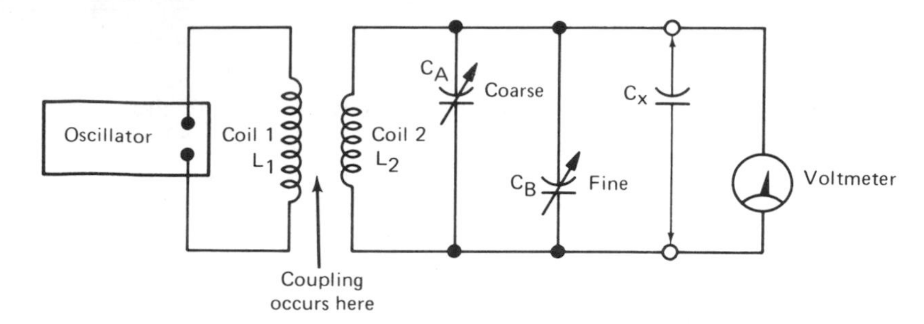

Figure 8-4. Resonance apparatus for dielectric constant measurements.

these conditions there corresponds a certain setting of the coarse and fine capacitors which brings the circuit into resonance. For case (1)

$$C_{A1} + C_{B1} = \frac{1}{L_T\omega^2} \tag{8-49}$$

For case (2)

$$C_{A2} + C_{B2} + C_{xe} = \frac{1}{L_T\omega^2} \tag{8-50}$$

and for case (3)

$$C_{A3} + C_{B3} + C_{xf} = \frac{1}{L_T\omega^2} \tag{8-51}$$

The right-hand sides of these equations are all identical and so the left-hand sides may be set equal. Subtracting (8-49) from (8-50) and from (8-51) furnishes

$$C_{xe} = (C_{A1} + C_{B1}) - (C_{A2} + C_{B2}) \tag{8-52}$$

and

$$C_{xf} = (C_{A1} + C_{B1}) - (C_{A3} + C_{B3}) \tag{8-53}$$

where C_{xe} and C_{xf} are the capacitances of the empty and full cell, respectively, and are to be used in Eq. (8-17) to calculate the dielectric constant ϵ.

Generally speaking the oscillator is in the rf region and ω is on the order of megahertz. The coarse and fine capacitors should sum to a maximum on the order of 300 pF (picofarads), and L_2 should be selected so that with a cell capacitance of the same order of magnitude the circuit can be brought into resonance (i.e., on the order of 30 pF).

Heterodyne Beat Method. In this method the circuitry, outlined in Fig. 8-5, consists of two separate oscillators, one held at a fixed frequency while the other, in which the dielectric cell appears, has an adjustable frequency. The outputs of these two oscillators are combined in a mixer which subtracts the frequencies of the two waves and gives an oscillatory output whose frequency is equal to the difference. The mixer output can be displayed on an oscilloscope or on a fre-

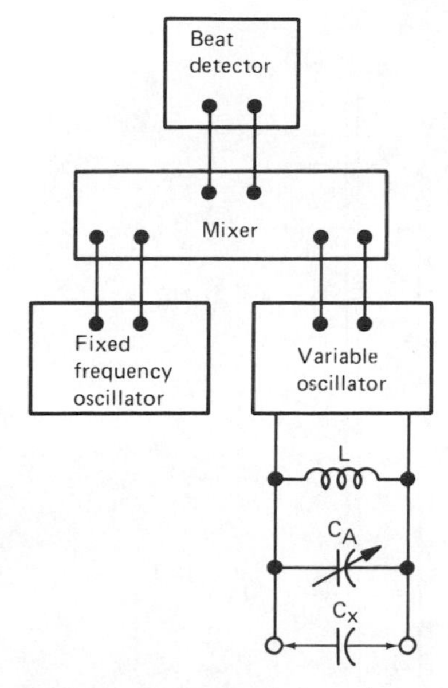

Figure 8-5. Heterodyne beat apparatus for measuring dielectric constants.

quency counter or detected as an audio frequency with a pair of earphones. Caution must be exercised because the mixer furnishes only the magnitude of the frequency difference and not its algebraic sign. Thus, the circuit containing the cell can be oscillating at a frequency above or below the fixed oscillator. Furthermore, harmonics (integer multiples) of the frequency often appear in the circuit containing the cell. These are, in fact, usually employed in the experiment, but care must always be exercised so that the same harmonic is used. The fixed frequency oscillator is generally 1 MHz, and the fundamental frequency of the oscillator containing the sample is likely to be near 0.25 MHz when the inductance is near 30 μH (microhenries). The frequency of the variable oscillator is

$$\omega = \frac{1}{\sqrt{L(C_A + C_X)}} \tag{8-54}$$

and if some harmonic of this frequency is adjusted to be 60 Hz or 60 $\times$ 2π radians/sec greater than the frequency of the fixed oscillator in three experiments as outlined for the resonance method, then Eqs. (8-52) and (8-53) apply again to furnish C_0 and C. An ordinary oscilloscope is a suitable detector for a 60-cycle frequency difference if normal ac line voltage is used to drive the horizontal scope axis while the mixer output drives the vertical axis. When C_A is adjusted to give a 60-cycle difference an elliptical Lissajous figure will appear on the scope.

Dielectric Constant Cell. Each of the methods requires a capacitance cell into which the dielectric can be reproducibly inserted. In this experiment liquids are used and a suitable dielectric cell can be constructed of two concentric cylinders electrically insulated from one another with the outer cylinder grounded. Figure

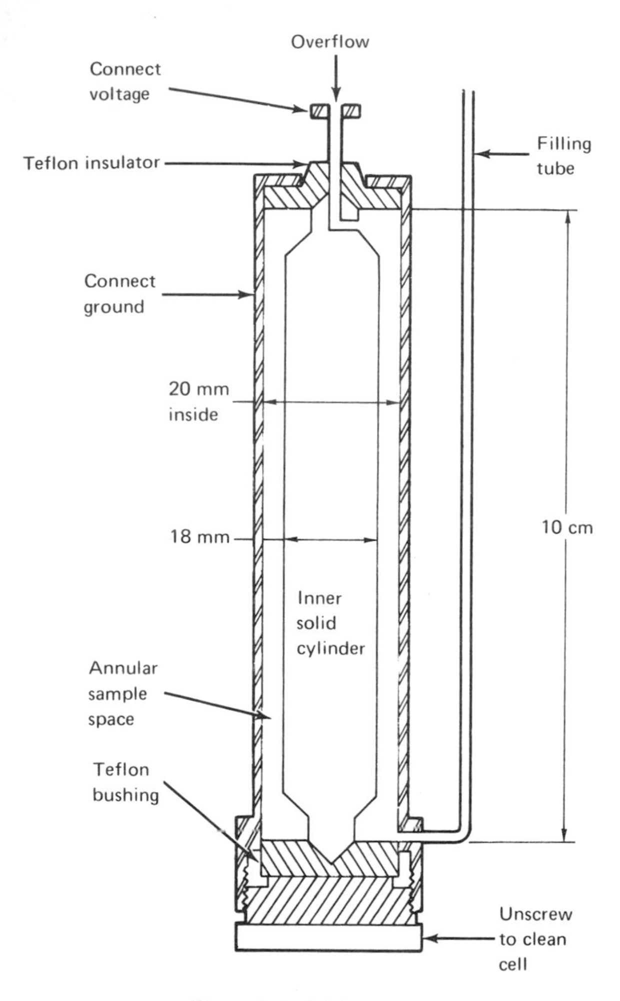

Figure 8-6. Dielectric cell.

8-6 shows one possible design which has a capacitance (filled with air) of about 100 pF. The capacitance can be estimated in picofarads using

$$C = 0.2416\ell \left[\log_{10} \frac{r_2}{r_1}\right]^{-1} \tag{8-55}$$

where ℓ is the length in centimeters and r_2 and r_1 are the inside diameter of the outer cylinder and the outside diameter of the inner cylinder in centimeters. Great care must be exercised in filling the cell in order to avoid air bubbles. The design of Fig. 8-6 can be filled and emptied most successfully with a syringe with a long needle.

EXPERIMENT. A wide variety of experiments can be done. One particular type is mentioned here in which a polar molecule existing in more than one isomeric form is dissolved in a nonpolar solvent. Using experimentally determined dielectric constants, an average of the square of the dipole moment is calculated, and an attempt is made to resolve this moment into parts contributed by different isomers, in this case rotational isomers. This assumes, of course, that the various isomers have different dipole moments. For example, a molecule such as *n*-propyl nitrite can exist in the *cis* and *trans* forms and each has a different dipole moment. The dipole moment determined from the combination of Eqs. (8-42) and (8-43) is thus some average over individual molecules in each of these forms. Assuming that a_2 from Eq. (8-42) is the same for both, we can write

$$\overline{\mu_2^2} = X_{cis}\mu_{2,cis}^2 + X_{trans}\mu_{2,trans}^2 \tag{8-56}$$

where X_i is the mole fraction of species i and $\mu_{2,i}$ is the dipole moment of species i. Another example is the normal paraffin molecules with one halogen atom attached to each end, for example, 1,3-dibromopropane, 1,2-dichloroethane, etc. In this case there are certain rotational configurations about the single-bonded carbon atom skeleton which are more stable than others. These rotational isomers each contribute to the observed average. In the case of 1,2-dichloroethane, for example, the *trans* and *gauche* forms are relatively stable, while the *eclipsed* is not.

For the purposes of the experiment outlined here we assume the use of 1,2-dibromoethane or 1,2-dichloroethane as the solute and benzene as the solvent. Since every apparatus is likely to have different operating characteristics, detailed instructions should be provided by the instructor. A general outline for the heterodyne beat procedure is given here:

1. Prepare four accurately known solutions by weight in the range 1–5% by weight of the solute. Make enough to fill the dielectric constant cell at least twice. Store these so that evaporative losses are minimized. Use volumetric flasks, or weigh both solvent and solute.
2. Determine the capacitance of the clean, dry, and air-filled dielectric cell by making two capacitance measurements, one with the cell in and the other with it out of the circuit. Adjust the beat frequency to 60 cycles in each case.
3. With the cell in the circuit and the apparatus tuned to the 60-cycle beat frequency, begin slowly adding solution to the cell. Begin the series of measurements with pure solvent and move upward in concentration. The beat frequency will change and the Lissajous figure will be destroyed. If a pair of earphones is used as a replacement for the scope, the beat frequency will be audible, and as solution is added the pitch of the sound will rise. In any case adjust the variable capacitance as solution is added to keep the beat frequency near 60 Hz and thus avoid problems of losing the particular harmonic that is being used. When solution reaches the top of the overflow, stop adding solution and record the capacitance and the temperature of the solution. The assumption is that the temperature is room temperature

and constant near 25°C throughout the experiment. For precise work or for studies involving the temperature dependence of the dielectric constant a controlled thermostat should be used around the cell.

4. Remove as much solution as possible with the syringe, and then clean and dry the syringe and the cell. Put the cell back together and repeat the measurements with the other concentrations. An alternative way of cleaning, but less successful, is to rinse with pure solvent and then dry with a stream of warm and dry air.
5. If time permits, repeat the whole set of measurements with the monosubstituted compound. If not, a dipole moment for the monosubstituted compound will be provided.

ANALYSIS

1. Calculate the dielectric constant for each solution.
2. Plot the dielectric constant versus composition according to Eq. (8-34) and determine b.
3. Using known (at 25°C) densities and indices of refraction, evaluate Eqs. (8-40) and (8-43).
4. Compute $\overline{\mu_2^2}$.
5. Assume that $\overline{\mu_2^2}$ is given by

$$\overline{\mu_2^2} = X_{2,trans}\mu^2_{2,trans} + X_{2,g^+}\mu^2_{2,g^+} + X_{2,g^-}\mu^2_{2,g^-} \tag{8-57}$$

where X_i is the mole fraction of rotational isomer, i. The labels trans, g^+, and g^- refer to the rotational forms of 1,2-dibromoethane or 1,2-dichloroethane.

6. Develop an expression for the dipole moments of each of these species in terms of the dipole moment of the monosubstituted species. Assume that each end of the disubstituted molecule has a dipole equal to that of the monosubstituted species and oriented along the C—Br bond at the tetrahedral direction (i.e., the C—C—Br bond angle is 109° 30′). Looking down along the C—C bond direction the *trans*, g^+, and g^- configurations look like

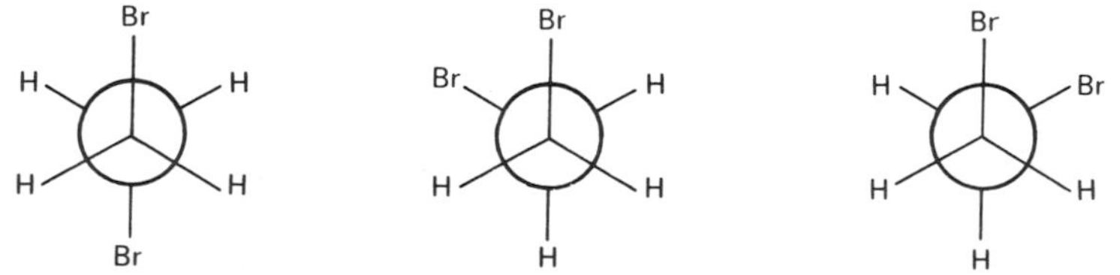

trans g^+ = *gauche* (120°) g^- = *gauche* (−120°)

From the side the *trans* form looks like

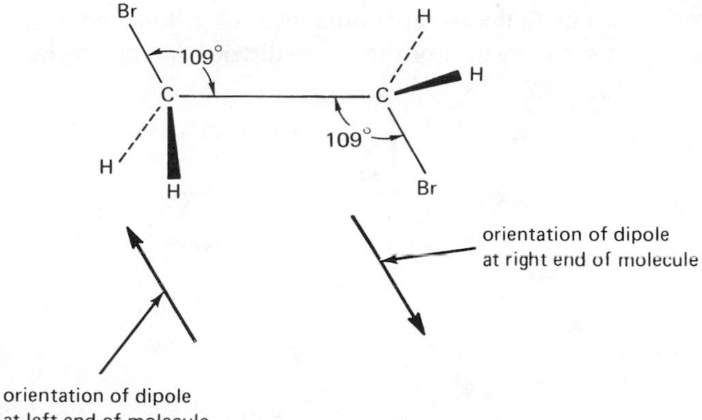

To get the net dipole resulting from both ends, break each dipole into components, one perpendicular to the C—C bond and the other parallel to the bond. Then sum the components vectorially to get the net dipoles parallel and perpendicular to the bond axis. One of these will generally go to zero.

7. Since the g^+ and g^- configurations are energetically degenerate their equilibrium concentrations will be equal, $X_{2,g^+} = X_{2,g^-}$. Further $1 - X_{2,trans} = X_{2,g^+} + X_{2,g^-}$. Use these results and those above for the dipole moments in Eq. (8-57) to evaluate $X_{2,trans}$, X_{2,g^+}, and X_{2,g^-}.
8. Using what you consider to be reasonable estimates of the uncertainties in the measured parameters, estimate the uncertainties of the measured dielectric constants and dipole moments.
9. Compare your results with literature values.

PROBLEMS

1. The temperature-dependent part of the polarizability, Eq. (8-25), can be calculated by computing, for a thermal distribution of energies, the average orientation of a dipole μ in an electric field $\mathbf{E}$. The potential energy, V, of such a dipole depends on the orientation with respect to the field and is given by $V = \mu \cdot \mathbf{E} = |\mu||\mathbf{E}| \cos\theta$, where θ is the angle between the two vectors. According to Boltzmann statistics the probability f of having a state of energy V to V + dV is

$$f(V)\, dV = A \exp\left\{-\frac{V}{kT}\right\} dV \tag{8-58}$$

where A is a constant to be fixed by normalizing f(V) to unity. This equation is valid for nondegenerate energies such as those we are dealing with here. We wish to compute the average component of μ along the field direction, namely $\overline{\mu \cos \theta} = \mu \overline{\cos \theta}$. For the above distribution this is given by

$$\mu \overline{\cos \theta} = \frac{\int_{v=0}^{\infty} (\mu \cos \theta) f(V)\, dV}{\int_{v=0}^{\infty} f(V)\, dV} \tag{8-59}$$

Using V as given above, show that

$$\overline{\cos \theta} = \frac{\int_{0}^{\pi} \cos \theta \exp \{-\frac{\mu E \cos \theta}{kT}\} \sin \theta \, d\theta}{\int_{0}^{\pi} \exp \{-\frac{\mu E \cos \theta}{kT}\} \sin \theta \, d\theta} \tag{8-60}$$

The right-hand side of this expression is known as the Langevin function, and it reduces to

$$\overline{\cos \theta} = \frac{e^{x} + e^{-x}}{e^{x} - e^{-x}} - \frac{1}{x} \tag{8-61}$$

where $x = \mu E/kT$. Show that for $x << 1$ this can be approximated as

$$\overline{\cos 0} = \frac{x}{3} = \frac{\mu E}{3kT} \tag{8-62}$$

and hence that

$$\mu \overline{\cos \theta} = \frac{\mu^2 E}{3kT} \tag{8-63}$$

2. Not every molecule is in one of the precisely defined configurations used in Eq. (8-57). There is some distribution of configurations about each of these angles. How might this effect contribute to the observed average dipole moment?

3. The derivation of the net dipoles assumes that the dipole moment observed in the monosubstituted species can be used as a *group* dipole in the disubstituted species. What interaction does this clearly neglect in the disubstituted species?

4. If an air bubble gets into one of the solutions in the dielectric cell, how will this influence the observed capacitance?
5. If the solution is frozen, how might this effect the capacitance?
6. How would you expect the average dipole moment for the molecule studied to vary with temperature?

REFERENCES

Section 8.1

P. W. Selwood, *Magnetochemistry* (Wiley-Interscience, New York, 1956). A classical text describing important consequences of magnetism for chemistry.

N. Davidson, *Statistical Mechanics* (McGraw-Hill, New York, 1962); T. L. Hill, *An Introduction to Statistical Thermodynamics* (Addison-Wesley, Reading, Mass., 1960). Statistical mechanics and statistical thermodynamics texts including the derivation of Curie's law and stressing chemical applications.

J. L. Deutsch and S. M. Poling, *J. Chem. Educ.* **46**, 167 (1969); W. C. Dickinson, *Phys. Rev.* **81**, 717 (1951); D. F. Evans, *J. Chem. Soc.*, 2003 (1959); W. D. Phillips, C. E. Looney, and C. K. Ikeda, *J. Chem. Phys.* **27**, 1435 (1957); E. N. Sloth and C. S. Garner, *J. Chem. Phys.* **22**, 2064 (1954).

Handbook of Chemistry and Physics (Chemical Rubber Publishing Company, Cleveland), any edition. Tabulates magnetic susceptibilities.

L. M. Mulay in A. Weissberger and B. W. Rossiter (eds.), *Techniques of Chemistry*, Vol. I, (Wiley-Interscience, New York, 1972), Part IV, Ch. VII. A full discussion of techniques for measuring magnetic susceptibility, including the widely used Gouy balance method.

Section 8.2

C. P. Smyth, *Dielectric Behaviour and Structure* (McGraw-Hill, New York, 1955); J. W. Smith, *Electric Dipole Moments* (Butterworth's, London, 1955); R. J. W. LeFevre, *Dipole Moments, Their Measurement and Applications in Chemistry* (Methuen, London, 1953). Full-length texts describing dielectric materials and techniques for investigating them.

H. B. Thompson, *J. Chem. Educ.* **43**, 66 (1966). A review of dipole moment measurements in solution.

I. F. Halverstadt and W. D. Kumler, *J. Am. Chem. Soc.* **64**, 2988 (1942). Outlines a method for finding experimentally the molar polarization of the solute.

G. K. Estok, *J. Phys. Chem.* **60**, 1336 (1956). Gives the suggestion for avoiding solution density measurements.

A. L. McLellan, *Tables of Experimental Dipole Moments* (W. H. Freeman, San Francisco, 1963); A. A. Maryott and E. R. Smith, "Tables of Dielectric Constants of Pure Liquids," *National Bureau of Standards Circular No. 514* (Superintendent of Documents, U.S. Government Printing Office, Washington, D.C., 1951). Give tables of dipole moments.

J. B. Moffatt, *J. Chem. Educ.* **43**, 74 (1966); P. Gray and M. J. Pearson, *Trans Faraday Soc.* **59**, 347 (1963). Deal with *n*-propylnitrite.

C. L. Braun, W. H. Stockmayer, and R. A. Orwell, *J. Chem. Educ.* **47**, 287 (1970); H. J. G. Hayman and I. Eliezer, *J. Chem. Phys.* **35**, 644 (1961). Deal with 1,2-dichloroethane or 1,2-dibromoethane.

C. P. Smyth in A. Weissberger and B. W. Rossiter (eds.), *Physical Methods of Chemistry,* Vol. I (Wiley-Interscience, New York, 1972), Part IV, Ch. VI. Deals with dipole moment measurements.

Chapter 9

Electrochemistry

9.1 Electrochemical Cells

The purpose of this experiment is to determine the standard half-cell potentials for several different kinds of electrodes and for several different solution concentrations. The theoretical expressions for these potentials are developed.

THEORY. Batteries are commonplace and depend for their operation on electrochemistry and electrochemical reactions, which involve both positive and negative ions in solution and the reactions among them which arise as a result of their inherent electrical properties. For example, it is well known that when a strip of metal is dipped into a solution containing ions of that metal a potential is developed at the liquid-solid interface. In attempting to measure this potential, say with a voltmeter, we find it necessary to make at least one additional metal-solution connection, thereby thwarting any attempt to find experimentally the potential of a single electrode. With this restriction we also get some freedom, namely, the freedom to choose the voltage of some standard reference single electrode and then to measure all other single electrodes with respect to it. The key words here are "with respect to"; they imply an arbitrarily selected scale origin. By international agreement the standard electrode is the standard hydrogen electrode, which is defined as a single electrode (half-cell) consisting of a hydrogen gas atmosphere at 1 atm and 298°K surrounding a platinum electrode coated with platinum black dipped in an acid solution of unit activity in H^+ ions.

This electrode is arbitrarily assigned a voltage of zero, which serves as the scale origin mentioned above. A complete cell made of this standard electrode and any other half-cell then possesses a measurable potential difference ascribable to the second half-cell. By convention all half-cell voltages are written as reduction reactions. For example, the standard hydrogen half-cell reaction is

$$H^+ + e^- \rightarrow \frac{1}{2}H_2(g), \qquad E^\circ = 0$$

This is one type of electrode: the gas electrode consisting of a gas and a solution of the ions it forms on removal of electrons. Another type is the metal-metal ion electrode, for example,

$$Zn^{2+} + 2e^- \rightarrow Zn, \qquad E^\circ = -0.763 \text{ V}$$

Still another is the metal-insoluble salt including the example

$$AgCl + e^- \rightarrow Ag + Cl^-, \qquad E^\circ = 0.222 \text{ V}$$

Finally, there are redox half-cells (electrodes) in which a chemically inert electrode is inserted into a solution containing two ions in different oxidation states, for example,

$$Fe^{3+} + e^- \rightarrow Fe^{2+}, \qquad E^\circ = +0.771 \text{ V}$$

Note that each of the above half-cells is characterized by a particular voltage E° measured with respect to the standard hydrogen electrode. These voltages, known as standard half-cell voltages, are taken to be measured at 25°C and under standard conditions of 1 atm and unit activity of all the ions. If these conditions are not met, the voltages will be different and given by the Nernst equation, which we shall develop later. For the moment we shall restrict ourselves to standard-state conditions and note that the combination of any two half-cells provides an electrochemical cell with some voltage between its two electrodes which can be calculated from a table of half-cell voltages. In operation such a cell generates an electric current as a result of a chemical reaction. This current can be made to do useful work. It is a well-known fact that, because there is internal resistance in the cell itself, the voltage between the two electrodes falls as current is drawn. To avoid this drop and to measure the voltage at thermodynamic equilibrium conditions, a potentiometric (no current flow) method is used. Potentiometer circuits are described in Section 2.4

If both half-cells are at standard-state conditions, it is a simple matter to calculate the expected cell voltage directly from readily available tables of half-cell voltages. To find the correct voltage certain conventions must be followed and certain pitfalls avoided. First, some older compilations write half-cells as oxidations rather than by the presently accepted reduction convention. If the table at hand happens to be for oxidations, all the reactions and the algebraic signs of the voltages should be reversed to construct a reduction table. The end result will always be the same provided two different kinds of tables are not used and attention is given to writing the reactions properly. It is also useful to remember that in electrochemical cells one half-cell involves oxidation and the other, reduction. This implies that one of the half-cell reactions taken from the table must always be inverted and the sign of its voltage reversed. The voltage of an electrolytic cell is normally taken to be positive except when standard cell potentials are being measured. In the latter case, the hydrogen half-cell is always taken as the ground terminal (zero voltage), and the sign of the other half-cell voltage is determined by whether it must be connected to the positive or negative side of the potentiometer to get a reading.

In most routine cases, the standard hydrogen electrode is not used because it is experimentally troublesome. Rather, some secondary standard is used whose voltage with respect to the standard hydrogen electrode is accurately known. A common example is the easily constructed saturated calomel electrode described in detail below. The electrode reaction is

$$Hg_2Cl_2(s) + 2e^- \underset{\text{saturated in KCl}}{\longrightarrow} 2Hg(l) + 2Cl^-, \qquad E^\circ = 0.2444 \text{ V}$$

We conclude therefore that when a calomel electrode and a standard hydrogen electrode are connected together in such a way that no significant additional potentials develop, the cell voltage will be 0.244 V under conditions of no current flow and that reduction will take place at the calomel electrode. If we now properly connect a saturated calomel electrode (0.2444 V) with a standard state silver-silver chloride electrode, we find a total cell voltage of 0.022 V, with the calomel electrode having the most positive voltage. The standard-state half-cell voltage for the silver-silver chloride portion is then calculable as

$$\begin{aligned} E^\circ_{cell} &= E^\circ_{calomel} - E^\circ_{Ag,AgCl} \\ E^\circ_{Ag,AgCl} &= E^\circ_{calomel} - E^\circ_{cell} = 0.244 - 0.022 \\ &= 0.222 \text{ V} \end{aligned} \tag{9-1}$$

Certain shorthand conventions are normally followed in denoting electrochemical cells. For example, the cell just considered would be denoted

$$Ag(s)|AgCl(s)||KCl(aq)|Hg_2Cl_2(s), Hg(l)$$

if the silver electrode were dipped in the same KCl solution as the calomel electrode. With the above scheme, we have the oxidation half-cell on the left and the reduction half-cell on the right.

Thus far our discussion has centered on standard-state electrodes. Quite naturally we shall inquire about voltages of cells not at standard-state conditions. Before undertaking the derivation of the appropriate equations we shall establish a connection between cell voltages and thermodynamics. As an example, consider the cell

$$Zn|Zn^{2+}||Cu^{2+}|Cu$$

Written out, this shorthand implies the half-cell reactions

$$Zn^{2+} + 2e^- \rightarrow Zn, \qquad E^\circ = -0.763 \text{ V}$$

$$Cu^{2+} + 2e^- \rightarrow Cu, \qquad E^\circ = 0.337 \text{ V}$$

Assume that the electrolyte also contains Cl^-. The total cell voltage is (at standard state) thus 1.100 V, with oxidation occurring at the zinc electrode. This voltage measured under reversible thermal equilibrium conditions represents the ability of the cell to do electrical work by transporting charge through the potential 1.100 V. Each ion in the above half-cell accommodates two electrons and so if 1 mole of copper ion were reduced under the above conditions, 2 moles of electrons would need to be transported across the potential gradient. The work involved in transporting a charge q through a potential V is qV. On a molar basis and referred to an electrochemical cell, we have the electrical work done on the surroundings given as

$$W_{elec} = nN_0eE = n\mathbf{F}E \tag{9-2}$$

where n is the number of electrons appearing in the balanced half-cell reactions, N_0 is Avogadro's number, e is the electronic charge, and $\mathbf{F} = N_0e$ is Faraday's constant, usually given as 9.6487×10^4 coulombs mole^{-1}.

From thermodynamics we have at constant temperature and pressure

$$\begin{aligned} G &= H - TS \\ dG &= dQ - dW + p\,dV - T\,dS \\ dG &= dQ - p\,dV - dW_{elec} + p\,dV - T\,dS \\ dG &= dQ - dW_{elec} - T\,dS \end{aligned} \tag{9-3}$$

Provided the reaction is carried out reversibly

$$dQ = T\,dS$$

and

$$\begin{aligned} dG &= -dW_{elec} \\ \Delta G &= -W_{elec} = -n\mathbf{F}E \end{aligned} \tag{9-4}$$

Equation (9-4) may be used in developing a mathematical relation for equilibrium cell voltages away from standard-state conditions. From the thermodynamics we have a general relation between ΔG and the equilibrium constant K_a (in terms of activities):

$$\Delta G = \Delta G^\circ + RT \ln K_a \tag{9-5}$$

Using Eq. (9-4) to eliminate both ΔG and ΔG° we arrive at the widely used Nernst equation:

$$E = E^\circ - \frac{RT}{nF} \ln K_a \tag{9-6}$$

For the cell described above, the following equilibrium is presumed to persist:

$$Zn + Cu^{2+} \rightleftharpoons Zn^{2+} + Cu$$

The equilibrium constant is given by

$$K_a = \frac{[a(Zn^{2+})]\,[a(Cu)]}{[a(Zn)]\,[a(Cu^{2+})]}$$

which reduces to

$$K_a = \frac{[a(Zn^{2+})]}{[a(Cu^{2+})]} \tag{9-7}$$

since the pure metal activities are each unity.

Thus, the voltage of this electrolytic cell when the ion activities are not at unit activity can be calculated using

$$E = E^\circ - \frac{RT}{nF} \ln \frac{[a(Zn^{2+})]}{[a(Cu^{2+})]} \tag{9-8}$$

where, for this case, $n = 2$. It should be noted in passing that E° depends on temperature and that the indiscriminate use of values directly from tables (usually for 25°C) must be avoided. Rigorously, only when the measurements are carried out at 25°C can the values of E° be used directly from the tables. Otherwise the variation of E° with temperature must be known. From a practical point of view the temperature variation of E° is often quite small and may be neglected.

We have written the equilibrium constant expression in terms of activities rather than mole fractions. For dilute solutions of nonelectrolytes we know the latter is a good approximation essentially because in this concentration region the solute molecules do not interact with one another. In electrolyte solutions, however, the Coulombic interaction is of sufficiently long range to make solute-solute interactions important even in quite dilute solutions. What we have is a situation where each positive ion is surrounded by a group of negative ions with some degree of geometric order in the arrangement–not much, but enough to preclude treatments based on a completely random geometric configuration of the ions. The extent of this kind of effect increases with concentration and leads to easily measurable deviations from Raoult's law at very low concentrations. It is therefore necessary when treating electrolyte solutions to deal with activities, not mole fractions or concentrations.

We shall now proceed to develop the appropriate thermodynamic relations describing electrolyte solutions in terms of molal activity coefficients. Customarily, molality m_i is used in electrochemistry to express concentration. Corresponding to this choice of concentration units is a set of molal activity coefficients γ_i which for every species i satisfy the relation

$$\gamma_i = \frac{a_i}{m_i} \tag{9-9}$$

The activity coefficient is unitless; thus, the activity a_i takes the units of concentration. Using this expression the equilibrium constant expression for the above electrochemical cell becomes

$$K_a = \frac{\gamma(Zn^{2+})m(Zn^{2+})}{\gamma(Cu^{2+})m(Cu^{2+})} \tag{9-10}$$

but $m(Zn^{2+}) = m(ZnCl_2)$ and $m(Cu^{2+}) = m(CuCl_2)$ and we assume that both of these molalities are calculable from the weights of solute and solvent used. The difficult part is now before us: How can we measure or calculate the activity coefficients?

From an experimental point of view, it is not possible to measure the activity of a single ion because both positive and negative ions are always present in electrolyte solutions. Hence, we are forced to use some form of mean value for each salt. Because of the form taken by the equation for the chemical potential ($\mu_i = \mu_i^\circ + RT \ln a_i$) and the fact that the chemical potentials of the positive and negative ions are additive, it is possible to preserve the form of the chemical potential equation for the mixture of positive and negative ions if a geometric mean activity coefficient is defined as

$$a_\pm \equiv \left(a_{Zn^{2+}} a_{Cl^-}^2\right)^{1/3} \tag{9-11}$$

The general form of this expression is

$$A \rightarrow \sigma^+ B^{+n} + \sigma^- C^{-n}$$
$$a_\pm \equiv [a^+ a^-]^{1/(\sigma^+ + \sigma^-)} \tag{9-12}$$

where the σs are stoichiometric coefficients and the a's activities. The mean molality is defined similarly:

$$m_\pm \equiv [m^+ m^-]^{1/(\sigma^+ + \sigma^-)}$$
$$\equiv m\left[\sigma^{+\sigma^+} \sigma^{-\sigma^-}\right]^{1/(\sigma^+ + \sigma^-)} \tag{9-13}$$

Combining Eqs. (9-12) and (9-13) we can now define a mean molal activity coefficient as the ratio of the mean activity to the mean molality:

$$\gamma_\pm \equiv \frac{a_\pm}{m_\pm} \tag{9-14}$$

Activities and thereby activity coefficients of this type are available experimentally from experiments involving freezing point depression, salt solubility, and cell potentials.

From a theoretical point of view the rigorous calculation of activity coefficients is a very difficult undertaking because of the number of particles and number of interactions which must be included. The most widely used approximate theory of activity coefficients of very dilute electrolyte solutions is due to Debye and Hückel. Debye-Hückel theory begins with a simplified tractable model which treats ions as point masses, ascribes all deviations from ideality to electrostatic interactions, assumes complete dissociation of the electrolyte at thermal equilibrium, and assumes a continuous distribution of charge.

At thermal equilibrium a classical statistical Boltzmann distribution of ions is formed about any given ion (the reference ion of charge q). This ion interacts with others, the potential V depending on how far the second ion is removed from the first, the dielectric constant of the solvent, etc. According to Boltzmann statistics, the fraction of ions i possessing potential energy V_j and charge q_i is given by the expression

$$f_i(V_j) = \frac{N_{ij}}{N_i} = \exp\{-\frac{q_i V_j}{kT}\} \tag{9-15}$$

where N_i is the total number of i-type ions per unit volume. Letting ρ_j be the charge density of ions whose potential energy is V_j we write for the total charge density

$$\rho_j = \sum_i N_i f_i(V_j) q_i \tag{9-16}$$

where the summation is over all types of ions. Requiring very dilute solutions so that the nearest ions to the reference are far removed and thus $q_i V_j << kT$ permits the Boltzmann factor to be simplified using a Taylor series expansion followed by a truncation. Expanding the negative exponential in powers of $y = q_i V_j/kT$ furnishes

$$f_i(V_j) = 1 - \frac{q_i V_j}{kT} + \frac{1}{2}\left(\frac{q_i V_j}{kT}\right)^2 - \frac{1}{6}\left(\frac{q_i V_j}{kT}\right)^3 + \cdots \tag{9-17}$$

Because $q_i V_j << kT$, the higher-order terms in this power series have magnitudes which approach zero very rapidly. Debye-Hückel theory uses only the first two terms and the total charge density thus becomes

$$\rho_j \cong \sum_i N_i q_i - \sum_i \frac{N_i q_i^2 V_j}{kT} \tag{9-18}$$

Since overall electrical neutrality in a unit volume must be preserved, $\sum_i N_i q_i = 0$ and we thus have the charge density at potential energy V_j given by

$$\rho_j \simeq - \sum_i \frac{N_i q_i^2 V_j}{kT} \tag{9-19}$$

From electrostatic theory the fundamental equation relating charge density and electric potential is Poisson's differential equation, which is written

$$\nabla^2 V_j = -\frac{4\pi}{\epsilon k T \rho_j} \tag{9-20}$$

where ϵ is the dielectric constant of the solvent. Using Eq. (9-19) to eliminate ρ_j we have the differential equation to be solved:

$$\nabla^2 V_j = -\frac{4\pi(\sum_i N_i q_i^2)}{\epsilon k T} V_j \tag{9-21}$$

Expressing ∇^2 in spherical polar coordinates and noting that the problem possesses spherical symmetry when ρ_j is regarded as a continuous variable, the angular parts can be immediately disposed of as integration constants. The radial equation to be solved is

$$\frac{d^2(rV_j)}{dr^2} = \left[\frac{4\pi}{\epsilon k T}(\sum_i N_i q_i^2)\right][rV_j] \tag{9-22}$$

subject to the boundary condition that $V_j \rightarrow 0$ as $r \rightarrow \infty$. The result is

$$V_j = \frac{q}{\epsilon r} \exp\left\{-\sqrt{\frac{4}{kT}(\sum_i N_i q_i^2)}r\right\} \tag{9-23}$$

where q is the charge on the reference ion to which the coordinate system is attached. Restricting ourselves for the moment to very small values of r and low concentrations (small $\sum_i N_i q_i^2$) the exponential in Eq. (9-23) can be expanded to furnish a power series for V_j which can be truncated as

$$V_j \cong \frac{q}{\epsilon r} - \frac{q}{\epsilon}\sqrt{\frac{4\pi}{\epsilon k T}(\sum_i N_i q_i^2)} \tag{9-24}$$

The first term on the right-hand side is the potential at r rising from the reference charge q, while the second term, independent of r, furnishes approximately the potential caused by all the remaining charge. It is important to keep in mind that this expression is valid only for $1/r << \sqrt{(4\pi/\epsilon kT)(\sum_i N_i q_i^2)}$.

Having now established the magnitude of the potential as a function of r in Eq. (9-23), we turn to a calculation of the electrical work required to place a charge q at the origin in the presence of all the other charges. The field at this point in the absence of q is just the second term of Eq. (9-24), which we denote as V', and the work required to add an increment dq of charge is just $V'dq$. The total work required to place a total charge q at the origin is thus

$$-W_{elec} = \int_0^q V'dq = -\frac{q^2}{2\epsilon}\sqrt{\frac{4\pi}{\epsilon kT}(\sum_i N_i q_i^2)} \qquad (9\text{-}25)$$

This work is related to the Gibbs free energy through an equation having the form of Eq. (9-2). If we dissect Eq. (9-5) into parts, one for each ion, we arrive at an equation for the chemical potential:

$$\begin{aligned} \mu_i &= \mu_i^\circ + RT \ln a_i \\ &= \mu_i^\circ + RT \ln m_i + RT \ln \gamma_i \end{aligned} \qquad (9\text{-}26)$$

If we now imagine a very dilute solution in which the charge on species i is reduced toward zero, $\gamma_i \rightarrow 1$ and $\mu_i \rightarrow \mu_i$ (ideal solution) $= \mu_i^\circ + RT \ln m_i$. The remaining term in Eq. (9-26) is then identified with the chemical potential caused by the formation of ions of charge q_i. Consequently the electrical work we have just calculated per ion may be set equal to $RT \ln \gamma_i$ divided by Avogadro's number. Rearranging furnishes

$$\ln \gamma_i = -\frac{q^2}{2\epsilon kT}\sqrt{\frac{4\pi}{\epsilon kT}(\sum_i N_i q_i^2)} \qquad (9\text{-}27)$$

At this point we are "stuck" with the experimental impossibility of measuring activity coefficients for either the positive or negative ion separately. Again use is made of the geometric mean of the activity coefficients:

$$\gamma_\pm^\sigma = \gamma_+^{\sigma^+} \gamma_-^{\sigma^-} \qquad (9\text{-}28)$$

We apply Eq. (9-27) to each ion to find

$$\begin{aligned} \sigma \ln \gamma_\pm &= \sigma^+ \ln \gamma_+ + \sigma^- \ln \gamma_- \\ &= \frac{-(\sigma^+ q_+^2 + \sigma^- q_-^2)}{2\epsilon kT}\sqrt{\frac{4\pi}{\epsilon kT}(\sum_i N_i q_i^2)} \end{aligned} \qquad (9\text{-}29)$$

This relation may be simplified by asserting electrical neutrality ($\sigma^+ q_+ = \sigma^- q_-$). This restriction leads to

$$\sigma^+ q_+^2 + \sigma^- q_-^2 = \sigma q_+ q_- \qquad (9\text{-}30)$$

and

$$\ln \gamma_\pm = -\frac{q_+ q_-}{2\epsilon kT}\sqrt{\frac{4\pi}{\epsilon kT}(\sum_i N_i q_i^2)} \qquad (9\text{-}31)$$

All the quantities in Eq. (9-31) are readily available except $\sum_i N_i q_i^2$. Its calculation requires the assumption of 100% dissociation in order to relate a known molality to a sum over all the ions of the square of their charge times their number density. If the solutions are very dilute, we may use the solvent density d in place of the solution density d_s. The molality m_i is then given in terms of the molarity ($M_i = 1000N_i/N_0$) and the density by the expression

$$m_i = \frac{1000N_i}{dN_0} \tag{9-32}$$

Solving Eq. (9-32) for N_i and substituting yield

$$\sum_i N_i q_i^2 = \frac{dN_0}{1000} \sum_i m_i q_i^2 \tag{9-33}$$

and thus

$$\ln \gamma_{\pm} = -\frac{q_+ q_-}{2\epsilon kT} \sqrt{\frac{8\pi e^2}{kT\epsilon} \frac{dN_0}{1000}} \sqrt{I} \tag{9-34}$$

where I is called the ionic strength and is defined as

$$I = \frac{1}{2} \sum_i m_i z_i^2 \tag{9-35}$$

z_i is the number of units of electric charge on ion i. In Eq. (9-34), e is the charge on the electron. This completes the formal development of the Debye-Hückel limiting law. It predicts that the activity of an ion in solution varies with the square root of the ionic strength, a relationship followed for many strong electrolytes in very dilute solution.

The effect of the solvent in Eq. (9-34) is only through the dielectric constant, and no explicit account is taken of ion-solvent interactions. The restrictions to very dilute solutions and 100% dissociation are also distressing. Extensions of the limiting law to encompass other conditions have met with some success, but their discussion is beyond the scope of this text.

As pointed out in the experimental section, salt bridges consisting of a concentrated salt solution (usually KCl or NH_4NO_3) are sometimes inserted between the liquid portions of two half-cells. The boundaries between these regions are maintained by porous glass discs. Salt bridges are necessary if very precise data are to be obtained from half-cells in which the electrolyte solution is not the same for both half-cells. When different solutions are present at the boundary, irreversible phenomena can occur; the most important of these is that the current is carried by different species when the flow through the cell is reversed. Because the electrical properties of these carriers differ, a liquid junction potential arises. If a salt bridge is introduced, say concentrated KCl, then the current across the junction will be carried mainly by K^+ and Cl^- irrespective of the direction of the current. There will, of course, be a residual irreversible effect at the boundary between the salt bridge and the half-cells, but hopefully it will be minimized. Another irreversible effect may be the occurrence of irreversible chemical reactions at the junctions.

EXPERIMENTAL. The saturated calomel electrode will be used as the reference electrode in all the experiments described here. The other half-cell will consist of a metal-metal ion electrode. While the glassware needed in the construction of these electrodes can be prepared relatively easily, it may be desirable to use commercially available electrode apparatus.*

The calomel electrode may be constructed in one of several ways. One of the simplest makes use of the apparatus depicted in Fig. 9-1. This half-cell consists of three glass tubes. The inner one contains the platinum electrode and electrical connections for making external voltage measurements. The intermediate tube contains the electrode material in contact with the platinum electrode. The outer tube contains saturated KCl solution, which contacts, through a hole in the intermediate tube, the electrode. Contact with the second half-cell is made through the porous glass junction. This porous glass disc maintains electrical contact but slows diffusion.

The materials required for the inner tube include a length of platinum wire, some Wood's metal, and a length of copper lead-in wire. The pellets of Wood's

*Kontes Glass Company, Vineland, N.J.

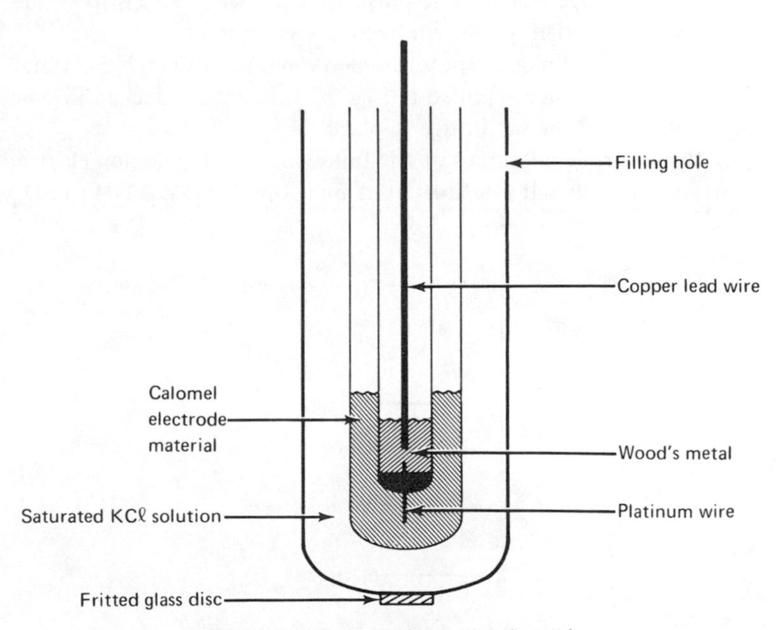

Figure 9-1. Saturated calomel electrode.

metal are placed inside the glass tube, and the two wires are inserted. Then the assembly is heated in boiling water to liquefy the Wood's metal after which it is cooled with the wires in their proper geometric positions.

The intermediate tube can be constructed by simply putting in a few drops of chemically purified and distilled mercury. Above the mercury, a few milligrams (enough to cover the surface of the exposed mercury) of finely divided Hg_2Cl_2 is added. Then a few crystals of purified KCl are added. The inner glass tube is then carefully inserted into the intermediate tube and the resulting assembly inserted into the outer tube. Saturated KCl solution is then added through the filler hole until the KCl crystals in the intermediate tube are covered.

The metal electrodes are easily constructed by soldering a piece of the electrode wire onto a lead-in wire and then sealing the electrode wire into the base of the glass holder as shown in Fig. 9-2.

The solutions called for in the following experiments should be made from chemically pure and dry reagents. As an experimental note, if standard taper electrode holders are used, it will be convenient to store the electrodes by inserting them in an outer standard taper joint. With the calomel electrode this outer joint should contain saturated KCl solution.

Some experiments require a salt bridge, while others make use of one as a matter of convenience. A salt bridge is simply a device which isolates the liquid portions of the half-cells but maintains electrical (ionic) communication. Two salts are in common use, KCl and NH_4NO_3, and they are present in concentrated solution in the salt bridge region. The purpose of the salt bridge is to ensure reversibility of the cell. This point will become clearer later.

The measurement technique is potentiometry and is outlined in Section 2.4. The complete cell is shown assembled in Fig. 9-3 into a three-necked Erlenmeyer flask. Figure 9-3 shows the salt bridge in place.

Measure the electrode potentials of the following metal-metal ion electrodes: Zn, Cu, Pb, and Ag. The salt solutions used are 0.1-M $ZnSO_4$, 0.1-M $CuSO_4$,

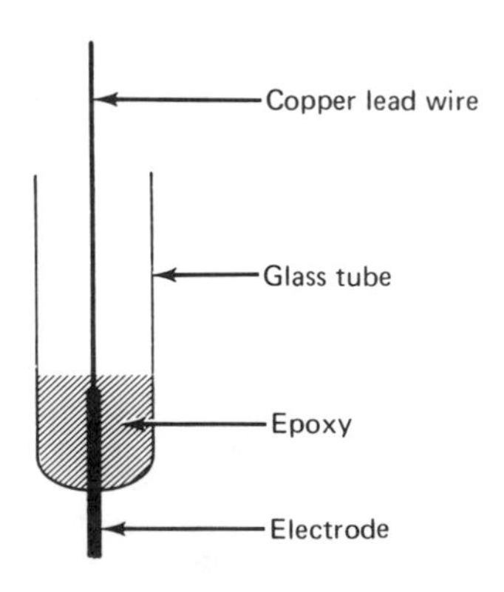

Figure 9-2. Metal electrode.

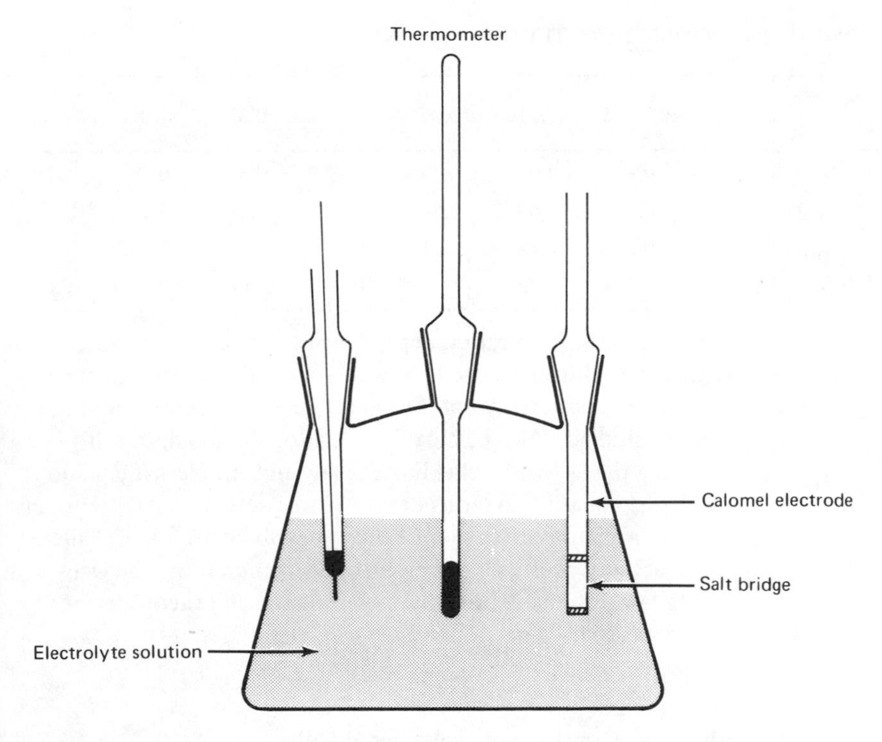

Figure 9-3. Electrochemical cell arrangement.

0.1-M $Pb(NO_3)_2$, 0.1-M $AgNO_3$, and 2.0-M NH_4NO_3. The equilibrium cell temperature should be 25°C. The Pb and Ag electrode potentials must be measured with the salt bridge in place. With the other two electrodes, the cell potential should be measured both with and without the salt bridge.

In a second set of experiments, measure the potential of one of the metal-metal ion electrodes with respect to the saturated calomel electrode at varying metal ion concentrations of 0.1, 0.050, 0.010, 0.005, and 0.001 M.

ANALYSIS

1. Using Table 9-1 as a source of activity coefficients, calculate the standard half-cell voltages of each of the cells using 0.1 M ions. Comment on their agreement with accepted literature values.
2. For two of the cells, measurements were made both with and without the salt bridge. What differences were observed? Comment on possible irreversible effects at the liquid junction when no salt bridge is present.
3. For the other two cells only the salt bridge was used. The solubility product constants for AgCl and $PbCl_2$ are 2.8×10^{-10} and 1.6×10^{-5}, respectively. Why does this imply the need for using the salt bridge?

Table 9-1. MEAN IONIC ACTIVITY COEFFICIENTS

Salt \ m	0.001	0.005	0.01	0.05	0.1	0.5	1.0	2.0
$AgNO_3$	–	0.92	0.90	0.79	0.72	0.51	0.40	0.28
$Pb(NO_3)_2$	0.88	0.76	0.69	0.46	0.37	0.17	0.11	–
$CuSO_4$	0.74	0.53	0.41	0.31	0.16	0.11	0.047	–
$ZnSO_4$	0.70	0.48	0.39	–	0.15	0.065	0.045	0.036

4. Using the data on the concentration dependence of the cell potential (remember that the calomel electrode is in its standard state), calculate a value for the mean ionic activity coefficients at each concentration of the electrolyte being studied. Also calculate a value for the standard cell potential E° using the Debye-Hückel limiting law and the Nernst equation. Express the Nernst equation in terms of mean ionic activity coefficients and mean molalities. An appropriate plot of the data will be useful. Calculate the mean ionic activity coefficient at each concentration using the Debye-Hückel limiting law. Compare both the experimental and theoretical results with the data from Table 9-1.

9.2 Thermodynamics and Electrochemical Cells

The purpose of this experiment is to measure the temperature dependence of the cell voltage and from the data calculate thermodynamic quantities attributable to the electrode reactions. The theory, mainly a review, is presented without considerable detail.

THEORY. In Section 9.1 a relation was established between the molar Gibbs free energy change of the electrode reactions and the cell voltage measured under conditions of no current flow:

$$\Delta G = -nFE \tag{9-36}$$

If the cell is in its standard state, we have

$$\Delta G^\circ = -nFE^\circ \tag{9-37}$$

and for any concentration

$$\Delta G = \Delta G^\circ + RT \ln K_a \quad \text{and} \quad E = E^\circ - \frac{RT}{nF} \ln K_a \tag{9-38}$$

We conclude that measurement of cell voltages under various conditions suffices to determine the molar Gibbs free energy, the standard-state molar Gibbs free energy, and the equilibrium constant in terms of activities.

In principle we can also measure the variation of cell potential with absolute temperature and thus find $\partial E/\partial T$, which from Eq. (9-36) furnishes

$$\frac{\partial E}{\partial T} = -\frac{1}{nF}\frac{\partial \Delta G}{\partial T} \tag{9-39}$$

From general thermodynamic considerations we know that

$$-\frac{\partial \Delta G}{\partial T} = \Delta S \tag{9-40}$$

Combining Eqs. (9-39) and (9-40) the molar entropy change is calculable as

$$\Delta S = nF\frac{\partial E}{\partial T} \tag{9-41}$$

For a process carried out isothermally (or very nearly so) the molar enthalpy change can be computed from the relation

$$\begin{aligned} \Delta H &= \Delta G + T\,\Delta S \\ &= -nFE + nF\frac{\partial E}{\partial T}T \end{aligned} \tag{9-42}$$

Equations (9-41) and (9-42) both apply to all equilibrium states including standard states.

EXPERIMENTAL. This experiment makes use of the saturated calomel electrode described in Section 9.1. The other half-cell is the metal-insoluble salt electrode

$$AgCl + e^- \rightarrow Ag + Cl^- \qquad E^\circ = 0.222 \text{ v}$$

The silver-silver chloride electrode is prepared by first constructing a metal electrode as described in Section 9.1. The resulting silver electrode is then treated with 3N nitric acid for about 2 min. after which it is washed with distilled water. The silver wire is then anodized in a solution of 0.1-M HCl at 2-mA current for about 1 hr. using an isolated platinum electrode.

The fully assembled cell is shown in Fig. 9-4. Clearly the saturated KCl electrolyte is common to both electrodes. Measure the cell voltages at at least five temperatures in the range 10°–60°C, repeating one of the measurements to ascertain reproducibility. A large-volume thermostatted water bath will be quite helpful.

As an extension of this experiment it is suggested that the $Cu\text{-}Cu^{2+}$ half-cell used in Section 9.1 be studied in conjunction with the calomel electrode. Similar voltage measurements over the same temperature range should be made.

ANALYSIS. Write the balanced half-cell reactions and the appropriate shorthand cell notation. From the relative signs of the two electrodes, ascertain at which electrode oxidation occurs. Write the Nernst equation corresponding to each half-cell and then combine the results to furnish an expression for the cell

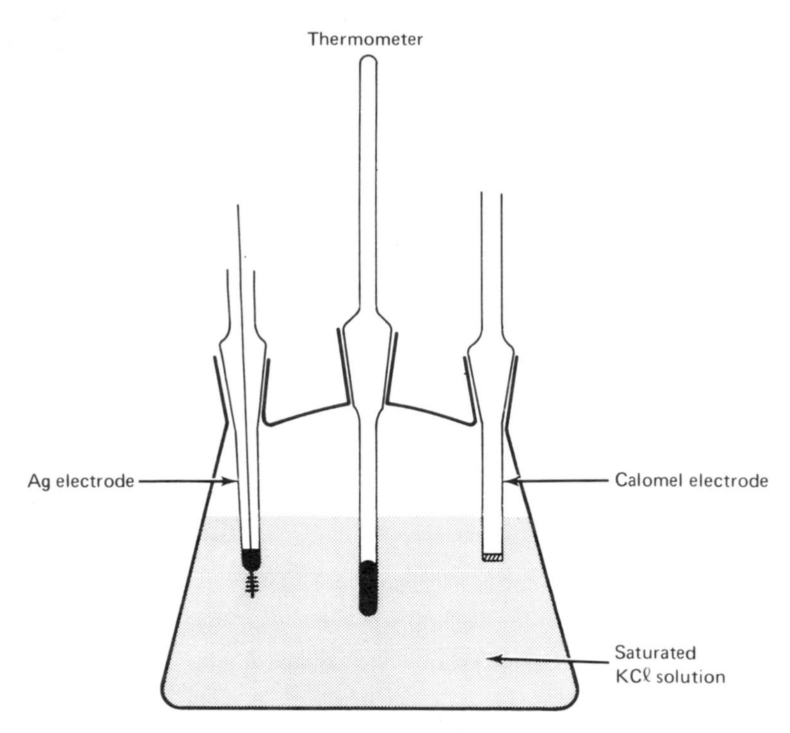

Figure 9-4. Electrochemical cell for measuring the variation of efm with temperature.

voltage in terms of the standard cell voltage and the ion activities.

From the data and a plot of E versus T, calculate ΔG, ΔH, and ΔS at each temperature used. To what chemical reaction do these thermodynamic parameters apply? Based on this reaction and tabulated standard heats of formation and absolute entropies, determine ΔH° and ΔS° at 298°K and compare them with the experimentally determined values.

If the suggested extension involving the $Cu\text{-}Cu^{2+}$ half-cell were performed, an analysis like that described in the previous two paragraphs would be appropriate. In addition the differences between the two cells should be considered and comments should be made regarding the experimental advantages and disadvantages of both.

9.3 Conductance Measurements in Electrolyte Solutions

The purpose of this experiment is to measure, using an ac bridge circuit, the ability of electrolyte solutions to conduct electric current. From the data, including calibration data, the equivalent conductance at each concentration is determined. By extrapolation, in the case of strong electrolytes, the equivalent conductance at infinite dilution is determined. Both strong and weak electrolytes are studied. In the theory section an introductory basis is given for understanding the behavior of electrolyte solutions.

THEORY. In this experiment the properties of a solution known to conduct an electric current are studied. Whereas in metal conductors the current is carried by electrons, the current of interest here is carried by both positive and negative ions immersed in a nonconducting solvent. The ability of a solution to conduct electric current can be stated, at least approximately, by Ohm's law:

$$I = \frac{E}{R} \tag{9-43}$$

where I is the current in amperes, E the electrical potential in volts, and R the resistance in ohms. Insofar as this experiment is concerned most of the interesting features lie buried in the resistance. What we seek then is some means of describing a solution which will provide a basis for the magnitude of R in various systems under various conditions.

To begin we shall dissect R into parts: one characterizing the geometry of the sample and the other characterizing the material making up the sample. This is written as

$$R = \rho \frac{\ell}{A} \tag{9-44}$$

This relation is valid for a length ℓ of a conducting medium whose cross-sectional area perpendicular to the axis of ℓ is uniform and of area A. The parameter ρ characterizes the material comprising the sample and its physical condition. The term *specific resistance* is used to denote this physical property. The inverse of ρ is called the *specific conductance,* κ. It measures the intrinsic ability of a solution to conduct electric current irrespective of the geometry of the solution. However, the specific conductance will vary with concentration, temperature, solvent, other ion concentrations, etc. Another quantity defined and used in conductance studies is the *equivalent conductance* Λ. It is defined in terms of concentration and specific conductance in a way which at infinite dilution removes the dependence of Λ on concentration C:

$$\Lambda = \frac{\kappa}{C} \times 1000 \tag{9-45}$$

where C is expressed in gram equivalents per liter of solution. For example 0.5-M $CuSO_4$ contains $0.5 \times 2 = 1$ g equivalent/liter, whereas 2-M NaCl contains $2 \times 1 = 2$ g equivalents/liter. The definition provided by Eq. (9-45) attempts to define a parameter which characterizes the conduction properties of a solution independent of the concentration of that solution. Only partial success is achieved because no account is taken of possible variations in the equilibrium concentrations of ions and in variations of ionic mobility as C changes. The former variation will be extremely important in weak acid, weak base, or slightly soluble salt solutions where a large fraction of the material may remain un-ionized except at very low concentrations. The latter concentration effect has to do with the ability of an ion to move through its local environment of solvent molecules and other ions toward an electrode. If the local concentration of oppositely charged ions is high, the mobility of a given ion can be significantly reduced.

Historically four famous names have become attached to the conductance of electrolyte solutions. Arrhenius is credited with proposing that ions are formed and account for the conductance in certain solutions. He suggested that the value of Λ should decrease with increasing concentration because the extent of ionization will drop, but he assumed that the mobilities of ions were unaffected by concentration changes. Debye and Hückel developed the theory of electrolytes on the basis of electrostatic considerations in the local environment of the ion. Some aspects of this theory are discussed in Section 9.1 and several important concepts are introduced there. The essential physical notions are that negative ions tend to cluster around positive ions and vice versa rather than being randomly distributed. Thermal energy tends to destroy this semiordered configuration so that at equilibrium there is a balance between the two. Furthermore, when a voltage is applied, the ions tend to drift in opposite directions and try to "drag along" oppositely charged ions. The fourth name attached to conductance is Onsager, who established a relation between ion concentration and equivalent conductance for very dilute solutions:

$$\Lambda = \Lambda_0 - (a\Lambda_0 + b)C^{1/2} \tag{9-46}$$

where a and b are constants depending on temperature and solvent properties (not C) and Λ_0 is the limiting value of Λ as C in Eq. (9-45) is reduced to zero. All the dependence of Λ on C is contained in the $C^{1/2}$ term according to Onsager's theory. An extension of this theory has been made by Onsager and Fuoss treating the ions as having finite size rather than point charges as in Debye-Hückel theory and Eq. (9-46). Equation (9-46) and its extensions are used predominantly in the treatment of dilute solutions of strong electrolytes.

In this experiment Λ is determined as a function of concentration in a variety of dilute electrolyte solutions and, by proper extrapolation, Λ_0 is found. Under conditions corresponding closely to Λ_0, the oppositely charged ions move essentially independently of one another. In this circumstance we may conceptually, although not actually, divide Λ_0 into two parts: one associated with the positive ion λ_0^+ and the other with the negative ion λ_0^-. The combination furnishes

$$\Lambda_0 = \lambda_0^+ + \lambda_0^- \tag{9-47}$$

The utility of this concept arises in connection with finding Λ_0 for weak electrolytes. Plots of Λ versus $\sqrt{C}$ or C for these systems are very nonlinear at low concentrations, making it difficult to extrapolate the results to C = 0. If values of Λ_0 can be determined for three strong electrolytes, one containing the positive ion of interest, the second containing the negative ion of interest, and the third containing the negative and positive ions, respectively, of the first two, then Λ_0

for the weak electrolyte can be calculated. Summarizing symbolically the above discussion we have four systems and four limiting equivalent conductances:

1. Weak electrolyte, A^+B^-: $\Lambda_{0AB} = \lambda^+_{0A} + \lambda^-_{0B}$
2. Strong electrolyte, A^+C^-: $\Lambda_{0AC} = \lambda^+_{0A} + \lambda^-_{0C}$
3. Strong electrolyte, D^+B^-: $\Lambda_{0DB} = \lambda^+_{0D} + \lambda^-_{0B}$
4. Strong electrolyte, D^+C^-: $\Lambda_{0DC} = \lambda^+_{0D} + \lambda^-_{0C}$

A value for Λ_{0AB} can be computed by the proper combination of the limiting equivalent conductances of the other three solutions. The result is

$$\Lambda_{0AB} = \Lambda_{0AC} + \Lambda_{0DB} - \Lambda_{0DC} \tag{9-48}$$

In weak electrolyte solutions the value of the limiting equivalent conductance determined as in Eq. (9-48) can be used with a measured value of Λ at some concentration C to estimate the degree of dissociation, α, at that concentration. This relation as originally expressed by Arrhenius is

$$\alpha = \frac{\Lambda}{\Lambda_0} \tag{9-49}$$

Values of α may be related to equilibrium constants in the usual way. Thus, measurements of α and C suffice in principle for the determination of the equilibrium constant. The relation between equilibrium constants and degree of dissociation is discussed in Section 5.7.

Mobility of ions has been introduced above in an intuitive fashion. Looking in more detail we shall define the mobility as the macroscopic nonrandom drift speed of an ion toward an electrode of the opposite charge divided by the voltage applied to the electrode. Operationally this definition tries to separate the speed of an ion into two parts: one caused by the intrinsic character of the ion and its local environment and the other caused by the influence of the outside perturbation Mobilities are defined for both positive and negative species, u^+ and u^-, and so we have

$$u^+ = \frac{v^+}{V} \tag{9-50}$$

where v^+ is the drift speed of the positive ion and V is the applied voltage. A similar equation applies to the negative ions. The mobility is determined by the local electrical and physical environment. At the molecular level this requires consideration of the number and kind of nearby ions and the physical properties of the solvent, in particular its dielectric constant and viscosity. For extremely dilute solutions, and in the case of the limiting equivalent conductance, neighboring ions can be neglected. Under these conditions the mobility can be written as

$$u^+_0 = \frac{\lambda^+_0}{\mathbf{F}} \quad \text{and} \quad u^-_0 = \frac{\lambda^-_0}{\mathbf{F}} \tag{9-51}$$

where **F** is the Faraday constant (amount of charge in one mole of electrons). The ability of an ion to move through the solvent under these limiting conditions is governed by several factors including the charge and size of the ion and the viscosity, η, and dielectric constant, ϵ, of the solvent. These influences are summarized approximately in one form of Stoke's law:

$$u_0^+ = \frac{\zeta^+\epsilon}{6\pi\eta} \quad \text{and} \quad u_0^- = \frac{\zeta^-\epsilon}{6\pi\eta} \tag{9-52}$$

where ζ^+ and ζ^- are called zeta potentials and describe the effective charge and size of the ions. The dielectric constant is treated thoroughly in Section 8.2. The viscosity measures the ability of a medium to retard the motion of a particle through it. The cgs unit of viscosity is the poise with dimensionality g cm^{-1} sec^{-1}. Tabulations of viscosities are generally given in centipoise (cP), which is equal to 0.01 P. Water at 25°C has a viscosity of 0.8937 cP, for example.

The zeta potential appearing in Eq. (9-52) may be thought of as the electrical potential that arises between a solvated ion and the bulk of the solution. A solvated ion is a species comprised of an ion and a group of more-or-less tightly bound solvent molecules which move through the solution as a unit. Electrical potentials arise because of the ion charge, the polarizability of the solvent, and the nonrandom orientation of the solvating molecules about the central ion.

Two of the more unusual ions with regard to mobility are H^+ and OH^- when they are present in protic solvents (i.e., those having protons which will participate in hydrogen bonding). These two ions have unusually large mobilities in such solvents in comparison with other ions. The explanation appears to be that protons move from one location to another in the solvent without dragging along the attending solvent molecules. This is accomplished by a proton transfer reaction. In aqueous acid solution as the proton transfers a new H_3O^+ ion is formed at another location and is equivalent mathematically to H_3O^+ migration. In aqueous basic solution proton transfer occurs from a water molecule to an OH^- and is equivalent to OH^- migration in the opposite direction.

EXPERIMENTAL TECHNIQUE. As pointed out in Eq. (9-43) the basic experimental measurement is the resistance of a solution from which the conductance and equivalent conductance can be calculated. This resistance must be measured under well-defined and essentially equilibrium conditions where no chemical reaction occurs. To this end conductance measurements are made under null (bridge) conditions of current flow. Section 2.4 describes Wheatstone bridge circuits which are used to measure unknown resistances. Essentially the same circuit is used in this experiment with modifications to allow for ac excitation and capacitance adjustments. A typical circuit employing 1000-Hz excitation is shown in Fig. 9-5. When properly balanced the unknown resistance of the cell is

$$R_x = R_3 \frac{R_2}{R_1} \tag{9-53}$$

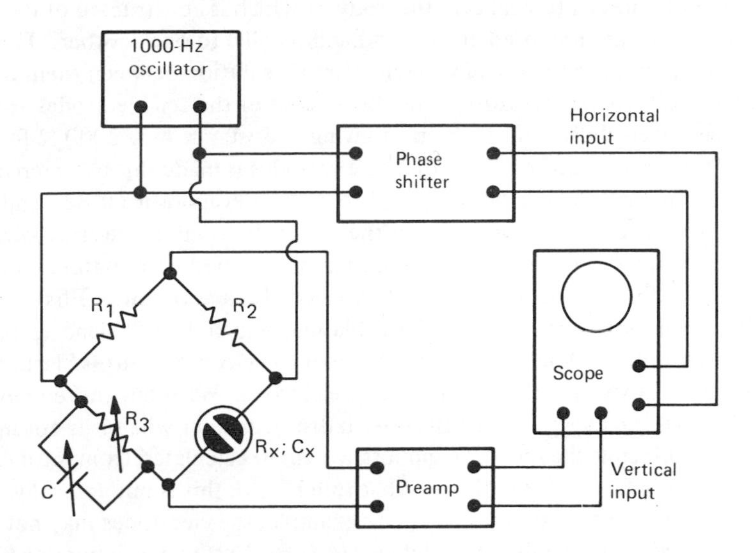

Figure 9-5. Circuit for conductance measurements.

Because the cell electrodes act as a capacitor, the bridge circuit includes a variable capacitor C which is adjusted to balance the capacitive reactance of the cell.

In operation the ratio R_2/R_1 can be varied by powers of 10 through the values 0.1, 1.0, 10.0, 100.0, etc., while R_3 is continuously variable. An oscilloscope is used as a detector and is more precise than the more traditional earphones. A small part of the 1000-Hz oscillator output is applied to the horizontal scope input after it passes through a phase shift network. The output of the bridge has a frequency of 1000 Hz, but the phase is shifted on passing through the bridge. The phase shifter in the circuit of Fig. 9-5 compensates for this so that the signal applied to the vertical scope input from the oscillator and that applied to the horizontal input from the bridge can be adjusted to some fixed phase relation. The bridge output is amplified because in searching for the null point the bridge output becomes very small. At the start of an experiment it is necessary to set the phase shift using one of the standard solutions described below. This procedure begins by using the scope as an ac voltmeter. The horizontal input of Fig. 9-5 is disconnected and replaced with a signal generated internally in the scope. Balance of the bridge is indicated when the vertical amplitude of the displayed signal is minimized. If the horizontal signal is now again connected as in Fig. 9-5, the pattern displayed will become a tilted ellipse when the phase shift is adjusted properly. Once the phase has been set, bridge balance is then indicated for other solutions when an adjustment of R_3 and C displays this same tilted pattern. The variable resistor R_3 should be an accurately calibrated device covering the range 1–200,000Ω.

Figure 9-6 shows a typical cell, the body of which is constructed of Pyrex, with platinum electrodes mounted at each end and parallel to one another. These two electrodes form a capacitor and with electrolytic solution between them form the boundaries of a *solution resistor.* The dimensions of the cell electrodes and the length have been chosen on the basis of giving a resistance near 2000 Ω for 0.01-M KCl at 25°C. Electrical contact with the electrodes is made through mercury contacts. The filling tubes can be stoppered to prevent evaporative losses, and they are located outside the region between the electrode to minimize stray currents.

The platinum electrodes should be *platinized* (covered with platinum black) prior to use. This can be accomplished by electrolysis as follows: First clean the cell with a solution containing 3% chloroplatinic acid and 0.02% lead acetate. Electrolyze with 5–10-mA current until one of the electrodes turns black. Then reverse the polarity and platinize the other electrode. Wash out thoroughly with distilled water, and keep the cell filled with distilled water when it is not in use.

While in principle the specific conductance can be calculated from the measured resistance using Eq. (9-44) with $1/\kappa$ substituted for ρ, this is not generally done because no cell is perfectly uniform. For example, the electrodes may not be precisely parallel. Generally an effective $1/A$ (cell constant) is determined by measuring the resistance of a standard solution, the specific conductance of which is presumed known. The standard is generally taken as potassium chloride solution of known concentration in the range 0.01–0.05 M. The specific conductance of these solutions is accurately known at 25°C in this range. Linear interpolation from Table 9-2 will suffice to determine κ for this experiment.

Table 9–2. SPECIFIC CONDUCTANCE OF KCl AT 25°C

C(M)	κ
0.01	0.0014127
0.02	0.0027668
0.05	0.0066685

The solvent used in conductance measurements should be essentially nonconducting when compared to the conductance of the solutions studied. This is easier said than done because small amounts of ionic impurities are often present— and always present in tap water. As a result special distilled water must be used from which salt ions and dissolved gases such as CO_2 and NH_3 have been removed. Ion exchange resins are probably satisfactory for the requirements of this experiment. The conductance of the solvent should be 10^{-6} or less.

EXPERIMENTS. A wide variety of experiments can be done, each illustrating one or more of the principles discussed above. Several are outlined below. In every case the cell constant should be determined using about 0.02-M aqueous KCl solution. It is, of course, not crucial that the concentration be 0.02-M, but

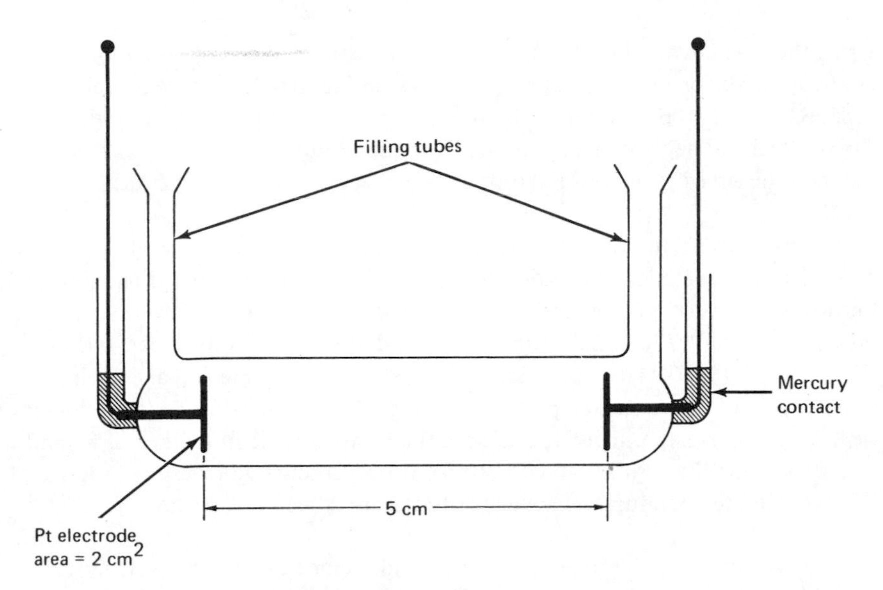

Figure 9-6. Conductance cell.

it is crucial that the concentration be accurately known. Further, in every conductivity measurement it is essential that the temperature be accurately regulated using a thermostat. Between measurements it is essential to rinse the cell thoroughly with distilled water and then with a little of the next solution to be studied.

Experiment 1. In this experiment the influence of concentration and solvent are investigated at 25°C. Prepare three accurately known solutions of KI in each of the solvents—water, methanol, and acetone—which cover the concentration range 0.005–0.05-M. The solubility at 25°C of KI in units of grams per 100-g solvent is 127.5, 17, and 1.0 in water, methanol, and acetone, respectively. Determine the specific conductance and equivalent conductance of each solution from resistance measurements. From an extrapolation using Eq. (9-46), determine Λ_0 for each case. Using Eq. (9-51), determine the sum of the mobilities for each solution. The viscosity and dielectric constant of the three solvents at 25°C are listed in Table 9-3.

Table 9–3. VISCOSITIES AND DIELECTRIC CONSTANTS OF SOLVENTS AT 25°C

Molecule	η(cP)	ϵ
H_2O	0.8937	78.54
CH_3OH	0.547	32.63
CH_3COCH_3	0.316	20.70

Using these values and Eq. (9-52), calculate the sum of the positive and negative zeta potentials for each type of solution. Assuming that λ_0^+ for the K^+ ion is 73.5, 52.4, and 80.6 in water, methanol, and acetone, respectively, calculate both the positive and negative zeta potentials for each solution. Discuss your experimentally observed variations in terms of the properties of the solvents.

Experiment 2. In this experiment the temperature dependence of the limiting equivalent conductance of a strong electrolyte is probed. Prepare three accurately known solutions of KI in water as in Experiment 1. For each of these solutions, measure the resistance and determine the conductance at four temperatures in the range 25°–40°C. Plot according to Eq. (9-46), determine Λ_0, and then construct a plot of Λ_0 versus temperature. Using the limiting equivalent conductance of K^+ as given in Experiment 1, calculate the limiting ionic mobilities of K^+ and I^- and examine their variation with temperature. Discuss your results in terms of the expected temperature dependence of relevant solvent properties.

Experiment 3. In this experiment the conductance of a weak electrolyte is studied as a function of temperature. Prepare a solution of accurately known composition near 0.03-M of acetic acid in water. Measure the resistance and calculate the equivalent conductance for this solution at four temperatures in the range 25°–40°C. Calculate the degree of dissociation at each temperature using $\Lambda_0 = 390.7$. Calculate an acid ionization constant $K_a = [H^+]\,[C_2H_3O_2^-]/[HC_2H_3O_2]$ at each temperature and try to determine an effective activation energy for dissociation. Note that the acid ionization constant here is in terms of concentrations, not activities.

Experiment 4. In this experiment the limiting equivalent conductance of a weak electrolyte $HC_2H_3O_2$ is determined by measuring the properties of three strong electrolyte solutions, NaCl, $NaC_2H_3O_2$, and HCl. Prepare three accurately known solutions of each of these electrolytes with concentrations in the range 0.005–0.03 M. Determine their resistance at 25°C, and by plotting Λ versus $C^{1/2}$, determine Λ_0. Calculate Λ_0 for $HC_2H_3O_2$.

Experiment 5. In this experiment the conductances of three strong electrolyte solutions are determined and the conductances of H^+ and OH^- are extracted. Prepare three accurately known solutions of each of the electrolytes NaCl, HCl, and NaOH in the concentration range 0.005–0.03-M. Measure their resistance at 25°C, calculate the equivalent conductance, and plot according to Eq. (9-46) to find the limiting equivalent conductance. Assuming that λ_0^+ for Na^+ is 50.1, calculate the limiting equivalent conductances of Cl^-, H^+, and OH^-. From these data, evaluate the mobilities of each of the ions.

9.4 The Determination of Transference Numbers of Electrolytes

The purpose of this experiment is to measure the transport of current by the positive ions and by the negative ions in an electrolyte solution. Coulometry is used to determine the total current involved, and chemical analysis of the solution near the cathode and the solution near the anode is used to distinguish between the current carried by the positive ions (cations) and negative ions (anions). This procedure is called the Hittorf method. A transference number for each ion is determined and is a measure of the fraction of the total current carried by that ion. In this particular experiment the transference numbers of Cu^{+2} and SO_4^{-2} are determined.

THEORY. The basic idea underlying the notion of a transference number is that when current is passed through an electrolyte the positive ions and negative ions do not participate equally in the conduction that occurs in the solution. These differences arise because the interactions between the ions and the solvent vary and thus change the ability of an ion to move under the influence of the voltage applied. Section 9.3 discusses many of the concepts employed here and should be read in conjunction with this experiment. As pointed out there, hydrogen and hydroxyl ions have unusually high mobilities, and thus in processes such as those involved in many biochemical reactions, transference numbers have some utility. Relative transference numbers are also of significance in understanding the behavior of galvanic cells.

If we define the transference number of an ion as the fraction of the total current carried by that kind of ion, it is important to recognize that a rigorous extraction of this number from experimental data is possible only for strong electrolytes. The essential reason for this is that laboratory measurements provide only the overall macroscopic current or overall transport of material. In the case of weak electrolytes or cases where complex ions are formed, account must be taken of the changes in total numbers of ions resulting from processes other than those taking place at the electrodes. For example, in weak electrolytes as one ion is removed at the cathode the local reduction in concentration is partially offset by further ionization of the weak electrolyte. These processes make it impossible to determine the transference numbers exactly.

In this experiment we employ the so-called Hittorf method, which can be analyzed with the help of Fig. 9-7. If the cell is filled with a solution of a strong electrolyte (100% ionized) and a current is passed through it from the current source, we expect negative ions (anions) to migrate toward the anode, while the cations migrate in the opposite direction. Depending on the electrode processes we may expect depletion of negative ion concentration in the cathode region and positive

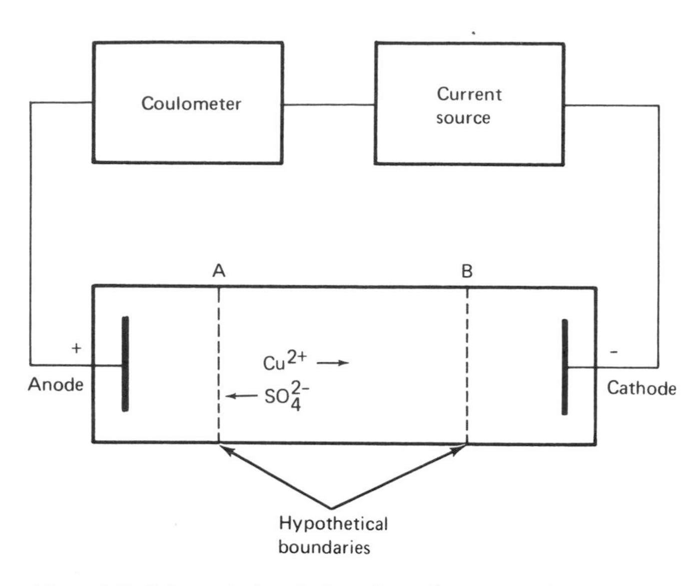

Figure 9-7. Schematic description of transference number apparatus.

ion depletion in the anode region. For example, if the electrodes are made of copper and copper sulfate solution fills the cell, then as current passes copper ions will migrate toward the cathode while sulfate ions will migrate away from it. At the anode the opposite occurs. For this situation we now examine three spatially separated regions in the cell: (1) the anode region, (2) the central region, and (3) the cathode region. In the anode region, defined in Fig. 9-7 by the left-hand end of the cell and the hypothetical boundary at A, Cu^{2+} enters the region at the anode and leaves the region by passing across the boundary at A. If n Faradays of charge are passed through the cell, as measured by the coulometer, then n equivalents (n/2 moles) of copper enter the solution at the anode. If the fraction of the total current carried by Cu^{2+} is symbolized as $t_{Cu^{2+}}$, then $nt_{Cu^{2+}}$ equivalents pass from the anode region into the intermediate region by passing across boundary A. As a result of these two processes the Cu^{2+} concentration in the anode region may change as current is passed unless the amount of copper coming into solution as Cu^{2+} exactly balances (in the anode region only) the amount of copper leaving the anode region. Generally the two will not balance each other and the net gain of copper ion in equivalents is

$$\Delta n^A_{Cu^{2+}} = n - nt_{Cu^{2+}} \tag{9-54}$$

where $n^A_{Cu^{2+}}$ is the number of equivalents of copper ion in the anode region.

Since the transference numbers are defined as fractions, the sum over all the ions must equal unity. For this case,

$$t_{Cu^{2+}} + t_{SO_4^{2-}} = 1 \tag{9-55}$$

Substituting Eq. (9-55) into Eq. (9-54) and rearranging, we obtain

$$\Delta n^A_{Cu^{2+}} = nt_{SO_4^{2-}} \tag{9-56}$$

but

$$nt_{SO_4^{2-}} = \Delta n^A_{SO_4^{2-}} \tag{9-57}$$

and so

$$\Delta n^A_{Cu^{2+}} = \Delta n^A_{SO_4^{2-}} \tag{9-58}$$

Equation (9-58) is a mathematical consequence of the assumption of overall electrical neutrality which is embedded in Eq. (9-55). It simply states that the change in the number of equivalents of positive ion is identical to the change in the number of equivalents of negative ion, with emphasis on the word *change.*

In the intermediate region of the cell, the net transport of positive and negative ions is such that the number of equivalents of each in the region does not change. The boundaries A and B shown in Fig. 9-7 are hypothetical, and in an actual experiment they can be selected so as to satisfy the above requirement. In practice this means that the actual apparatus must somehow provide for the physical separation of the three regions shown in Fig. 9-7.

In the cathode region Cu^{2+} ions pass from solution to the cathode, while the sulfate ions tend to migrate toward the anode (across boundary B moving to the left). From considerations identical to those used in describing the anode region the following expression arises:

$$\Delta n^C_{Cu^{2+}} = \Delta n^C_{SO_4^{2-}} \tag{9-59}$$

where the superscript C denotes the cathode region. There is no reason that the quantities appearing in Eq. (9-59) must be equal to those appearing in Eq. (9-58). As a matter of fact we generally expect them not to be equal since the anode and cathode processes differ. This example using copper sulfate can be generalized to any strong electrolyte solution.

EXPERIMENTAL TECHNIQUE. Several problems occur in transference number determinations. One of the most important ones is how to define and preserve the boundaries between the three regions outlined in Fig. 9-7. One way is to construct a cell with three compartments in it of the style shown in Fig. 9-8(a), each compartment having a volume of about 30 cc. A second method is to simply use a vertical tube, as shown in Fig. 9-8(b) and drain off portions immediately after stopping electrolysis. The volume of this device should be about 100 cc, and it should be at least 50 cm long. In either case one must be careful to avoid diffusion and mixing of electrolyte after electrolysis is stopped. Mixing is more troublesome with the straight tube than with the three-compartment cell but not terribly serious.

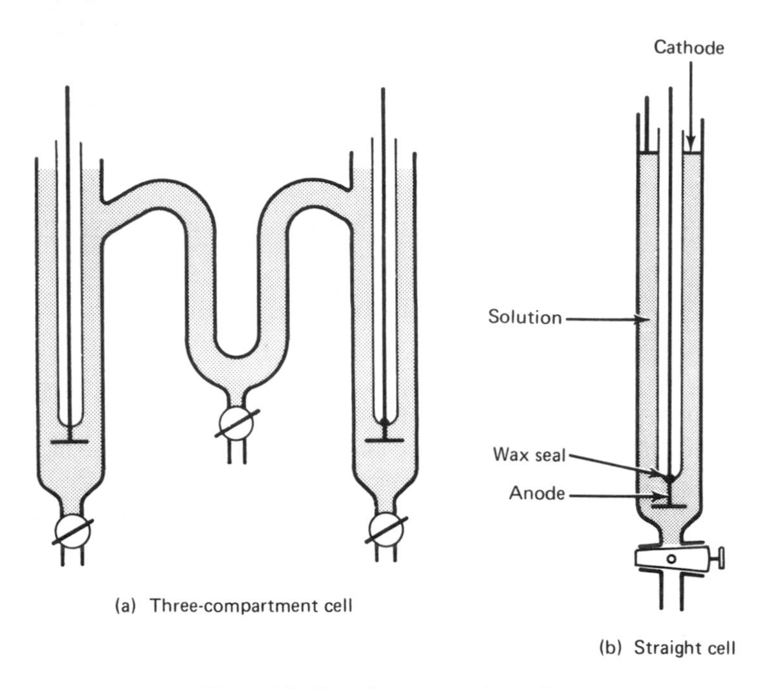

Figure 9-8. Transference number cells.

The anodes and cathode in this experiment are both made out of copper wire (20 gauge is satisfactory). For apparatus (a) of Fig. 9-8 both may consist of a few turns of clean copper wire sealed with sealing wax (Apiezon) into glass tubes. In the straight tube apparatus (b) of Fig. 9-8 the anode should be a spiral of several turns of wire, keeping the coil diameter to about 0.5 cm. It should be sealed into a glass tube as shown. The cathode is a single flat spiral of copper wire, and when in place the axis of the spiral coincides with the axis of the tube. The stopcocks shown should have Teflon plugs.

Coulometry (in this experiment, electrolysis of copper) is used to determine accurately the total number of electrons used. One possible arrangement is shown in Fig. 9-9. It consists of a pair of 50-cc glass tubes connected by a syphon, the longer arm of which is in the anode cavity. These are filled with an aqueous solution of 0.25-M Na_2SO_4 and 0.05-M H_2SO_4. When connected into the circuit of Fig. 9-7 as shown and when current passes, Cu passes from the anode into solution as Cu^{2+} ions, and if 2 moles of electrons is used, 1 mole of Cu^{2+} appears in solution. To determine an unknown number of moles of electrons (in Faradays) we must analyze the anode region for Cu^{2+} and determine its concentration increase. As described below this is done by chemical means. It is important at the end of the experiment to take the syphon out of the coulometer slowly and in a way which allows the solution in it to flow into the anode region. The syphon should then be rinsed off with a small amount of water to assist in ensuring that all the Cu^{2+} ions formed are retained in the anode container.

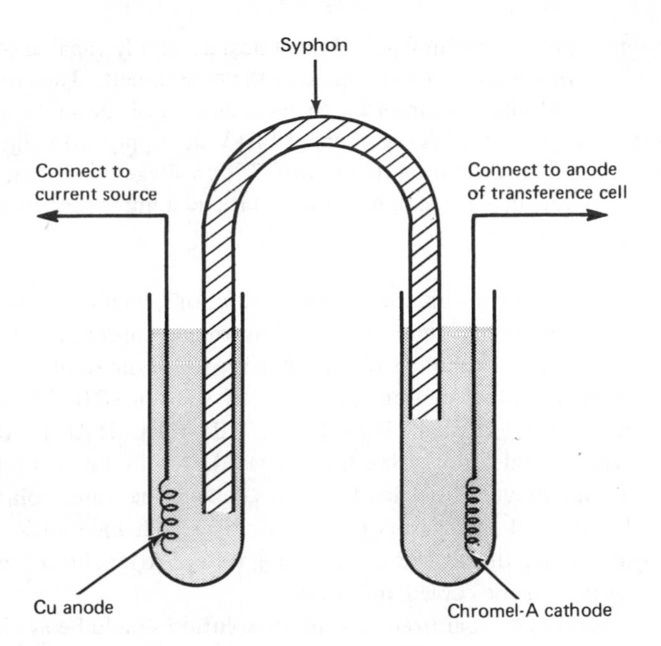

Figure 9-9. Coulometer.

As mentioned above some means is required for determining the amount of Cu^{2+} in the coulometer and in each of three regions of the transference cell. In this experiment chemical means are employed to this end. In particular, KI is added to each of the resulting solutions. As the KI dissolves the following reaction occurs, liberating I_2:

$$4I^- + 2Cu^{2+} \rightarrow 2CuI + I_2 \tag{9-60}$$

This reaction proceeds very nearly to completion if excess KI is used. After the KI is dissolved the solution is stabilized with acetic acid and the I_2 is then titrated with standardized thiosulfate until the yellow color of the solution disappears. Then starch solution is added, and the end point is titrated by noting the disappearance of the blue color. It is easy to pass by the end point and so the thiosulfate must be added slowly and with good mixing. Knowledge of the thiosulfate ion concentration and the volume used can then be used in conjunction with the stoichiometric equation

$$I_2 + 2S_2O_3^{2-} \rightarrow S_4O_6^{2-} + 2I^- \tag{9-61}$$

as a means of determining the number of equivalents of Cu^{2+} in each of the samples. If this same procedure is carried out on a sample of the solution used in the transference number apparatus, then by difference the change in the number of equivalents can be determined for use in Eqs. (9-54) and (9-55). The number of Faradays used is determined by titration of the anode portion of the coulometer.

The current supply shown in Fig. 9-7 must be a dc supply capable of delivering about 20 mA continuously over the course of the experiment. The current should not markedly exceed this value in order to avoid heating of the solution and undesirable electrode processes. A commercial 110-V dc supply with adjustable current output up to 25 mA is quite satisfactory. If dc voltage is available in the laboratory but is not current-regulated, a rheostat and a milliammeter must be connected into the circuit.

EXPERIMENT. In preparing the following solutions, high-resistance distilled water should be used. Prepare an accurately known and approximately 0.05-M copper sulfate solution of volume sufficiently large to fill the transference tube twice. Prepare an adequate amount of coulometry solution (0.025-M Na_2SO_4 and 0.05-M H_2SO_4) and of stabilizing solution (0.01-M $HC_2H_3O_2$). About 10 cc of the latter will be needed for each sample titrated. About 4 g of solid KI and 5 cc of starch solution will be needed for each sample. The starch solution can be prepared by first making a paste of 1 g of soluble starch and a little cold water and then slowly adding this to 100 cc of boiling water. After about 2 min of boiling the solution can be cooled and used.

A stock solution of standardized thiosulfate solution should be available. The concentration should be about 0.02 M. Standardization can be against potassium iodate or potassium dichromate as described in quantitative analysis texts.

PROCEDURE.

1. Set up the transference number apparatus as shown schematically in Fig. 9-7 following the techniques described above. Make certain the polarity of the current supply is proper, and then turn on the apparatus. At least 2 hr. and likely longer depending on the conditions will be required for this part of the experiment. It should be stopped when the solution near the cathode becomes light blue and copper tendrils can be seen growing on the cathode. At this point, stop the current flow and withdraw the samples immediately into previously weighed flasks.
2. While the above experiment is running, set up the titration apparatus and titrate two known weights (near 30 g) of the stock copper sulfate solution. The procedure is as follows:
   a. Weigh out to 0.1 g two samples of stock solution.
   b. Add 4 g of solid KI.
   c. After the KI has dissolved, add 10 ml of acetic acid solution.
   d. Titrate with thiosulfate until the yellow color is nearly gone.
   e. Add 5 ml of starch solution.
   f. Titrate to the end point (disappearance of blue) with good mixing.
   g. Record the volume of thiosulfate solution required.

   Repeat for the second sample. The two titrations should check to better than 1%. If they do not, your technique is poor or one of the reagents is bad.

3. At the end of part 1, withdraw the anode, cathode, and intermediate regions into three separate flasks of known weight. Weigh again with the solutions present. Titrate each one as described in part 2.
4. Titrate the anode portion of the coulometer.

ANALYSIS. For each set of data the number of equivalents of Cu^{2+} per gram of solution should be determined. The values obtained for the stock solution and the intermediate region of the transference apparatus should agree. If they do not, some adjustment of the transference data will need to be made.

Using the coulometer results, the known anode and cathode solution weights, and the data from the previous paragraph, calculate $\Delta n^{A}_{Cu^{2+}}$ and $\Delta n^{C}_{Cu^{2+}}$.

Calculate, using Eqs. (9-55) and (9-56), $t_{SO_4^{2-}}$ and $t_{Cu^{2+}}$. Repeat the calculation for the cathode region data. If necessary, work out a correction for changes in the intermediate region.

Estimate the uncertainties in your results.

REFERENCES

Section 9.1*

W. M. Latimer, *The Oxidation States of the Elements and Their Potentials in Aqueous Solutions* (Prentice-Hall, Englewood Cliffs, N.J., 1952); B. F. Conway, *Theory and Principles of Electrode Processes* (Ronald, New York, 1965). Full-length texts on electrode potentials and reactions.

Section 9.2

The References for Section 9.1 are useful.

Section 9.3*

H. S. Robinson and R. H. Shokes, *Electrolyte Solutions,* 2nd ed. (Academic Press, New York, 1955); H. S. Harned and B. B. Owen, *The Physical Chemistry of Electrolyte Solutions,* 3rd ed. (Van Nostrand Reinhold, New York, 1958). Deal with the behavior of electrolyte solutions.

*All of the references in Appendix XI treat electrochemical cells. See also the references to to other laboratory texts and physical chemistry texts given in Appendix XI.

T. Shedlovsky in A. Weissberger (ed.), *Technique of Organic Chemistry,* Vol. 1 (Wiley-Interscience, New York, 1960), Part IV, Ch. XLV; E. Edelson and R. M. Fuoss, *J. Chem. Educ.* **27**, 610 (1950). Helpful references on experimental techniques.

E. Price in J. J. Lagowski (ed.), *The Chemistry of Non-Aqueous Solvents,* Vol. I (Academic Press, New York, 1966), Ch. 2. Reviews the solvation of electrolytes.

P. Debye and E. Hückel, *Physik. Z.* **24**, 305 (1923); L. Onsager, *Physik. Z.* **28**, 277 (1927); L. Onsager and R. M. Fuoss, *J. Phys. Chem.* **36**, 2689 (1932). References to the original literature.

Section 9.4*

M. Spiro in A. Weissberger (ed.), *Technique of Organic Chemistry,* Vol. I (Wiley-Interscience, New York, 1960), Part IV, Ch. XLVI. A summary of transference number measurement methods.

*The References for Section 9.3 are appropriate.

Chapter 10

Macromolecular Physical Chemistry

10.1 Viscosity of Polyvinyl Acetate-Acetone Solutions

The purpose of this experiment is to measure the intrinsic viscosity of polyvinyl acetate in acetone and methanol and, using the empirical Mark-Houwink relation between intrinsic viscosity and molecular weight, calculate an average molecular weight for the polymer.

THEORY. Viscosity, one of the transport properties, arises because of intermolecular attractive and relatively long range forces. In the present experiment we are interested in using viscosity measurements as a tool for determining the average molecular weight of a polymer solution. An average molecular weight is calculated because the polymer molecules do not all have the same mass.

The viscosities will be calculated by measuring the time required for a known volume of solution to flow through a capillary tube. We shall begin the quantitative description of this process by restricting the discussion to Newtonian and laminar flow. Newtonian flow occurs when velocity-dependent molecular orientation effects are absent. Intuitively we might expect long rod-shaped molecules, for example, to be partially oriented as they fall through a narrow tube (the rod axis oriented preferentially along the tube axis) under the influence of gravity. As the flow velocity increases the orientation effect would be expected to increase. In Newtonian flow these effects are either absent or negligible. Laminar flow implies that the velocity vector of any molecule in solution is pointed very nearly parallel to the axis of the capillary tube. As a result material appearing in

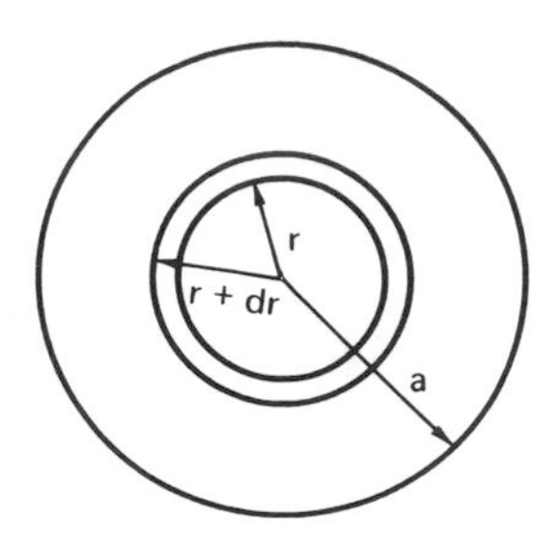

Figure 10-1. Cross section of capillary flow tube.

the cross-sectional region $r \rightarrow r + dr$ of Fig. 10-1 remains in this radial region throughout the course of passage axially along the tube. Generally speaking, laminar flow is achievable at relatively low flow velocities.

At the molecular level viscosity arises because of (1) attractive interactions between the capillary tube walls and both the solvent and solute molecules and (2) attractive interactions between the solution molecules themselves. This attraction retards the flow, and the extent is measured by the coefficient of viscosity. The latter is defined macroscopically in terms of the frictional force between two volume elements as they pass over one another. For example, visualize two small blocks of wood. The resulting force in a solution is proportional to the contact area dA between the volumes and to the rate at which the velocity changes, dv/dx, as one passes from one element to the other. See Fig. 10-2. That is, the frictional force dF is

$$dF \propto \frac{dv}{dx} dA \tag{10-1}$$

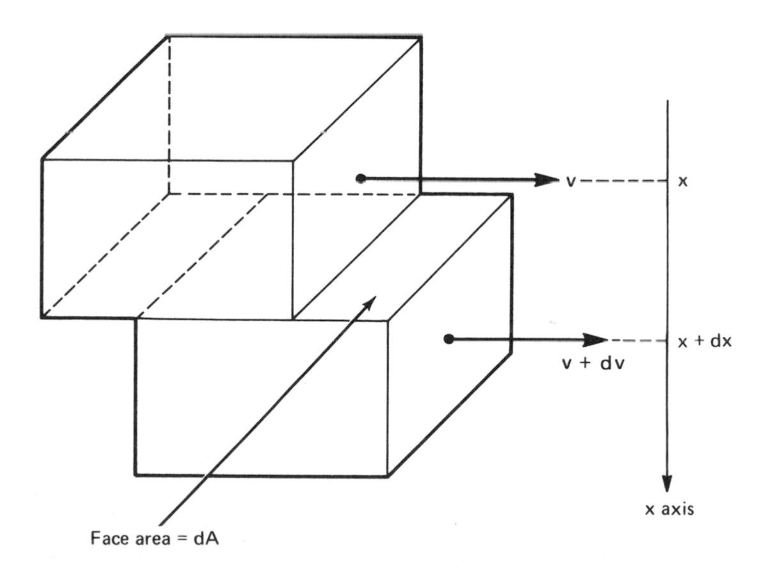

Figure 10-2. Flow of one volume element past another.

The proportionality constant η is the coefficient of friction. Hence, η is defined as

$$\eta = \frac{|dF/dA|}{dv/dx} \tag{10-2}$$

For flow through a uniform capillary of length l and radius a (the cross section is shown in Fig. 10-1) we shall consider the frictional forces on the volume element lying between r and r + dr. Just outside this element the fluid is moving more slowly than it is just inside it. Hence, the fluid within the element tends to be accelerated by the fluid at smaller values of r and retarded by the fluid at larger values of r. The net force under conditions of equilibrium flow is given by the difference between the inner and outer forces, and the result is equal to the applied force (pressure times the cross-sectional area of the element between r and r + dr). Figure 10-3 illustrates the side view of the capillary and the thin imaginary (mathematical) tube being described here. The applied force is

$$dF_a = 2\pi pr' \, dr' \tag{10-3}$$

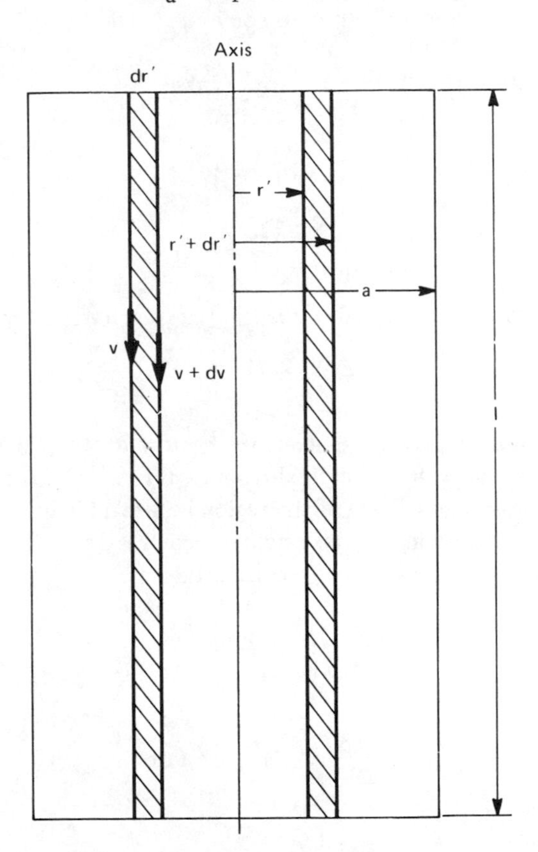

Figure 10-3. Axial cross section of capillary.

where p is the pressure and $2\pi r'\,dr'$ the area of the ring in Fig. 10-1 to which p is applied. From Eq. (10-2) the viscous force is

$$dF = 2\pi r'\eta\ell\left(\frac{dv}{dr}\right)_{r=r'} - 2\pi(r' + dr')\ell\eta\left(\frac{dv}{dr}\right)_{r=r'+dr'} \qquad (10\text{-}4)$$

The first term in this equation is the net frictional force on the inner area element of the small mathematical tube of Fig. A7-3, with $2\pi r\ell$ furnishing the total area. η is the coefficient of viscosity, and $(dv/dr)_{r=r'}$ is the slope of the velocity profile measured perpendicular to the axis of the tube. The second term in Eq. (10-4) represents the force on the outer surface of the mathematical tube in a similar way. Progress toward a solution is made by expanding the velocity gradient in a Taylor series abour $r = r'$. That is,

$$\left(\frac{dv}{dr}\right)_{r'+dr'} = \left(\frac{dv}{dr}\right)_{r=r'} + \left(\frac{d^2v}{dr^2}\right)_{r=r'} dr' + \frac{1}{2}\left(\frac{d^3v}{dr^3}\right)_{r=r'} (dr')^2 \qquad (10\text{-}5)$$

Neglecting terms beyond dr' and then combining (10-5) with (10-4) we obtain

$$dF = -2\pi r'\ell\eta\left(\frac{d^2v}{dr^2}\right)_{r=r'} dr' - 2\pi\ell\eta\left(\frac{dv}{dr}\right)_{r=r'} dr' \qquad (10\text{-}6)$$

Using Eq. (10-3)

$$dF_a = dF$$

or

$$-\frac{pr'}{\ell\eta} = \frac{dv}{dr} + r\frac{d^2v}{dr^2} = \frac{d}{dr}\left(r\frac{dv}{dr}\right) \qquad (10\text{-}7)$$

This equation may be integrated to furnish the velocity as a function of r:

$$v = \frac{p}{4\ell\eta}(a^2 - r^2) \qquad (10\text{-}8)$$

Normally, the total volume passing out the bottom of the tube per unit time is measured. This is given by an integral from $r = 0$ to $r = a$ of the volume passing through the cross-sectional area of the region between r and $r + dr$. If the velocity is v, the amount passing by this ring per second is $2\pi vr\,dr$. The volume V leaving the bottom of the tube per second is then

$$V = 2\pi\int_0^a v(r)r\,dr$$

$$= \frac{\pi p}{2\ell\eta}\int_0^a (a^2 - r^2)r\,dr$$

$$= \frac{\pi p a^4}{8\ell\eta} \qquad (10\text{-}9)$$

Equation (10-9) is known as Poiseuille's law, and it suggests a strong dependence of volumetric flow rate on the radius a of the capillary tube and weaker dependences on the viscosity, capillary length, and pressure.

The integral over time of the volume flow per second provides the time required for the viscous fluid to fall through a certain height. Figure 10-4 illustrates the Ostwald viscometer which is used in these experiments. It is used by measuring the time required for a fixed volume of fluid contained between h_1 and h_2 to flow through the capillary. Since the height of the column of material varies with time, so does the mass of the column and therefore the pressure. At any point the pressure at the top of the capillary is $p = \rho gh$, where ρ is the fluid density, g the gravitational constant, and h the column height above position h_2. Then the integral over time is

$$\int_0^t dt' = \int_{h_1}^{h_2} \frac{8\eta\ell}{\pi g\rho a^4} \frac{dV}{h} \tag{10-10}$$

or

$$t = \frac{8\eta\ell}{\pi g\rho a^4} \int_{h_1}^{h_2} \frac{dV}{h} \tag{10-11}$$

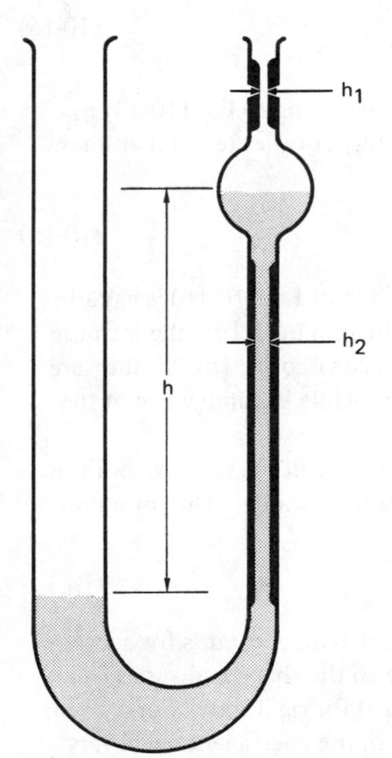

Figure 10-4. Ostwald viscometer.

The quantity indicated by the quadrature in Eq. (10-11) may be approximated as a constant D of the apparatus to give

$$t = \frac{8\eta\ell}{\pi g \rho a^4} D \tag{10-12}$$

The apparatus constant D may be determined by measuring t for a solution of known viscosity and density.

In experimental work a reduced viscosity, called the specific viscosity, η_{sp}, is defined as

$$\eta_{sp} = \frac{\eta_i - \eta_0}{\eta_0} \tag{10-13}$$

where η_i is the viscosity of solution i and η_0 the viscosity of pure solvent. In terms of the approximate equation (10-12), the specific viscosity can be expressed in terms of measured times t_i and densities ρ_i as

$$\eta_{sp} = \frac{t_i\rho_i - t_0\rho_0}{t_0\rho_0} \tag{10-14}$$

Intuitively we expect, at a given temperature, η_{sp} to depend on the solute concentration, c, and the geometric configuration of the solute. The concentration dependence of η_{sp} may be written as a Taylor series in c expanding about c = 0:

$$\eta_{sp} = \eta_{sp}^{\circ} + c\left(\frac{d\eta_{sp}}{dc}\right)_{c=0} + \frac{c^2}{2}\left(\frac{d^2\eta_{sp}}{dc^2}\right)_{c=0} + \cdots \tag{10-15}$$

where η_{sp}° is the specific viscosity when c = 0. According to Eq. (10-13) $\eta_{sp}^{\circ} = 0$ and Eq. (10-15) may be rearranged, neglecting higher-order terms than those shown, to a form

$$\frac{\eta_{sp}}{c} = [\eta] + kc[\eta]^2 \tag{10-16}$$

where $k[\eta]^2 = \frac{1}{2}(d^2\eta_{sp}/dc^2)_{c=0}$. The left-hand side of Eq. (10-16) is experimentally measurable and plots of η_{sp}/c versus c furnish both $[\eta]$, the intrinsic viscosity, and k, the Huggins constant. The intrinsic viscosity $[\eta]$ is a measure of the magnitude of the solute-solvent interaction, while k is indicative of the solute-solute interaction.

Empirically it has been found that the intrinsic viscosity is sensitive both to the shape and molecular weight of the macromolecular solute. The form normally used in the Mark-Houwink relation:

$$[\eta] = KM^a \tag{10-17}$$

where M is the average molecular weight and K and a are constants for a given solvent-solute system. The constant a is sensitive to the shape and varies from zero for hard spheres to 0.5 for random coils to 2.0 for rigid rods. For a given polymer and a fixed molecular weight distribution, the coefficient a will vary from solvent to solvent because of different solvent-solute interactions which tend to alter the shape of the solute. In this context a good solvent is one which causes the solute to unravel extensively (big a), while a poor solvent causes "balling up" of the solute (low a). The viscosity average molecular weight from Eq. (10-17) is really a very complex average and not rigorously either the weight average M_w or the number average M_n. By definition these two averages are

$$M_w = \frac{\sum_{\text{all types}} C_i M_i}{\sum_{\text{all types}} C_i} \tag{10-18}$$

$$M_n = \frac{\sum_{\text{all types}} n_i M_i}{\sum_{\text{all types}} n_i} \tag{10-19}$$

where M_i is the molecular weight of the ith-type particle, C_i the weight concentration (grams per cubic centimeter) of the ith-type particle, and n_i the number concentration (particles per cubic centimeter) of the ith-type particle. The viscosity average molecular weight is approximately equivalent to the weight average molecular weight.

Throughout the course of this presentation we have assumed η to be constant for a given solute-solvent system. The coefficient of viscosity varies with the experimental temperature and, for liquids, only very slightly with the system pressure. Generally η decreases exponentially with temperature; hence, in the experiment performed here, the temperature cannot be allowed to vary significantly during the time when the set of measurements is taken.

EXPERIMENTAL. Solutions of polyvinyl acetate in acetone and methanol containing approximately 1 mg/ml will either be furnished or should be prepared by dissolving the polymer in the solvent and filtering as described in Section 10.3. By dilution three other concentrations covering the range 0.2–0.8 mg/ml should be prepared. Clean and dry volumetric flasks and pipettes should be used, and the solutions should be kept capped at all times when not in use to minimize loss of solvent by evaporation.

Using a clean, dry viscometer, determine at room temperature the times required for solution to pass from h_1 to h_2 as shown in Fig. 10-4. This is achieved by pipetting bubble-free solution into the left-hand column of the viscometer and drawing it through the capillary column to a point slightly above h_1.

In deriving Eq. (10-12) the apparatus constant D is used to replace an integral appearing in Eq. (10-11). This integral is evaluated between two heights h_1 and h_2, which are distances between the fluid levels in the left and right sides of the viscometer (Fig. 10-4). To ensure that h_1 and h_2 are identical throughout the course of a set of measurements, the same total volume of material should be used in each experiment and should be selected so that when the solution is drawn up through the bulb, good-sized column height appears but some fluid remains in the left-hand tube of Fig. 10-4. For solutions of very nearly identical densities this procedure is satisfactory. However, since the flow rate is governed by the pressure difference in the two columns of liquid and in turn this is governed by the product of density times height, the density variation must be accounted for if it is significant.

As the column drops, a timer is started on passing h_1 and stopped on passing h_2. Determine t_i for pure solvents and each of the solutions. Between measurements, clean the viscometer with cleaning solution (sulfuric acid-potassium dichromate), rinse with solvent, and dry thoroughly.

Using a pycnometer as outlined in Section 5.6, determine the density of each of these samples. Using Eq. (10-14), calculate η_{sp} and η_{sp}/c from these measurements.

ANALYSIS. For both acetone and methanol solutions, plot η_{sp}/c versus c and calculate, by least-squares fitting, $[\eta]$ and k. Using the empirical formula (10-17) and the constants from Table 10-1, determine the molecular weights of solute in these two solvents. Make certain to use c in grams per milliliter. Comment on the spatial extension of polyvinyl acetate in these two solvents. Also comment on any differences in molecular weight.

Table 10-1.

Solvent	K	a
Acetone	21.4×10^{-5}	0.68
Methanol	38×10^{-5}	0.59

10.2 Gel Chromatography of Macromolecules

The purpose of this experiment is to investigate the diffusional partitioning of solute macromolecules in a gel chromatographic column. Partition coefficients will be calculated from retention times or retention volumes, and the concentration profiles will be examined under various conditions. A brief theoretical introduction is also given.

THEORY. Biological polymer molecules as well as other macromolecules are often amenable to characterization and separation by the technique of gel chromatography. This technique utilizes porous gels of various types in a vertical column through which solvent and solute flows. The solvent occupies two readily distinguished volumes: (1) the volume outside the porous gel particles called the void volume V_0 and (2) the available interior volume of the gel particles called the internal volume V_i. The total volume of the column V_t is given by

$$V_t = V_0 + V_i + V_g \tag{10-20}$$

where V_g is the volume occupied by the gel material itself. Under operating conditions the gel and the two solvent regions are at thermal equilibrium, with solvent passing through the column (being eluted) and being collected at the lower end, as shown in Fig. 10-5. To start the chromatographic experiment a small amount of dilute solution of macromolecules is injected at the top of the column just above the gel. As the solvent and sample move downward under the influence of gravity, the dilute solute molecules find themselves in a region containing both gel and solvent. Figure 10-6 illustrates the environment schematically. Assuming that the interaction between the gel matrix and the solute is characterizable as a hard sphere potential it becomes clear on a little reflection that solute molecules "see" two distinctly different solvent environments: one characterized by very narrow spatial boundaries (internal) and the other by essentially no boundaries (void). Since passage through the column axially (from top to bottom) is measured, those solute molecules which enter the interior of the gel will require longer times to pass from the top to the bottom than those which do not. Crudely speaking, whether a molecule passes into the gel interior or not is determined by the size and shape of the gel pores and the size

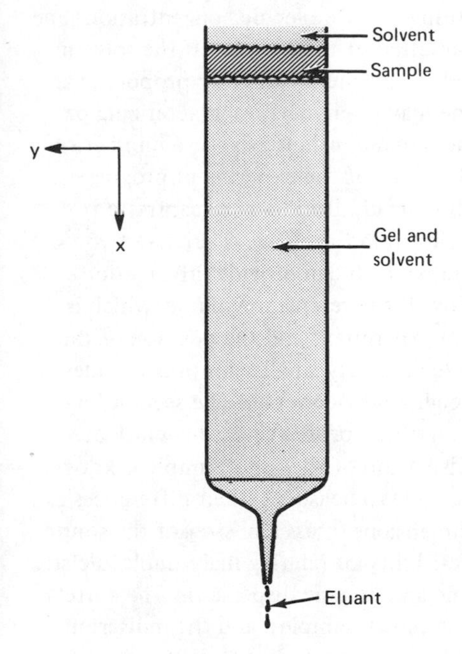

Figure 10-5. Gel chromatograph column.

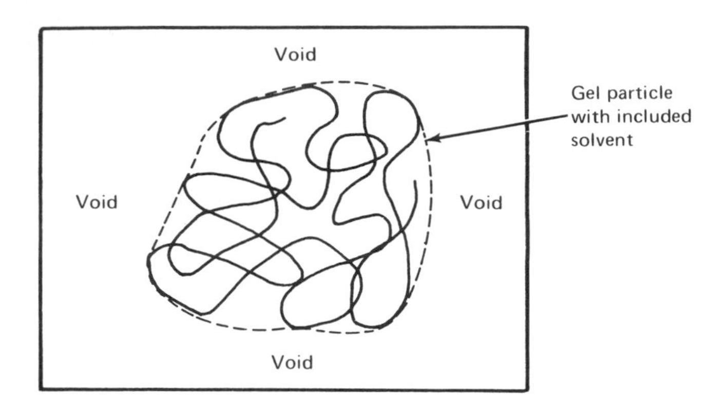

Figure 10-6.

and shape of the solute molecules. In actual fact, the physical phenomena is not dissectable as a "yes" or "no" to entry into the interior of the gel but rather as a fractional partitioning ("yes" with some probability and "no" with some probability) between V_0 and V_i at every increment dx along the axis of the column. The partition coefficient is defined to provide a measure of the partitioning between the two environments. Complete thermodynamic equilibrium is generally not achieved at any position along the column because the flow of solvent and solute perturbs the equilibrium.

The measured quantities in an elution chromatography experiment of the type described here are solute concentration (macromolecule concentration) and total volume of solvent eluted after introduction of the sample. If the solvent flow rate is constant, eluted volume and elapsed time are directly proportional, and elution time instead of elution volume may be used as a characterizing parameter. Solute concentrations can be measured continuously, or a number of fractions of the total eluted volume can be taken as the experiment progresses. For a single component the fraction method would lead to a concentration-volume profile (chromatogram) like that shown in Fig. 10-7. There are two basic features of the resulting chromatogram which can provide information about the molecular size of the solute. First is the retention volume, which is the volume eluted between the start of the experiment and the position of the maximum concentration of the profile. Qualitatively small retention volumes correspond to large molecular size and weight and vice versa. The second feature of importance is the volume dispersion of the profile (e.g., the width at half-height in Fig. 10-7). If the volume dispersion of the input sample is known and the volume dispersion of the eluted sample is measured, their differences can in principle be related to the molecular dimensions (mass and size) of the solute. The above discussion has suggested the possibility of relating macromolecule size and weight to the measured elution volume and volume dispersion. The correlation requires an empirical calibration using known samples, and the indiscriminate use of these calibrations can often lead to erroneous conclusions.

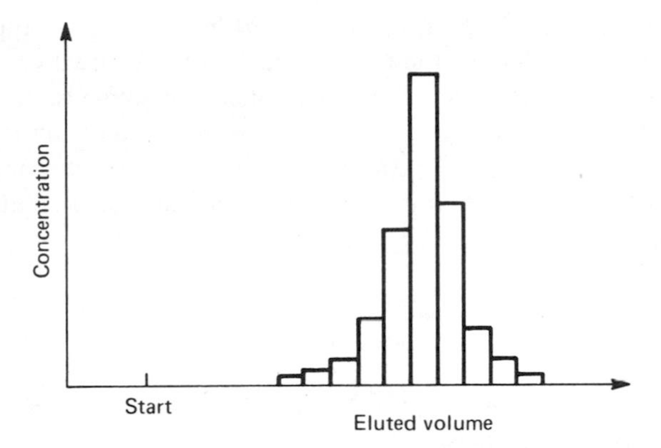

Figure 10-7. Concentration-volume profile.

With this brief qualitative introduction we shall proceed now to a mathematical description of column action, keeping in mind two basic underlying assumptions: (1) Partition coefficients depend on molecular size and shape, not on electrical charge or localized molecular features, and (2) columns are very nearly linear chromatographic devices. Deviations from these two assumptions often occur but seldom are of primary significance.

At the outset we wish to emphasize the two fundamental overall processes which must be treated. First is the solute partitioning phenomenon, which, within every increment dx along the axis (lengthwise direction) of the column, leads to at least a slight momentary retention of solute. This is essentially a diffusional partitioning in the yz plane of Fig. 10-5. Conceptually we can devise a column in which only solute partitioning occurs. The resulting concentration-volume profile is characterized by a retention volume for each species and a volume dispersion which is identical to the volume dispersion of the input sample.

The second overall process is axial dispersion, which in general leads to a spreading out (larger half-width) of the solute volume profile as elution occurs. Several phenomena contribute to this process; among them are axial diffusion of solute into the solvent ahead and behind the profile, nonuniform flow velocities around the surface of gel particles, and disequilibrium of the solute exchange between the two solvent environments.

We shall begin the mathematical development* with a phenomenological description and define a partition coefficient σ which relates the total weight of solute within the gel interior Q_i to both the concentration C_e (grams per cubic centimeter) of solute in the void volume and the internal volume V_i. By definition the partition coefficient is

$$\sigma = \frac{Q_i}{C_e V_i} \tag{10-21}$$

*Several different concentrations and volumes are defined in this development. It is important for the reader to gain a clear understanding of their meaning.

In general σ is a function of both C_e and V_i, but in ideal cases it approaches a constant value independent of these two variables. Conceptually, at any rate, we can examine the ideal case where the concentration of solute within the volume it occupies in the gel interior is equal to the concentration of solute in the void volume. The volume occupied by the solute within the gel interior is not equal to V_i but is rather some fraction of it. If we define the occupied interior volume at equilibrium to be V_{0c}, then the ideal conditions require

$$C_{0c} = \frac{Q_i}{V_{0c}} \tag{10-22}$$

$$C_{0c} = C_e \tag{10-23}$$

and from Eq. (10-21) we obtain

$$\sigma_{ideal} = \frac{V_{0c}}{V_i} \tag{10-24}$$

That is, under the above-defined ideal conditions the partition coefficient is simply the fraction of the total internal gel volume occupied by solute molecules. For small deviations from ideality we may expand the partition coefficient in a power series in C, the solute concentration in the injected sample, using $C = 0$ as a point about which to expand. The result is

$$\sigma = \sigma_{ideal}[1 + KC + \cdots] \tag{10-25}$$

and all terms beyond the linear one may generally be neglected.

Experimentally we assume that a small sample, whose volume is much less than V_t, the total column volume, is injected at the top of the column and that volumes eluted to the position of maximum solute concentration are measured. The resulting volume V_{el} is found to be related to previously defined parameters as

$$V_{el} = V_0 + \sigma V_i \tag{10-26}$$

This basic operational equation provides a means of determining σ if the three volumes V_{el}, V_0, and V_i are available. V_{el} is directly measured in the elution experiment. V_0 can be determined by measuring V_{el} for a solute whose σ is very nearly zero (e.g., a so-called *totally excluded* solute). V_i is more difficult but may be determined by measuring the volume of solvent V_s taken up by a known weight W of anhydrous gel-forming material. The volume taken up per unit weight may then be written

$$v = \frac{V_s}{W} \tag{10-27}$$

and

$$V_i = vW_c \tag{10-28}$$

where W_c is the weight of anhydrous gel-forming material used to prepare the column. Having values for the necessary volumes we can operationally determine σ for various concentrations C of injected solute. By extrapolation we can determine σ_{ideal}, and from the variation of σ with C, K of Eq. (10-25) is calculable. If a given column is suitably calibrated with species of known shape and weight, then measurement of σ for an unknown sample provides information about its shape and weight. As suggested previously such correlation must be done with caution since there is no absolute guarantee of its uniqueness (e.g., other sizes and shapes may lead to the same retention data). In passing we note that in the operational equation (10-26) all the interesting physics is suppressed into the partition coefficient σ. Quantitative theories relating σ to molecular properties arc as yet undeveloped.

The above discussion deals only with retention volumes and does not include axial dispersion. Both may be included in a theoretical equation of continuity which is developed, like Fick's laws of diffusion, on the basis of solute flux through an area perpendicular to the axis of the column. The resulting continuity equation is, following Ackers (see the References)

$$\frac{\partial C}{\partial t} + \frac{F}{\xi}\frac{\partial C}{\partial x} = L\frac{\partial^2 C}{\partial x^2} \tag{10-29}$$

where L is the axial dispersion coefficient, F the volume flow rate, and ξ the average cross-sectional area which solute molecules could occupy at any position x along the column axis. ξ is made up of two parts: one part resulting from void area, α, and the other resulting from area which can be occupied within the gel. If β is defined as the average total cross-sectional area inside the gel, then $\beta\sigma$ is the fraction which can be occupied by solute. Thus,

$$\xi = \alpha + \beta\sigma \tag{10-30}$$

Equation (10-29) furnishes both the time and axial variations of the solute concentration, C. Solutions require statement of boundary conditions and initial conditions. Assuming that σ is constant and that the conditions on the eluted volume V_{el} are

$$V_{el} = 0, \quad \begin{cases} C = M \text{ if } x = 0 \\ C = 0 \text{ if } x \neq 0 \end{cases} \tag{10-31}$$

$$V_{el} > 0, \quad C = 0 \text{ if } x = 0 \tag{10-32}$$

the solution for the volume dependence of the eluted concentration is

$$C = \frac{M}{\sqrt{4\pi L V_{el}/F}} \exp\left[-\frac{[\ell - (V_{el}/\xi)^2 F]}{4LV_{el}}\right] \tag{10-33}$$

where ℓ is the length of the column. The mathematical form of (10-33) is approximately Gaussian in V_{el} but skewed and shifted slightly because of the appearance of V_{el} in the radical coefficient. The maximum in Eq. (10-33) will occur very near the position

$$\ell - \frac{V_{el}}{\xi} = 0 \tag{10-34}$$

and using Eq. (10-30) we find that

$$V_{el} = \ell\xi = \ell\alpha + \ell\beta\sigma \tag{10-35}$$

$\ell\alpha$ is the product of the average void area, at any position x along the axis, with the column length or simply the void volume V_0. Similarly, $\ell\beta = V_i$. Thus, Eq. (10-35) becomes

$$V_{el} = V_0 + \sigma V_i \tag{10-36}$$

which is identical with the operational equation (10-26). Thus, we have a theoretical basis for (10-26) and its use to determine σ, namely, the continuity equation (10-29) and the conditions imposed on it.

By numerical curve-fitting procedures similar to those outlined in Section 1.5 the measured C versus V_{el} profile can be fit to Eq. (10-33) and both L and σ determined. In principle this should provide more reliable values of σ than those determined directly from the maximum of the profile as in Eq. (10-34). Generally, however, using (10-34) is adequate.

The parameter L, defined as the axial dispersion coefficient, is related to the width of the volume profile. The processes which contribute to the magnitude of L include nonuniform solvent flow velocity around gel particles, axial diffusion, and disequilibrium between internal and void volume regions. Treating these as distinct additive effects furnishes

$$L = Fl_p + \xi D + \frac{qd^2(1 + \alpha)F^2}{\beta\sigma\xi^2 D} \tag{10-37}$$

where l_p reflects the contribution of nonuniform flow to axial dispersion. The magnitude of l_p is to a first approximation independent of F but increases with gel particle size. For a given column it is therefore a constant at low flow rates. The second term in Eq. (10-37) accounts for axial diffusion. In any chromatography experiment axial solute concentration gradients exist and diffusion occurs along these directions. The coefficient D is known as the axial diffusion coefficient in these experiments. The parameter ξ enters because the diffusion occurs in both the void and internal regions. The final term in Eq. (10-37) accounts for the extent of nonequilibrium between the void and internal regions. q is a gel particle packing factor, and d is the gel particle diameter. The mathematical development of this term is quite complicated and will not be given here. It depends on consideration of the mean time required for equilibration as compared with the time required for a unit of solute to pass through an area perpendicular to the flow direction.

EXPERIMENTAL. The experiments to be carried out employ porous Sephadex* gels. A wide variety of other gel materials is also available. The Sephadex gels are cross-linked dextran gels which have found considerable utility in biological applications because they provide a means of molecular weight

*Pharmacia Fine Chemicals, Inc., 800 Centennial Avenue, Piscataway, N.J.

determination which does not, in general, alter the macromolecule conformation. This is very important in biological applications because macromolecules of biochemical significance often have very labile features. The cross-linked dextrans are prepared by reacting epichlorohydrin with partially hydrolyzed native dextran to form the cross-linked three-dimensional porous network. The extent of cross-linkage is variable, and smaller internal volumes are associated with larger extents of cross-linkage. Other gel materials of utility in gel chromatography include polyacrylamide (Biogel), agar, porous glass, and polystyrene (Styragel).

The column and eluant reservoir are shown in Fig. 10-8. The column should be cleaned and sealing rings greased prior to use according to instructions provided in the laboratory. A support net which provides a base for the gel material is located at the bottom of the column. It should be cleaned and properly inserted. The column itself should be set up vertically using a plumb bob. Add a small quantity of 0.1-M NaCl solution from the laboratory stock to the empty column and allow it to pass through the support net until no air bubbles can be seen below or in the net. At this point the exit tube of the column should be stoppered with a column of eluant solution 1 cm high in the column above the net.

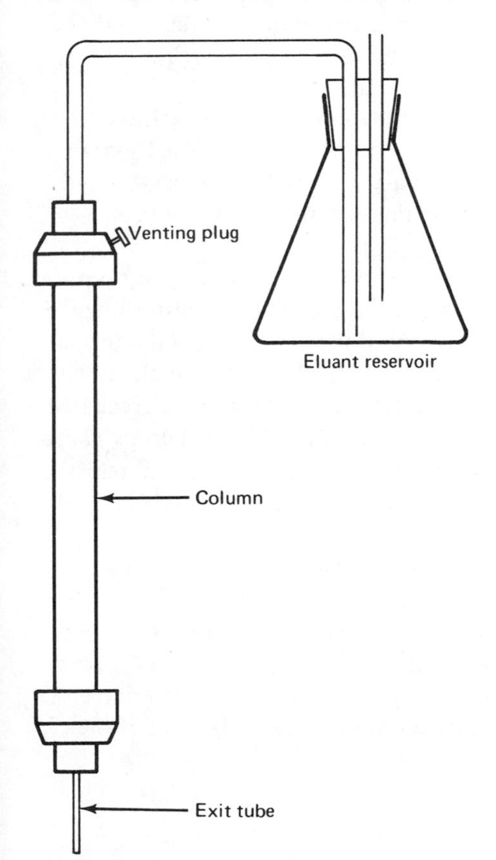

Figure 10-8. Gel permeation chromatography apparatus.

Sephadex G-25 column material will be used in the separations outlined below and the gel preparation procedure is as follows. Bring to boil about 50 ml of 0.1-M aqueous NaCl eluant solution. Place about 15 ml of this hot solution in a 25-ml graduated cylinder and add slowly and with stirring an accurately weighed sample (approximately 3 g) of Sephadex G-25. Add additional hot NaCl solution to give a total volume of 25 ml. Stir for about 15 min and then allow the gel to settle until enough clear supernatant liquid can be removed to leave 18 ml of gel and eluant behind. Remove the supernatant liquid with an eyedropper or pipette. Slurry the gel with rapid circular motion and pour the material into the chromatographic column. If air bubbles remain below the surface of the gel slurry and do not rise to the top of the column, pour the contents back into the original container, add 3 or 4 ml of eluant, reslurry, and add to the column. If it is not possible to add all of the gel slurry, wait 5 min and then unstopper the column exit tube, and as the material settles add the remaining gel. At this point attach the cap to the top of the column and connect the reservoir. A slight external pressure applied to the open tube of the reservoir will cause solution to pass to the column. Air in the connecting tubing must be removed by opening the venting plug at the top of the column. When the air is removed, replace the venting plug and adjust the height of the reservoir above the exit tube so that a flow rate near 0.1 ml/min is obtained. Allow this flow to continue for 30 min to provide stabilization of the column bed material.

The sample to be analyzed is prepared by dissolving 9 mg of Blue Dextran 2000 and 2.7 mg of vitamin B_{12} to 3 ml of hot distilled water. Blue Dextran 2000 has a partition coefficient, σ, of very nearly zero for Sephadex G-25 and will be used to determine the void volume in this experiment. Vitamin B_{12} alone is yellow colored in saline solution.

To add the sample solution to the top of the column, remove the top cap and with a pipette take off nearly all the liquid above the gel without disturbing the gel. Let the eluant continue to exit the bottom of the column until the top surface appears to be dry and then stopper the exit. Add 0.3 ml of sample dropwise to the bed, unstopper the exit, and begin collecting eluant in a 25-ml graduated cylinder. When the sample has entered the bed add about 12 small drops of eluant and then fill the column with eluant and reconnect the top cap and reservoir. Keep a constant level of eluant in the reservoir by adding stock 0.1-M NaCl solution through the open reservoir tube.

As the sample passes down the column distinct blue and yellow regions should appear. When the blue layer approaches the bottom, remove the graduated cylinder and begin collecting 5-min fractions of eluant in small test tubes. Record the volume of eluant in the graduated cylinder. Continue to collect samples until the yellow vitamin B_{12} color has completely vanished. Determine the volume of each fraction using the flow rate and time or by calibration of the test tubes.

After ascertaining the volumes, dilute each fraction by an amount sufficient to make optical absorption analysis on a spectrophotometer possible. Instructions regarding this will be provided. Visually find the sample whose color appears the deepest blue, and on the spectrophotometer find the wavelength of its maximum absorption in the 4000–5000-Å region. Record the percentage absorption at this maximum and then record the percentage absorption for all the other blue samples. In a similar way, record the percentage absorption of the yellow samples using a wavelength selected as above but in the 5400–6400-Å region.

Repeat the experiment using 0.6 ml of sample. If time permits, dilute the remaining sample by a factor of 2 and perform the elution experiment again.

ANALYSIS. Plot percentage absorption as a measure of concentration versus eluted volume from the start of each experiment. Determine the void volume V_0 from the Blue Dextran 2000 elution profile and V_{el} for vitamin B_{12} from its elution profile. Comment on the reproducibility of V_0 from experiment to experiment. Calculate V_i assuming that $v = 2.5$ ml/g for Sephadex G-25. Determine the partition coefficient and volume widths at half-height for vitamin B_{12} in each experiment. If σ shows any concentration dependence, determine an approximate value for K of Eq. (10-25). Examine the variation of peak width at half-height and try to ascertain why any observed variations occur. Comment on the symmetry or the lack of it in the elution profiles. What fraction of the internal gel volume is available to vitamin B_{12} if this system is assumed to be ideal?

10.3 Light Scattering of Polyvinyl Acetate Solutions

The purpose of this experiment is to measure the intensity of light scattered by a dilute solution of macromolecules and from the data calculate the weight-average molecular weight of the dilute solute. Theoretical background is given describing the interaction of the incident light with the sample and its attendant scattering into various angular regions.

THEORY. Light scattering arises because of the classical interaction between the electromagnetic radiation from a source and the molecular or atomic electron density of a sample. The interaction produces an oscillating electric dipole in each molecule. In turn these locally accelerated charges emit radiation of the same frequency but in different directions. As an example, consider Fig. 10-9, which diagrams the scattered radiation emanating from a dipole which has been excited by polarized incident rays whose electric vector is in the z direction. In part (a) of the figure the incident polarized ray is shown propagating along the x direction and interacting with a molecule. Assuming that the molecule is opti-

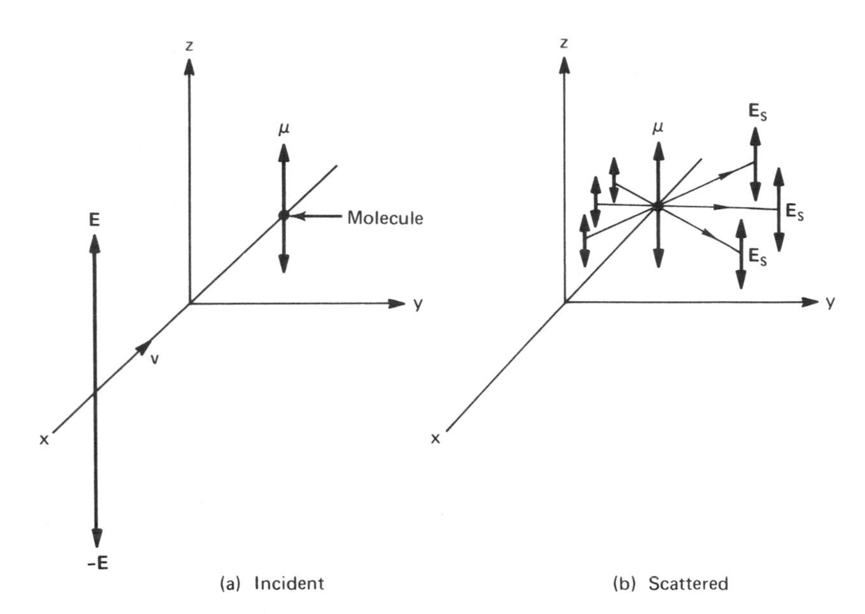

Figure 10-9. Excitation of and scattered light from an oscillating dipole.

cally isotropic, the induced electric dipole moment is, by definition of the term isotropic, restricted to be parallel to the incident electric field vector **E**. The accelerating electron density emits electromagnetic radiation as spherical waves, that is, in all directions. The intensity of the emitted radiation, however, depends on the direction taken with respect to the electric dipole axis. No radiation is emitted along the axis itself, and the maximum intensity appears at 90° to this axis (e.g., in a plane perpendicular to the dipole). Mathematically, for molecules whose dimensions are small compared to the wavelength of the incident light (say $1/20\lambda$ or less), the scattered intensity may be developed by considering the molecule as a point-scattering source as follows. A plane-polarized light beam has an electric field E describable as

$$E = E_0 \cos\left(\omega t - \frac{2\pi x}{\lambda}\right) \qquad (10\text{-}38)$$

where ω is the angular frequency in radians per second, λ the wavelength, t the time, and x the propagation distance. The interaction of this electromagnetic wave with an optically isotropic medium is characterized by the polarizability, α, of the medium, and the resulting induced electric dipole moment p is

$$p = \alpha E \qquad (10\text{-}39)$$

or

$$p = \alpha E_0 \cos\left(\omega t - \frac{2\pi x}{\lambda}\right) \qquad (10\text{-}40)$$

Equation (10-40) expresses the oscillatory nature of the dipole moment. The electric field E_s of the radiation scattered by the dipole is proportional to $|d^2p/dt^2|$ or

$$E_s \propto \omega^2 \alpha E_0 \cos\left(\omega t - \frac{2\pi x}{\lambda}\right) \tag{10-41}$$

Since the emitted radiation can be described as spherical waves, the intensity passing through a square centimeter of a sphere drawn around the dipole must decrease in proportion to $1/r^2$, where r is the radius of the sphere. Intensity is proportional to E_s^2 and hence E_s must fall off as $1/r$. Furthermore, because electric fields arise from both the positive and the negative ends of the dipole and tend to cancel one another, there is, as mentioned above, an angular dependence in E_s. If the angle between the dipole axis and a line from the dipole to the observer is called θ, E_s is proportional to $\sin\theta$.

The net result for scattering from small molecules is

$$E_s = \frac{\omega^2 \alpha E_0 \sin\theta}{c^2 r} \cos\left(\omega t - \frac{2\pi x}{\lambda}\right) \tag{10-42}$$

where c is the speed of light and r the distance from the scattering center to the detector. The scattered radiation from an optically isotropic molecule then has the same frequency ω as the incident radiation and has an electric field strength proportional to the square of the incident frequency.

The most easily measured property of the scattered radiation is its intensity (energy per square centimeter per second), which is proportional to the time average of E_s^2 over one period of the ray. The ratio of the scattered light intensity to the incident intensity is then

$$\frac{I_s}{I_0} = \frac{\int_0^{1/\nu} \omega^2 \alpha^2 E_0^2 \sin^2\theta \cos^2[\omega t - (2\pi x/\lambda)]\, dt}{\int_0^{1/\nu} E_0^2 \cos^2[\omega t - (2\pi x/\lambda)]\, dt} \tag{10-43}$$

Performing the quadrature furnishes

$$\frac{I_s}{I_0} = \frac{\omega^2 \alpha^2}{c^4 r^2} \sin^2\theta$$

$$= \frac{16\pi^4 \alpha^2}{\lambda^4 r^2} \sin^2\theta \tag{10-44}$$

where $\omega/c = 2\pi/\lambda$ has been used.

Equation (10-44) describes what is known, after its developer, Lord Rayleigh, as Rayleigh scattering. It accounts for the strong inverse wavelength dependence of scattered light. A good example is the scattering of visible sunlight in the upper atmosphere which makes for a blue sky.

Equation (10-44) applies only to small optically isotropic molecules. Often systems are studied in which the molecules are neither small compared to the wavelength of the incident radiation nor optically isotropic. A molecule is said to be optically anisotropic if the induced dipole moment is not parallel to the electric vector of the incident radiation. In this circumstance the polarizability becomes a tensor quantity. Anisotropy can be detected by measuring the polarization of light scattered through the angles $\eta = \pi/2$, $\theta = \pi/2$ in Fig. 10-10. If the molecule is isotropic and unpolarized incident light is used, the dipole radiation may be described as having two independent electric field components, both perpendicular to the incident light. One of these points is along the x axis in Fig. 10-10 and the other is along the z axis. The combination of these two furnishes the scattered radiation. When viewed at an angle of $\theta = \pi/2$ and $\eta = \pi/2$ the only contribution to the scattered light must come from the electric vector component which points along the z axis because the direction of propagation and the electric vector must be perpendicular. At this viewing position we therefore expect that a detector which responds only to light polarized in the xy plane will find no radiation, while one which responds to light polarized in the xz plane will indicate that light is present. Contrariwise, if the molecule is anisotropic, the above polarization arguments are no longer valid, and some radiation will be scattered which has a component in the xy plane. The ratio of intensities of the xy component and xz component is called the depolarization ratio ρ and is a measure of the optical anisotropy. The formulas for scattered intensity developed on the basis of optically isotropic materials can be corrected approximately by multiplying by the factor $(6 - 7\rho)/(6 + 6\rho)$.

Equation (10-44) arose from considerations of the scattering of incident polarized radiation. If unpolarized light is used and the medium is isotropic, the scattered light can be regarded as the superposition of two rays: one originating from that component of the incident light polarized in the yz plane of Fig.

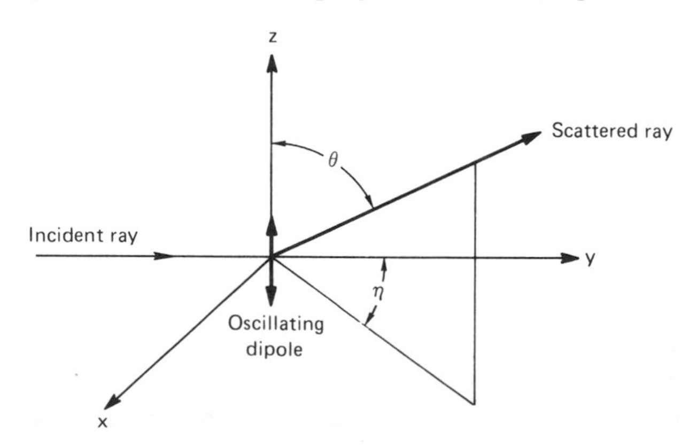

Figure 10-10. Important angles in light-scattering measurements.

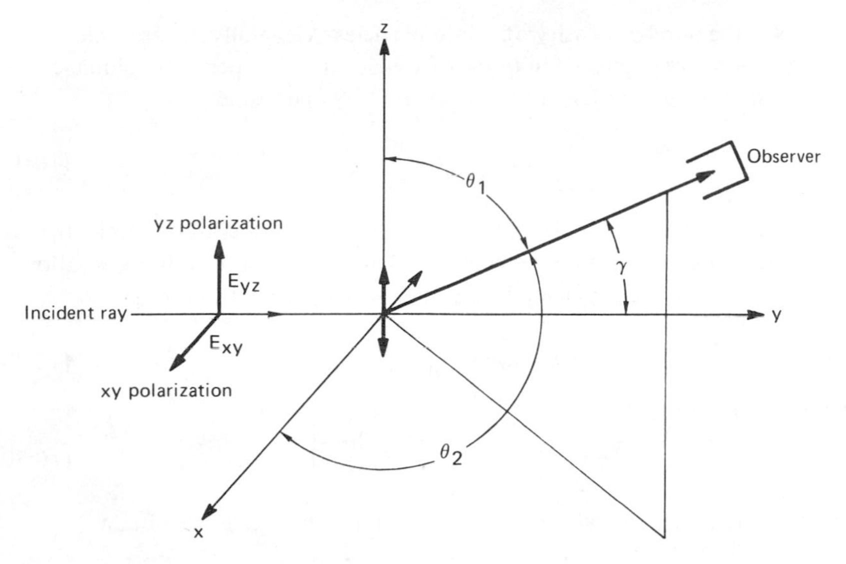

Figure 10-11. Superposition of two polarized rays to form an equivalent unpolarized ray.

10-11 and the other from the component polarized in the xy plane. Because the incident light is completely unpolarized, the magnitudes of the electric vectors associated with these two components, and therefore their intensities, are equal. An observer stationed as indicated in Fig. 10-11 observes an intensity given by

$$I_s = \frac{16\pi^4\alpha^2 I_0}{\lambda^4 r^2} [\sin^2 \theta_1 + \sin^2 \theta_2] \tag{10-45}$$

or

$$I_s = \frac{16\pi^4\alpha^2 I_0}{\lambda^4 r^2} (1 + \cos^2 \gamma] \tag{10-46}$$

where γ is the angle between the y axis and the line from the dipole to the observer.

At the outset of this discussion a relationship was alluded to between light-scattering intensity and molecular weight. This relation arises from the dependence of the polarizability α on molecular weight M and index of refraction n. We shall consider the case of scattering from a dilute solution (refractive index n) of macromolecules in a solvent of refractive index n_0. The macromolecules are assumed to be independent scattering centers. The difference between the squares of the refractive indices of the solution and solvent is related to the polarizability of the macromolecules by the classical formula

$$n^2 - n_0^2 = 4\pi\alpha N \tag{10-47}$$

where N is the number density of solute particles. Generally, macromolecule concentrations are expressed in terms of weight of solute per unit volume c rather than number density. In these terms (10-47) becomes

$$n^2 - n_0^2 = 4\pi\alpha \frac{c}{\overline{M}_w} N_0 \tag{10-48}$$

where N_0 is Avogadro's number. With this choice of concentration units the index of refraction depends on molecular weight. In the limit of dilute solutions $n \rightarrow n_0$ and n may be expanded in a Taylor series in c about n_0 to give approximately

$$n = n_0(1 + \frac{dn}{dc} c) \tag{10-49}$$

and squaring furnishes

$$n^2 = n_0^2 \left[1 + 2c \frac{dn}{dc} + c^2 (\frac{dn}{dc})^2\right] \tag{10-50}$$

Neglecting the last term within the brackets and rearranging, we obtain

$$n^2 - n_0^2 = 2c \frac{dn}{dc} n_0^2 \tag{10-51}$$

and together with Eq. (10-48) this results in

$$\alpha = \frac{\overline{M}_w (dn/dc) n_0^2}{2\pi N_0} \tag{10-52}$$

With the choice of grams of solute per cubic centimeter as the concentration variable and measurement of dn/dc, the variation of refractive index with concentration in these units, the polarizability becomes a direct measure of the molecular weight. Under these circumstances measurement of the scattered light intensity becomes a measure of the molecular weight. That is, from Eqs. (10-46) and (10-52) we find that

$$\frac{I_s}{I_0} = \frac{2\pi^2 n_0^2 (dn/dc)^2 \overline{M}_w^2}{\lambda^4 r^2 N_0^2} [1 + \cos^2 \gamma] \tag{10-53}$$

which furnishes the scattering from a single macromolecule in the above circumstances. With the independent particle assumption stated above the scattering from a unit volume of solution is the sum of the scattering from each particle in the unit volume or N_0c/M times Eq. (10-53):

$$\frac{I_s}{I_0} = \frac{2\pi^2 n_0^2 (dn/dc)^2 \overline{M}_w c^2}{\lambda^4 r^2 N_0} [1 + \cos^2 \gamma] \tag{10-54}$$

All quantities on the right-hand side of Eq. (10-54) are known from measurements other than light scattering except the solute molecular weight $\overline{M}$, which is thus calculable from measurement of the light-scattering ratio I_s/I_0. Often, especially in polymer solutions, the solute species do not all have identical molecular weights. In this circumstance $\overline{M}$ from Eq. (10-54) is the weight average molecular weight $\overline{M}_w$. It is a weight average because of the concentration units chosen for c.

Commonly, the scattering of light by macromolecules is expressed in terms of the turbidity τ, defined by the expression

$$\tau = \frac{1}{\ell} \ln \frac{I_0}{I} \tag{10-55}$$

where I is the intensity not scattered by the sample and ℓ the path length of light within the sample. By conservation I and therefore τ are related to the summation of all the scattered light. From this point of view τ is given by

$$\tau = 2\pi \int_0^{\pi} R_\gamma{}' \sin \gamma \, d\gamma \tag{10-56}$$

where the Rayleigh ratio $R_\gamma{}'$ is defined as

$$R_\gamma{}' = \frac{I_s r^2}{I_0} \tag{10-57}$$

Another so-called Rayleigh ratio is defined below [Eq. (10-60)]. Generally the context will make clear which definition is to be used. For Rayleigh scattering I_s/I_0 is given by Eq. (10-54).

If the criterion for Rayleigh scattering (the particle size is small compared to the wavelength of light) is not met, the scattering particle can no longer be considered a single point-scattering center. Rather the macromolecule must now be regarded as an assembly of scattering centers whose positions with respect to one another are not random but at least semiordered. Hence, in this kind of scattering interference effects are noted similar to those encountered in the X-ray diffraction of liquids and electron diffraction of gases. The net result of the theoretical analysis is a change in the angular dependence of the scattering, with more intensity generally appearing at small angles γ. The precise angular dependence varies with the molecular geometry and thus measurement of I_s versus γ can in principle furnish some indication of the molecular shape. Scattering of this type is known as Debye scattering, and the angular dependence is described by the particle scattering factor $P(\gamma)$, which is defined as the ratio of the scattered intensity for a large particle including interference effects to the scattered intensity of the same molecule neglecting interference. The resulting function depends on molecular shape and as γ goes to zero it approaches unity. As a result an extrapolation of measured values of I_s/I_0 to $\gamma = 0$ furnishes an intercept described by Eq. (10-54) if the macromolecules are independent.

Thus far we have extended our thinking to large noninteracting solute molecules. These molecules are randomly arranged in space and behave like a liquid-phase ideal gas. In actual fact, however, solute molecules do interact with one another and as a result are nonideal. Just as for nonideal gases, the nonideality in the present case can be expressed in terms of a virial equation in powers of c which in the limit as $c \rightarrow 0$ furnishes the ideal relation. Thus, we have

$$\frac{I_s}{I_0} = \frac{2\pi^2 n^2 (dn/dc)^2 (1 + \cos^2 \gamma) c P(\gamma)}{N_0 \lambda^4 r^2} \left[\frac{1}{(1/\overline{M}_w) + 2Bc + 3Cc^2 + \cdots}\right] \tag{10-58}$$

The coefficients B and C are the second and third virial coefficients. The development of Eq. (10-58) is on the basis of fluctuation theory. From an experimental point of view it will prove convenient to plot, at a given concentration c, Qc/R_γ versus γ where

$$Q \equiv \frac{2\pi^2 n^2 (dn/dc)^2}{N_0 \lambda^4} \tag{10-59}$$

and

$$R_\gamma = \frac{r^2 I_s}{I_0 (1 + \cos^2 \gamma)} \tag{10-60}$$

At the intercept of such a plot $P(\gamma)$ of Eq. (10-58) is equal to unity, and the intercept b(c) is thus given by

$$b(c) = \frac{1}{\overline{M}_w} + 2Bc + 3Cc^2 + \cdots \tag{10-61}$$

Additional plots of Qc/R_γ versus γ at different concentrations provide additional intercepts b(c). These intercepts can in turn be plotted versus c, and according to Eq. (10-61) the intercept is $1/\overline{M}_w$ and the limiting slope is given by the second virial coefficient B. The magnitude of B is interpretable in terms of interactions between solute molecules leading to nonideal behavior. The angular dependence of Qc/R_γ is interpretable in terms of molecular geometries. For the purposes of this experiment we are only interested in measuring $\overline{M}_w$. The kind of plotting described is often combined into a single plot known as a Zimm plot.

EXPERIMENTAL. A discussion of the features of light-scattering equipment is given in Section 2.3 and the student should be familiar with the material there. Specific instructions for operating the particular apparatus in the laboratory will be given by the instructor.

In Fig. 10-12 a schematic description of a light-scattering apparatus is given which incorporates the essential ingredients. The source is generally a medium pressure mercury arc emitting many lines in the visible spectrum. Commonly the 436- and/or 546-nm lines are used, these being selected from the full spectrum by a narrow-band optical filter. The condensed beam is transmitted into a light-tight housing enclosing the sample cell and the detector, the latter being

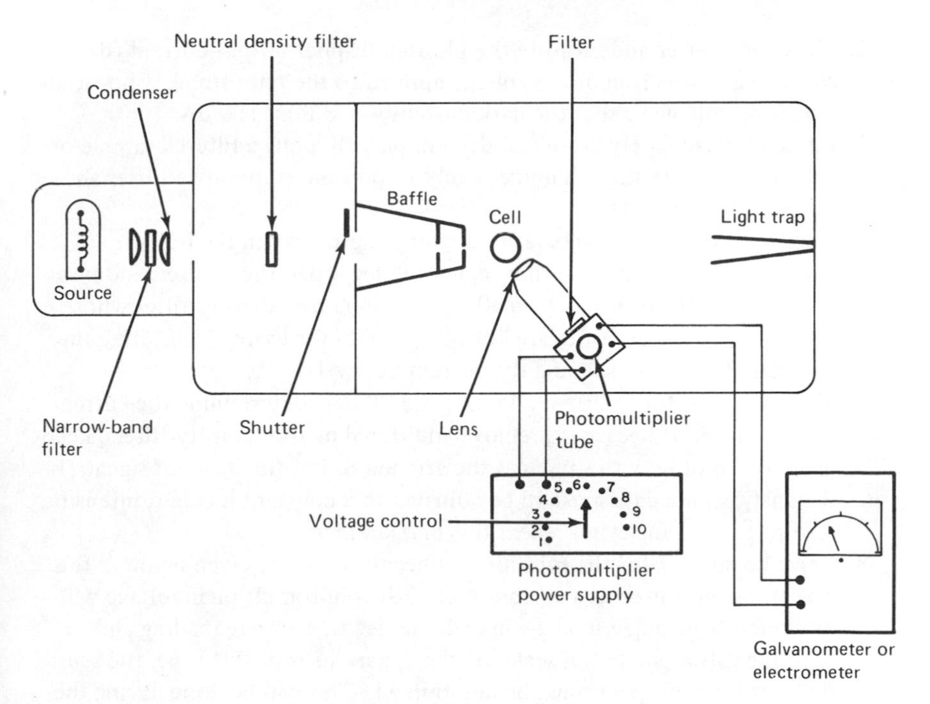

Figure 10-12. Light-scattering apparatus.

contained in its own housing to optically isolate it from the rest of the system. Several cell designs are available. One is a cylindrical design with two opposed flat regions which are aligned perpendicular to the beam. This combination of flat and cylindrical geometry reduces multiply reflected scattered light (flat part) and preserves the "natural" geometry desired when rotating the detector about the cell (i.e., the round part keeps the distance traversed by the scattered light uniform over all observation angles). The light trap captures unscattered radiation, thus preventing its return to the cell and its straying into the detector. The shutter provides a means of determining the dark current of the photomultiplier and nulling it out of the measurements by adjusting the current detector. The detector housing may contain a filter which allows only radiation of the proper wavelength to enter the detector.

With the advent of lasers the optical source, lens, and filter assembly shown in Fig. 10-12 can be greatly simplified because of the monochromatic and collimated nature of the laser output.

A typical set of measurements for samples, solvent, and solution would go something like the following:

1. With the neutral density filters in place to ensure that the photomultiplier is not current-saturated, turn on the source, the electrometer (zero it), and finally the photomultiplier supply.

2. Close the shutter and measure the photomultiplier output current (dark current) as a function of the voltage applied to the phototube. Choose an operating voltage so that the dark current is less than 100 μA.
3. Into an immaculately clean and dry sample cell, pour a filtered sample of acetone. Stopper the cell immediately to prevent evaporative losses and accumulation of dust.
4. Place this cell into the apparatus, aligning it properly in the holder.
5. Set the detector at a scattering angle of 0 deg, open the shutter, and measure the current. If it is very small, remove a neutral density filter whose attenuation characteristics are known. Avoid overdriving (saturating the photomultiplier) by keeping the current below 10^{-4} A.
6. Move the detector to 30, 35, 45, 60, and 90 deg, determining the current at each angle. If necessary, remove additional neutral density filters, keeping a record of how this reduces the attenuation of the incident signal. In the analysis, the data must all be adjusted to a constant incident intensity.
7. Return to zero angle and repeat this measurement.
8. Repeat steps 3–7 for the solutions (concentrations are given below). If a galvanometer is used as a current meter, the photomultiplier voltage will be altered from angle to angle in order to get an accurate reading and to keep the galvanometer on scale. If the apparatus is of this type, the sensitivity of the multiplier must be determined. This can be done during the course of the measurements by taking, at each angle, galvanometer readings with two different voltages on the multiplier. The current-voltage relations should be approximately linear.

Calculation of R_γ from the above measurements and Eq. (10-60) requires knowledge of r, the distance from the scattering center to the detector. While this can be measured, it is often incorporated into an instrument constant which is the r^2/I_0 part of Eq. (10-60). This constant is determined using a standard solution at a standard angle. For example, benzene has an R_γ of 16.3×10^{-6} at $90°$ and 546 nm and $R_\gamma = 46 \times 10^{-6}$ at $90°$ and 436 nm (Stacey, p. 86, in the References.) By measuring I_s at $90°$ for these solutions, r^2/I_0 can be calculated. One incorporates I_0 into the instrument constant because the ratio of intensities of the scattered and incident beams is on the order of 10^{-6}, making it difficult to determine the ratio accurately. The calibration procedure offered here presents one way of avoiding the instrumental problems associated with a direct and accurate measurement.

Samples of polyvinylacetate in acetone will either be supplied or prepared. The solute concentrations should range from 0.1 to 0.5 g/100 ml. It is essential that dust and other solid material be removed from the sample. This may be accomplished by filtration through high-quality filter paper through which solvent has been passed prior to the filtration of samples. At least four samples with concentrations spanning the above range and in addition a sample of pure solvent should be prepared. Appropriate cells for use in the actual scattering measurement will be provided.

The light-scattering intensities for each of these samples should be recorded as a function of angle γ. In addition the refractive index of each solution should be measured. The method is described in Section 2.3.

ANALYSIS. Using $dn/dc = 0.1040\ cm^3/g$ and the measured quantities described above, construct plots of Qc/R_γ versus γ for each sample. From the intercepts, plot b(c) versus c and determine the average molecular weight $\overline{M}_w$ and the second virial coefficient B. Comment on the magnitude of B and its relation to the properties of the solution. Deviations of $P(\gamma)$ from unity are reflected in the plots of Qc/R_γ versus γ. Comment on the direction of this deviation and, on the basis of the definition of $P(\gamma)$, comment on the significance of interference effects in the polyvinyl acetate-acetone solutions.

REFERENCES

Section 10.1

C. Tanford, *Physical Chemistry of Macromolecules* (Wiley, New York, 1961). A well-known text on the physical chemistry of macromolecules.

S. N. Chinai, P. C. Scherer, and D. W. Levi, *J. Polymer Sci.* **17**, 117 (1955); W. R. Moore and M. Murphy, *J. Polymer Sci.* **56**, 519 (1962); R. H. Wagner, *J. Polymer Sci.* **2**, 21 (1947). Deal with the polyvinyl acetate viscosity problem.

Section 10.2

G. K. Ackers, *Adv. Protein Chem.* **24**, 343 (1970). A review article discussing the separation of biologically important molecules.

P. Andrews, *Biochem. J.* **96**, 595 (1965); J. Cazes, *J. Chem. Educ.* **43**, A567 (1966); D. D. Bly, *Science,* **168**, 547 (1970). Literature articles of interest.

C. Tanford, *Physical Chemistry of Macromolecules* (Wiley, New York, 1961). A text treating the physical properties of macromolecules.

Section 10.3

B. H. Zimm, *J. Chem. Phys.* **16**, 1099 (1948). A detailed description of an early light-scattering apparatus.

J. Kremen and J. J. Shapiro, *J. Opt. Soc. Am.* **44**, 500 (1954); W. H. Anghey and F. J. Baum, *J. Opt. Soc. Am.* **44**, 833 (1954); R. Speiser and B. A. Brice, *J. Opt. Soc. Am.* **36**, 364 (1946); B. A. Brice, M. Habnew, and R. Speiser, *J. Opt. Soc. Am.* **40**, 768 (1950); G. Oster, *Anal. Chem.* **25**, 1165 (1953). Describe instruments available commercially.

G. Oster in A. Weissberger (ed.), *Technique of Organic Chemistry*, (Wiley-Interscience, New York, 1960), Volume I, Part III, Ch. XXXII; K. A. Stacey, *Light Scattering in Physical Chemistry* (Academic Press, New York, 1956); C. Tanford, *Physical Chemistry of Macromolecules* (Wiley, New York, 1961). Give a full discussion of light scattering.

S. N. Chinai, P. S. Scherer, and D. W. Levi, *J. Polymer Sci.* **17**, 117 (1955). Reports light scattering of polyvinyl acetate solutions.

Chapter 11

Molecular Structure

11.1 Crystal Structure Determination by X-ray Diffraction (Powder Method)

The purpose of this experiment is to obtain the X-ray diffraction pattern of a powdered crystalline material belonging to the cubic system. From the experimental data, including a separate density measurement, the interplanar spacing and the lattice type are determined.

THEORY. One of the most important chemical problems is that of molecular structure and X-ray diffraction affords one of the most powerful techniques for determining the position of atoms in molecules and for determining the relative positions of the molecules in a crystal. The characteristic features of X-rays and crystalline materials that are essential to the structure determination include (1) the wave character of the X-rays, which implies that they will show interference effects, and (2) the regular repetitive nature of molecular arrangements in the crystal.

The power of the X-ray method is illustrated by reports of the structure of myoglobin, ribonuclease, hemoglobin, and other large biologically significant molecules containing enormous numbers of atoms. It is important to note here one of the limitations of the X-ray method: It cannot be used to locate hydrogen atoms for reasons that will be made clearer below. Another limitation of the X-ray method is its resolving power. High-quality X-ray analysis will resolve bond distances differing by more than 0.01 Å, whereas optical and microwave spectroscopic techniques are capable of much higher resolution. X-ray techniques are also applied to the study of liquids where there is some local orderly molecular arrangement.

At the outset it is worth looking for a qualitative explanation of how X-rays interact with matter (electrons and nuclei). X-rays are one type of electromagnetic radiation characterized by short wavelengths on the order of 1 Å and high energies near 12,000 eV. As all other electromagnetic radiation, X-rays may be characterized in part by an alternating electric field which interacts with charges. This is the language used to describe the X-ray–matter interaction. The electric field of the X-ray provides a means of accelerating the electrons and nuclei. In turn these accelerated particles radiate electromagnetic energy as spherical waves (i.e., in all directions). These radiated waves have the same frequency as the incident waves. Because of their relatively large mass, nuclei are very ineffective scatterers of X-rays when compared to electrons. In the discussion that follows we shall tacitly assume that all X-ray scattering is due to electrons. Because X-rays have such large energies, they easily influence even the most tightly bound electrons of an atom. The ability of an atom to scatter electrons increases as its number of electrons increases (i.e., there are more radiating electrons in a heavy atom than in a light one). For this reason it is virtually impossible to locate accurately hydrogen atoms by X-ray scattering; they have only one electron, and it is generally shared to form a chemical bond. Hydrogen atoms do in fact appear in the analyses of X-ray data, but their positions cannot be reliably determined.

Visible light (4000–8000-Å wavelengths) is another form of electromagnetic radiation. Its wave properties are very much the same as X-rays except these properties are exhibited on a scale of different dimensions. Whereas visible light can be diffracted with a grating whose lines are several thousand Angstroms apart, X-rays require a grating whose lines are at most a few Angstroms apart. The spacing between molecular species in a crystal is of this order, and since the same spacing is repeated over and over to generate one plane in a crystal, such materials make suitable diffraction gratings for X-rays.

Having discussed qualitatively the interaction between X-rays and electrons and the basic feature of crystals responsible for diffraction we shall now look into models used to represent crystals and then examine the interference patterns generated when X-rays interact with these models. The three-dimensional periodicity of a crystal is generally represented by a series of lattice points, each one of which bears a definite relation to all the others. This relation is embodied in the crystal translation vector **T** as

$$\mathbf{T} = n_1\hat{a} + n_2\hat{b} + n_3\hat{c} \tag{11-1}$$

where $\{\hat{a}, \hat{b}, \hat{c}\}$ define the crystal axes and n_1, n_2, and n_3 are integers or zero. Beginning with an arbitrary lattice point all other lattice points are located at the terminus of some **T** vector generated by using various combinations of n_1, n_2, and n_3. It is important to note that the triad $\{\hat{a}, \hat{b}, \hat{c}\}$ need not be at right angles to each other nor of the same length. In the cubic system, into which the crystals studied here fit, the axes are of equal length and at right angles to each other. For the five other crystal systems, hexagonal, tetragonal, monoclinic, orthorhombic, and triclinic, such is not the case. In some cases, depending on how the unit cell is defined, the set $\{n_1, n_2, n_3\}$ may be fractions. Later in this discussion two examples, body-centered cubic and face-centered cubic unit cells, will be discussed where this is true.

While the crystal lattice provides a model for the three-dimensional periodicity, it does not provide a means of locating individual atoms. In fact, each lattice point, in the case of large molecules, represents in some average way the location of many atoms in a crystal. Atom positions are located with respect to each lattice point in terms of a set of position vectors which define what is called a basis. In the present experiment we shall not be concerned with the basis vectors, but in a full-fledged structure determination the basis vectors must always be determined. There is generally more than enough X-ray data (line positions and intensities) to make such a determination.

X-rays scatter from crystals because electrons in the sample are accelerated and radiate in all directions (spherical waves). The resulting *observed* intensity of diffracted (scattered) radiation is the result of superimposing the scattered wavelets from a large number of electrons from various lattice points. We shall treat the problem in an approximate way by considering that all the electrons associated with a given lattice point are concentrated at the point rather than distributed in space. For X-rays of wavelength λ the commonly employed Bragg law is developed by assuming specular (mirror-like) reflection from an array of lattice points as diagrammed in Fig. 11-1. The spacing between planes is d, and a search is made for reflections leading to constructive interference at the detector. The inbound wave is a plane wave which can be decomposed as a set of individual rays whose amplitudes are in phase over the space between the source and the crystal (if these rays were out of phase, no radiation would appear at the crystal). The rays leaving the crystal are a superposition of a huge number of spherical waves which, when many wavelengths away from the surface, result in an approximately plane wave passing to the detector. If we reduce the problem to its simplest terms, scattering from the two lattice points A and B shown in Fig. 11-1 is considered. The angles and distances of interest are shown in expanded form in the inset. Lattice points A and B are located in two planes perpendicular to the figure, and like A and B these planes are a distance d apart. The inbound plane wave, decomposed here into two rays whose amplitudes are in phase, interacts with electrons at A and B. Assuming that a maximum in the intensity is observed at the detector, we know that the outbound rays must also be in phase. The ray interacting with point A travels a shorter distance by PBQ than the ray interacting with point B. Thus, if the outbound rays are to be in phase, the distance PBQ must be an integral number of wavelengths:

$$PBQ = n\lambda \tag{11-2}$$

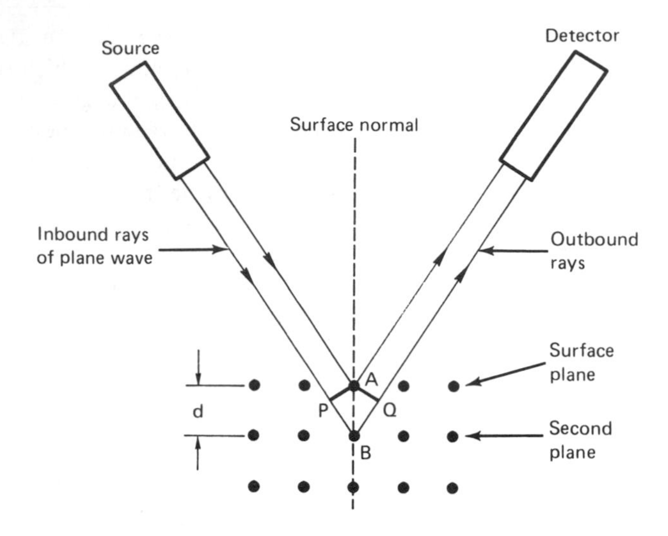

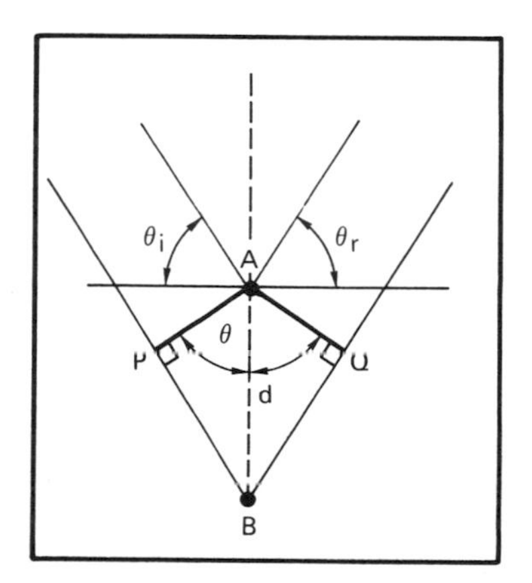

Figure 11-1. Reflection of X-rays from a lattice.

Using simple geometric considerations, the incident angle θ_i and the internal crystal angle, defined as shown in the inset, are equal. As a result PBQ is also given by

$$PBQ = 2d \sin \theta \tag{11-3}$$

Combining (11-2) and (11-3) gives the Bragg diffraction condition for observing an intensity maximum:

$$n\lambda = 2d \sin \theta \tag{11-4}$$

where n = 1, 2, 3, ... Thus, by determining angles for which the intensity observed is maximized the ratio d/n can be calculated. The parameter n is called the *order* of the diffraction, and only through experience can it be properly assigned.

The Bragg equation (11-4) is justified because it works, not because of its sophisticated physical content. More rigorous derivations of diffraction conditions are possible considering the scattering from a large number of lattice points and incorporating the fact that the electrons responsible for the observed scattered rays are not confined to the latttice point. The development of these equations is beyond the scope of this discussion. They show, among other things, that specular (mirror-like) reflection arises quite naturally and need not be invoked as an ad hoc postulate as is done in the derivation of Bragg's law.

In the so-called powder method employed in this experiment a finely ground crystalline powder contained within a small capillary tube is placed in the path of a fixed-position narrow-pencil X-ray beam. Figure 11-2 schematically shows a powder diffraction camera. The source provides a nearly monochromatic beam of X-rays which are limited spatially by the collimator. The sample intercepts these rays and is located at the center of a cylindrical camera. The axis of the capillary sample holder is perpendicular to the plane of Fig. 11-2. Most of the incident rays pass directly through the sample and out of the camera body through a second collimator. These undeviated rays are of no interest, and their removal from the camera eliminates film overexposure problems. Some of the incident energy is absorbed and reemitted as scattered radiation. These scattered rays contain the structural information desired. Any microcrystal in the powder whose orientation with respect to the incoming ray is an angle satisfying the Bragg equation will emit X-rays constructively at an angle of $\pm 2\theta$ measured from the incident beam direction. In addition constructive interference occurs at an angle of $\pi - 2\theta$ measured with reference to the incident beam direction. In any crystal a variety of orientations (crystal planes) exist each of which has a characteristic d separating it from the next plane of the same type. Constructive interference occurs from each type of plane, but since the characteristic ds vary, the associated scattering angle varies also. The purpose of using a powder is to provide a randomly oriented sample. That is, all possible crystal orientations are supposedly equally likely. In practice this may or may not be accomplished. To

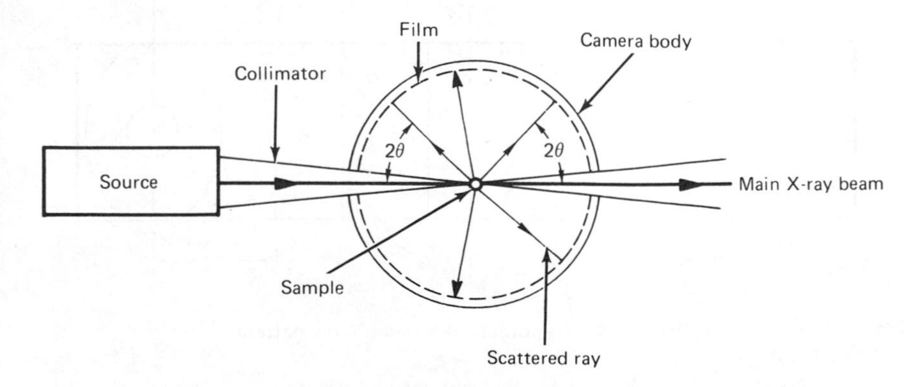

Figure 11-2. X-ray powder diffraction camera.

assist in achieving a fully random sampling of all orientations the capillary sample holder is sometimes rotated about its axis during the X-ray exposure.

The output of an X-ray diffraction experiment is an exposed piece of photographic film which on development provides a series of darkened lines like those of Fig. 11-3 in which the film is unrolled and approximately half of it is left off, namely a replica of the part shown between the collimator holes.

The record is symmetric about the line labeled 0, which corresponds to a scattering angle of 90°. The distances labeled s_2 should be equal and represent forward and backward scattering from some plane in the crystal. To find the distance d between planes of this type, the scattering angle corresponding to s must be calculated. The parameter s is the length of arc from the beam direction around the circumference of the camera to the second dark line of the spectrum. The radius R of the camera is assumed known and the angle $2\theta_2$ is calculable as

$$2\theta_2 = \frac{s_2}{2\pi R} \times 360 \qquad \text{(degrees)} \tag{11-5a}$$

or

$$2\theta_2 = \frac{s_2}{R} \qquad \text{(radians)} \tag{11-5b}$$

There is an arc length s_i corresponding to each line on the film and thus a θ_i satisfying the conditions of the Bragg equation. Measurement of the s_i values is done with a densitometer for precise work, but in this experiment a good millimeter scale is satisfactory. The interplanar spacing d_i is given by

$$d_i = \frac{n\lambda}{2 \sin \theta_i} \tag{11-6}$$

After analyzing the film in this way one has a set of interplanar spacings, $\{d_i\}$. From these and the measured density of the material one derives the structure.

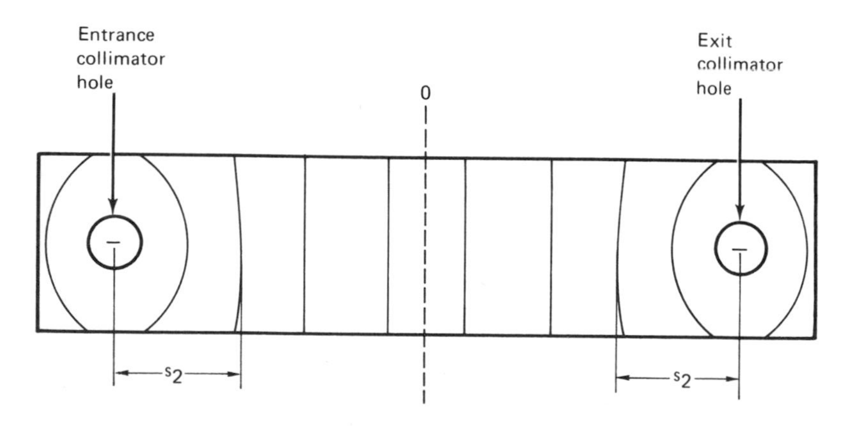

Figure 11-3. Hypothetical powder X-ray pattern.

We shall now discuss the relationships between the set $\{d_i\}$ and the geometry of the lattice. As a simple example suppose that the lattice is simple cubic; that is, the lattice points all occupy the corners of a cube as shown in Fig. 11-4. Stacking together mathematically these small cubes generates the full three-dimensional crystal. Now we ask what distinct planes are there in this crystal, the characteristics ds of which should appear in the experimentally observed set if, in fact, the actual material is simple cubic. To answer we shall study the model. From a view of the front face there is clearly a set of planes separated by a distance a. Lattice points 4, 5, and 8 define one member, while 10, 11, and 13 define a second member. From this we conclude that d = a should appear in the set $\{d_i\}$. A view of each face reveals a similar set of planes of the same spacing and so these views lead to nothing new. A different set is achieved by considering the planes defined by lattice points 1, 5, and 11 and by 4, 8, and 13. The distance between these planes is just the perpendicular distance between the line

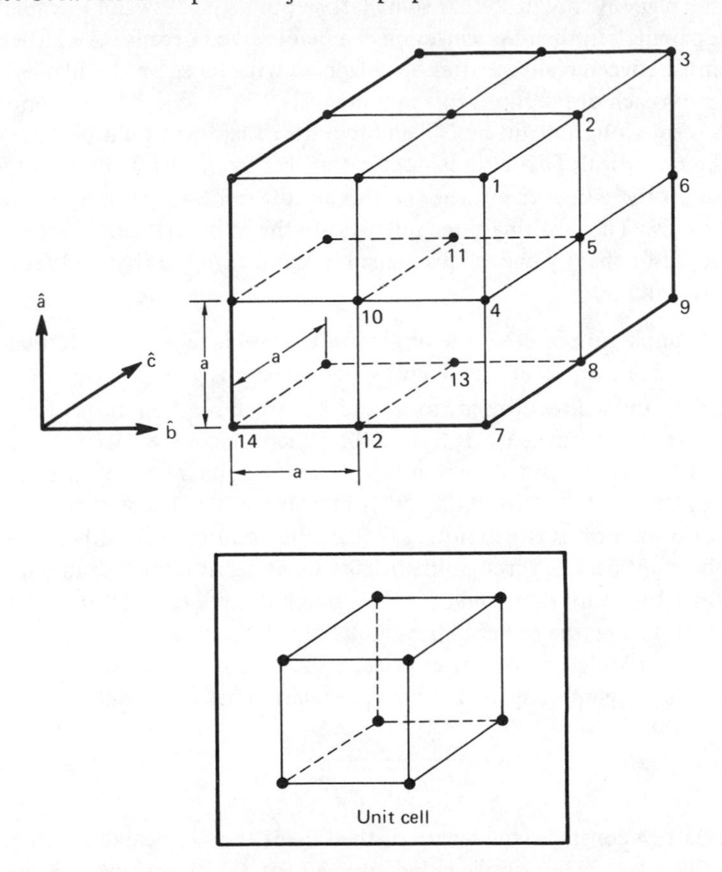

Figure 11-4. Simple cubic lattice.

joining points 1 and 5 and the point labeled 4. Thus, we expect the second distance, $d = a/\sqrt{2}$, to appear in the experimentally observed list. Proceeding in this way it is possible to generate a set of ds for the simple cubic geometry and for all other possible geometries.

Systematizing the above procedure is accomplished using what are called Miller indices. These indices are sets of three integers which denote various planes. To find the set of Miller indices (hkl) associated with a given crystal lattice structure, the following procedure is used:

1. Using the lattice point model, locate various planes by selecting three lattice points not on the same line. The plane thus generated will intercept the crystal axes $\{\hat{a}, \hat{b}, \hat{c}\}$ at three points. In selecting the three lattice points to generate the plane, one precaution must be exercised. The full set of planes parallel to the one selected must contain all the lattice points. If the plane is chosen so that some lattice points lie between members of the parallel set, then no scattering can be observed because the lattice points between planes scatter out of phase with those on the planes.
2. Express each of the intercepts in units of $|\hat{a}|$, $|\hat{b}|$, and $|\hat{c}|$. For example, the intercept along $\hat{a}$ might be $2|\hat{a}|$ and in units of $|\hat{a}|$ this would be 2.
3. Take reciprocals of the three numbers arising from part 2 and find the smallest three-integer set whose ratios are the same as the ratios of the reciprocals. The resulting three integers are the Miller indices of the plane.
4. Proceed further by choosing a plane not parallel to the first and repeat steps 2 and 3.

As an example suppose that the origin in Fig. 11-4 is taken to be lattice point 14. One possible plane is that defined by points 10, 11, and 12. The intercepts along the $\hat{a}$, $\hat{b}$, and $\hat{c}$ directions are ∞, 1, and ∞, respectively, in units of $|\hat{a}|$, $|\hat{b}|$, and $|\hat{c}|$. The reciprocals are (0, 1, 0) and the Miller indices are (010).

Consider the plane defined by points 4, 5, and 7. The intercepts are $(\infty, 2, \infty)$, the reciprocals are $(0, \frac{1}{2}, 0)$, and the Miller indices are (010), just as for the other plane. At this point it is worth noting that the procedure outlined here works for any plane not passing through the origin. Consider as a third example the plane defined by points 4, 8, and 13. The intercepts are $(1, \infty, 1)$, the reciprocals are (1, 0, 1), and the Miller indices are (101).

The utility of Miller indices arises because they are related to the set $\{d_i\}$. In the cubic crystallographic system the proper relation has the form

$$d = \frac{a}{\sqrt{h^2 + k^2 + l^2}} \tag{11-7}$$

where a is lattice constant (the length of the side of the simple cubic unit cell) shown in Fig. 11-4. For a simple cubic unit cell any set of positive integers (hkl) is satisfactory because the planes generated from any one of these sets will, as a group, pass through every lattice point. For the other two types of cubic lattices some restrictions must be placed on the integers in order to avoid omitting some lattice points from the planes. In the body-centered cubic structure shown in Fig. 11-5, the unit cell has a lattice point occurring at $(\frac{1}{2}a, \frac{1}{2}b, \frac{1}{2}c)$. In this case any triad (hkl) whose sum is *even* is a satisfactory set of Miller indices. For the face-centered cubic structure, also shown in Fig. 11-5, the lattice has points at $(0, \frac{1}{2}b, \frac{1}{2}c)$, etc. A satisfactory set (hkl) in this case is one in which h, k, and l are either *all even* or *all odd.* Why do these two rules arise? Because an experimentally observed d calculated from Eq. (11-6) involves scattering from a set of planes containing all lattice points. Therefore in making a list of d values from a model, we want to include only those which satisfy this experimental restriction. The above two rules guarantee that the list of ds calculated from Eq. (11-7) will generate sets of planes containing every lattice point.

Rearranging Eq. (11-7) and squaring, we obtain

$$h^2 + k^2 + l^2 = \left(\frac{a}{d}\right)^2 = \left(\frac{1}{D}\right)^2 \tag{11-8}$$

where D is a reduced distance expressed in units of the lattice constant a. The left-hand side of Eq. (11-8) is an integer since h, k, and l are all integers. Using the above restrictions on h, k, and l, a sequence of integers for $(1/D)^2$ values can be developed for each of the crystal types, and these can be compared with the experimental sequence. We need to look a bit further to see how to do this because we have no a value from the experimental data. However, since the parameter a is a constant, the experimental data can be reduced to a relative sequence by dividing every experimental 1/d value by the smallest one. Dividing every sum $(h^2 + k^2 + l^2)$ by the smallest one achieves the same thing for the model. Comparison of the model and experiment then establishes the lattice type. If the experimental relative sequence does not fit any of the sequences generated from the three cubic models, the crystal possesses non cubic symmetry. Table 11-1 shows the first few elements of the sequences for the three cubic lattice types. Comparing the relative sequence columns it is apparent that face-centered lattices are easily distinguished from the other two types. It is more difficult to distinguish between simple cubic and body-centered cubic lat-

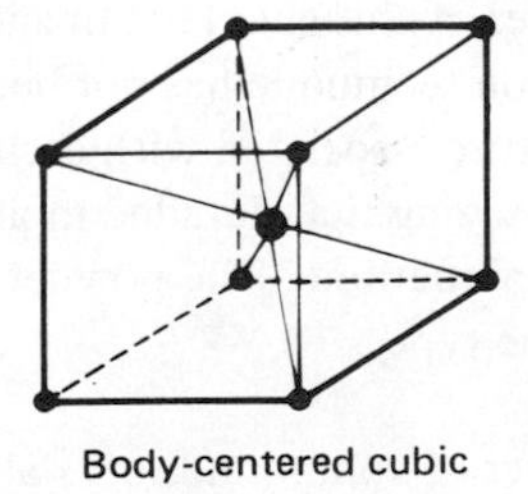

Body-centered cubic

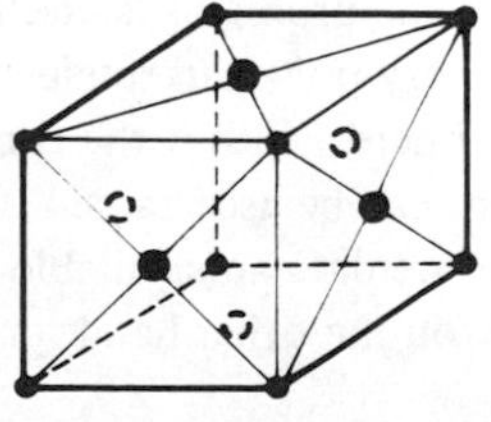

Face-centered cubic

Figure 11-5. Unit cells for cubic lattices other than simple cubic.

tices. The only difference through the first several lines is that 7 is missing from the simple cubic sequence.

Table 11-1. SEQUENCES OF $(1/D)^2$ VALUES FOR CUBIC LATTICES

Simple cubic			Body-centered cubic			Face-centered cubic		
hkl	$(1/D)^2$	Relative sequence	hkl	$(1/D)^2$	Relative sequence	hkl	$(1/D)^2$	Relative sequence
100	1	1	110	2	1	111	3	1
110	2	2	200	4	2	200	4	1.33
111	3	3	211	6	3	220	8	2.67
200	4	4	220	8	4	311	11	3.67
210	5	5	310	10	5	222	12	4
211	6	6	222	12	6			
–	–	–	321	14	7			
220	8	8	400	16	8			
221, 300	9	9	330, 411	18	9			
310	10	10	420	20	10			
311	11	11	332	22	11			
222	12	12	422	24	12			

Once the lattice type is determined the lattice constant can be related to the density by

$$a^3 = \frac{nM}{\rho N_0} \tag{11-9}$$

where N_0 is Avogadro's number, M the formula weight of the species associated with each lattice point, ρ the density, and n the net number of lattice points per unit cell in the macroscopic crystal. For simple cubic systems $n = 1$, for body-centered $n = 2$, and for face-centered $n = 4$.

In this brief presentation, we have passed over a systematic development of symmetry operations, symmetry elements, point groups, and other group theoretical concepts employed in treating crystal structure. These properties are discussed in Section 7.4 as they apply to gas-phase (or isolated) molecules. Their use in crystal problems is discussed in the References of this chapter. In addition the very powerful single-crystal X-ray diffraction technique has not been discussed here. With it the three-dimensional structure associated with each lattice point can be ascertained. This is equivalent to saying that detailed molecular structure data are available from the single-crystal method. The powder method, on the other hand, provides only the lattice type.

EXPERIMENTAL MATERIALS. A large number of compounds crystallize in the cubic system including many commonly available alkali halides and metal oxides. Examples include KCl, NaCl, NH_4Cl, MgO, LiF, FeO, FeS_2, PbS, and *bis*-benzene chromium. The *bis*-benzene chromium compound is an interesting example of a simple organo-metallic material. Its preparation and handling require a dry box, and its study is recommended only as a special project. A comparison of KCl and NaCl diffraction patterns illuminates the importance of isoelectronic character in chemistry. Both these salts are ionic crystals and have their ions arranged in a geometrically identical fashion. However, in the case of KCl the K^+ and Cl^- ions are isoelectronic (i.e., they both have identical numbers of electrons in the same type atomic orbitals). For NaCl the Na^+ and Cl^- are not isoelectronic. As a result the X-rays see KCl as simple cubic and NaCl as face-centered.

PROCEDURE. General procedural guidelines will be given here; the instructor will furnish details regarding which compounds to use, how to operate the X-ray instrument, and how to develop the exposed film.

The sample should be pure (to avoid extraneous spectral lines) and finely ground. Use an agate mortar and pestle. Very thin-walled glass capillary tubes of a diameter near 0.2 mm will be provided. These should be sealed off at one end, then filled with the finely ground crystalline material, and finally sealed off with picein wax at the other end. Depending on the kind of X-ray apparatus used, it may be necessary to mount the capillary on a metal rod which fits into the camera body and locates the sample on the camera cylinder axis. Detailed instruction will be provided by the instructor for mounting the sample.

The camera must be loaded with unexposed film in a photographic darkroom. Some types of film can be handled in red light with no damage, whole others can not. Again, specific instructions will be given. If the film must be handled in darkness, it is a good idea to practice the necessary manipulations with a piece of exposed film. In general the film must have holes punched in it to accommodate the collimators and must also fit smoothly around the body of the camera.

Once the camera is loaded with film it can be remounted on the X-ray apparatus and prepared for exposure. It is common knowledge that X-rays damage human tissues and thus exposure is to be avoided. The required time of exposure and proper safety precautions should be discussed with the laboratory instructor. Generally an exposure of 7 or 8 hr is required for a powder spectrum.

After exposure and shutdown of the X-ray source, the film must be developed, thoroughly washed, fixed, thoroughly washed again, and dried. For the present experiment in which only distances to lines are measured, these procedures need not be accurately controlled with respect to temperature, chemical concentration, development time, etc. For accurate work, and especially where line intensities are determined by measuring the relative densities of the lines, the film-handling procedures must be carefully worked out and controlled. Most developers used with X-ray film require about 5 min after which the film should be washed in water for about 1 min and then fixed for 15 min. Once fixed, the film can be exposed to light and should be washed in running water for about 30 min after which it can be hung up to dry.

If the density of the material you are studying is not available, it can be determined using a pycnometer as outlined in Section 5.6. The procedure is as follows using a pycnometer of about 10-ml volume, shown schematically in Fig. 11-6:

1. Weigh the clean and dry pycnometer to 1 mg.
2. Weigh the pycnometer filled with pure CCl_4. The carbon tetrachloride should completely fill the capillary of the stopper when it is inserted.
3. Empty the pycnometer, add about 1 g (accurately weighed to 1 mg) of the material studied in the X-ray experiment, refill the pycnometer with CCl_4, and weigh.
4. Use the above four weights and correct for the buoyancy of air to calculate the density of the sample.

The dimensions of the camera (its radius) and the wavelength of the X-rays will be provided. Typically the Cu(K_α) X-rays are used which have a wavelength of 1.5418 Å. A typical X-ray camera has a radius of about 60 mm.

ANALYSIS. Using an accurate millimeter scale, make measurements on the dried film determining a full set of s_i values. From these calculate an experimental set of $(1/D)^2$ values and an experimental relative sequence of $(1/D)^2$ values. Compare the experimental sequence with Table 11-1 and determine the cubic lattice type. From the density, determine the lattice constant. Finally convert the $(1/D)^2$ values obtained experimentally into a set of interplanar spacings, d, associated with a set of Miller indices. Report these spacings in a table labeled d_{hkl}.

PROBLEMS

1. Qualitatively, what do X-rays do to human tissue?
2. Why is it impossible to determine the local three-dimensional arrangement of atoms about a lattice point using the powder method?

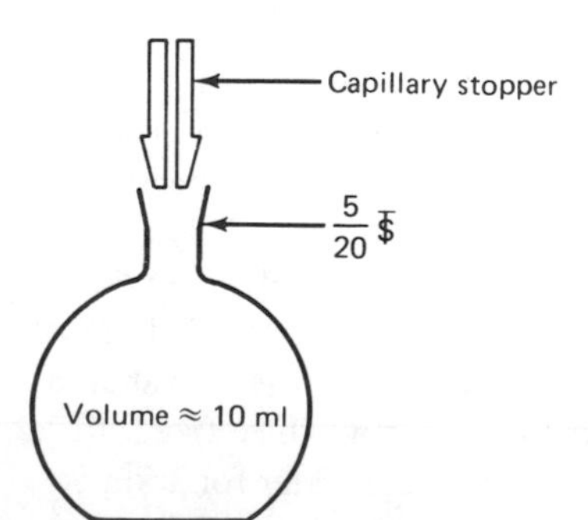

Figure 11-6. Pycnometer for determining the density of solid materials.

3. The momentum of a photon is h/λ, where h is Planck's constant. If a photon is completely absorbed by an electron, how much does the electron velocity change? How does this change compare with the average velocity of an electron in an atom?

4. Suppose that it were possible to introduce a few Na atoms in place of K atoms in an otherwise pure KCl crystal. How would this change the observed X-ray pattern?

REFERENCES

Section 11.1

J. D. H. Donnay (ed.), *Crystal Data, Determinative Tables,* 2nd ed. (American Crystallographic Association, New York, 1963). References the literature dealing with X-ray structure determination.

H. Lipson and H. Steeple, *Interpretation of X-ray Powder Diffraction Patterns* (Macmillan, New York, 1970); R. W. M. D'Eye and E. Wait, *X-ray Powder Photography* (Butterworth's, London, 1960); L. V. Azároff and M. J. Buerger, *The Powder Method in X-ray Crystallography* (McGraw-Hill, New York, 1958). Deal with the powder method.

J. H. Fang and F. D. Bloss, *X-ray Diffraction Tables* (Southern Illinois University Press, Carbondale, 1966). A set of X-ray diffraction tables for converting 2θ values into lattice spacings for different common X-ray wavelengths.

J. M. Bijvoet, N. H. Kolkmeyer, and C. H. Macgillavry, *X-ray Analysis of Crystals* (Wiley-Interscience, New York, 1951); B. D. Cullity, *Elements of X-ray Diffraction* (Addison-Wesley, Reading, Mass., 1956); H. Lipson and W. Cochran, *The Determination of Crystal Structures* (G. Bell & Sons Ltd., London, 1953); M. J. Buerger, *X-ray Crystallography* (Wiley, New York, 1942); G. L. Clark, *Applied X-rays,* 4th ed. (McGraw-Hill, New York, 1955); E. Nuffield, *X-ray Diffraction Methods* (Wiley, New York, 1966); M. Woolfson, *An Introduction to X-ray Crystallography* (Cambridge University Press, New York, 1970); G. H. Stout and L. H. Jensen, *X-ray Structure Determination; A Practical Guide* (Macmillan, New York, 1968); A. J. C. Wilson, *Elements of X-ray Crystallography* (Addison-Wesley, Reading, Mass., 1970). Deal with X-ray analysis in general.

W. N. Lipscomb and R. A. Jacobson in A. Weissberger and B. W. Rossiter (eds.), *Techniques of Chemistry,* Vol. I (Wiley-Interscience, New York, 1972), Part IIID, Ch. I. A discussion of experimental X-ray techniques applied to large molecule structure problems.

F. Jellinek, *J. Organometal. Chem.* 1, 43 (1963). Gives the X-ray data for the *bis*-benzene chromium mentioned in the text.

Appendices

Appendix I

Thermocouple Voltages for Chromel-Alumel Junction Assuming that the Reference Junction is at 0°C

T (°C)	V (mV)	T (°C)	V (mV)	T (°C)	V (mV)
-80	-2.87	100	4.10	280	11.39
-70	-2.54	110	4.51	290	11.80
-60	-2.20	120	4.92	300	12.21
-50	-1.86	130	5.33	310	12.63
-40	-1.50	140	5.73	320	13.04
-30	-1.14	150	6.13	330	13.46
-20	-0.77	160	6.53	340	13.88
-10	-0.39	170	6.93	350	14.29
00	0.00	180	7.33	360	14.71
10	0.40	190	7.73	370	15.13
20	0.80	200	8.13	380	15.55
30	1.20	210	8.54	390	15.98
40	1.61	220	8.94	400	16.40
50	2.02	230	9.34	410	16.82
60	2.43	240	9.75	420	17.24
70	2.85	250	10.16	430	17.67
80	3.26	260	10.57	440	18.09
90	3.68	270	10.98	450	18.51

Appendix II

General Physical Constants

Constant	Symbol	Value	Units (SI)	Units (cgs)
Speed of light	c	2.998	$\times 10^{8}$ m/sec	$\times 10^{10}$ cm/sec
Electron charge	e	1.6022	$\times 10^{-19}$ coulomb	
		4.803		$\times 10^{-10}$ esu
Avogadro's number	N_A	6.022	$\times 10^{23}$ mole^{-1}	$\times 10^{23}$ mole^{-1}
Atomic mass unit	amu	1.661	$\times 10^{-27}$ kg	$\times 10^{-24}$ g
Electron rest mass	m_e	9.110	$\times 10^{-31}$ kg	$\times 10^{-28}$ g
Proton rest mass	m_p	1.673	$\times 10^{-27}$ kg	$\times 10^{-24}$ g
Faraday constant	F	9.649	$\times 10^{4}$ coulomb/mole	
Planck constant	h	6.626	$\times 10^{-34}$ J/sec	$\times 10^{-27}$ erg/s
Charge per mass (electron)	e/m_e	1.759	$\times 10^{11}$ coulomb/kg	
Rydberg constant	R	1.097	$\times 10^{7}$ m^{-1}	$\times 10^{5}$ cm^{-1}
Gyromagnetic ratio for proton	γ_p	2.675	$\times 10^{8}$ rad sec^{-1} T^{-1}	$\times 10^{4}$ rad sec^{-1} G^{-1}
Bohr magneton	μ_B	9.274	$\times 10^{-24}$ J T^{-1}	$\times 10^{-21}$ erg G^{-1}
Gas constant	R	8.314	$\times$ J K^{-1} mole^{-1}	$\times 10^{7}$ erg K^{-1} mole^{-1}
Boltzmann constant	k	1.381	$\times 10^{-23}$ J K^{-1}	$\times 10^{-16}$ erg K^{-1}

cgs: centimeter-gram-second
esu: electrostatic unit
G: gauss
g: gram
J: joule
K: kelvin
kg: kilogram
m: meter
rad: radian
SI: Systeme International
sec: second
T: tesla

Appendix III

Conversion Factors

1. Energy: 1 erg/molecule $= 1.4394 \times 10^{13}$ kcal/mole
$= 5.0345 \times 10^{15}$ cm^{-1}
$= 6.2418 \times 10^{11}$ eV
$= 7.2436 \times 10^{15}$ °K
$= 2.294 \times 10^{10}$ a.u.
$= 10^{-7}$ J/molecule

2. Distance: 2.54 cm = 1 in.
3. Mass: 453.6 g = 1 lb.
4. Current: 3.336×10^{-10} A = 1 statamp (esu)
5. Charge: 3.336×10^{-10} coulomb = 1 statcoulomb (esu)
6. Gas constant: R = 8.3142 J/deg/mole
   = 1.9872 cal/deg/mole
   = 0.082053 liter-atm/deg/mole
7. Prefixes for fractions and multiples*:

deci = 10^{-1} (d)	deka = 10^{1} (da)
centi = 10^{-2} (c)	hecto = 10^{2} (h)
milli = 10^{-3} (m)	kilo = 10^{3} (k)
micro = 10^{-6} (μ)	mega = 10^{6} (M)
nano = 10^{-9} (n)	giga = 10^{9} (G)
pico = 10^{-12} (p)	tera = 10^{12} (T)

Appendix IV

Common Circuit Symbols Used in Schematic Diagrams

Resistors: Fixed, Variable

Capacitors: Fixed, Variable

Inductors (choke): Fixed, Variable

Ground

Transformers: Air core, Iron core, Variable

Switches: Single pole, Double pole

*Standard symbols are shown in parentheses.

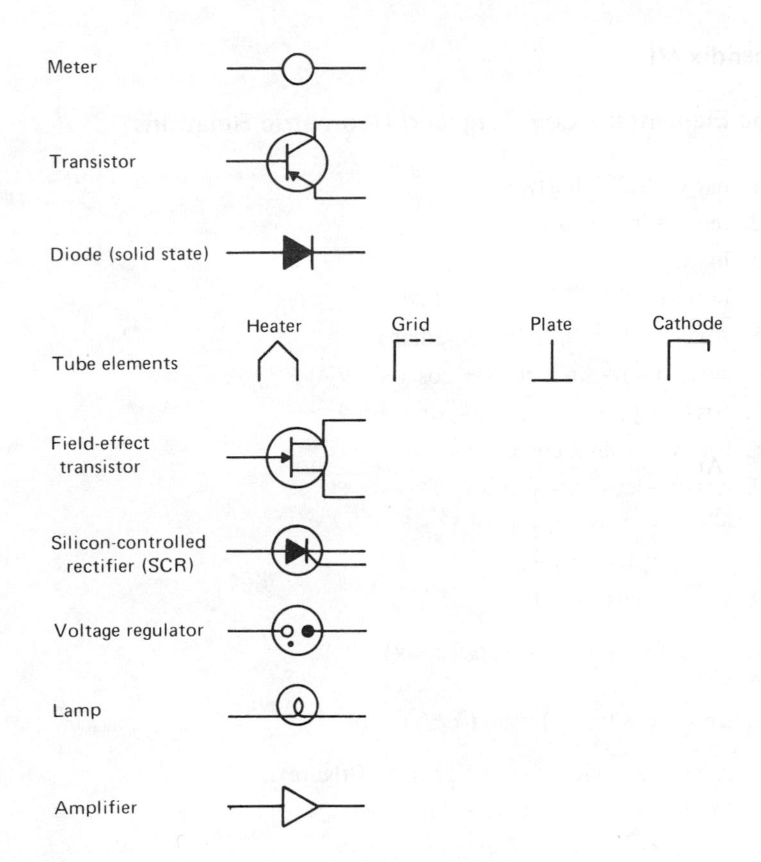

Appendix V

Basic Thermodynamic Expressions*

1. $H = E + pV$
2. $G = H - TS = E + pV - TS$
3. $A = E - TS$
4. $dH = T\,dS + V\,dp$
5. $dE = T\,dS - p\,dV$
6. $dA = -S\,dT - p\,dV$
7. $dG = -S\,dT + V\,dp$
8. $\left(\frac{\partial S}{\partial V}\right)_T = \left(\frac{\partial p}{\partial T}\right)_V$
9. $\left(\frac{\partial S}{\partial p}\right)_T = -\left(\frac{\partial V}{\partial T}\right)_p$
10. $C_p = \left(\frac{dQ}{dT}\right)_p = T\left(\frac{\partial S}{\partial T}\right)_p$
11. $C_V = \left(\frac{dQ}{dT}\right)_V = T\left(\frac{\partial S}{\partial T}\right)_V$

*See W. J. Moore, *Physical Chemistry* (Prentice-Hall, Englewood Cliffs, N.J., 1972) and A. Tobolsky, *J. Chem. Phys.* **10,** 644 (1942), for an indication of how to employ these 11 equations to develop a wide variety of other thermodynamic expressions for particular situations.

Appendix VI

Some Elementary Log, Trig, and Geometric Relations

1. $\ln_e x = 2.303 \log_{10} x$
2. $\ln xy = \ln x + \ln y$
3. $\ln (x/y) = \ln x - \ln y$
4. $\ln (e^x) = x$
5. Angle of $1° = 0.0174533$ radians
6. $\sin (x + y) = \sin x \cos y + \cos x \sin y$
7. $\cos (x + y) = \cos x \cos y - \sin x \sin y$
8. $\sin 2x = 2 \sin x \cos x$
9. $\cos 2x = \cos^2 x - \sin^2 x$
   $= 2 \cos^2 x - 1$
   $= 1 - 2 \sin^2 x$
10. $\sin^2 x + \cos^2 x = 1$
11. $\cos x = \sin (\frac{\pi}{2} - x) = -\cos (\pi - x)$
12. $\sin x = \cos (\frac{\pi}{2} - x) = \sin (\pi - x)$
13. Area of a circle $= \pi r^2$ (r = radius of figure)
14. Circumference of circle $= 2\pi r$
15. Volume of sphere $= \frac{4}{3}\pi r^3$
16. Surface area of a sphere $= 4\pi r^2$
17. $\pi = 3.14159265$

Appendix VII

Some Integrals

1. $$\int_0^\infty x^{2n} \exp \{-a^2x^2\}\, dx = \frac{(2n - 1)(2n - 3) \cdots (1)\sqrt{\pi}}{2^{n+1}a^{2n+1}}; \qquad n = 1, 2, 3, ...$$

2. $$\int_{-\infty}^\infty x^{2n} \exp \{-a^2x^2\}\, dx = 0 \qquad \text{for n odd}$$

3. $$\int_{-\infty}^\infty x^{2n} \exp \{-a^2x^2\}\, dx = 2 \int_0^\infty x^{2n} \exp \{-a^2x^2\}\, dx \qquad \text{for n even}$$

4. $\displaystyle\int_0^{\infty} x^{2n+1} \exp\{-a^2x^2\}\, dx = \frac{n!}{2a^{2n+2}}; \quad n = 1, 2, 3, \ldots$

5. $\displaystyle\int_0^{x} \exp\{-x^2\}\, dx = \frac{\sqrt{\pi}}{2} - \frac{\exp\{-x^2\}}{2x}\left[1 - \frac{1}{2x^2} + \frac{3}{4x^4} - \frac{15}{8x^6} + \cdots\right]$

6. $\displaystyle\int_0^{\pi} \sin^2 mx\, dx = \int_0^{\pi} \cos^2 mx\, dx = \pi/2$

7. $\displaystyle\int_0^{\infty} \exp\{-ax\}\, dx = \frac{1}{a}$

8. $\int \ln x\, dx = x \ln x - x$

9. $\int x \ln x\, dx = \frac{x^2}{2} \ln x - \frac{x^2}{4}$

10. $\int x \sin x\, dx = \sin x - x \cos x$

11. $\int x^2 \sin x\, dx = 2x \sin x - (x^2 - 2) \cos x$

12. $\int x \cos x\, dx = \cos x + x \sin x$

13. $\int x^2 \cos x\, dx = 2x \cos x + (x^2 - 2) \sin x$

14. $\int x \sin^2 x\, dx = \frac{x^2}{4} - \frac{x \sin 2x}{4} - \frac{\cos 2x}{8}$

15. $\int x \cos^2 x\, dx = \frac{x^2}{4} + \frac{x \sin 2x}{4} + \frac{\cos 2x}{8}$

Appendix VIII

Relations Involving Complex Numbers

1. A complex number can be considered as an ordered pair of real numbers (x, y).
2. Complex numbers are often represented as a point in a plane with the x axis being the real axis and the y axis being the imaginary axis.

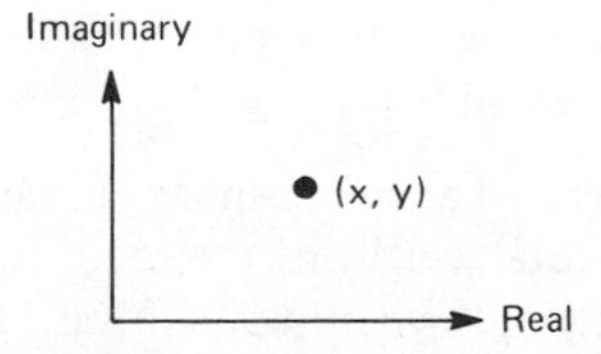

3. Two complex numbers are equal if and only if their real parts are equal and their imaginary parts are equal.
4. In Cartesian form the complex number z is written $z = x + iy$, where x is

the real component and y is the imaginary component. The quantity i is algebraically equivalent to $\sqrt{-1}$.

5. Complex numbers are also written in polar form: $z = r\exp(i\theta)$, where r is the distance from the origin to the complex point in the figure accompanying 2 above and θ is the angle between the x axis and the line connecting the origin with the point. Another polar form is $z = r[\cos\theta + i\sin\theta]$.
6. The addition of two complex numbers z_1 and z_2 is defined as the addition of their real parts and the addition of their imaginary parts as

$$z_1 + z_2 = (x_1 + x_2) + i(y_1 + y_2)$$

7. Multiplication of two complex numbers is defined as

$$z_1 z_2 = (x_1 x_2 - y_1 y_2) + i(x_1 y_2 + x_2 y_1)$$

Multiplication is more straightforward if the complex numbers are written in polar form. Then the above product becomes

$$z_1 z_2 = r_1 r_2 \exp(i\theta_1 + i\theta_2)$$

8. The magnitude of a complex number is just r in the polar form and is calculable as

$$r = \left(x^2 + y^2\right)^{1/2}$$

9. The complex conjugate of the complex number z is written z^* or $\bar{z}$ and has the form

$$z = x + iy$$

$$z^* = x - iy$$

10. The sine and cosine of the real angle θ take the form

$$\sin\theta = \frac{1}{2i}(e^{ix} - e^{-ix}) \qquad \cos\theta = \frac{e^{ix} + e^{-ix}}{2}$$

Appendix IX

Concentration Scales

Several different methods are used to express concentrations of mixtures. They all describe the same thing so one must be able to convert among these systems. Six systems are described here:

1. *Weight percent* is defined as the number of grams of one constituent present divided by the total weight of the solution with the result multiplied by 100%. For example, if we mix 5 g of NaCl and 50 g of water, the weight percent of salt is $(5/55) \times 100\% = 9.1\%$ and the weight percent of water is $(50/55) \times 100\% = 90.9\%$.

2. *Mole fraction* is defined as the number of moles of one constituent divided by the total number of moles in the system. The mole fraction of salt in the above solution is

$$X_{NaCl} = \frac{\frac{5}{58}}{\frac{5}{58} + \frac{50}{18}} = 0.030$$

3. *Molarity* is defined as the moles of one constituent divided by the volume of the solution in cubic centimeters with the result multiplied by 1000. In the above example 50 g of water mixed with 5 g of salt will yield about 51 cc = 51 ml of solution. The molarity of salt is then

$$M_{NaCl} = \frac{\frac{5}{58}}{51} \times 1000 = 1.690$$

while the molarity of the water is

$$M_{H_2O} = \frac{\frac{50}{18}}{51} \times 1000 = 54.47$$

4. *Molality* is defined as the number of moles of one constituent (called the solute and generally the species present in the smallest amount) divided by the number of grams of the other constituent (solvent) with the result multiplied by 1000. For example, if salt is solute and water the solvent, the molality of the salt in the above example is

$$m_{NaCl} = \frac{\frac{5}{58}}{50} \times 1000 = 1.724$$

5. *Formality* is defined just like molarity except the words *formula weight* replace the word *moles.* This concept has utility when situations arise in which a molecule in the normal sense of the term is difficult to define. For example, solid salts are crystals composed of distinct ionic entities, not molecules in the normal sense.

6. *Normality* is defined as the number of equivalent weights of solute divided by the volume of the solution in milliliters with the result multiplied by 1000. Remember that an equivalent weight is that amount of a species which is associated with the gain, loss, or sharing of 1 mole of electrons in any chemical process.

Appendix X

Operation of Two-Stage Cylinder Regulators

The figure below illustrates the very commonly used two-stage gas cylinder regulator valve, which has two pressure gauges: One reads the total gas pressure in the main cylinder which is applied to the valve by opening the cylinder valve. The second gauge reads the pressure delivered to the system or at least to the needle valve. This pressure is variable and is controlled by the regulator valve. Clearly, one of the purposes of this type of valve is to reduce the cylinder pressure to a more useful region. Operation of the system is described below.

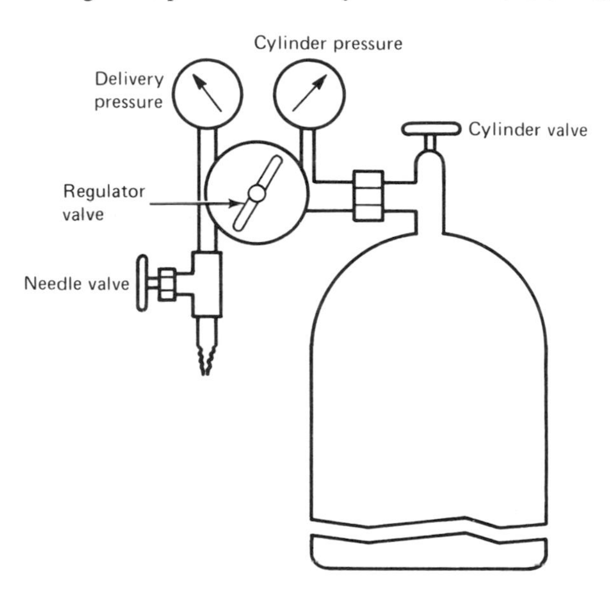

Operation of Regulator Control

1. Make certain the needle valve is closed and that the regulator control valve is closed by turning it counterclockwise until the handle is loose.
2. Open the cylinder valve by turning counterclockwise. The pressure in the cylinder is indicated on the cylinder pressure gauge.
3. Turn the regulator control valve clockwise until a pressure of about 10 psi is indicated on the delivery pressure gauge.
4. Admit gas into the system by opening the needle valve slowly.
5. When finished, close the cylinder valve and open the needle valve until the gauges both indicate zero pressure. Make certain that the system is open to the atmosphere during this process. When both gauges indicate zero pressure, close the needle valve and the regulator control valve, remembering that this valve is closed by counterclockwise rotation of the handle.

Appendix XI

A List of Generally Helpful Texts

Physical Chemistry Texts

W. J. Moore, *Physical Chemistry,* 4th ed. (Prentice-Hall, Englewood Cliffs, N. J., 1972).

G. M. Barrow, *Physical Chemistry* (McGraw-Hill, New York, 1960).

S. M. Glasstone and D. Lewis, *Elements of Physical Chemistry* (Van Nostrand Reinhold, New York, 1960).

S. H. Maron and C. F. Prutton, *Principles of Physical Chemistry,* 4th ed. (Macmillan, New York, 1965).

G. W. Castellan, *Physical Chemistry,* 2nd ed. (Addison-Wesley, Reading, Mass., 1971).

F. Daniels and R. A. Alberty, *Physical Chemistry,* 2nd ed. (Wiley, New York, 1961).

D. F. Eggers, N. W. Gregory, G. D. Halsey, Jr., and B. S. Rabinovitch, *Physical Chemistry* (Wiley, New York, 1964).

E. Hutchinson, *Physical Chemistry* (Saunders, Philadelphia, 1962).

E. A. Moelwyn-Hughes, *Physical Chemistry*, 2nd ed. (Pergamon, Elmsford, N.Y., 1961).

Thermodynamics Texts

K. Denbigh, *The Principles of Chemical Equilibrium,* 2nd ed. (Cambridge University Press, New York, 1966).

R. Dickerson, *Molecular Thermodynamics* (Benjamin, Reading, Mass., 1961).

E. A. Guggenheim, *Thermodynamics,* 5th ed. (North-Holland, Amsterdam, 1967).

H. B. Callen, *Thermodynamics* (Wiley, New York, 1960).

I. Koltz, *Chemical Thermodynamics* (Benjamin, Reading, Mass., 1961).

G. N. Lewis and M. Randall (revised by K. S. Pitzer and L. Brewer), *Thermodynamics* (McGraw-Hill, New York, 1961).

F. T. Wall, *Chemical Thermodynamics* (W. H. Freeman, San Francisco, 1965).

Laboratory Texts

F. Daniels, R. A. Alberty, J. W. Williams, C. D. Cornwell, P. Bender, and J. E. Harriman, *Experimental Physical Chemistry* (McGraw-Hill, New York, 1970).

D. P. Shoemaker and C. W. Garland, *Experiments in Physical Chemistry* (McGraw-Hill, New York, 1967).

W. C. Oelke (ed.), *Laboratory Physical Chemistry* (Van Nostrand Reinhold, New York, 1969).

F. A. Bettelheim, *Experimental Physical Chemistry* (Saunders, Philadelphia, 1971).

H. W. Salzberg, J. I. Morrow, S. R. Cohen, and M. E. Green, *Physical Chemistry, A Modern Laboratory Course* (Academic Press, New York, 1969).

H. W. Salzberg, J. I. Morrow, and S. R. Cohen, *Laboratory Course in Physical Chemistry* (Academic Press, New York, 1966).

O. F. Steinbach and C. V. King, *Experiments in Physical Chemistry* (American Book, New York, 1950).

Appendix XII

Equipment and Chemicals List for Each Experiment

Section 3.1

1. Mechanical vacuum pump
2. Timer (seconds)
3. Four-mm bore stopcocks, seven each
4. Joints for connecting capillaries, two each
5. Several pieces of capillary tubing
6. Glass vacuum apparatus shown in Fig. 3-1

Section 3.2

1. Potentiometer
2. Dewar flasks, three each
3. Liquid nitrogen
4. Trichloroethylene
5. Copper-constantan thermocouple
6. Standard cell
7. Three-volt dc power source

Section 3.3

1. Electronic parts called for in Fig. 3-2

Section 3.4

1. Oscilloscope
2. Signal generator of variable frequency
3. Audio amplifier
4. Load for audio amplifier; 3-kΩ resistor or transformer-coupled speaker

Section 3.5

1. Oscilloscope
2. Electronic parts specified in list for B^+ supply and the triode amplifier
3. Load for B^+ supply; 10-kΩ resistor
4. Load for triode amplifier; 3-kΩ resistor or transformer-coupled speaker

Section 4.1

1. Mechanical pump
2. Dewar flasks, three each
3. Mercury manometer
4. Four-mm bore stopcocks, seven each
5. Three-way stopcock
6. Steam generator
7. Glass connecting tubing shown in Fig. 4-2

Section 4.2

1. Oscilloscope
2. 1000-Hz oscillator
3. Microphone
4. Radio speaker
5. Audio amplifier
6. Enclosed tube shown in Fig. 4-5
7. Teflon supports and rings

Section 5.1

1. Bomb calorimeter
2. Oxygen supply cylinder
3. Ignition power supply
4. Ignition wire
5. Oxygen regulator
6. Pellet press
7. Benzoic acid pellet
8. Unknown pellet
9. Thermometer (0.01° divisions, range 20°–31°C)

Section 5.2

1. Cottrell boiling point apparatus shown in Fig. 5-6
2. Beckmann thermometer
3. Carbon tetrachloride
4. Naphthalene
5. Benzoic acid
6. Ethanol
7. Boiling chips
8. Five to 10-junction copper-constantan thermocouple

9. Potentiometer
10. Light mineral oil
11. Large beaker for heat bath

Section 5.3

1. Freezing point apparatus shown in Fig. 5-10
2. Thermometer–100° in 0.2° divisions
3. A pair of chemicals such as naphthalene-biphenyl or others given in the experimental section
4. Strip chart recorder
5. Copper-constantan thermocouple

Section 5.4

1. Sample flask and condenser tube shown in Fig. 5-12
2. Mercury manometer
3. Ballast volume, 5–10 liters
4. Three-way stopcock
5. Four-mm bore stopcock
6. Thermometer–100°C in divisions of 0.2°
7. Mechanical vacuum pump
8. Glass tubing connections shown in Fig. 5-12
9. Pure liquid such as ethanol, acetone, carbon tetrachloride, toluene, etc.

Section 5.5

1. Refractometer
2. Low voltage-high current power supply (6 V at 8 A)
3. Boiling apparatus shown in Fig. 5-16
4. Thermometer–100°C in 1° divisions
5. Laboratory barometer
6. Chloroform
7. Acetone
8. Eyedropper
9. Pipettes
10. A few small beakers
11. A few small test tubes

Section 5.6

1. Pycnometer
2. Balance–0.1 mg
3. Thermometer–100°C in 1° divisions
4. Aspirator or rubber suction bulb

5. 250-ml beakers, five each
6. Sodium chloride or potassium iodide or sodium acetate
7. *n*-Propanol or some other nonelectrolyte

Section 5.7

1. Nitrogen dioxide–lecture bottle

Three methods are outlined in the text. For the method outlined in Fig. 5-18 the following items are needed:

2. Pressure transducer–0–15 psi
3. Thermometer–100°C in 1° divisions
4. Mercury manometer
5. 2000-ml flask
6. Water bath and heater
7. Four-mm bore stopcocks, four each
8. Sample vessel shown in Fig. 5-18
9. Balance–100-g capacity weighing to 0.1 mg
10. Bourdon tube pressure gauge–vacuum to 1-atm range
11. Dry ice
12. $\frac{5}{20}$ ₮
13. Glass tubing to make connections shown in Fig. 5-18 and 5-20
14. Mechanical vacuum pump

For the method outlined in Fig. 5-21 the following items are needed:

2. Mechanical vacuum pump
3. Four-mm bore stopcocks, three each
4. Teflon plug stopcock
5. 250-ml sample flask
6. Glassware for vacuum line shown in Fig. 5-21
7. Balance–two pans capable of weighing 300 g to a precision of 0.1 mg
8. Thermostat controlled to 1°C over the range 30°–60°C

For the method outlined in Fig. 5-22 the following items are needed:

2. Mechanical vacuum pump
3. Vacuum line shown in Fig. 5-22
4. Optical absorption cell
5. Air thermostat
6. Two 5-cm focal length Pyrex lenses
7. Interference filter–5460 Å = 546 nm
8. Mercury arc source
9. Phototube

10. Microammeter
11. Iron-constantan thermocouple
12. Potentiometer
13. Bromobenzene

Section 5.8

1. Ammonium carbamate
2. Mechanical vacuum pump
3. Pressure transducer
4. Bellows-sealed metal valves, two each
5. Four-mm bore stopcocks, three each
6. Mercury manometer
7. Thermometer–100°C in 1° divisions
8. Air thermostat–30°-60°C
9. Glass-to-metal seals, two each
10. Heater–500 W
11. O-ring connector to connect sample flask to vacuum line of Fig. 5-23
12. 1000-ml, flat-bottomed flask

Section 5.9

1. Thermal conductivity cell
2. Sample injection ports
3. Chromatography columns, two each
4. Mercury manometers, two each
5. Cylinder of carrier gas
6. Air oven, thermostatted
7. Wheatstone bridge
8. Strip chart recorder
9. Soap film flow meters, two each
10. Samples of benzene, toluene, phenol, cyclohexane, and cyclohexene
11. 10-μl syringe

Section 5.10

1. Mechanical vacuum pump, two each
2. Glass-type oil diffusion pump
3. Vacuum line shown in Fig. 5-26
4. Mercury manometer
5. Oxygen thermometer
6. Sample of porous Vycor or carbon black, etc.
7. Dewar flask
8. Electrical flasher circuit for constant-volume manometer
9. Cathetometer

Section 5.11

1. Mechanical vacuum pump
2. Metal liquid nitrogen Dewar flask
3. Copper-constantan thermocouple
4. Valving arrangement shown in Fig. 5-29
5. Sample container shown in Fig. 5-29
6. Argon cylinder
7. Mercury manometer
8. Potentiometer
9. Nitrogen cylinder
10. Liquid nitrogen
11. Glass connecting tubing shown in Fig. 5-28

Section 6.1

1. Thermostatted water bath
2. Pipettes
3. Burettes, two each
4. Potassium persulfate
5. Potassium iodide
6. Potassium nitrate
7. Sodium thiosulfate
8. Starch (soluble)
9. Thermometer–100°C in 1° divisions
10. Timer (seconds)
11. Storage vessels for reagents

Section 6.2

1. UV spectrophotemeter
2. One cm liquid sample cells (quartz), two each
3. Two cc syringes, two each
4. 1,10-Phenanthroline
5. Mercuric nitrate
6. Cobalt nitrate
7. Nickel nitrate
8. Copper nitrate
9. Variable temperature thermostat for sample cell
10. Variable temperature thermostat for reagent solutions

Section 6.3

1. Furnace diagrammed in Fig. 6-6
2. Mechanical vacuum pump
3. Diffusion pump
4. McLeod gauge

5. Mercury manometer
6. Chromel-alumel thermocouple
7. Potentiometer
8. Dewar flask
9. Four-mm bore stopcocks, four each
10. Ten-mm bore stopcocks, two each
11. Sample cell shown in Fig. 6-6
12. Timer (seconds)
13. Sample such as 2,5-dihydrofuran

Section 6.4

1. Vacuum line described in Fig. 6-11
2. Mechanical vacuum pumps, two each
3. Four-mm bore stopcocks, nine each
4. Ten-mm bore stopcock
5. Diffusion pump (oil)
6. Pressure transducer
7. Mercury manometer
8. Quartz reaction vessel
9. Helium
10. Hydrogen bromide
11. Spectrophotometer (210–420 nm)
12. Low-pressure mercury lamp
13. Dewar flask
14. Liquid nitrogen

Section 6.5

1. Computer

Section 7.1

1. Hydrogen discharge tube
2. Helium discharge tube
3. Spectroscope
4. Spectrograph
5. High-voltage ac supply for discharge tubes

Section 7.2

1. Franck-Hertz apparatus shown in Fig. 7-6
2. Furnace (160°C)
3. dc voltage supply
4. Microammeter
5. Franck-Hertz tube

Section 7.3

1. Infrared spectrometer
2. Infrared cell
3. HCl-DCl or HBr-DBr mixture
4. Small vacuum line for filling cell

Section 7.4

1. Apparatus same as Section 7.3
2. SO_2
3. CO_2
4. NH_3

Section 7.5

1. Visible-ultraviolet spectrophotometer
2. Pyrex sample cell for gases
3. Bromine
4. Iodine
5. Heater for iodine sample

Section 7.6

1. Fluorescence cell
2. Hg lamp
3. Visible spectrometer
4. Didymium filter
5. Pyrex convex lenses, two each

Section 7.7

1. NMR apparatus shown in Fig. 7-24
2. NMR sample tubes
3. Triflouracetic acid
4. Copper nitrate

Section 8.1

1. NMR spectrometer
2. Coaxial sample tube
3. Thin-walled glass sample tube
4. *t*-Butyl alcohol
5. Nickel chloride
6. Copper sulfate
7. Ferric sulfate
8. Potassium ferricyanide
9. Variable temperature NMR sample holder

Section 8.2 Three methods are outlined in the text.

Bridge method:

1. Schering bridge circuit described in Fig. 8-3
2. Dielectric cell for sample described in Fig. 8-6
3. Samples described below

Resonance method:

1. Oscillator and coupling coils described in Fig. 8-4
2. Dielectric cell shown in Fig. 8-6
3. Samples described below

Heterodyne beat method:

1. Fixed 1-MHz oscillator
2. Variable frequency oscillator
3. Mixer
4. Dielectric cell shown in Fig. 8-6
5. Oscilloscope
6. Inductor and variable capacitor shown in Fig. 8-5

Each of the above techniques requires a dielectric sample. Several different experiments are suggested in the text using the following chemicals and manipulative equipment:

1. *n*-Propyl nitrate
2. 1,3-Dibromopropane
3. 1,2-Dichlorocthane
4. Benzene
5. Syringe

Section 9.1

1. Calomel electrode
2. Salt bridge
3. Zn, Cu, Pb, and Ag metal electrodes
4. Zinc sulfate
5. Copper sulfate
6. Lead nitrate
7. Silver nitrate
8. Ammonium nitrate
9. Thermometer–0°–100°C in divisions of 1°
10. Potentiometer
11. Standard cell

Section 9.2

1. Saturated calomel electrode
2. Silver-silver chloride electrode
3. dc current source (2mA)
4. Platinum electrode
5. Flask assembly shown in Fig. 9-4
6. Potentiometer
7. Standard cell
8. Water bath covering range 10°–60°C

Section 9.3

1. ac bridge circuit shown in Fig. 9-5
2. One-MHz oscillator
3. Preamplifier
4. Oscilloscope
5. Phase shifter
6. Conductance cell shown in Fig. 9-6
7. Potassium chloride
8. Methanol
9. Acetone
10. Potassium iodide
11. Acetic acid
12. Sodium acetate
13. Sodium chloride
14. Hydrochloric acid
15. Sodium hydroxide

Section 9.4

1. Transference number cell like one shown in Fig. 9-8
2. Coulometer like that shown in Fig. 9-9
3. Copper wire
4. Copper sulfate
5. Sodium sulfate
6. Sulfuric acid
7. Potassium iodide
8. Sodium thiosulfate
9. Soluble starch
10. dc current supply capable of delivering 25 mA
11. Acetic acid

Section 10.1

1. Viscometer
2. Polyvinyl acetate
3. Acetone
4. Methanol
5. Pycnometer
6. Timer (seconds)

Section 10.2

1. Sephadex gel
2. Chromatographic column shown in Fig. 10-8
3. Sodium chloride
4. Blue Dextran
5. Vitamin B_{12}
6. Spectrophotometer

Section 10.3

1. Light-scattering apparatus
2. Polyvinyl acetate
3. Acetone

Section 11.1

1. X-Ray powder apparatus
2. Sodium chloride
3. Potassium chloride
4. Other compounds may be used; some are mentioned in the text

Appendix XIII

Suppliers of Equipment and Chemicals

1. Ammonium carbamate
   Apache Chemicals, P.O. Box 17, Rockford, Ill. 61105
2. Barometers, laboratory
   a. Central Scientific, 2600 S. Kostner Ave., Chicago, Ill. 60623.
   b. Sargent-Welch, 4647 Foster Ave., Chicago, Ill. 60630.
   c. Wallace and Tiernan, 25 Main St., Belleville, N.J. 07109.
3. Capacitance bridge circuitry
   a. General Radio, 300 Baker Ave., Concord, Mass. 01742.
   b. Leeds and Northrup, Sumneytown Pike, North Wales, Pa. 19454.

4. Cathetometers
   a. Ealing, 2225 Massachusetts Ave., Cambridge, Mass. 02140.
   b. Eberbach, Box 1024, Ann Arbor, Mich. 48106.
   c. Gaertner Scientific, 1201 Wrightwood Ave., Chicago, Ill. 60614.

5. Chromatograph column materials
   a. Johns-Manville, 22 E. 40th St., New York, N.Y. 10016.
   b. Varian Aerograph, 2700 Mitchell Dr., Walnut Creek, Calif. 94598.
   c. Beckman Instruments, 2500 Harbor Blvd., Fullerton, Calif. 92634.
   d. Analabs, 80 Republic Dr., North Haven, Conn. 06473.

6. Combustion bombs
   a. Parr Instrument, 211 53rd St., Moline, Ill. 61265.
   b. American Minechem, Coraopolis, Pa. 15108.

7. Connectors, O-ring, glass-to-metal
   Cajon, 32550 Old South Miles Rd., Solon, Ohio. 44139.

8. Deuterated reagents
   a. Merck, Sharp and Dohme, 350 Selby St., Montreal 6, Quebec., Canada.
   b. Stohler Isotope Chemicals, 49 Jones Rd., Waltham, Mass. 02154.

9. Dewar flasks
   a. Union Carbide, 270 Park Ave., New York, N.Y. 10017.
   b. Superior Air Products, 132 Malvern St., Newark, N.J. 07105.
   c. Refer to stopcock suppliers.

10. Electrochemical Cell Glassware
    a. Kontes Glass, Spruce St., Vineland, N.J. 08360.
    b. Ace Glass, P.O. Box 688, Vineland, N.J. 08360.

11. Electronic equipment (general)
    a. Keithley Instruments, 28775 Aurora Rd., Cleveland, Ohio. 44139.
    b. Heath, Hilltop Rd., Benton Harbor, Mich. 49022.
    c. Hewlett-Packard, 1501 Page Mill Rd., Palo Alto, Calif. 94304.
    d. John Fluke, P.O. Box 7428, Seattle, Wash. 98133.
    e. Kepco, 131-38 Sanford Ave., Flushing, N.Y. 11352
    f. Lambda Electronics, 515 Broad Hollow Rd., Melville Long Island, N.Y. 11746.

12. Filters, glass color
    a. Bausch and Lomb, 820 Linden Ave., Rochester, N.Y. 14625.
    b. Corning Glass Works, Science Products Div., Corning, N.Y. 14830.

13. Filters, interference
    a. Baird-Atomic, 125 Middlesex Tnpk., Bedford, Mass. 01730.
    b. Corion Instrument, 23 Fox Rd., Waltham, Mass. 02154.
    c. Optics Technology, 901 California Ave., Palo Alto, Calif. 94304.

14. Franck-Hertz apparatus
    Leybold-Heraeus, 200 Seco Rd., Monroeville, Pa. 15146.
15. Gas discharge tube, low pressure
    a. Central Scientific, 2600 S. Kostner Ave., Chicago, Ill. 60623.
    b. Sargent-Welch, 4647 Foster Ave., Chicago, Ill. 60630.
16. Gases (lecture bottles)
    a. Matheson Gas Products, Box 85, East Rutherford, N.J. 07073.
    b. Union Carbide, Specialty Gas Office, 223 Highway 18, New Brunswick, N.J. 08816.
17. Gauges, Bourdon
    a. Marsh Instrument, Box 190, Wilmette, Ill. 60091.
    b. Wallace and Tiernan, 25 Main St., Belleville, N.J. 07109.
    c. Honeywell, 1100 Virginia Dr., Ft. Washington, Pa. 19034.
18. Gauges, McLeod
    a. See Vacuum equipment
    b. Eck and Krebs, 27-09 40th Ave., Long Island City, N.Y. 11101.
    c. Kontes Glass, Spruce St., Vineland, N.J. 08360.
    d. Bendix, 1775 Mt. Read Blvd., Rochester, N.Y. 14603.
19. Gel chromatography apparatus
    a. Pharmacia Fine Chemical, 800 Centennial, Piscataway, N.J. 08854.
    b. Waters Associates, 61 Fountain, Framingham, Mass. 01701.
20. Heaters, cartridge
    Emerson Electric, 7500 Thomas Blvd., Pittsburgh, Pa. 15208.
21. Laser, ruby
    a. Crystal Optics Research, 3680 S. State, Ann Arbor, Mich. 48104.
    b. Union Carbide, 270 Park Ave., New York, N.Y. 10017.
    c. Spacerays, NW Industrial Pk., Burlington, Mass. 01803.
    d. Hadron, 800 Shames Dr., Westbury, N.Y. 11590.
22. Lenses, Pyrex
    a. Oriel Optics, 1 Market St., Stamford, Conn. 06902.
    b. Edmund Scientific, 801 Edscorp Bldg., Barrington, N.J. 08007.
    c. Ultra-violet, 5114 Walnut Grove, San Gabriel, Calif. 91778.
23. Light-scattering apparatus
    a. Phoenix Precision Instrument, Rt. 208, Gardiner, N.Y. 12525.
    b. American Instrument, 8030 Georgia Ave., Silver Spring, Md. 20910.
    c. C. N. Wood, Newton Industrial Commons, Rt. 332, Newton, Pa. 18940.
24. Light sources, Mercury
    a. Hanovia Lamp, 100 Chestnut, Newark, N.J. 07105.
    b. Ealing, 2225 Massachusetts Ave., Cambridge, Mass. 02140.
    c. PEK, 825 E. Evelyn Ave., Sunnyvale, Calif. 94806.
    d. American Ultraviolet, 64 Commerce St., Chatham, N.J. 07928.

25. NMR apparatus
    a. Varian, 611 Hansen Way, Palo Alto, Calif. 94303.
    b. JEOL USA, 477 Riverside Ave., Medford, Mass. 02155.
    c. Central Scientific, 2600 S. Kostner Ave., Chicago, Ill. 60623.

26. NMR sample cells

    Stohler Isotope Chemicals, 49 Jones Road, Waltham, Mass. 02154.

27. Oscilloscopes
    a. Hewlett-Packard, 1501 Page Mill Rd., Palo Alto, Calif. 94304.
    b. Tektronix, Box 500, Beaverton, Or. 97005.
    c. Heath, Hilltop Rd., Benton Harbor, Mich. 49022.

28. Ovens, circulating warm air
    a. Blue M Electric, Blue Island, Ill. 60406.
    b. Lindberg Hevi-Duty, 2450 W. Hubbard, Chicago, Ill. 60612.
    c. See Stopcocks.

29. Photomultiplier tubes
    a. Varian/EMI, 80 Express St., Plainview, N.Y. 11803.
    b. Bendix, Galileo Park, Sturbridge, Mass. 01518.
    c. RCA Electronic, 415 S. 5th St., Harrison, N.J. 07029.

30. Phototubes
    a. RCA Electronic, 415 S. 5th St., Harrison, N.J. 07029.

31. Potentiometers
    a. Leeds and Northrup, Sumneytown Pike, North Wales, Pa. 19454.
    b. James G. Biddle, Plymouth Meeting, Pa. 19462.
    c. Brinkmann Instruments, Cantiague Rd., Westbury, N.Y. 11590.

32. Pressure transducers
    a. CGG/Datametrics, 127 Coolidge Hill Rd., Watertown, Mass. 02172.
    b. MKS Instruments, 25 Adams St., Burlington, Mass. 01803.
    c. Whittaker, 12838 Saticoy St., North Hollywood, Calif. 91605.
    d. Validyne Engineering, 18819 Napa St., Northridge, Calif. 91324.

33. Quartz glassware
    a. Coleman-Delmar Glass, 318 Madison St., Maywood, Ill. 60153.
    b. Quartz Products, Box 628, Plainfield, N.J. 07061.
    c. Ruska Instrument, Box 36010, Houston, Texas 77036.
    d. Quartz Scientific, 34602 Lakeland Blvd., Eastlake, Ohio 44094.
    e. Engelhard Industries, Amersil Quartz Div., 685 Ramsey Ave., Hillside, N.J. 07205

34. Recorders, strip chart
    a. Heath, Hilltop Rd., Benton Harbor, Mich. 49022.
    b. Hewlett-Packard, 1501 Page Mill Rd., Palo Alto, Calif. 94304.
    c. Houston Instrument, 4950 Terminal Dr., Bellaire, Texas 77401.

  d. Esterline Angus, Box 24000, Indianapolis, Ind. 46224.
  e. Leeds and Northrup, Sumneytown Pike, North Wales, Pa. 19454.
  f. Sargent-Welch, 4647 Foster Ave., Chicago, Ill. 60630.

35. Refractometers
  a. Bausch and Lomb, 820 Linden Ave., Rochester, N.Y. 14625.
  b. Carl Ziess, 444 5th Ave., New York, N.Y. 10018.
  c. Fisher Scientific, 711 Forbes Ave., Pittsburgh, Pa. 15219.

36. Seals, glass-to-metal
  a. Cajon, 32550 Old South Miles Rd., Cleveland, Ohio 44139.
  b. Eck and Krebs, 27-09 40th Ave., Long Island City, N.Y. 11101.
  c. Scientific Glass Apparatus, 2375 Pratt Blvd., Elk Grove Village, Ill. 60007.

37. Signal Generator
  See Oscilloscope suppliers.

38. Spectrophotometer (IR)
  a. Beckman Instruments, 2500 Harbor Blvd., Fullerton, Calif. 92634.
  b. Perkin-Elmer, 800 Main Ave., Norwalk, Conn. 06852.
  c. Bausch and Lomb, 820 Linden Ave., Rochester, N.Y. 14625.

39. Spectrophotometer (UV-Vis-IR)
  a. Bausch and Lomb, 820 Linden Ave., Rochester, N.Y. 14625.
  b. Cary Instruments, 2724 S. Peck Rd., Monrovia, Calif. 91016.
  c. Coleman Instruments, 42 Madison St., Maywood, Ill. 60153.
  d. Heath, Hilltop Rd., Benton Harbor, Mich. 49022.
  e. McKee-Pederson Instruments, Box 322, Danville, Calif. 94526.
  f. Beckman Instruments, 2500 Harbor Blvd., Fullerton, Calif. 92634.
  g. Perkin-Elmer, 800 Main Ave., Norwalk, Conn. 06852.

40. Standard cells
  a. Eppley Laboratory, 12 Sheffield Ave., Newport, R.I. 02840.
  b. Julie Research Labs, 211 W. 61st., New York, N.Y. 10023.

41. Stopcocks
  a. Eck and Krebs, 27-09 40th Ave., Long Island City, N.Y. 11101.
  b. Sargent-Welch, 4647 Foster Ave., Chicago, Ill. 60630.
  c. Curtin Scientific, Box 1546, Houston, Texas 77001.
  d. Kontes Glass, Spruce Street, Vineland, N.J. 08360.
  e. Ace Glass, Box 688, Vineland, N.J. 08360.

42. Thermal conductivity cell
  Gow-Mac Instrument, 100 Kings Rd., Madison, N.J. 07940.

43. Thermistor bridge circuit
    a. Sargent-Welch, 4647 Foster Ave., Chicago, Ill. 60630.
    b. Yellow Springs Instrument, Box 279, Yellow Springs, Ohio 45387.
    c. Tri-R Instruments, 48 Merrick Rd., Rockville Centre, N.Y. 11570.
    d. Victory Engineering, Victory Rd., Springfield, N.J. 07081.

44. Thermocouples
    a. Pyro Electric, Walkerton, Ind. 46574.
    b. Omega Engineering, Box 4047, Springdale Sta., Stamford, Conn. 06907.

45. Timers
    a. Eagle Signal, 736 Federal St., Davenport, Iowa 52803.
    b. Chrono-Log, 2583 W. Chester Pike, Broomall, Pa. 19008.
    c. Lab-line Instruments, 15th and Bloomingdale Ave., Melrose Park, Ill. 60160.

46. Vacuum equipment
    a. Varian-NRC, 160 Charlemont St., Newton, Mass. 02161.
    b. Veeco Instrument, Terminal Dr., Plainview, N.Y. 11803.
    c. Aero Vac, Box 448, Troy, N.Y. 12181.
    d. Sargent-Welch, 4647 Foster Ave., Chicago, Ill. 60630.
    e. Curtin Scientific, Box 1546, Houston, Texas 77001.
    f. Bendix, 1775 Mt. Read Blvd., Rochester, N.Y. 14603.

47. Valves, cylinder regulator
    See Gases (lecture bottle).

48. Valves, metal
    a. Hoke, 1 Tenakill Pk., Cresskill, N.J. 07626.
    b. Matheson Gas Products, Box 85, East Rutherford, N.J. 07073.
    c. Veeco Instrument, Terminal Dr., Plainview, N.Y. 11803.

49. Water baths
    a. Arthur H. Thomas, Box 779, Philadelphia, Pa. 19105.
    b. See Ovens, circulating warm air.

50. Wheatstone bridges
    a. Sargent-Welch, 4647 Foster Ave., Chicago, Ill. 60630.
    b. Leeds and Northrup, Sumneytown Pike, North Wales, Pa. 19454.
    c. General Radio, 300 Baker Ave., Concord, Mass. 01742.

51. X-Ray diffractometers
    a. Philips Electronics, 750 S. Fulton Ave., Mt. Vernon, N.Y. 10550.
    b. General Electric, 40 Federal St., Lynn, Mass. 01910.
    c. Picker Scientific Apparatus, 1020 London Rd., Cleveland, Ohio 44110.

Index

D

E

F

Q

R

S

V

W

X

Y

Z